W9-BCU-969

YALE AGRARIAN STUDIES SERIES
James C. Scott, Series Editor

The Agrarian Studies Series at Yale University Press seeks to publish outstanding and original interdisciplinary work on agriculture and rural society—for any period, in any location. Works of daring that question existing paradigms and fill abstract categories with the lived-experience of rural people are especially encouraged.
—JAMES C. SCOTT, *Series Editor*

James C. Scott, *Seeing Like a State: How Certain Schemes to Improve the Human Condition Have Failed*

Deborah Fitzgerald, *Every Farm a Factory: The Industrial Ideal in American Agriculture*

Stephen B. Brush, *Farmers' Bounty: Locating Crop Diversity in the Contemporary World*

Brian Donahue, *The Great Meadow: Farmers and the Land in Colonial Concord*

J. Gary Taylor and Patricia J. Scharlin, *Smart Alliance: How a Global Corporation and Environmental Activists Transformed a Tarnished Brand*

Michael Goldman, *Imperial Nature: The World Bank and Struggles for Social Justice in the Age of Globalization*

Arvid Nelson, *Cold War Ecology: Forests, Farms, and People in the East German Landscape, 1945–1989*

Steve Striffler, *Chicken: The Dangerous Transformation of America's Favorite Food*

Parker Shipton, *The Nature of Entrustment: Intimacy, Exchange, and the Sacred in Africa*

Alissa Hamilton, *Squeezed: What You Don't Know About Orange Juice*

Parker Shipton, *Mortgaging the Ancestors: Ideologies of Attachment in Africa*

Bill Winders, *The Politics of Food Supply: U.S. Agricultural Policy in the World Economy*

James C. Scott, *The Art of Not Being Governed: An Anarchist History of Upland Southeast Asia*

Stephen K. Wegren, *Land Reform in Russia: Institutional Design and Behavioral Responses*

For a complete list of titles in the YALE AGRARIAN STUDIES SERIES, visit www.yalebooks.com.

BENJAMIN R. COHEN

Notes from the Ground

SCIENCE, SOIL, AND SOCIETY IN THE AMERICAN COUNTRYSIDE

YALE UNIVERSITY PRESS NEW HAVEN & LONDON

Published with assistance from the foundation established in memory of Philip Hamilton McMillan of the Class of 1894, Yale College.

Copyright © 2009 by Benjamin R. Cohen.
All rights reserved.
This book may not be reproduced, in whole or in part, including illustrations, in any form (beyond that copying permitted by Sections 107 and 108 of the U.S. Copyright Law and except by reviewers for the public press), without written permission from the publishers.

Illustration on p. xiii: Plate from Virgil's *Georgics* showing farmers testing the soil. Courtesy of Special Collections, University of Virginia Library.

Set in Scala Roman types by Keystone Typesetting, Inc.
Printed in the United States of America.

Library of Congress Control Number: 2009928178
ISBN 978-0-300-13923-5 (hardcover : alk. paper)

A catalogue record for this book is available from the British Library.

This paper meets the requirements of ANSI/NISO Z39.48-1992 (Permanence of Paper).

10 9 8 7 6 5 4 3 2 1

This book is for C, of course

In the course of the period marked by the birth of modern scientific discourse the map has slowly disengaged itself from the itineraries that were the condition of its possibility.
—MICHEL DE CERTEAU, *The Practice of Everyday Life*

CONTENTS

PREFACE

This book is about how and why dirt became an object of scientific interest. It is, to that end, a story about defining the modern landscape with scientific means. I began the research leading to it interested in how people come to know what they know. With a course of study in history and eventual doctorate in science and technology studies (STS), a concurrent identity as an environmental historian, and a wayward background in chemical engineering (somehow), I aimed that question at the combination of science and nature. How did science become the dominant way to know nature? Today we take it for granted that scientific knowledge is the most politically legitimate and ecologically relevant kind for directing action in nature, but, as a matter of historical inquiry, how and when did that start to become so obvious? It turned out that the nineteenth century was the place to look for answers. To make my questions manageable, I examined a subset of this huge combination, science and nature, aiming in particular at agricultural science and agroenvironmental nature.

Agrarian sites present a particularly potent and timeless forum for questions about knowing nature. Everyone has access to the ground. Everyone eats the products of the agricultural process. How do we know how to do that, to produce food? Confucius touched on this, observing, "The best fertilizer on any farm is the footsteps of the owner." Two or so millennia later, during the era when agricultural chemistry was first named and debated, a writer in *The Southern Planter* in Richmond, Virginia mused similarly—to know the soil, he wrote in 1842, "the philosopher must exchange his laboratory for the open field." Untold numbers of others have noted the same thing, that practices of living in nature form the connection through which people come to know nature—be it dirt, forest, water, city, or more.

Notes from the Ground offers a similar focus on knowing and then working in nature. It examines the historical and cultural basis from which agriculture and science first came together in America, a story that begins in the later eighteenth century and becomes fully manifest by the mid-nineteenth. By looking to farmers, planters, politicians, publishers, natural philosophers, chemists, and other advocates and critics, this book examines the moral and material bases from which our agricultural (i.e., food-making) practices became directed through scientific principles. The study takes soil identity, soil fertility, and the cultural reasons to seek more systematic knowledge of both as the basis of its narrative. These issues of soil identity and fertility interest me not only in the perhaps Nixonian way of wondering what rural citizens knew and when they knew it, but also at the level of questions about credibility and authority: when advocates claimed to know something new about the soil, why did anyone else believe them? Taking the knowledge and credibility questions together, in its larger ambition *Notes from the Ground* is a study of how science became a culturally credible means for humans to interact with the environment.

This book began earlier in the decade as a dissertation. In its first form I was naively undeterred from asking overly lofty questions, questions that clearly would take a career if not several to approach feasibly. Through the process of researching the work, and in the intervening years of revision, emendation, and reconfiguration, I pared down the vastness of those questions with the help of a great range of friends and colleagues. First and foremost has been my wife Chris, who insisted that talking about some remote historical era have some purpose for all of us living now, not then. Whitt was born just before I embarked upon the research, and now that he's seven perhaps he can practice reading by searching for local references in this text. Alex joined us at almost the exact moment graduate school was officially completed, and her encouragement through unbridled enthusiasm for life has since been a sincere motivator. My parents, Lieba and Gerald; in-laws, Janis and Cesar; sister and brothers- and sister-in law, Shosh, Kurt, Denning, and Kimsey—though they never really knew what I was doing, were ever supportive even so, and can now put a copy of this on their shelves. If pressed, I may even sign their copies.

As for academics, now also friends, I thank Mark Barrow for advising me as I defined the original scope of this work and for pushing me to consider more effectively whom it was for and why. Ann LaBerge, my first academic adviser when I became a graduate student, provided me with the background in and fascination with Enlightenment and post-Enlightenment studies that got me to consider issues of knowledge production, culture, and philosophy together. Tim Luke, Richard Burian, Matthew Goodrum, and Saul Halfon not only pushed me to clarify and round out earlier versions, but did so with interest and sincere scrutiny. Among my peers, Wyatt Galusky read and reread almost everything more than once so that his intellectual contributions to this work are in fact unparalleled amongst everyone in this paragraph. From Blacksburg, conversations with and help from Jody Roberts, Jane Lehr, Brent Jesiek, Steve Bennet, Lois Sanborn, Piyush Mathur, Joe Baker, Gary Downey, Joe Pitt, Marco Esquandolas, Chikako Takeshita, Brian Britt, Barbara Reeves, Joseph Knecht, Peter Wallenstein, and many others formed the background of what would become this book; in Charlottesville, I thank Ed Russell, Peter Onuf, Jack Brown, Bernie Carlson, Alex Checkovich, Gwen Ottinger, Eric Stoykovich, and my other colleagues for their conversation and feedback about parts of the work. And I want to acknowledge the various kinds of help offered from afar by Arnold Davidson, Jason Delborne, Michael Egan, Deborah Fitzgerald, Chris Henke, Paul Lucier, David Nye, Emily Pawley, Michael Pollan, Alan Rocke, Laura Sayre, Steven Stoll, Paul Sutter, Emily Thompson, and Jeremy Vetter. Along the way, I was fortunate to have two non-academic editors for work I've done beyond the academy, John Warner and Vendela Vida, who had little to do with the specific content of this book but a great deal to do with its writing and form in general. At Yale, I thank Jean Thomson Black, Matthew Laird, Susan Laity, James Scott, Kay Scheuer, and three anonymous manuscript reviewers for their help, support, and encouragement. Institutionally, I thank the libraries and special collections staffs at the University of Virginia, Virginia Tech, the University of Pennsylvania, and MIT. I also want to acknowledge the Virginia Historical Society, where I was an Andrew

Mellon Fellow in 2003, and the Library of Virginia, where I consistently found congenial help. An earlier version of chapter 5 appeared as "Surveying Nature: The Environmental Dimensions of Virginia's First Geological Survey, 1835–1842," *Environmental History* 11 (2006): 38–69. I thank Marc Cioc for his editorial guidance there.

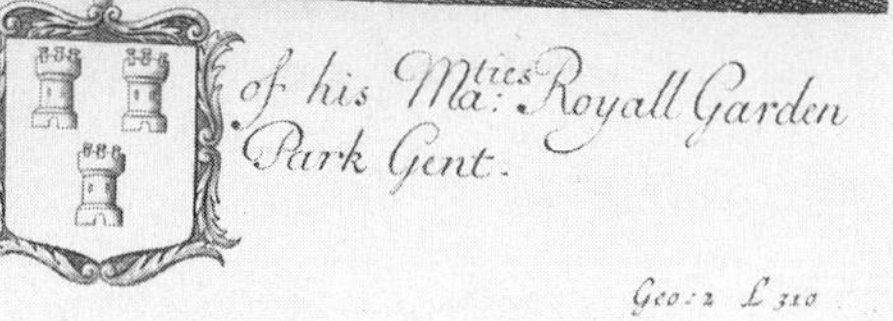

To George London of his Ma:ties Royall Garden in St James's Park Gent.

Geo: 2 L 310

Introduction

> Listen sometimes to the moans of an educated man of the nineteenth century suffering from toothache . . . not simply because he has a toothache, not just as any coarse peasant, but as a man affected by progress and European civilization, a man who is "divorced from the soil and the national elements," as they express it now-a-days.
>
> —Fyodor Dostoevsky, *Notes from Underground*, 1864

IN *NOTES FROM UNDERGROUND*, Fyodor Dostoevsky's unnamed narrator decries Enlightenment rationalism and the progressive ideology ushered in by a materialist scientific worldview. It was a mid-nineteenth-century commentary on the disenchantment of a world fostered by eighteenth-century Enlightenment values, a view that was already echoing from the Romantic movement. Dostoevsky's audience considered quantification, technical analysis, disembodied logic, and the values of efficiency and rationality to be hallmarks of modernity, guarantors of progress, and the underpinnings for political and economic success. The Industrial Revolution and the rise of the professional scientific class helped forward those values and confirm the alliance between science, progress, and improvement. At least, this was the outlook offered by the psychically fragmented author as he glossed a standard view of the rise of modernity. By the mid-nineteenth century, most people accepted and promoted the association of science with improvement, so much so that Dostoevsky's critique was recognizable only because he spoke against a clear and unquestioned historical phenomenon. He questioned whether the thoroughly cultivated improvement ethic of the previous century had in fact made the world better.

In another work downplayed as mere "notes," but written a century before and a hemisphere away from Dostoevsky, Thomas Jefferson looked ahead to envision a natural world of unified description made common by the language and practice of science. Rather than speaking at a high level of abstraction, as Dostoevsky had, Jefferson evoked the possibilities of science at the level of a practical and concrete subject: the physical land of Virginia. *Notes on the State of Virginia* (1787) was a pioneering survey of his native state, a veritable compendium of the natural features of the Old Dominion intended to present in systematic form its extant identity and future possibilities. An embodiment of the American Enlightenment, Jefferson dreamed of the very world—scientifically described, mathematically certain, rationally codified—that Dostoevsky's Underground Man rejected. From the start, he homed in on the land as the site for those future possibilities. In many passages of *Notes,* Jefferson in fact premised his argument for a great nation on assumptions of composition and soil identity made evident by science. Arguing with the French natural historian Buffon, for example, he appealed to the truth of universal scientific descriptions to make his case that America was as fertile, lush, and ripe for improvement as anywhere in the world—"[As] if a soil of the same chemical composition was less capable of elaboration into animal nutriment." Simply put, the American continent had the material means and the moral promise, vested in "the chosen people of God," he wrote, to justify the idealism of the American experiment.[1]

Jefferson's enthusiasm for a scientifically identifiable landscape was ultimately borne out, albeit through avenues less direct and more culturally situated than he could have ever imagined. Dostoevsky had already observed with qualms the success of scientific perspectives by the mid-nineteenth century; farmers, planters, politicians, and the newly identifiable "scientists" shared the observation, though generally without Dostoevsky's misgivings. In the twentieth century, agriculturalists could look back to the pivot from pre-modern farming to modern scientific agriculture as a nineteenth-century transformation occurring after Jefferson and before Dostoevsky. Assuming the science-improvement relationship, they often located that pivot within the 1840s contributions of one man, the chemist Justus von Liebig. It was Liebig who first codified the mechanistic philosophy of soil identity—arguing that soil was comprised of nitrogen (N), phosphorous (P), and potassium (K), and that problems of fertility were thus problems of maintaining the proper quantities of those elements (now known as the NPK paradigm). In this, he set "agriculture on its industrial path," Michael Pollan has written, breaking down "the quasi-mystical concept of fertility in soil with a straightforward inventory of the chemical elements plants required for growth."[2] He opened the door for the artificial fertilizers being produced by the 1850s, provided the conceptual basis for governments to make policies that were measurable and direct, and overcame dogmatic farmers to offer the final break from organic agriculture and, perhaps, an organic worldview. It was Liebig who solved the problems of the first famous agricultural chemist, Humphry Davy, the leading British gentleman of science who had forwarded organic, not mechanistic, theories of soil and fertility in the 1810s. Such has been the received view.

Today, the authority and credibility of science for observing and defining elements *of* the land and then prescribing a plan of action *on* that land are accepted without question. In public discourse, the terms "science" and "scientific practice" are also treated generally and with the assumption that they are solid and immutable. For the purposes of achieving some specific goal, be it economic, environmental, or political, citizens and policy-makers alike take science as the obviously best way to know and alter our world. Early twenty-first-century citizens discuss the problems and possibilities of industrial or organic or industrial organic agriculture, arguing over what's best for the soil, for food, for nature, for our society—that is, for nature and culture. They defer to a scientific and technical calculus all the while. But it took a lot of work for this to make sense, work that cannot so easily be summarized as the inevitable triumph of rational scientific progress over stubborn, resistant tradition. It took more than Liebig, who was fortunate to leverage the conditions already being produced in the American agrarian landscape. The received view, a straight line from Davy to Liebig, is incomplete and misleading, concealing the work of local, non-expert citizens that paved the way for the contributions of scientific writers and their codified texts. It ignores the strong cultural context of working agrarians, while presuming that a clear and stable concept of science was waiting in the wings for its eventual entrance onto the world stage. I have found the post-Enlightenment setting of the early American Republic a fruitful site for coming to understand the work that made those accomplishments, in retrospect, seem neat. At a time when the moral and material dictates of improvement were of a piece—so that to improve the land was to improve the virtuous agrarian culture—when systematic and scientific practices were being developed and fitted as a new prism through which to see the land, and when the prevailing environmental ethic within which those improvements could be set was a georgic, or experience- and work-based, ethic—at that time and under those conditions, the sciences of agriculture began to cultivate and grow authority.

Notes from the Ground examines the cultural conditions from which science and agriculture first came together in America. It looks to the early Republic combination of scientific practices, improvement values, and agrarian virtues for the roots of why and how Americans began to accept scientific practices as legitimate practices in their agricultural lives. I argue that agricultural science became politically significant and ecologically relevant by mid-century, first, because it fit into the complex improvement ethic of that period and, second, because it drew from contributions by local citizens. It was their land, after all, that was being redefined and reconceptualized. It was also their improvement ethic that framed debates about the value of science, an ethic at once cultural and material. Thus, the new systematic practices that would soon be called scientific had to be understood as bettering both community values and soil properties at the same time. The nonscientists, the agrarian citizens, therefore played a crucial role in granting credibility to the new practices. Their views provide the bulk of discussion in this book.

The early American Republic provides a unique setting to think about when and how scientific agriculture began to become part of a set of new rural practices. It was certainly not the only place improvers were attempting to formulate more systematic methods of identifying and manipulating the soil. In settings across the world, from at least eastern Europe to Britain, from the range of states in the United States to Brazil, farmers and governments were working to experiment with their lands with greater methodological rigor and to codify their results in text. In the United States, though, these efforts were more tightly integrated with cultural and moral goals, where something as seemingly mundane as dirt was given great prominence, situated as a centerpiece of political economy for both North and South.[3]

For centuries, humans grew food following basic practices of tilling and plowing and seeding, practices understood through the slow cycle of the seasons from experience and tradition. Experience provided the most central means through which people came to understand the nature most relevant to their lives—by working the fields, they knew what the land was. In the late eighteenth century, as improvement advocates began to agitate for "systematic" fertilizing and experimenting with the soil, there was no such thing as science or agricultural chemistry in the sense we consider it today. What is more, most people who would write about it were natural philosophers, men like Davy in London or Liebig in Giessen, men who knew little about the soil from personal experience. Yet here were early Americans, agrarians in culture and economy, in community and political structure, struggling with new ways to work their land in ways very similar to Davy and Liebig.

We might consider the differences in rural meeting halls between the early and the mid-nineteenth century to gain a sense of the shifting authority of approaches to understanding that soil. First picture Jefferson and his central Virginia neighbors gathering to discuss strategies for improving their land's productivity. It's the early 1800s, and they've huddled in Charlottesville or Richmond—or their peers to the north have done so in Philadelphia or Albany—trading results from individual experiments in their fields. Let's say it's a sunny day, with carriages parked outside, probably a slave tending to the horses (in Virginia, at least), dusty floors, and men in post-Revolutionary homespun attire with an almanac under arm. They were model agrarians, the very symbols of rural virtue. What technique for crop rotation had they used? How much lime or gypsum did they apply to their soil? How often did they add fertilizer? Were there new fertilizers available? These were the particular, direct questions at the table. The answers came by reference to actual experiences on the ground, by the leading authority of hands-on, practical knowledge of farming. The questions may not sound all that riveting to a twenty-first-century sensibility. Yet the records of farmers, planters, and early agricultural societies speak to the vibrancy of such discussions. More intriguing is that the gentlemen sat inside a larger context to ask those questions, as Jefferson and his ilk well knew. They were also asking: What did improvement mean? What was their goal—improvement toward what? How did the material aspects of agriculture meld with the cultural ideals of agrarianism? What was the connection between everyday practices of working the land and intellectual patterns of understanding what that land was? And what was all of this intellectual and moral work for?

By the mid-nineteenth century, the carriages might have been sturdier, the roads better built, the locomotive apparent in the distance, the market revolution receding into the past, and the political sky much cloudier with the storm of Civil War looming on the horizon. The same kinds of conversations were going on, but the individuals now at the table had the means to seek answers by reference to a newer kind of authority, with explicitly scientific principles and techniques. This was true in several senses: the concept of improvement was attached to the idea of, and in fact growing to be synonymous with, scientific knowledge and technical practice; the means for improving productivity on the land was scientific or, in a term more widely used, systematic agriculture; the way to know which kinds of fertilizers were best was through systematic or chemical experimentation. What new artificial fertilizers were being produced? How closely had the application followed chemical ideals? What theory of fertility and nutrition had the farmer employed in choosing his fertilizer? Instead of the rural home, these men met at the Agricultural Society or maybe even a state office, where an official chemist or a competent reader of a scientific text weighed in on the matter of land management. Instead of sole deference to experiences on the land, they made policies based on reference to codified, "scientific" techniques.

The gentlemen used new mechanical and scientific metaphors and turns of phrase to understand and describe the land, language that helps readers today see the shift from more general to more specific, and scientific, descriptions. They captured those changes by promoting the new view that science and agricultural improvement were of a piece. In the 1850s, rural citizens as little known to historians of science and the environment as the Virginian Richard Eppes would write that "a farm is another name for a chemical laboratory." Franklin Minor, a fellow Virginian speaking to the Virginia Agricultural Society in 1855, offered a clear cultural critique by stating his astonishment at the "stupid prejudice against scientific agriculture." That same year, Thomas Ewbank, the former U.S. Commissioner for Patents, published his view in *The World a Workshop* that the earth was but a site for "the cultivation and application of chemical and mechanical science." Politicians of the period offered similar perspectives. The Pennsylvania congressman James Campbell, as one example, asked his audience to "let science, applied to the culture of the earth, go hand in hand with practical labor." In the 1860s, the professional chemist James Nichols wrote in his *Chemistry of the Farm and the Sea*, "Let us . . . feel assured that chemistry, which holds the key which has unlocked so many rich chambers in the storehouse of nature, will open others fully capable of supplying all the wants of the husbandman."[4] Soil and the possibility of a soil science formed the central locus of antebellum political economy, highlighting as they did the agrarian, cultural, political, and environmental character of society. The newly authoritative sciences of agriculture were more than disembodied laboratory endeavors, as the gentlemen mentioned above knew, instead serving as signifiers of broader goals of improvement, knowledge, and political economic organization in the early Republic.

We could understand the difference between the two eras—speaking here to changes in agricultural practice—in at least two ways. The first would be by examining the productive capacity and sheer quantity of land under cultivation. How did it look different? How much more was there? How much more food did it produce? Those kinds of queries would speak to the well-trodden topic of soil exhaustion and rebirth, of declension and recovery. They are questions about modes of *production*. Essentially, did all that systematic attention to agricultural practice save the soil or not? These kinds of questions have been common in the literature on the early Republic. I see a second way to approach the possible differences between the earlier and later eras, a way that has us understanding how agrarians interacted with and thus could understand that land. How did they develop concepts of it? What did they come to know that led them to treat it differently? Were scientific principles considered of a piece with improvement? This set of queries hits the conceptual side of things more squarely, although, given the political, cultural, and economic contexts of agrarian America, saying "the conceptual side" does not divorce the issue from the others. They are questions about modes of *interaction* between people and their land. My interest here is in this latter issue, especially because I want to put the role of connection to the land into the foreground so that those modes of interaction can be recognized as sites of contested authority.

To be fair, I may have overstated the differences the half century made, suggesting a stark before and after picture, one without science, one with.

That isn't completely accurate. The stakes were not all-or-nothing, nor were hands-on forms of knowledge tied to heritage and family work displaced forever after because science had trumped them. A quick conversation with any farmer of the last 150 years would make this abundantly clear. What is more, students of scientific history would note my over-generality with the term "science." But if I introduce the differences too sharply here, it is only to clarify the two sides of a shifting locus of authority, the first about tradition-laden, experience-based, locally derived knowledge of the land, the second about systematic, codified, and universal knowledge. The government's decision to form the U. S. Department of Agriculture in 1862 may serve best to underscore that by then Americans were organizing and legislating a different kind of knowledge of the land, one meant to be installed in fields, accessible and profitable in ways dictated by scientific principles of soil identity and manipulation theretofore only loosely understood or practiced. It led to a change not just in how Americans produce food but in how we interact with agrarian nature in ways that remain with us today. I draw out the contrast in ways to know the soil because I want to understand how practices on the land and the ideas that gave them meaning developed. In the main, I want to understand, in the sense given by the philosopher Michel de Certeau, the itineraries that were the conditions of possibility for the emergence of a scientific and technical discourse about agricultural environments. My approach is to understand the early Republic as a place where the conditions for making a scientific agricultural site were first developed, not to suggest that scientific primacy and credibility were thereafter unquestioned.

All of this leaves me with at least one basic question: Why did agrarian citizens come to seek systematic—later, scientific—strategies for improvement? In retrospect, public understandings of science consider the association of science with progress as obvious and uncontentious. Farmers' use of scientific techniques for agricultural improvement should thus be understood as, perhaps, a "natural" course of action. In fact, if we do the opposite, if we question the assumed relationship, that understanding is the odd thing. There Dostoevsky sits, skeptical about the mathematical certainty of mechanistic science. Yet, the story of what happened and what made it possible is not so simple. How could that transition from the enthusiasm of Jefferson to the antipathy of Dostoevsky take place? Where did science fit in the countryside of the early national period? How did

Americans move from the early years of a new nation to the onset of the Civil War, from the pre-Davy years of the early 1800s to the post-Liebig years of the 1850s, from a world known by daily working experience to one known through scientific codification? What lessons can we garner from that transition for today for the sake of environmental ethics and policy? And, most basically, what can the changes tell us about how people come to understand their environments, the very nature under their feet, in new and more commanding ways? These are my guiding questions in *Notes from the Ground.* For answers, I have dug deeply into the cultural dimensions of agricultural improvement.

In many ways this study of science and agriculture is about the relationships between ideas of and practices in the environment. Yet, environmental studies scholars have left underdeveloped the viable environmental ethic that best captures the kind of relationship agrarian citizens had with their land. For this reason, I see the environmental ethics context of that agrarian world as providing a useful window onto the period. The pastoral ethic certainly stands as the most prominent, most studied variant of a descriptive environmental ethic in that time. It is born of Virgil's poetry and evocative of the comfortable imagery of humans fitting easily and unobtrusively into their landscapes. Grand portraits of eighteenth- and nineteenth-century America, of flowing fields and frolicking planters, speak to the idyll of a pastoral landscape. With the pastoral's imagery of sublime appreciation, leisure is life. It is an abstraction, and not surprisingly poets, artists, and philosophers provide our sources for it. But, curiously, it does not represent the actual relationship most Americans had to their land in the early Republic and, thus, it limits our access to that era.

Virgil also offered another view of rural virtue, that of his *Georgics.* The *Georgics* finds the land as a site of labor. As one literary scholar put it, the essence of the georgic is "not a Golden Age where apples drop freely from the boughs but a fallen world of hardship and toil where one lives by the sweat of one's brow."[5] And in that sweat comes the famous virtue of Jefferson's chosen people of God, the farmers. To address the questions of "what happened" in the countryside of the early Republic and "what made it possible," I draw out the importance of this experience- and work-based georgic ethic for the rhetorical value it had in debates about right practice on the land as an explanatory means for those conversations.

For Americans, that georgic ethic was a more basic credo, one that understood labor as life. The overwhelming majority of Americans worked in agrarian settings or for agrarian purposes for the entirety of the antebellum era, even as the Industrial Revolution was finding new ways to convert the products of the farm into salable and tradable goods. The soil provided a central organizing principle in the early Republic, serving as an ideological, even if contested, basis for wealth, livelihood, and community. Twentieth-century scholarship may have drawn stereotypical differences between industrial and agricultural pursuits—perhaps between John Adams and Jefferson, or the archetypal pro-industrial Federalists Alexander Hamilton and Tench Coxe against the farmers—but such contrasts are generally overplayed. For early Republic actors, those differences were more about how to convert the products of the land into industrial products than about whether one should stay and the other should go. Hamilton himself considered industry a way "to improve the State of Agriculture." Adams thought "the Enthusiasm for Agriculture like Virtue will be its own reward. May it run and be glorified."[6] The georgic ethic that helps explain this cultural focus on the land is thus both widely applicable and surprisingly absent in the current literature on the early Republic.

To explore the basis from which science and technical practices became new forms of interaction with the land, I want to regain the sense of experiential knowledge and connectedness early Americans had to that land. The georgic improvement ethic helps this cause. Agrarians made sense of their land through the labor and experience of working it; they also drew from the cultural value of labor even if the wealthier among them were distanced from the fields through the use of hired labor, overseers, and slaves. To understand, then, how forms of interaction were also forms of knowledge—that is, how the way people see something informs what they end up thinking about it—I rely on notions of work, activity, practice, and process.[7]

Consider, for example, the space between humans and the land. The space is not empty; in fact, it is stuffed and overflowing with ideas, values, beliefs, and tools, with other individuals and other families, with political arrangements both local and beyond. All of those factors form the terms of the relationship between humans and their land. When scientific practices started to fill that space, they offered a new means with which people could see the land beneath their feet. That lens, more a prism than a clean piece of glass, was a tool. It may even have been a physical tool like a surveyor's kit or a chemist's fertilizing apparatus, but it was also a way of seeing, a conceptual tool. To push the point in what follows, I treat science as an activity, as something people do. Science is certainly also a set of ideas, a community of styles, and a body of rules. For the rural citizenry I examine, though, it was something that moved and was actively shaped, an activity that was produced and reproduced in everyday practice.

Georgic experience and the overtures toward promoting its value provide the setting for early Republic views of nature. "Work itself is a means of knowing nature," the environmental historian Richard White has shown, so that experience, labor, and practice are not issues of cultural identity alone, but also, as forms of interaction, markers that show the production of environmental knowledge.[8] In that sense, the georgic also provides the framework from which we can see science as a form of work and experience. Fortunately, recent work in literary environmental studies has begun to put the work-based georgic ethic into the foreground, bringing it out from the shadows of its dominant partner, the pastoral. Timothy Sweet's *American Georgics,* in particular, introduces the improvement-based underpinnings of a georgic ethic in America. As it was defined through agricultural work and agrarian virtue, the georgic held together the moral and the material. Much as the historian Drew McCoy notes that the notion of "political economy" at the time did not distinguish between economic and ethical facets, so too did the georgic avoid distinguishing between moral and material pursuits.[9] The two were of a piece: improving the land was consonant with improving the moral stature of the individual and the community. Modern readers, taking for granted a now-common perception, look at the eventual rhetorical distinction between morality and materiality and consider them as separate areas. To regain the sense of experiential knowledge and connectedness early Republic citizens had to that land, this book maintains the non-distinction between moral and material dimensions of the improvement ethos. Sweet's work goes far in helping make such interconnection apparent. The point makes further sense once we recover the basic observation that agricultural work and the agrarian life were fundamental elements of the Republican experiment.

The chapters that follow examine how science aided agricultural improvement, how such science-based improvement was promoted or resisted by local farmers, and in what ways it became a credible practice for interpreting and interacting with agrarian nature. Rather than taking what could be perceived as the inevitable solidity of science for granted—as if farmers were waiting in the wings for its *sui generis* arrival—the discussion considers that term in flux over the early Republic, understood, crafted, and defined differently in the thick world of agrarian practices. Part I of the book discusses the moral, practice-based, and instrumental place of this contested term, agricultural science, in the American countryside; Part II asks instead how people used so-conceived scientific practices to interpret the land, looking first at the county and then at the state level. While the first half of this book examines farmers from South Carolina to New Hampshire, the second half is geographically and culturally focused on the case of Virginia, a state with a rich historiography but which has nevertheless suffered from relatively little attention to its environmental history.[10] By putting the circulation of agricultural science in the context of improvement-minded agents, I hope to better locate agrarian American culture in a post-Enlightenment setting while developing a more robust picture of how morality, materiality, and theory were wedded in the much-revered principles of practice (experience) and practicality (utility).

The example of the agricultural sciences stands as an approachable case that speaks to the larger transition from Jefferson to the outbreak of the Civil War. The rising credibility of agricultural science provides a particularly useful and precise case of a new form of interaction with the land coming into prominence for agrarian citizens. It also allows me to draw from a well-established body of literature in science studies and environmental history, a corpus that, although vast, generally speaks to the most famous, professional, or powerful agents, usually chemists and geologists and other "men of science" without attention to the actual farmers, planters, and laborers whose changing worldviews were the true mark of a shifting sense of nature. Because those nonspecialized and lesser-known agents were already experimenting, theorizing, and promoting notions of chemical descriptions of their own, they provide a view into how the working context of the era shaped how and in what way new means for defining nature could become credible. In agrarian America, improvement advocates had built the conditions that helped set a scientifically mediated view of rural land in its basic form by the mid-nineteenth century. Not just retrospective scholars, but nineteenth-century actors alike speak to the authoritative status of science for knowing nature by that time. I pointed earlier to Richard Eppes, Franklin Minor, James Campbell, Thomas Ewbank, and James Nichols, not because they were leaders of the time, but because their views were representative of a common perception.[11]

Understanding how the basis for the scientization of nature was produced is, at one level, an unfathomable task. The very idea of science is to study nature, meaning that this book's goal would be to explain the whole of the historical scientific enterprise. But at a more concrete scale, the one where Americans lived, worked, and formed their identities, we can look to the introduction of systematic technical practices onto the farm because they offer a recognizable, not just abstract, focus on altering the ways people understood their lands and their lives. Something interesting went on over the course of the early American Republic, something that started with unorganized, disparate attempts to study the land, its productivity, and its potential, and ended with the possibility for codified, organized, and economically and politically reliable means for acting upon the land. New environmental policies of the mid-nineteenth century—what in retrospect came to mark the first promise of modern agroscience—were aided by scientific and technical terms because those practices were shaped by a moral and cultural context that promoted experience and virtue.

Dostoevsky's view of a rationalized, materialistic world relies on the conditions developed in the half century prior to his penning of his words. Steven Stoll summarized it well when he wrote: "Reformers of the early Republic . . . should be considered progenitors of a general regard for nature that would, in the next century, become a policy of government. They provide good evidence that what ultimately became codified by professional scientists and conservationists had roots in amateur practice."[12] The research in this work is rooted in sources from the early American Republic. But those examples from that time provide a useful and relevant situated case of several themes that resonate today. Issues of credibility and authority, of lived experience in the field versus speculation from afar, and of cultural production from below versus hierarchical dissemination from above speak to a context beyond only the early Republic. In addition, *Notes from the Ground* elucidates some recurring historical themes: that science has been built into the land with practices of working experience, that noncentralized agents contribute to and define the terms of scientific authority, and that forms of human connection to nonhuman nature have always been the arbiter of what we think of that nature and what we then do as members of it. To change concepts of nature today to create the possibility of different environmental activity, we must pay attention to and shift the forms of interaction that produce those concepts. We must, that is, be aware of the possibility of new kinds of science and technology that promote connection, engagement, and work.

PART I THE PLACE OF SCIENCE

CHAPTER 1

Distinguishing the Georgic

The Father himself hardly / willed that agriculture would be easy when he called forth / the field with his art, whetting human minds with worries, / not letting his kingdom slip into full-blown laziness.

—Virgil, *The Georgics*

JOHN SPURRIER WAS "an old experienced farmer" and a late eighteenth-century British transplant living in America. He farmed the Brandywine region of Delaware in the decades surrounding the Revolution, seeking the well-advertised fertile land of the Americas and the cultural improvement the agrarian world offered. He recorded years of rationally planned experimental observations in cultivating his soil. In 1793, he finally published *The Practical Farmer* as a compendium of those experiments "to contribute to the public good" and to appeal in "plain language" to neighbors and friends alike. "I trust that such observations as tend to promote the public welfare will be deemed worthy of some attention," he wrote, in a demure style common to such public treatises. In the compendium, page after page presented the fruits not just of tilling the land but of applying "reason and experience" to the enrichment of soil and, by virtue of his goals for public welfare, culture alike. Discussing the use of different manures for fertilizing and the process of treating different fields to various degrees of tilling, noting the variety of "mechanical, chemical, and philosophical elements of agriculture" available to him, his account showed a life spent learning about the soil from working it. *The Practical Farmer* was not a remarkable treatise for its novelty of form—Spurrier wrote in a mode common to the Englishman Jethro Tull's (1733) *The Horse-hoeing Husbandry*, Jared Eliot's (1762) *Essays Upon Field Husbandry in New England*, and a range of eighteenth-century personal works on land practice. It was remarkable, rather, as an example of a late Enlightenment synthesis of agriculture, natural philosophy, and cultural goals, showing that natural philosophy—and chemistry—held a visible place in the farmer's life.[1]

As with many others in the early Republic, Spurrier's relationship to the land was georgic, a kind of relationship named in reference to the Roman poet Virgil's *Georgics*. That georgic outlook was, and is, an agrarian and environmental ethic combining cultural and material values. For Spurrier, it was premised on the value of work, of human labor in nature. The georgic ethic stands as a means to understand the land that relies upon the lived experience of the laboring individual. In this, it runs in tension with a modern mode of agricultural understanding that reduces the field to a laboratory and prioritizes the minimization of work. As we look back onto the fields of the late Enlightenment world from the twenty-first century, it offers us a description of the orientation Spurrier and his early Republic neighbors had to their land. From lived experience, they understood that the improvement of the soil—increased fertility and health—was synonymous with the improvement of society—a healthier, stronger, and more virtuous culture. *The Practical Farmer* was an example of holding those together, soil and society, and seeking improvements of both with the benefit of "natural philosophy," or, the sciences of agriculture.

Virgil had crafted classical expressions of farming virtue in *The Georgics*. Written in 29 B.C.E, before his *Aeneid* and after the *Eclogues*, the Roman poet's verse was didactic, aimed at inspiring Romans to return to the labor of the land and away from their transition to militarized state. Within, he gave the terms for an environmental ethic integrating the goals of moral, cultural, and agricultural improvement. In kind, Spurrier and his neighbors did not reduce farming to a mere material activity, nor did they speak of it only as a way of life. It was always both at once. For them, to be sure, agriculture and the farming life were not easy. As Virgil said, "Father himself hardly / willed that agriculture would be" so. But this was not entirely problematic—agriculture's taxing physical demands provided the basis for esteemed cultural values of labor and experience. It defined an austere work ethic, a community value, and the cultural promise of increased public welfare. The English agricultural advocate Henry Home (Lord Kames) had expressed this sentiment by tying it to nationalist goals in 1776 when he wrote, "In a political view, [agriculture] is perhaps the

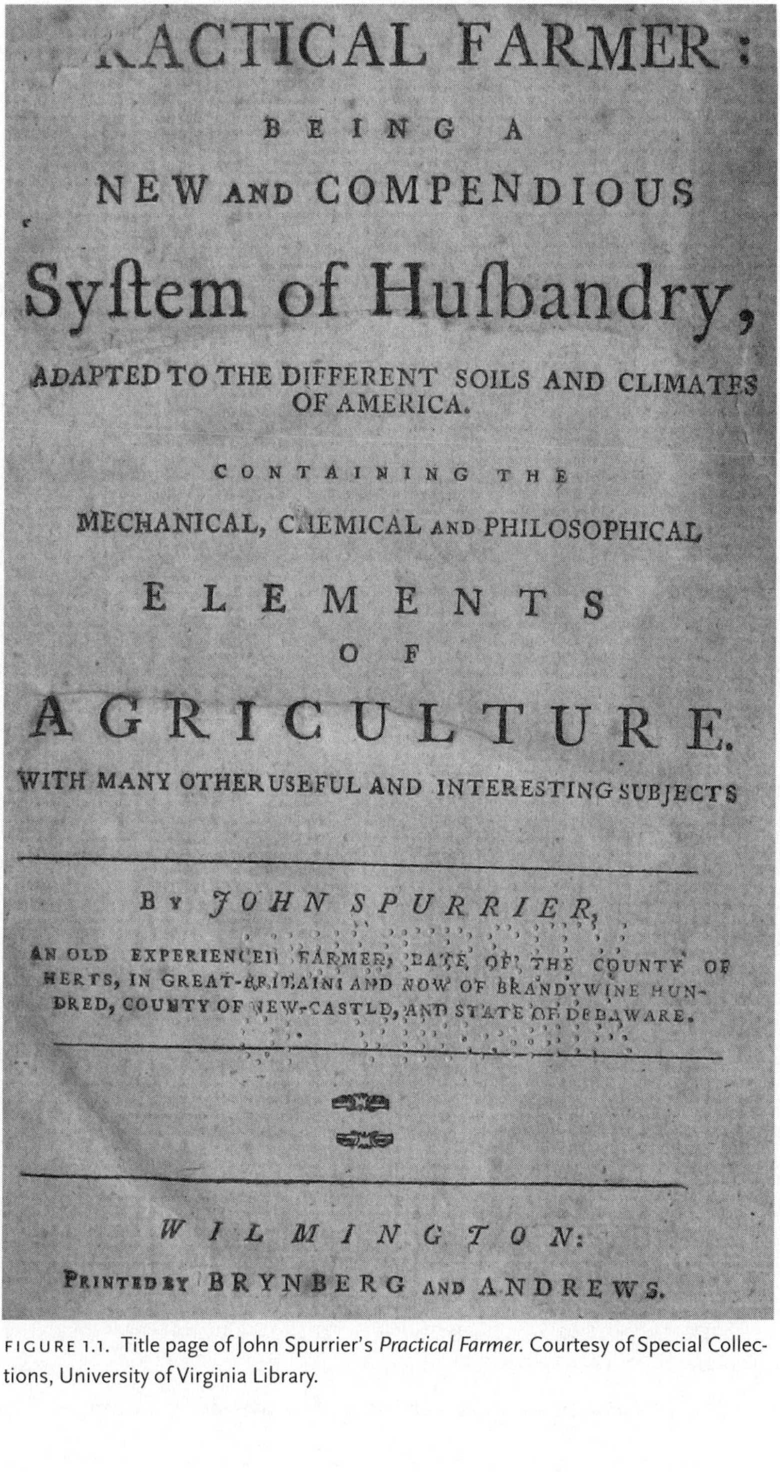

ACTICAL FARMER:

BEING A

NEW AND COMPENDIOUS

Syſtem of Huſbandry,

ADAPTED TO THE DIFFERENT SOILS AND CLIMATES OF AMERICA.

CONTAINING THE

MECHANICAL, CHEMICAL AND PHILOSOPHICAL

ELEMENTS

OF

AGRICULTURE.

WITH MANY OTHER USEFUL AND INTERESTING SUBJECTS

By *JOHN SPURRIER*,

AN OLD EXPERIENCED FARMER, LATE OF THE COUNTY OF HERTS, IN GREAT-BRITAIN; AND NOW OF BRANDYWINE HUNDRED, COUNTY OF NEW-CASTLE, AND STATE OF DELAWARE.

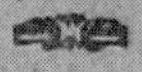

WILMINGTON:

PRINTED BY BRYNBERG AND ANDREWS.

FIGURE 1.1. Title page of John Spurrier's *Practical Farmer.* Courtesy of Special Collections, University of Virginia Library.

only firm and stable foundation of national greatness. As a profession, it strengthens the mind without enervating the body. In morals, it has been well observed, it leads to increase of virtue, without introducing vice. In religion, it naturally inspires devotion and dependence on Providence." George Washington would echo this view two decades later and in a different country when, in a 1796 address to Congress, he stated, "With reference either to individual or national welfare, Agriculture is of primary importance. . . . By diffusing information, [we could] encourage and assist a spirit of discovery and improvement." Spurrier sought to promote the same ethos.[2]

The georgic ethic helps frame this complex agrarian identity. With it, we can see that the value of hard work and the fertility of the land were tightly bound to the virtues of the agrarian citizen. The ethic provides a way to understand human relationships to the soil that stands as an alternative to a more passive, more idyllic, and pastoral one. It keeps humans close to and in touch with the dirt under their feet. However, this georgic ethic has received limited attention in literature on agroenvironmental history and ethics, and as a valuable lens for understanding how Americans knew their world through work. Literary scholars have focused on it for its poetic and narrative expression, American and agricultural historians have generally tended to give it short shrift, and environmental scholars have typically taken notice of it, when they notice it at all, as a "hard" pastoral. In the present study, I take a bit from each, but go further to explore and champion the georgic ethic in its relation to science, agriculture, and improvement. Spurrier was but one of a host of author-farmers in America's early years, individuals who would prepare the patterns of recording agricultural experiments that later held together that science-improvement nexus on the land. His book's 1793 publication helps set the early marker for a transition from qualitative and unsystematic accounts of agricultural practice to a systematic, chemistry-aided *georgic* science that held together the dual moral and material mandates of American improvement.

GEORGIC ROOTS: WALKING AND TOURING

John Spurrier based his compendium on the value of years of experiments in his Delaware fields. But from where did Spurrier's experiments come? What conceptual basis was his work drawn from to convey a systematic view of "properties ascertained by repeated experiment," a practice "enabled from my own experience" based on "precept, reflection and study"? We have little record of Spurrier's biography but for that gleaned from the record and methods of his experiments in *The Practical Farmer.* That book reveals an approach common to the eighteenth-century treatise, one based on knowledge "the agriculturalist" gains by "riding or walking around his farm."[3] This was the spirit of practice Cato and Virgil offered at the height of the Roman Empire, and the basis from which those Romans had elevated the dignity and civic stature of the farmer. It was also the basis from which a new form of agricultural improvement literature was born in the later decades of the eighteenth century in Britain, the Georgic Tours genre. I describe those Tours below, but first speak to the georgic's presence in studies of late Enlightenment scholarship.

The georgic context has been somewhat buried in historical and environmental scholarship because of the dominance of the more Romantic pastoral ethic. The pastoral, like the georgic at once an environmental ethic, a poetic construction, a narrative form, and an aesthetic sensibility, has served the literature on environmental studies well. Its analytical utility comes perhaps from its relevance for a wide range of scholarly approaches. It helps define a middle ground along a spectrum from wilderness to civilization, as Roderick Nash expresses it in his *Wilderness and the American Mind;* it stands as an ideal, an organic, life-affirming antipode to dehumanizing technology, as Leo Marx explained it in his landmark *The Machine in the Garden;* it suggests an ideal of literature that emphasizes, as the ecocritical scholar Lawrence Buell says, "an ethos of rurality or nature or wilderness over against an ethos of metropolitan." Donald Worster, in *Nature's Economy,* draws clear distinctions between the Arcadian (as synonymous with Pastoral) and imperial studies of nature. That Arcadian view represents a peaceful relationship to the world within which humans live (as with, for example, Thoreau), while the imperial school of thought is understood by its goal of controlling and dominating nature (as with, for example, Linnaeus and Bacon).[4]

As a way to express a sense of contrast in studies of environmental thought, of nature writing, and of technological history, the pastoral offers a clear tool demarcating one view of nature from another—civilization from wilderness, culture from nature, city from country, mechanical from organic. It is born of Virgil's *Eclogues* and of the aesthetic and emotional

FIGURE 1.2. Thomas Cole's *Pastoral, or Arcadian State*. This picture is the second of five in Cole's *Course of Empire* series. It follows *The Savage State* and precedes *The Consummation of Empire*, in the process exemplifying the middle ground of the pastoral between wilderness and civilization. The painting also depicts the passive, leisurely placement of humans in the landscape—note the philosopher seated to the left, pondering the world around him, and the shepherd in the center, easily tending his flock in the sun. Reproduced with permission from the collection of The New-York Historical Society.

response to timeless, gentle, and leisured cultivation. It situates humans as part of the natural world, not outside it; even as they cultivate, herd, and develop their resources, they do so within the constraints of a world greater than themselves. The pastoral also exemplifies an aesthetic and narrative form—in particular an aesthetic of passive contemplation, of staff-holding shepherds resting in mountain valleys, providing the narrative thrust that characterizes humans as frolicking leisurely in a nature well suited for Romantic values. As the classicist Bruno Snell framed it a half century ago, the Arcadia of Virgil's *Eclogues* is set in "a far away land overlaid with the golden haze of unreality."[5] In the 1830s, Thomas Cole, the founder of the Hudson River Valley School of landscape painting, portrayed the pastoral as the prelude to civilization in his magisterial sequence *The Course of Empire*. His landscape shows humans gently and almost passively placed within the contours of the mountains, valleys, streams, and fields of his view. People frolic.

But Virgil gave us more than the pastoral: he also presented the land as a site of labor. In this georgic world, people work.[6] The environmental literary critic Timothy Sweet lends further clarity to the definition contrasting the georgic to the pastoral as two distinct modes of orientation to the land: for the georgic, labor is life; for the pastoral, leisure is life.[7] Human intervention is a central tenet of the georgic ethic, not a problematic relationship to be explained away. It is an aesthetic and narrative form, like the pastoral, but one which casts the image of difficult engagement beyond Romantic impulses to glorify the agrarian life as one of comfort. The distinction between the two ethics, then, is not that the georgic elides nature/culture or wilderness/civilization differences while the pastoral keeps them separate; nor is it that the one places humans in the landscape while the other keeps us out. In fact, the georgic has often been subsumed within the pastoral, sometimes as a "hard" pastoral, and as such is often conflated with it. The distinction I want to note is one of emphasis, with the georgic highlighting the ways humans *interact* with their world and the pastoral highlighting the two sides of that interaction instead of the mediation between them.

That element of interaction introduces moral demands, since it emphasizes relations between different people and between people and their land. Virgil understood this. Rather than defining his work reductively, he combined moral instruction with practical advice in the person of the farmer. In so doing, he enfolded moral and material elements: promoting the occupation of farming was part and parcel of modeling the practice of agriculture as civic virtue. In American historical lore, Thomas Jefferson's 1780's view that "those who labor in the earth are the chosen people of God" stands as the early national expression of agriculture's virtuous identity. Today, Wendell Berry's neogeorgic call to return human labor to the core of an environmental ethic, Wes Jackson's similar focus on agrarian contexts, and recent work by pragmatist-oriented environmental ethicists, who promote an approach that asks us to debate environmental matters based on practices in nature, not just abstract reflection about it, stand as part of the legacy of this lost ethic.[8]

The georgic offers a valuable perspective from which to interpret how Americans have understood their land, especially in the early decades of the nation's history. Spurrier's writing, for example, is cast more fairly as georgic than pastoral. Because Americans of his time were predominantly agricultural in their political economic lives, our understanding

To Sr Thomas Trevor
His Majestys
of the Inner Temple Knight
Attorny Generall.

FIGURE 1.3. Plate from Virgil's *Georgics,* showing active farmers working the land, applying labor to the fields. In some ways the imagery is similar to the strictly pastoral (the farmers here also live in a middle ground between uncultivated wilderness and controlled civilization) but in others it is importantly different—here they work, rather than sitting in passive contemplation. Courtesy of Special Collections, University of Virginia Library.

today of environmental perceptions at that time cannot rely solely on pastoral, sublime, or wilderness-based forms of experience. The land is where people lived; their views of nature were born of those lives. Even in the important narrative of settlement, Americans were coming to understand their environs through the development and protection—and the difficulty in both—of farmland. Any discussion of the environmental sensibilities of the early American Republic must not just *make reference* to agriculture, but should be embedded *within* it. Through the georgic, we can study agriculture for its moral connotations as much as its economic and material meaning. Through the georgic, I mean to underscore the human *connection to* nature, though in an agricultural era just coming to deal with further separations from working the land that industrialization would bring.[9]

Because my focus is initially on the early decades of the United States, the revival of the georgic in the genre of the eighteenth-century agricultural tour is more relevant here than a faithful reference to the original. Taking its cue from the Roman agrarian poetic tradition of which Cato's *De Agricultura* was also a member, the ethic of Virgil's *Georgics* reappeared in rhetorical expression and gained new cultural currency in that eighteenth-century form.[10] There, the term was taken up by Scottish and English rural improvement advocates, finding expression in the "Georgic Tours" literature of that time.

The Georgic Tour was a literary genre that sought to promote Enlightenment ideals of improvement through observations of farming. The purpose of the Georgic Tour, then, was to promote an admittedly malleable ideal of "improvement"; its premise was to refine that generic ideal and to suggest that improving the future possibilities and current productivity of the countryside required an intimate knowledge of that place, knowledge best gained by direct experience on the land. By collecting observations from the land, putting them in writing, distributing them as an eighteenth-century version of "best practices," and doing so with the intimation that "science" would have some role to play in the process, the Georgic authors were inventing and directing the goal of agricultural improvement.

The historian and agricultural writer Laura Sayre, in her study of the prominent eighteenth-century rural British tourists Arthur Young and William Marshall, has observed that the georgic "suggested new ways of reading and writing the rural landscape, establishing an essential connection between the intellectual work of the gentleman and the physical work of the laborer."[11] A more fine-grained approach to tilling, cultivating, fertilizing, and managing the soil would lead to more productivity and better control of the land, the men posited. Virgil had advised farmers, "you who work the land," to test soil variety by feel, and by taste, and by crumbling it in one's hands.[12] These were all tactile measures that relied on direct observation and required immersion in the experience. One cannot speculate about feel; one has to actually touch something to gauge feel.

The Tours sought to highlight this hands-on ethos of agrarian life. In his *Six Months Tour through the North of England* (1771), for example, Young detailed crops, experiments, and farm practices (including prices sought and wages paid) in four volumes of methodical accounting, quantifying that which had theretofore been only qualified; that same year he published his *Farmer's Tour through the East of England;* he later followed these reports with *Travels During the Years 1787, 1788, & 1789*. While explicitly dedicated to Virgil's kind of direct contact with working the land, these efforts were admittedly an upper-class—and thus somewhat sanitized—thrust at observing rural practice firsthand, recording observations, and offering a somewhat codified set of observations as the basis for future improvement practices. But Young—"the climax of the late eighteenth-century improvers," as one agricultural historian has noted—Marshall, and others traveled the countryside taking notes and reporting observations for the sake of recrafting an agrarian ideal based on experience, not speculation.[13] In the person of such improvement-minded tourists, the georgic was a way to knot together the civic virtue of labor with the demands of agricultural improvement.

This georgic approach emphasized the dual development of intellectual work and physical work as another way to treat moral and material pursuits as of a piece. Scotland's Earl of Dundonald, Archibald Cochrane, in a treatise "shewing the Intimate Connection that Subsists between Agriculture and Chemistry," believed that "[Agriculture is a science] morally and politically conducing to the true happiness of man . . . whence flow health, social order and obedience to lawful authority." He there summarized the georgic's multifaceted value set by announcing its simultaneous and irreducible promotion of health, wealth, and social order, furthering the genre to which Henry Home before him had also contributed. Alexander Hunter, a Scottish physician and later (1770) founder of the Agricultural Society of York, similarly and more explicitly captured the importance of the ethic, publishing eight volumes of his *Georgical Essays*. Hunter wrote the essays as a way to "reduce [York's farmers] thoughts and observations into writing."[14]

Spurrier's work was set within this walking, working, and touring context. His plea that agricultural "reasoning . . . founded on science combined with experiments minutely attended to" would lead to "conclusions of real utility" was a product of the same kind of encouragement for systematic observations that had driven Young and Marshall into the fields of Britain.[15] Yet Spurrier, who owned and worked the land on which he reported, aimed for a distinctively American view on the matter, one not just materially dissimilar but culturally too. The meaning of the georgic, that is, was not circumscribed only in the Scottish Highlands or English rural settings.

On the other side of the ocean, agricultural and political leaders of the new American nation, including Washington, Jefferson, and Madison, participated in transatlantic correspondences with the georgic tourists and crafted views of improvement in America that transcended mere material comforts. Although they did not travel as widely as Young and Marshall, the Americans forwarded the ideals of the Georgic Tour as part of their formulation of culture and economy. A vibrant correspondence record and civic spirit of community engagement also encouraged many in the young nation to share local land experiments with one another in a manner in keeping with the Georgic Tourists.

From the late eighteenth century, the importance of work and connection to the land had been threaded through with expressions of agrarian virtue in general. Home and Cochrane, two self-labeled improvers of the georgic genre, claimed political, professional, moral, and religious greatness from agriculture; they saw the future of healthy, wealthy, and wise men as related to the land; they associated the virtues of the soil with the possibilities of cultural strength. That common vision of virtue translated easily and then grew more forcefully across the ocean as colonists expanded land holdings, as colonies became states, so that Washington's statement of individual and national welfare as intimately related to agriculture echoed the broad view that an age of improvement was born within the young nation. However, material conditions as well as the heightened cultural value of improvement were different in the Americas than England. Planters of the time were well aware of differences in local conditions, an observation they believed required more explication in treatises and speeches. Spurrier's work was meant to do so, to speak to "the different soils and climates of the Americas," as his subtitle indicated. His was a work dedicated to the promise of systematizing agriculture with local, experiential knowledge for the purpose of building a stronger culture. The new Republic offered the venue for those possibilities most clearly.

THE GEORGIC IN AMERICA

The American ideal of improving agriculture was consistent with English and Scottish Enlightenment values of progress.[16] From the start, attempts at material advance through improved agricultural practice were also matters of moral order; for the founders, agrarianism meant that the virtue of the farmer and the moral basis for the new nation were synonymous. This ideal rested on a much broader conception of "the farmer" than the strict occupational one we hold today. As the historian Alan Marcus has noted, in the early nineteenth century, "farming was a state of mind and its practitioners included virtually all Americans regardless of where they lived, how they made a living, or how much wealth they had accrued. . . . Farmers," Marcus writes, giving substance to the dual moral and material discourse of improvement, "extolled the virtues of orderliness and Godliness [and of] working with one's hands." They were the "self-reliant souls upon whom the democratic republic was based."[17] The term "farmer" was broad enough to include a wide range of citizens, though narrow enough to define a specific moral code. Improvement of the farming class—a socioeconomic reference too vague today to hold up under scrutiny—was in the first decades of the country commonly understood as a social, material, and political goal of national significance. It was within this moral and material mandate for improvement that systematic study, experiment, and natural philosophy had a place.

In pursuing their multivalent goal, Americans of the early Republic had not yet made clear distinctions between agricultural, commercial, and industrial interests. The three elements were integrated, not autonomous. After the fact, such a view has instead been reduced to one of opposition in common historical lore: agriculture stands on one side, industry on the other. It's a view that makes sense within the dominant pastoral studies noted above that posit the idyll of the pasture against the corruption of the city. It makes less sense within a working georgic context. American historical reconstructions, for example, and in keeping with a romanticized rural-urban split, pit the agrarian Republicans of Jefferson's day against the industrial Federalists of the early National period. The

Jeffersonian Republicans believed more firmly in the views of the Physiocrats, a group of French political economists who proposed that the wealth of a nation derived from its agricultural base, than did the competing Federalist policy-makers. That distinction is worth noting.[18] But even in disagreement with the manufacturing-based views of arch-Federalist Alexander Hamilton, the agrarians were safe in asserting that factories were more properly the end product of agricultural origins, not an opposing principle.

Differences in political economy were differences in prioritizing the three coordinated elements of agriculture, commerce, and industry. For Hamilton, industry was a way to improve agriculture; "manufacturing establishments," he wrote, would promote a "more certain and steady demand for the surplus produce of the soil."[19] Because of this common understanding, Americans most often forwarded political economies based on *how* to encourage cultivation rather than *whether or not* to encourage it. Commerce, the third political economic component, was simply the means by which those products could move from agricultural origins to their industrial or manufacturing ends.[20] The farmer and the state of agriculture were the core of nearly all the policy arguments.

It was in his *Notes on the State of Virginia,* a survey in the 1780s of the vast expanse of Virginia's territory, that Jefferson called farmers the chosen people of God. His popular framing of the issue set agriculture as the "deposit for substantial and genuine virtue." That Americans "should be employed in its improvement," Jeffersonian agrarians believed, was thus the highest calling of the citizenry.[21] Washington, even if not as forcefully, expressed a sentiment similarly touting agriculture when he suggested that a national Board of Agriculture would be beneficial for both individual and national welfare. It would "assist a spirit of discovery and improvement."[22] Both men represent the perspective that improvement was understood as a multifaceted goal.

In brief, the terms of a georgic ethic, even if not the specific terminology, were found deeply embedded in the American context. Spurrier and his *Practical Farmer* espoused all the elements of a georgic ethic. His goal was one of cohesive moral and material progress, a goal sought through the cultivation of the soil. Improvement, land, and economy were intertwined, running through the identity of the farming—the young American—class. The American improvement conversation may have inherited a georgic tradition from Britain, but in the new nation that ethic encompassed moral and material duality far more directly. In one way, the strong cultural narrative of unbounded land and the burgeoning cultural ethic of the practical Yankee made the appeal to promoting a georgic attitude especially forceful and clear.[23] In a second way, it dovetailed with a set of social values based on the practice of farming, where practice was understood as experience.

DELINEATING A GEORGIC SCIENCE

Samuel Deane, in composing *The New England Farmer, or Georgical Dictionary* (1790), performed a popular task by presenting the "ways and methods" of husbandry. By 1822, his well-endorsed compendium was in its third, expanded edition. In Deane's "ways and methods" was an overture toward greater attention, technique, and systematic study of soil treatment and land management. The encyclopedic work made obvious reference to the georgic tradition not just in its title, but through the moral dictates in the text. It also drew in part on Alexander Hunter's *Georgical Essays* from the 1770s, which had already been well received in Europe and America.[24] Deane's dictionary aimed to put the "science of agriculture," as so many farmers called it, more centrally into the conversation on soil and society. Although American uses of the science of agriculture saw various invocations in the early Republic, rarely were they divorced from the larger georgic goal of moving toward a more virtuous society predicated on values of experience.

Americans were pursuing the core of the georgic ethic with the science of agriculture and with particular attention to chemistry. In the early national period, farmers began to associate the georgic orientation to the land with both the practical use of science and the variously conceived goal of improvement (moral, material, social). This led Americans to what might best be called "georgic science." Georgic science entailed a set of virtues consistent with the celebrated moral order of the early national period, among them hard work, diligence, industriousness, and individual experience. Science, for farmers of the age, denoted a specific form of systematic investigation beginning with praxis-oriented studies of the land. Georgic science, as a tool for ends other than itself, was not abstract or laboratory-based, instead suggesting direct attention to daily activity. Its terms referred not to scientific principles in general, or as definitive of

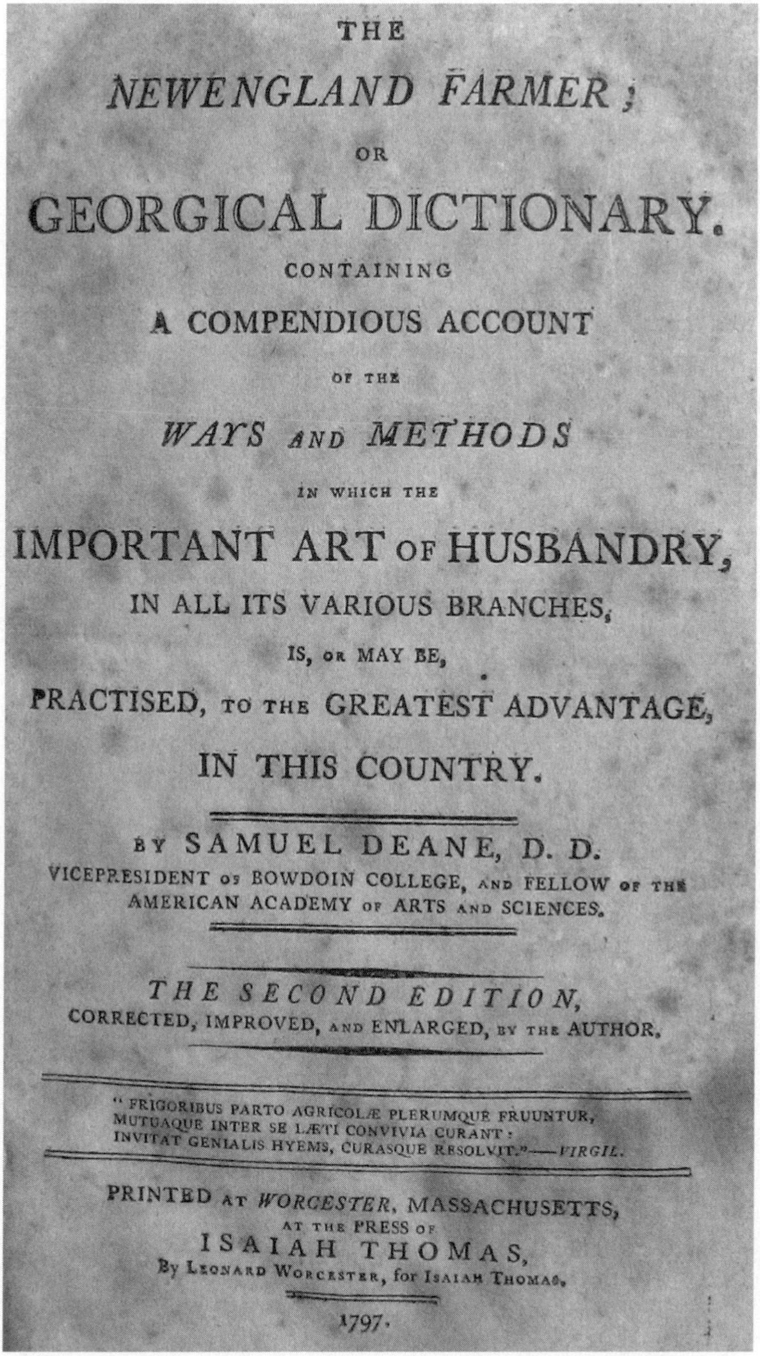

THE

NEWENGLAND FARMER;

OR

GEORGICAL DICTIONARY.

CONTAINING

A COMPENDIOUS ACCOUNT

OF THE

WAYS AND *METHODS*

IN WHICH THE

IMPORTANT ART OF HUSBANDRY,

IN ALL ITS VARIOUS BRANCHES,

IS, OR MAY BE,

PRACTISED, TO THE GREATEST ADVANTAGE,

IN THIS COUNTRY.

BY SAMUEL DEANE, D. D.

VICEPRESIDENT OF BOWDOIN COLLEGE, AND FELLOW OF THE AMERICAN ACADEMY OF ARTS AND SCIENCES.

THE SECOND EDITION,

CORRECTED, IMPROVED, AND ENLARGED, BY THE AUTHOR.

"FRIGORIBUS PARTO AGRICOLÆ PLERUMQUE FRUUNTUR,
MUTUAQUE INTER SE LÆTI CONVIVIA CURANT:
INVITAT GENIALIS HYEMS, CURASQUE RESOLVIT."—*VIRGIL.*

PRINTED AT *WORCESTER*, MASSACHUSETTS,

AT THE PRESS OF

ISAIAH THOMAS,

By LEONARD WORCESTER, for ISAIAH THOMAS.

1797.

FIGURE 1.4. Cover page of Samuel Deane's *New England Farmer, or Georgical Dictionary.* Courtesy of Special Collections, University of Virginia Library.

all "science," but as they addressed the dual discourse of improvement within an agrarian political economy, bridging agriculture and science with the values of rural virtue.

Deane's *Georgical Dictionary* spanned topics from seed preparation, manuring techniques, and livestock feeding suggestions to questions about artificial fertilizers, crop rotations, and different soil conditions for different crops (clayey, loamy, sandy, for example), in the process giving substance to georgic science. Deane was a Congregationalist clergyman from Maine as well as a member of the American Academy of Arts and Sciences. In those capacities, his contributions to improvement as a cultural value bridged the clerical and agricultural.[25] He spent years farming and dutifully recording his own experiences and experiments, seeking in the same spirit as Washington and Spurrier to "promote the improvement of agriculture" for the sake of individual and culture alike. In the *Dictionary,* Deane suggested that attempting a systematic approach to all of these elements at once would be unreasonable; the natural economy of the farm was too complex to be reduced to one system. But approaching them piecemeal was not altogether out of place, and the clergyman encouraged a systematic study of each element. Significantly, Deane promoted the value of writing and reporting individual experience as a way to work through this tension between ungainly complexity and manageable system, just as Young and Marshall had made attempts to codify the "ways and methods" in the context of their georgic agricultural tours in the decades before.

Another popular text of the antebellum period, Daniel Adams's *Agricultural Reader,* deferred to Dr. Deane's georgic testimony throughout its pages. Adams spent his life in New Hampshire and Massachusetts and was himself a physician as well as a "farmer" in the classical antebellum sense. He lauded Deane's work not just for its bald agricultural facts, but also for the insistence that through its advice "civilization, with all the social virtues, would, perhaps, be proportionably promoted and increased."[26] He encouraged the participation of farmers in a world of improvable land, supporting Deane's georgic orientation and motivating his readers with the overriding moral cause inherent in agricultural improvement. Along the way, he introduced and promoted the idea of systematic studies on the farm, forcing the matter of diligence and method onto his rural audience. Deane and Adams together, along with Spurrier to the south and Young, Marshall, and Hunter across the Atlantic, were pursuing systemization.

Theirs was a process measured by the mechanical and visual features of the soil—appearance, texture, feel, color. That is, they were working toward rigor in method, but remained at a level of superficial soil observation. Their works simply promoted a more systematic, less idiosyncratic and unaccounted for process for doing so.

More specific to the call for a science of agriculture was the frequent attention to chemistry in georgic-based texts of the era. Connecting chemistry to agriculture was a fashionable and popular topic of the early nineteenth century, just as agriculture itself was perceived as culturally virtuous.[27] The chemistry-agriculture connection, in fact, formed a sort of mini-genre. This approach occurred at a time when the so-called chemical revolution was spreading throughout Europe with the new identification and conceptualizations of elements, experimental practices, and theoretical contributions. In the arena of the rural improver, this topic of chemistry and agriculture also occupied common conceptual ground between earlier Scottish and English improvers and the young United States.

Spurrier wrote *The Practical Farmer* on an agrochemical basis, just a few years after Deane's *Georgical Dictionary*. By the 1790s, the uses of chemistry and philosophy were so prevalent, their reference so common, that authors began to define the terms with greater specificity. In his book intended for neighbors, friends, and other interested improvers (Jefferson had several copies in his library), Spurrier defined chemistry as an art by which the "several properties of soils and manures are discovered." Natural philosophy was knowledge "founded on reason and experience" that, when used, would "enrich the earth, and . . . promote vegetation."[28] Both concepts carried an active, engaged tone; both were presented as directly related to, in fact almost invented for, the cultivation of the soil. Spurrier, like Deane and Adams, considered the merit of chemistry by its relation to lived experience. It was a tool of the farmer conceived in terms similar to natural philosophy more generally.

Deane's *Dictionary*, also serving as an example of this chemical attention, carried entries on "Air," "Experiments," "Improvement," "Manure," and "Marle"—crushed shell deposits useful as calcareous manures—that all considered the arguments of chemistry and "natural philosophy." Deane encouraged readers to "ascertain the composition of sterile soils" for the sake of improvement. He also made the importance and relevance of chemistry more explicit by referring to the European chemists Carl

Scheele, Joseph Priestley, and Antoine Lavoisier when detailing for New England's rural classes "the essential qualities of atmospheric air." By the 1822 edition of his *Georgical Dictionary,* Deane could defer to the Englishman Humphry Davy's *Elements of Agricultural Chemistry* (1813) to explain "general principle[s] in chemistry."[29] Where scholars may be apt to wonder how the practice-based chemistry of rural improvers was distinct from the more theoretical accomplishments of Europe's finest philosophers, it seems that the improvement advocates themselves made no distinction. Spurrier and Deane highlight the ease and ubiquity of the chemical reference. In their georgic context, such contrasts between practice and theory had little meaning.

In Europe, too, of course, amateurs, gentlemen of science, and philosophers alike had been devoted to connecting agriculture and chemistry for the sake of social and individual welfare. It was only that the American context provided a richer cultural basis for such a connection. Not just cloistered and abstract analysts trying to define the meaning of an element, chemists were also extending an earlier workshop and pharmacy-based tradition to appeal to the practical problems of contemporary life. Part of observing and codifying farming practices was directed at and built in response to chemists' practical directives, to the application of chemistry to agriculture. Thus, although surely chemical studies of the time cannot be summarized as being of one kind or directed toward one end, the chemical community aiming to provide practical solutions to everyday problems had common cause with the goals of georgic tourists and with the chemistry-agriculture connection.

Cochrane's (1795) *Treatise Shewing the Intimate Connection that Subsists between Agriculture and Chemistry* was just one contribution to the genre. Samuel Parkes, a British chemist, offered another. His oft-reprinted *Chemical Essays* of 1815 were particularly popular in the United States for the fourth essay, "On the Importance of Chemistry as Connected With Agriculture." The piece was later excerpted in agricultural papers like *The New England Farmer,* providing an early example of bringing chemistry into rural circulation with the nascent agriculture press that I examine in the next chapter. Other work on "the application of chemistry to agriculture" and "the chemical history of vegetables" spoke to the promise of unlocking the potential of agricultural improvement to aid the concurrent goal of cultural progress.[30]

All of this brings us back to the place of chemistry and the science of agriculture in the United States and the role the georgic ethic played in that placement. The combination of agrochemical pursuits with a georgic environmental ethic meant that whatever science was being promoted or discussed was being understood through the agrarian ethic of practice, work, and virtue. It would be a georgic science. In America, by seeking to align chemistry with farming, rural economists had a precise example of a practical science for practical benefits. Jefferson complained, in fact, that his peers were wasting time with abstractions while the rural foundation of the country languished for want of more practical studies. He suggested the chemistry of cooking and household economies as examples of practical measures, but also applied chemical studies of soil fertility as an arm of his improvement ethos. He also made his point by negation, in one instance chastising chemistry as "useless" as compared to "botany, natural history, comparative anatomy, etc" because "for chemistry you must shut yourself up in your laboratory."[31] The very fact of its distance from the field cast it in a negative light: the place of the study mattered not just in physical terms (its location), but for cultural reasons (its practitioners). Jefferson meant his sense of "practical" in reference to chemistry's role as a practice on the farm rather than merely as something beneficial. For him, the location of chemistry determined its value. He would be pleased to laud chemistry, were it based on field conditions.

Like many works of the day, Thomas Ewell's (1806) *Plain Discourses on the Laws or Properties of Matter*—dedicated to Jefferson, no less—aimed to use chemical principles, no matter how uncodified, to achieve agricultural gains. As with his contemporaries, Ewell saw his work as a discourse on modern chemistry "connected with domestic affairs," believing that "agriculture is the most intimately connected with chemistry."[32] In the coming decades, even a yeoman-like, otherwise unknown farmer such as South Carolina's Eli Davis could write to the agricultural press about "the importance of chemistry as connected with Agriculture."[33] To be sure, chemistry was not the only science considered useful for the farm, but here it provides an interesting and clear case of the means for georgic improvement. It was, as a practice employed on the land, considered a tool useful for achieving a higher cultural and material goal.

Chemistry circulated throughout the farming community. But, at a time when science and improvement were not yet synonymous, what can we say about the specific correlations between them that hinged on the moral dimensions demanded from the georgic? On this question it is vital, though almost simplistic, to recognize that for the culture of the late Enlightenment in America, agriculture came first and the goal of improving it second. That is, the baseline for cultural identity was agriculture; the call for improvement relied on the existence of that preexisting baseline. Scholars in the history of science and science and technology studies have done encouraging work expressing the dynamics of amateur scientists and gentlemen scientists, but as such have led the conversation on science in society (especially before the "modern" era beginning in the early nineteenth century) to view amateurs and gentlemen as seeking or competing with professional status. The agrarians of the early Republic, however, sought no such status. They instead considered how they might make their working patterns more effective for their own sake, not for the purpose of becoming scientific.[34]

A starting point that takes agrarian identity as the baseline, and improvement as an augmentation of that baseline, suggests a great deal about how science could be credibly aligned with that improvement. It shows that the values used to interpret and judge the merits of improvement practices—georgic science—were based *in the existing culture,* not coming from the social sphere of a chemical community. This means that the questions of scientific credibility had to be answered by nonscientists (a term I use out of place, since it would not have been recognized in this era). On the face of it, that should not be surprising. The dawn of professional chemistry was still decades in the future. Thus, there was no significant American chemistry community and no set of values within such a community that could be promoted outside it. But, for the sake of studying how the science of agriculture gained legitimacy in rural culture we need to sit within that culture to see that science had to fit within an existing social sphere, defined by certain values and virtues that could not simply be imposed from the outside. I mean then for the term "georgic science" to stand as shorthand for the conversation on moral and material goals of the American agrarian culture that debated the value of chemistry. Americans rooted their perceptions of chemistry for improvement, like georgic science itself, in an appreciation of virtue and values. The early Republic conjoined the place of chemistry, the georgic ethic, and systematic agricultural practices in conversations about its broad catchword, improvement. Fertilizing techniques, specifically, highlight the focus on soil and science brought about through attention to the virtue and value of improvement.

The diaries of both George Washington and Thomas Jefferson, America's most famous planters, provide interesting touch points for issues of chemistry, fertilizing, writing, systematic attention to improved agricultural practices, and the confluence of material and moral improvement. Simply put, they offer a gloss on georgic science. Reading the diaries, one gets the sense that the planters were almost more interested in strategies of cultivation than policies of government. As founders of the new nation, both men had a clear interest in promoting virtuous behavior and moral righteousness with the agroeconomic tenets of the new government.

To do this, Washington long sought to improve his Virginia property by experimenting with crop rotation systems, crop variety, and fertilizers. By materially improving his own lot, and doing so with experiments, he would also stand as an example of the virtuous American. Demonstrating the value and practicality of experiments was an important element of his broader success, and it defined Washington georgically. Those "experiments" were in fact widespread, extending to the use of "animal dung, marl, green crops plowed under, and in at least one instance mud from the Potomac River bottom."[35] He corresponded frequently with Arthur Young, considering him the foremost British agricultural improver even before Young became the first secretary of the new British Board of Agriculture in the 1790s.[36] With Young, Washington discussed the techniques and merits of manuring, exchanged plant seeds, and compared test plots of crops throughout the late 1780s. Jefferson was also in touch with the British improvers, trading seeds and ideas with John Sinclair, president of the British Board of Agriculture, on numerous occasions. He introduced crop varieties from afar to aid the cause of local improvement.[37]

In the late eighteenth century, the georgic tourists took part in British improvements integrated within the broader culture of the Enlightenment. It was hardly by chance that Arthur Young became secretary of the Board of Agriculture after his years of tourism; the one followed logically from the other. The ideals of progress captured in high Enlightenment rhetoric took specific form in Britain's political economy with agricultural developments, even though examples of industrialization more often command historical attention in that regard. The tourists contributed to the Enlightenment milieu—one thinks of the systematic knowledge-collecting endeavor of Diderot and D'Alembert's *Encyclopédie,* a didactic effort in keeping with the Tour reports—by reporting the best practices from across the countryside, intending along the way to provide a common stock of farming knowledge. They took local practices and distributed them in literary form to the far reaches of the nation, in the process introducing a semblance of order and system to agricultural practice. Certainly the readership for those Tour accounts was limited, perpetuating the gentlemanly nature of the pursuit by their very format. But this early foray into systematic investigation was one specific means by which "improvement" could be achieved. It suggested an understanding of the science of agriculture that was synonymous with systematization.

Scotland made the intersection of agriculture, improvement, and georgic values most evident. In fostering the Scottish Enlightenment, the academic culture at Edinburgh promoted in particular the medical and chemical studies that made possible studies of agriculture and chemistry. Prominent philosophers, chemists, and patrons such as James Hutton, William Cullen, Henry Home, and the Reverend Dr. John Walker took part in sustained studies of agriculture, further forcing the integration of chemical pursuits with Enlightenment ideals.[38] In the end, the Scots and the British Georgics may not have all gotten their boots as muddy as farmers—in fact Marshall would criticize Young for his view from the carriage, rather than the understanding he could gain from letting the dirt sift through his fingers, so to speak—but their goals kept the multifaceted sense of improvement in view. Especially interesting for georgic science, the agricultural tourists and commentators like Young and Marshall sought to bring together the labor of both farming and writing with their literary strategy by explicit reference to Virgil's *Georgics.*

In America, and as correspondents with their European contemporaries, both Washington and Jefferson represented the model of the improvement-minded planter for more reasons than I can address here. Viewing Jefferson as the Enlightenment paragon, for example, goes some way in helping us see the connection between his practices and his philosophy. Although the two gentleman farmers may have differed in ideology, age, and agricultural success, they shared an interest in new fertilizing strategies and in participating in both foreign and domestic correspondence networks.

Both men were devoted followers of Judge Richard Peters and his experiments with plaster of Paris (gypsum) near Philadelphia. Jefferson was a frequent correspondent of Peters who, as the second president of the influential Philadelphia Society for Promoting Agriculture (hereafter, PSPA), had written *Agricultural Enquiries on Plaister of Paris* in 1797. The PSPA had been founded in the 1780s by wealthy Philadelphians in a spirit quite similar to that of the georgic writers in Britain. The members sought to advance agricultural practice by collecting reports of experience and offering public addresses. Peters placed himself precisely in the middle of the organization with his experiments and reports. He also espoused the ideals of the era by forwarding Virgilian notions of labor, virtue, and science—and not surprisingly, he did this by textual reference to the *Georgics*.[39] In 1816, Jefferson wrote to him that "plaister . . . is become a principal article of our improvement, no soil profiting more from it than that of the country around this place."[40] According to Washington, with whom Peters was also close friends, the *Agricultural Enquiries* treated "more extensively on Gypsum as a manure than any I have seen before."[41]

Peters was not alone in promoting plaster, nor were his prescriptions for applying fertilizers based on "systematic" principles uncommon. Jefferson found similar advice in the work of a northern Virginia farmer, John Binns. Binns's (1803) *Treatise on Practical Farming* reported on nineteen years of field experiments to suggest that the plaster acted as a manure and that his honed techniques of applying the manure promoted improved wheat growth.[42] Spurrier's *Practical Farmer* presented twenty-two pages of discussion on varieties of fertilizing manure. His treatise was notable for its omission of plaster, though he emphasized the value of a range of other lime and shell-based manures. Washington bought ten copies of the treatise; Jefferson, five.[43] It's worth repeating that Spurrier's aspiration, much like Peters's and Binns's, was to offer a compendium of the "mechanical, chemical, and philosophical elements of agriculture." Like his contemporaries, he dedicated the treatise to Jefferson by promising to "promote and increase upon the most rational principles, the real strength and wealth of the commonwealth."[44] As a summary statement of

ENQUIRIES, FACTS,

OBSERVATIONS and CONJECTURES,

ON

PLAISTER OF PARIS.

LETTER *of* R. PETERS, *and* ANSWERS *to* QUERIES *on* PLAISTER *of* PARIS, *by Mr.* WM. WEST, *of Darby Townſhip, Delaware County.*

SIR,

THE Gypſum, or Plaiſter of Paris, according to a late analyſis of its component parts, as declared in an Engliſh work, is ſaid to be compounded of a mineral acid, and a calcareous earth: the firſt an enemy, the ſecond friendly to vegetation. According as the one or the other prevails, it is ſaid to be good or bad. It is ſaid there to

B operate

FIGURE 1.5. The first page of Richard Peters's *Agricultural Enquiries on Plaister of Paris.* The excerpt indicates Peters's analytical interests in his response to questions posed by a fellow member of the area's farming culture. Courtesy of Special Collections, University of Virginia Library.

a georgic ethic, these works bound cultural improvement to principles of rationality on the farm.

These letters on farming were a rural and local subset of the broader Republic of Letters.[45] Participating in correspondence networks, the seed trade, and economic planning across the Atlantic world, improvement advocates encouraged attention to method and process in farming practice. The practical emphasis in this literature was indeed on *process*. Better methods would lead to better practices, better yields, and an improved agronomy. Washington and Jefferson provided a crucial rhetorical connection between eighteenth-century European improvers and nineteenth-century American farmers with their georgically oriented practices. They encouraged and read these works, promoting new fertilizing strategies and reports of experiments on their own land because of it. Their rhetorical prescriptions were backed up by experience.

As Laura Sayre has written, the georgic writers approached that labor-agriculture-writing junction "by stressing the time and effort of authorial production, by basing the authority to write about farming on a résumé of farming experience, and above all by insisting that the experience of farming could only be fully realized through habits of writing and reading."[46] The strong deference to authority, experience, and writing was common among the georgic agriculturists, in fact almost definitive of them, establishing a union that would later pervade the antebellum agricultural press. Despite a confusing assemblage of interpretive elements—farming, writing, morality, improving—lived experience stood as the common factor binding them together and foregrounding the moral component of the rhetoric. When Jefferson and Washington contributed to the goal of improvement through experiments and fertilization—as written in tours, treatises, and letters—they were following through in practice with the connection of agriculture to the success of an agrarian and republican social system. What came next was the further edification of the intellectual underpinnings for those experiments and processes.

The georgic science of early Republic authors brought together political, environmental, and aesthetic judgments based on an active philosophy of praxis—doing was knowing. Thus, discussing literary expressions of farming practice fell within a broader framework of establishing and enforcing value systems, of how people should live and work in their communities. Aesthetic and economic values, comprehending the bounty of nature for its beauty and as a resource, for example, mattered for the way early Americans and their British correspondents perceived the world around them. Most tellingly, and to summarize, the ostensible goals of addressing farming practice—to improve crop production and promote fertilization—were enmeshed in far deeper commentary about structuring social life, political economy, and perceptions of nature.

How did Americans shift the value locus of systematic studies of the land from within individual, scattered agricultural treatises—where georgic science was first built—to agricultural science as debated by a broader community of working farmers and rural improvement advocates in the early Republic? John Taylor of Virginia, a U.S. senator, veteran of the Revolutionary War's Battle of Yorktown, prominent statesman, and agrarian citizen, is an excellent transition figure in this regard. He fits the profile of the georgic American well, earnest, diligent, respected by his community, and politically and agriculturally successful. Taylor was also a wealthy, slave-owning, interested, and active planter, contributing to local farming issues by authoring agricultural treatises and later leading the Agricultural Society of Virginia (hereafter ASV, though it was also referred to as Society of Virginia for Promoting Agriculture in its day).[47] He was interested in "fitting ideas to substances, and substances to ideas," exemplifying that approach toward a dual moral and material discourse on agriculture in his famous collection of 1813, *Arator.*[48]

Taylor was ostensibly interested in the matter of soil identity—what was it, what could he do with it, and how were his answers to those questions formed through years of practical farming experience? He wrote *Arator* in response to the English tourist William Strickland's (1801) *Observations on the Agriculture of the United States,* accepting despite his nonetheless patriotic character Strickland's main agricultural point that American farming exhausted soil. The problem of soil exhaustion so well canvassed in agricultural history fits into the current georgic story because it provided the specific object—soil—to be improved.[49] It also forced attention to the material problem with farming, which was the *quality* of that soil, the means for assessing that quality, and the proposals for amending it. In *Arator,* Taylor set out to address these facets of the exhaustion problem. Put simply, his solution was to establish a program of improvement and rational agriculture that integrated philosophical principles and direct experience. In this, his work both leveraged the similarly minded treatises of the late eighteenth century—like Spurrier's and Deane's—and moved beyond them with more precise attention to soil exhaustion, fertilization, and soil identity, and with the greater public posture of a cultural leader only slightly less famous than his neighbors Washington and Jefferson. Despite their various interpretations of his character, historians agree that Taylor's contributions were intended to improve soil fertility and agriculture with appeals to a moral constitution fostered by rationality and systematization. In essence, he was an early Republic georgic writer.[50]

The early national literature on science and agriculture of which *Arator* was a part was often the product of civic societies established to promote local values and agricultural virtue. This organizational level helped points about soil and society to carry greater cultural cachet and to reach a potentially broader audience. The organizations were gentlemanly groups formed to sponsor calls for progress. The PSPA and, in Virginia, the ASV stand as two strong examples in this regard. They were the agents of agricultural improvement in America and they were georgically defined. Considered from an environmental perspective, they were agents of increased attention and change to the soil—in its concept, physical and chemical constitution, and augmentable possibilities—of the young nation.

The individual members of those societies and those members' practices were of course instrumental in promoting local policies. Judge Peters, for example, spoke from the platform of the PSPA. He used the *Memoirs of the PSPA* as an outlet for publishing reports of experiments and comments on prevailing theories of agricultural improvement.[51] Daniel Adams, the author of *The Agricultural Reader* mentioned above, had been the president of the Hillsborough (New Hampshire) Agricultural Society. He spoke at length from that podium about the civic virtues of "scientific agriculture" to his constituency, explaining that to become a "practical, scientific agriculturalist . . . it will be necessary [that the farmer] should have some acquaintance with the principles of natural philosophy, and especially of *agricultural chemistry*."[52] The full role of those organizations as outlets for the advancement of scientifically based improvement strategy is a key subject, as it were, but it lies beyond my aims in this chapter.[53] My interest here instead is their position as a bridge from

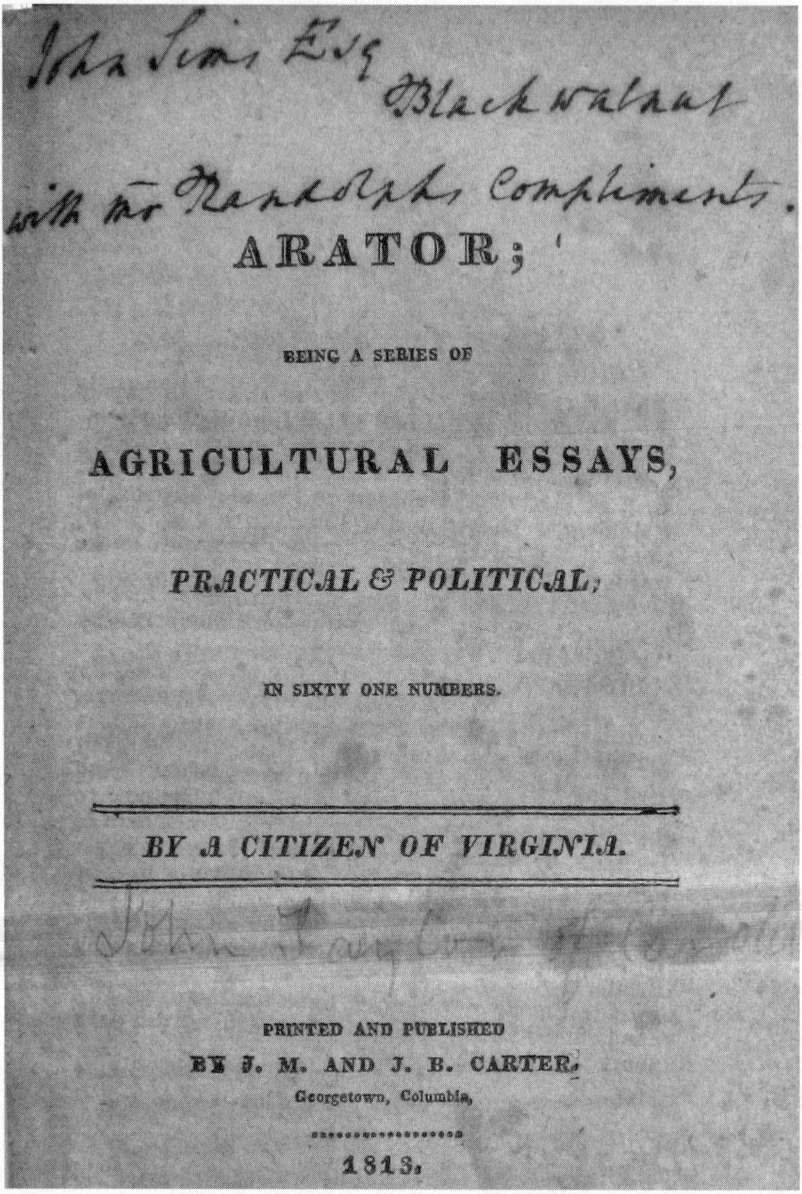

ARATOR;

BEING A SERIES OF

AGRICULTURAL ESSAYS,

PRACTICAL & POLITICAL;

IN SIXTY ONE NUMBERS.

BY A CITIZEN OF VIRGINIA.

PRINTED AND PUBLISHED
BY J. M. AND J. B. CARTER,
Georgetown, Columbia,

1813.

FIGURE 1.6. Title page of John Taylor's *Arator.* Fitting ideas to substances and substances to ideas, as Taylor wrote, his text was widely reprinted, with numerous editions in just the first few years following its publication. The copy shown here, a first edition of the book, was originally in the library of John Randolph of Roanoke, friend of Taylor and one of Virginia's prominent early Republic politicians. Courtesy of Special Collections, University of Virginia Library.

individual eighteenth-century studies in the "science of agriculture" to nineteenth-century outlets for promoting written work on science and farming through the rural press.

This is where Taylor's work stands out. Farmers and planters eagerly read and favorably received *Arator.* Its essays indeed bore testimony to moral and social order as much as, if not more than, agricultural order. As the historian John Grammer has written, the essays in *Arator* provided "not just an image of virtue but also a description of the daily practical labor by which virtue, in the real world, must be maintained."[54] A Jeffersonian Republican, Taylor closely followed the ideal that God preferred farmers, claiming thirty years after Jefferson's similar line that those who opposed agriculture dangerously preferred "a system which sheds happiness, plenty and virtue . . . fosters vice, breeds want, and begets misery" (316).

Inasmuch as *Arator* was a collection of farming advice, it was so only in relation to the commentary it offered on the moral structure of labor. The eleven essays on the "Political State of Agriculture," as well as essays on "Labour," rights, economy, and militia, were thus not peculiar for their inclusion in a collection on agricultural improvement, but definitive of it. As a "reason for uniting agriculture and politics," Taylor characterized "agriculture as the guardian of liberty, as well as the mother of wealth."[55] As another *Arator* scholar has written, Taylor's writings involved "a prescriptive wisdom, based on experience, and embodie[d] a practice of good manners toward the gods," a feat accomplished by associating "virtue and the proper order of human life with the disciplines of the farmer and the stockman."[56] Hard work and diligence, Taylor repeated in a neo-Virgilian spirit, were the foundation of agriculture and the guarantor of its virtuous calling. When he promoted fertilizer experiments for agriculture he associated the same virtues with those agricultural pursuits. Experiments too had to be diligent and born of hard work. In this way the science of agriculture could be an arm of improvement and not simply a shortcut to easy answers.

The georgic ethic that understood labor as life was both implicit and explicit in Taylor's *Arator.* Science and farming were both to be pursued with equally thorough and attentive rigor; both were meant to help each other define and promote improvement in society and improvement in the fields. With its 1813 publication, the book stood at a crossroads in American agriculture between earlier—and more individually guided—

Washingtonian and Jeffersonian efforts at improvement and later regional and communal efforts.

The year 1813 was a good one for agricultural texts. Humphry Davy first published *The Elements of Agricultural Chemistry* at the same time. Like *Arator,* Davy's book was a collection of past experience, a compilation of thoughts, experiences, and experiments on agricultural production and soil fertility.[57] But Davy's work was the product of a different social atmosphere. This was not only because it was British, but because Davy's hands-on experiments were not actually those of a working farmer. From the perspective of most Americans, the differences between Davy and Taylor were less that the former was a chemist and the latter was not, than that the latter could make overtures to farming experience while the former could not.

Taylor's experiences through the 1810s kept him on his Virginia farm and brought him new notoriety as a successful agricultural improver.[58] As a recognized improver, he soon shifted from his position as an individual author of distinction to a more participatory regional leader. As part of this shift, he also carried on more forcefully the dual discourse of georgic improvement. In 1817, he became the first president of the ASV, a title, he claimed, that was preferable to that of president of the United States. When he spoke to other planters and farmers of the Virginia countryside as a respected social and political figure and as the ASV's president, he brought with him the views of writing, working, and agrarian virtue that had already been developed in *Arator* and the earlier tour literature. Taylor's biography maps nicely onto the transition from a late eighteenth-century agricultural spirit derived from British roots to an early nineteenth-century agrarian ethic beginning to fit a unique American context. His emphasis on virtue and moral character did well to integrate the common georgic sense of agrarian ethics with what he cast as a more refined science of agriculture.

Consistent with his focus on the morally purifying virtues of labor and work, Taylor continued to trumpet the values of an agrarian citizen (this, even while managing a slave population to do the actual labor) and to plead against the vices of idleness, habit, and prejudice. With respect to science, he argued that his notion of improvement would come about as a result of his notion of successful scientific methods, namely, circulated reports of experience. In this, he did not associate science with standards of an admittedly unassembled scientific community, but presented it as a tool geared toward the virtuous and economic profits of agriculture. Thus, he encouraged farmers to circulate reports of experience, since that *experience* of farming bound virtue and material improvement together and countered perceptions of "error, rudeness, and vice."[59] His view of virtue and vice was sincere and deliberate. In a late 1818 address, he again championed improvement-minded farmers who, rather than falling "under the banner of vice," exemplified the "execution of a strong, virtuous, and patriotic mind." The vice-president of the ASV, Wilson Cary Nicholas, seconded the plea for a social dynamic of "health, happiness and virtue" based on the positive values of strength, patriotism, and labor. Be it physical or mental labor, each was defined in opposition to idleness. In all cases, the virtues of the agrarian citizen were the virtues of improvement.[60]

Taylor is the culmination of the early years of agricultural writing, systematic experiments, and reports of experience that were part of the late Enlightenment's georgic context. For him and his peers, moral and political economy were produced together. Their use of a systematic—later, scientific—approach to cultural goals was common to both aspects of that production. Throughout the first decades of the Republic, Americans were inconsistent and unclear in their references to "science"—sometimes meaning it as a method, sometimes as a category, sometimes as an ideal of study. Improvement advocates, though, offered a clear and consistent association of labor-based virtues with the perceived value of experiment and observation. John Spurrier, Samuel Deane, Daniel Adams, Thomas Jefferson, George Washington, and John Taylor represent this goal of associating virtue and experiment. Although shaped in part by Virgil's resurgence in the eighteenth-century British Georgic Tours literature, the georgic outlook represented by those men was soon part of a distinct American context thoroughly defined through agrarian virtue and culture. Their treatises on systematic agriculture, "the science of agriculture," or what I have called georgic science were meant to offer agricultural improvement the benefits of sustained and diligent experimental and experiential attention. But if experience was key, then the kind of experience best suited for improvement would also be an important point of debate. By the early decades of the nineteenth century, then, a georgic science was one that enabled moral and material improvement through the practice and experience of farmers. Altogether, this kind of improvement gave meaning to the importance of place (location) in the later construction of the authority of science, because these georgic virtues relied upon a direct connection to the land. They provided the terms for agricultural improvement, laying the necessary groundwork for a scientific way of knowing the land.

Georgic science provided new forms of interaction, and thus understanding, between Americans and their natural environment. "The practice of agriculture," John Spurrier had contended, "is in general confined, either to persons who have not had a sufficient education to enable them to keep regular accounts, or who are inattentive and do not choose to take that trouble without which, it is impossible to ascertain the profits and loss thereon—hence they cannot possibly attain the truth."[61] No longer could "the farmer" rely only on the undigested, unaccounted for, unquestioned experience that defined centuries of traditional knowledge practices. Keeping accounts of systematic experiments, doing so with rational deliberation, and reporting them to neighbors should become the basis for daily practice on the farm. This was also the context in which the rural press developed in the 1820s and within which arguments for and against the value of science for agricultural improvement gained meaning.

CHAPTER 2

"The Science of Agriculture and Book Farming"

HOMESPUN VIRTUE, DANDY VICE, AND THE CREDIBILITY OF "CHIMICAL MEN" IN RURAL AMERICA

Who are to be believed in this discussion, either the observing, practical farmers, who have ocular demonstrations of their own experiments, or chimical [*sic*] men, who know more about eating wheat than growing it?

—Gideon Ramsdell, *Genesee Farmer*, 1832

IN THE EARLY AMERICAN REPUBLIC, book farming—the practice of guiding field management by reference to written works on agriculture—was either a problem or a solution.[1] The dispute between those two positions turned on the perceived role of science for the cause of improvement. For those who favored it, book farming represented the pinnacle of modern thought and the very underpinning of improvement. The best methods and most detailed studies could be published and distributed for all farmers, equally and at the same time, to see. Also, sharing knowledge with one's neighbors was an important sort of communication, simultaneously fostering community and bettering each individual's land. By codifying practice and theory in text, book farming advocates thought they could negate certain harmful features of the agricultural community, offering a break from dogma, tradition, and the resistance born of ignorance. One anonymous writer, representative of the advocate spirit, wrote, "The *mere* clodhopper, the contemner [despiser] of *'book-larnin'* tells his ill-fated progeny to . . . put their trust in their mules and their oxen, and for the rest to watch the changes of the moon, and the shifting of the winds . . . as more important than all the philosophy that ever was promulgated." Another considered the resistance of dirt farmers, who "will neither take an agricultural paper, read it when given them, nor believe its contents if by chance they hear it read," a position of mere stubbornness.[2]

For those who opposed book farming, however, it was just another quick, easy, and ill-considered solution to the problems of land management. A rural press contributor using the pseudonym "Anti-Philosopher" considered it simply "the rage of the day." He condescendingly remarked that the "desire to explain every thing upon *philosophical* principles" was only a fad.[3] On the contrary, he asserted, personal experience was the best guarantor of agricultural knowledge. A direct acquaintance with one's own plot of land, anti-philosophers everywhere contended, could hardly be superseded by advice from beyond the farm. Book farming was bad farming promoted by "men with silk gloves on," the product of inexperienced agents of improvement who, rather than help, were likely to damage the fields.[4] Whom should they believe, Gideon Ramsdell asked in the *Genesee Farmer,* "chimical men, who know more about eating wheat than growing it" or the experienced farmers themselves? It turns out that the improvement trope could in fact make things worse, with plenty of evidence from generations of farming to know that miracle solutions were often wasteful. "Our *first* wish," John Skinner wrote as editor of Baltimore's *American Farmer,* "is to communicate the *experience* of the sun-browned practical Farmer, in preference to the fine spun lubrications of the Philosopher." And thus the argument that science and improvement were synonymous did not come across naturally to the minds of the working agriculturalists.

Ideas of and debates over book farming are fascinating examples of classic philosophical posturing, of contesting whose knowledge is credible. The term "book farming" carried well into the twentieth century; the cultural problem it stands in for is with us still—compare, from one view, street-wise urbanites to country bumpkins, or, from the opposite view, the old farmhand with his homespun wisdom to the clueless city slicker. The loose debates have long been a direct instantiation of the age-old problem of armchair philosophy that had the testimony of hands-on workers on one side, the physically disengaged pontificators, the "men with silk gloves on," on the other. The first legitimate version of that debate in a scientific and environmental forum came in the early nineteenth century. It reveals a great deal about environmental knowledge, land practices, and cultural values.

While the book farming debate continued far beyond the antebellum period, its terms and the position of science within it shifted. Whereas the early years of the debate saw "science" (in quotes at that time) as either better or worse for the farm—was it progress or not? progress toward what?—later years accepted the validity of scientific agriculture and argued instead how it should be circulated and funded. I use the early years of the book farming debate to explore how the place-based virtues and vices of the rural community set the terms for evaluating science's utility or, put another way, how science gained a place of social and epistemic authority in agricultural settings.

The concept of georgic science helps situate the salient issues of book farming. It provides a frame through which to examine perceptions of agricultural science in rural social experiences. Thinking georgically, we can treat the concern over book farming's validity as a question of promoting the most morally legitimate way to define the nature under one's feet. Again, which kind of testimony carries more weight, that of the experienced farmer or the man who knows more about eating wheat than growing it? The tension worth underscoring from the cultural dynamic of book farming is a georgic one, highlighting just that difference between knowledge from lived experience and knowledge from beyond the field. John Taylor's *Arator,* for example, taught that truth inheres in "books but also in experience"; John Spurrier's *Practical Farmer* was guided by the view that "theory without practice, is similar to a shadow without substance."[5] In this context, experience in farming, producing place-based knowledge, was the measure by which rural citizens could gauge the value of philosophical or scientific advice.

Ultimately, the fundamental issue for farmers was whether or not the science was virtuously pursued, not whether it was scientific or not. This shift to the terms of a new science of agriculture was more than an abstract issue of ideas. Rather, it had ecological consequences, since the shift led to an acceptance of scientific analysis that brought a new form of interaction between humans and their land to agrarian America.[6] The same John Skinner who sided with "the *experience* of the sun-browned practical Farmer," for example, believed that "no man can be a good farmer, and make the most of his land and his means, without some acquaintance with chemistry." Skinner was not trapped in a dichotomy of pro- or anti-science. Rather, his concern was more nuanced, questioning *whose* science was valued, the farmer's or the philosopher's. Again associating the moral and material, Skinner claimed that the farmer correctly using chemistry would become in "*society*, a more accomplished gentleman."[7]

CULTIVATING THE LAND IN THE RURAL PRESS

Book farming was bound up with the rural, or agricultural, press. The press began in earnest by the 1820s, resulting from a confluence of factors including the rising prominence of agricultural societies (which were interested in communicating their meeting minutes and public addresses), diligent local editors, new publishing and mailing opportunities that made serialized literature possible, and the generic plea for improvement that pervaded American culture at the time. Its purpose was to advocate rural economy by providing a forum for presenting and debating the issues of agriculture. It worked as a complement and counterweight to urban newspapers, listing market prices for farm goods, advertising rural products, and commenting on agricultural development. *The American Farmer* of Baltimore, which began publication in 1819, generally receives credit for being the first successful agricultural paper. By its fourth volume, the paper had 1,500 subscribers. In one year alone, 1829, it added another 300. *The New England Farmer,* published in Boston from 1822, claimed 16,000 subscribers by the 1850s. By the Civil War, 400 different agricultural papers had appeared, at least one in every part of the country.[8] Overall, readership was significant but not overwhelming, as judged by bare circulation statistics—though the use of such statistics to gauge influence is problematic. Nonetheless, subscription figures steadily increased over the years as the rural press became a staple of American agrarian life.

The American Farmer, Albany's *Plough Boy* (also founded in 1819), and Boston's *New England Farmer* provide three prime examples of the early wave of the rural press. Each paper participated in debates about book farming, and the editor of each acted as a pro-book farming advocate by invoking georgic values. In terms of rhetorical style, the papers were part of the same spirit of "improvement or discovery," as the Philadelphia improver Richard Peters put it, being fostered in the transactions of agricultural societies and public addresses.[9] The editors, John Skinner (1788–1851) at *The American Farmer,* Solomon Southwick (1773–1839) at *The Plough Boy,* and Thomas Fessenden (1771–1837) at *The New England Farmer,* were farmers in the general sense of the term. Their identity was

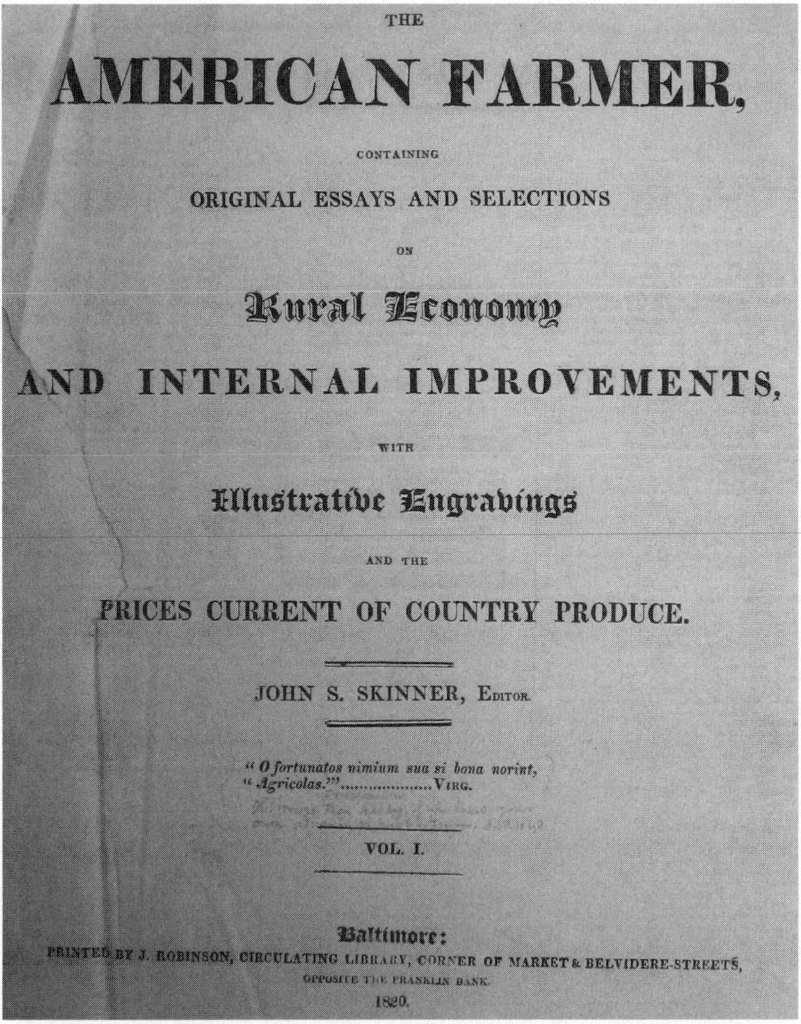

THE
AMERICAN FARMER,
CONTAINING
ORIGINAL ESSAYS AND SELECTIONS
ON
Rural Economy
AND INTERNAL IMPROVEMENTS,
WITH
Illustrative Engravings
AND THE
PRICES CURRENT OF COUNTRY PRODUCE.

JOHN S. SKINNER, EDITOR.

"O fortunatos nimium sua si bona norint,
"Agricolas."...................VIRG.

VOL. I.

Baltimore:
PRINTED BY J. ROBINSON, CIRCULATING LIBRARY, CORNER OF MARKET & BELVIDERE-STREETS,
OPPOSITE THE FRANKLIN BANK.
1820.

FIGURE 2.1. Title page of *The American Farmer*'s first issue, published in Baltimore in 1819 and bound as a volume the next year. Courtesy of Special Collections, University of Virginia Library.

embedded within the common culture of land cultivation even if they did not all or always engage in daily farming activity. Skinner, a lawyer and Baltimore's postmaster, parlayed his War of 1812 experience and friendship with James Madison into a successful publishing career. *The American Farmer* was only the first of a series of rurally directed papers he edited from Baltimore in his four decades of editorship.[10] Southwick founded and edited *The Plough Boy* after editing the Albany *Register* and another paper, *The Christian Visitant*. He was a charismatic but divisive figure, embedding a strong and overt tone of moral advocacy—from the value of improvement to the necessity for temperance—in his papers.[11] Fessenden brought an already successful career in writing and publishing to *The New England Farmer* as well as firsthand knowledge of the farm from young adult experiences and his lifelong work on the annual harvest. He was well known for his satires, many of which, like *The Modern Philosopher, or Terrible Tractoration!*, were attacks on the folly of what he considered speculative knowledge. Jefferson, for one, was the target of many of Fessenden's early Federalist-based critiques.[12]

Despite this diversity of editorial backgrounds, the press in the early years shared more features than it was divided over, not the least of which was concern for the issue of book farming. In this regard, the editors and their papers exhibited a distinct lack of overt political sectionalism, a strong though mostly informal affiliation with regional agricultural societies, and a propensity to couch book farming in terms of the specific problem of soil exhaustion and its converse, fertility. Each paper participated in bringing the science of agriculture into public debate even though none of the editors had any chemical or scientific training. They were all, however, determined to act as purveyors of method, system, and rationality.

In the 1820s, with political sectional strife still young, the first wave of papers had an air of commonality, especially with their various invocations for improvement. Differences that might have been expected from the geographical space that separated them were minimal. *The American Farmer*, for instance, published reports and commentaries from Maryland, Pennsylvania, Virginia, Tennessee, South Carolina, New York, and Europe in its first volume.[13] The material was sometimes original, but just as often reprinted from such geographically diverse sources as the Albany *Argus*, the *Memoirs of the Philadelphia Society for Promoting Agriculture*, *The Richmond Enquirer*, and *The Nashville Whig*. *The Plough Boy* and *The New England Farmer* used similar patterns of publishing and reprinting. Common editorial outlooks also added to the similarity in regional approaches. Both Skinner in Baltimore and Fessenden in Boston had the same concern for preventing emigration with their pleas for local improvement. Fessenden, in an editorial titled "The Science of Agriculture and Book Farming," saw that the "practical farmer . . . must understand, and in some degree practice these improvements," such as systematic rotation of crops and the use of plaster of Paris, or else "he must go . . . either to a poor house or to the state of Ohio."[14] Virginia, Massachusetts, and New York had obvious differences of culture and political economy; Fessenden's motivations and decisions could thus be distinguished from those of Southwick or Skinner in a broader cultural mapping. But as focused on the subject of encouraging more systematic attention to agricultural practice, those differences were less prominent. Despite growing rifts between political economies in the South and North, the papers in common presented a view of agriculture in need of attention and improvement.

The press forged a tight relationship with regional agricultural societies as both forums shared a focus on the salient issues of improvement and economy. Organizations played a key role in producing acceptable knowledge, vetting new practices, and encouraging innovation; the agricultural societies were no exception. The Philadelphia Society for Promoting Agriculture (PSPA), Agricultural Society of Virginia (ASV), and Agricultural Society of Albemarle (ASA) each had its own *Memoirs* or *Proceedings*. For even greater influence and reach they could also count on the press to spread their organized advice. The press and societies worked together toward improvement in a few key ways: by publishing reports on society prizes awarded for the best studies of farming experiments, by detailing the activities on the society members' lands, by fostering a generally cooperative spirit between the members of those societies and the readers of the press, and by publishing the transactions of the societies. These factors stood as early and georgic markers of closer attention to systematic, experimental, and then literary expressions of promoting rational agriculture.[15]

Turn-of-the-century treatises were given common cause by their concern for soil improvement through attention to fertilization; those concerns were echoed and reinforced in the later rural press. *Arator*'s John Taylor, for example, believed that "the first necessity of agriculture is fertility," with his peers and followers taking similar views.[16] To be sure, issues of animal husbandry (like proper feeding patterns and breeding strategies) and mechanical equipment (like new threshers and plows) also came under debate, but the brunt of book farming was more specific to fertilization and soil nutrition.

THE

NEW ENGLAND FARMER.

CONTAINING

ESSAYS, ORIGINAL AND SELECTED,

RELATING TO

AGRICULTURE AND DOMESTIC ECONOMY,

WITH

ENGRAVINGS, AND THE PRICES OF COUNTRY PRODUCE.

BY THOMAS G. FESSENDEN.

VOL. I.

BOSTON,

PUBLISHED BY THOMAS W. SHEPARD, ROGERS' BUILDINGS, CONGRESS STREET.

1823.

FIGURE 2.2. Title page of *The New England Farmer*'s first issue, published in Boston in 1822 and bound as a full volume the next year. Courtesy of Special Collections, University of Virginia Library.

At the time, a fertilizer was thought of as a natural agent added to the dirt, a product of the farm or the land such as animal dung, vegetable manure, lime, marl, and plaster of Paris (or gypsum).[17] To fertilize a field was to perform a simple, routine task that generations of farmers had enfolded into their daily practices. The dung heap was long a mainstay of farm life; there was certainly nothing new about using various manures to help vegetation grow.

The *methods* for using fertilizers, however, were coming into development as was an increased understanding of variety.[18] At the same time, earlier references to "science" for the sake of agriculture were losing their vagueness as a result of calls for systematic approaches and particular invocations of an agricultural chemistry. Following the tone set by works like Deane's *Georgical Dictionary,* the plea for systematization was everywhere. The New Hampshire physician and farmer Daniel Adams, in his *Agricultural Reader,* asserted that "the *'era of systematic agriculture'* has actually commenced."[19] Improving the moral structure and material capabilities of agricultural America was becoming a matter of method and system.

The rural press was an extension of earlier georgic writing.[20] It was dedicated to the georgic values of agricultural progress with the principle that it would "improve the soil and the mind," as the masthead of Albany's *Cultivator* announced with each issue. It carried on discussions about soil exhaustion and fertility that the georgic authors had made their modus operandi. And the press pursued with weekly and monthly attention the possibilities of improvement through rational practice. Editors and contributors alike were promoters, prescribing certain modes of action and discouraging others as viewed through the lens of experience. Certainly there was never a univocal voice in the press. (This lack of unity, in fact, helps explain the continuing debates over book farming and the prevalent contestation of advice.) But despite a diversity of opinion, the various papers advocated a clear value system in their form and content, one understandably favoring rural virtue and disparaging urban sophistry.

The contrast worth examining in the story of book farming, then, is not between an upper-class planter and a lower-class tenant, or an educated country squire and a day laborer. The one to explore is between the farming class and the philosopher class; it is georgic, between those who lauded the value of connection to the land and those who did not. Despite the many differences in class status, education, and social power of the antebellum American populace—and these were many, from tenants, yeoman, and slaves to market-oriented gentleman farmers and plantation owners—the majority of the agrarian society shared the position of the nonphilosophical citizen (later, nonscientists and, later still, lay people) wrestling with issues of new knowledge and advice from outside their community. The contrast that book farming debates represent was drawn between lived experience and contemplative inexperience. In kind, the story of book farming is not as much about the straightforward reception and diffusion of knowledge as it is about the thriving conversation and debate concerning principles of experimentation, codification, and trustworthy advice within the dominant agrarian culture.

In the forum of the rural press, readers understood the value of advice through a georgic prism, a perception defined through the virtues of labor, discipline, and practice. Such attitudes came across plainly with the juxtaposition, time and time again, of the sun-browned, practical, observant, and experienced farmer to the closet-bound, fine-spinning concocters called philosophers, "chimical men," or, eventually, scientists. Contributors to papers from South to North reinforced those contrasts, many times on the side of book farming, but just as often to the disparagement of such speculative practice. As with Gideon Ramsdell, the farmer debating the merits of botanical theory by asking who was to be believed, the case was one pitting "the observing, practical farmer" against the man more versed in eating wheat than growing it.[21] Farther south, an anonymous contributor in Virginia lauded an agricultural survey while noting that "knowledge . . . sanctioned by ocular demonstration . . . could have none of the disadvantages attributed to *book farming*." Many others followed the same tack, indicating their awareness of the pitfalls of written works while forwarding the salience of local, lived experience.[22]

The georgic reference point so commonly evoked at the turn of the century eventually gave way to new language in the rural press—as it were, uses of the specific term "georgic" were soon rare. By the 1820s and 1830s, a new ethos characterized by public politics and democratic participation had gained momentum. Civic agricultural societies then became examples of participation and rural advocacy, fighting for local rights amidst the new security and American identity of the post–War of 1812 era. When Alexis de Tocqueville visited the country in the 1830s to take stock of the democratic experiment fifty years on, he toured a nation that was visibly distinct from its British forbears in ways more complicated than just patterns of language and political philosophies. The georgic reference had also subsided (as it had in Britain), even as the importance of agrarian identity was becoming contrasted with more industrial, less land-based political economies in several of the nation's sections.[23]

During the rising Jacksonian Age, we can follow the same issues of experience, connection, and place by highlighting a different set of key terms. One prominent contemporary way to formulate the issue was by juxtaposing the *homespun* ethic to that of the *dandy*, the former virtuous and true, the latter lazy and despised. The term itself—homespun—referred first to homemade clothes and had political significance in the years of the social movements favoring American-made products over imported ones. In rural discourse, it came more clearly to signify simple or plain. Noah Webster's (1828) *American Dictionary of the English Language* gave definitions for adjective and noun: "Plain; coarse; . . . as a *homespun* English proverb" and "A coarse, rustic, unpolished person." In standard parlance, the values inhering in the label "homespun" were the same as those of the georgic—diligent, hard-working, industrious, practical. As a way to contrast with dandyism, the salient issues were still place-based. Homespuns lived where they worked; dandies gave the negative image of urban sophistry (not to mention their more recognizable gender-based representation as effeminate and dainty).

The concept of georgic science soon became one of homespun science, the homespun confirming the georgic salience of place—geographical and cultural—in the acceptance of agricultural chemistry. That is, in the agricultural press and throughout rural America, the debate shifted away from the perception of an essentialized science as good or bad, useful or wasteful, to the reconceptualization of two kinds, good science and bad science. That shift was echoed even in the language—"science" began to take on more specialized meaning in rural communities, just as it was doing in the broader history of science.

Historians of science have studied the early decades of the nineteenth century as the era of professionalization, as the second Scientific Revolution, and as the early years of a laboratory revolution. As science itself became defined in a way far more recognizable to readers today than the natural philosophy that preceded it—with professional organizations, university curricula, and even a new name (the word "scientist" was coined in 1833 by the British philosopher William Whewell)—so too did the science of agriculture become more heavily addressed in the rural press and for the sake of agricultural improvement. The same era has been understood as the Second Great Awakening, an epoch in American history when religious fervor affected the public conversation on morality and behavior as well as the social mores of the day to an even greater degree. As homespun science was about both the production of new knowledge and the morally apt way to do it, the method of identifying and then acting upon agricultural lands became a topic of vigorous debate.[24] A few contemporary editorials illustrate the point.

DANDIES AND HOMESPUNS, VIRTUE AND VICE

Thomas Fessenden wrote a long editorial called "The Science of Agriculture and Book Farming" in the first volume (1822) of his *New England Farmer.* It was the start of a long conversation on the value of science. The very title is worth attention, showing as it does that science and book farming were not the same thing. Fessenden's purpose was to discuss just how they were related. In the editorial, he combined economy, fertility, morality, and scientific credibility and equated them with the merits of book farming. Unlike Ramsdell writing to the *Genesee Farmer,* or the anonymous contributor to *The Southern Planter* who suggested philosophers leave their closets for the fields, Fessenden was a forceful proponent of book farming. He outlined existing qualms about the science of agriculture while carefully explaining its virtues in a commentary that urged prudence, reason, and dutiful observation. He said that while practical farmers could not be smitten with all the "theories not sanctioned by actual and repeated experiments," or afford to be "full of notions," they should still seek to separate speculation from fact. His view, he suggested, was even-handed, allowing for the criticisms of skeptics while tempering them with appeals to dominant values. But he presented the equitable view to transcend it. Those farmers who "never knew any good result from what they call *book-farming*" had misplaced their criticisms. They could improve their land by recognizing that "*book knowledge . . . is power.*"[25]

Fessenden did not approach the merits of book farming with accusations of ignorance against anti-book farmers or claims for the unbridled acceptance of all novelties, but he did associate the merits of book farming with the value of a systematic science. Quite simply, he said, he did not want to "check enterprise, but [rather,] inspire caution, and teach us that every novelty may not be an improvement, altho' every improvement was

NEW ENGLAND FARMER.

BOSTON:—SATURDAY, AUGUST 10, 1822.

THE SCIENCE OF AGRICULTURE AND BOOK FARMING.

Agriculture, the oldest of the arts, considered as a science, is still in its infancy. It is, we believe, not fifty years since chemistry was brought to the aid of agriculture, and this will eventually prove one of its principal pillars. Systematic Rotations of crops—Im-

FIGURE 2.3. Excerpt from *The New England Farmer.* This editorial by Thomas Fessenden about science and agriculture appeared on August 10, 1822, in the paper's first year. Courtesy of the Annenberg Rare Books and Manuscript Library, University of Pennsylvania.

once a novelty." Despite toeing the line so carefully in his effort to appeal to the range of his audience, he *was* pro-book farming. The editorial was directed at the proposition that anti-book farmers believe all theories and "whim-whams" derive from those "who know nothing about farming but what they get out of libraries."[26] Farmer "B," Fessenden's foil, rejected book advice because it was "not worthy of the attention of real, genuine, practical farmers." But if the knowledge observed and recorded in print was based on the testimony, observation, or experience of practical husbandmen, Fessenden explained, absurdities would "fast [yield] to reason and the lights of science."

Throughout his commentary, Fessenden emphasized that the sanctioning of these matters—who did it?—was what counted. His comments to that end were couched in a homespun framework. In so doing he presented a view of science where chemistry and philosophy, the terms he used synonymously for "the science of agriculture," were evaluated for their relevance to working farmers. His was not a strictly pro- or anti-science vision. Nor did Fessenden himself call for unbridled deference to the science of agriculture, but for a deference which asked for the use of reason by average citizens, the practical husbandmen. Book knowledge would aid the georgic and homespun goals of social stability, morality, and truth if sanctioned by industrious farmers. In fact, he was saying nothing more than John Taylor and Richard Peters before him, but with a stronger

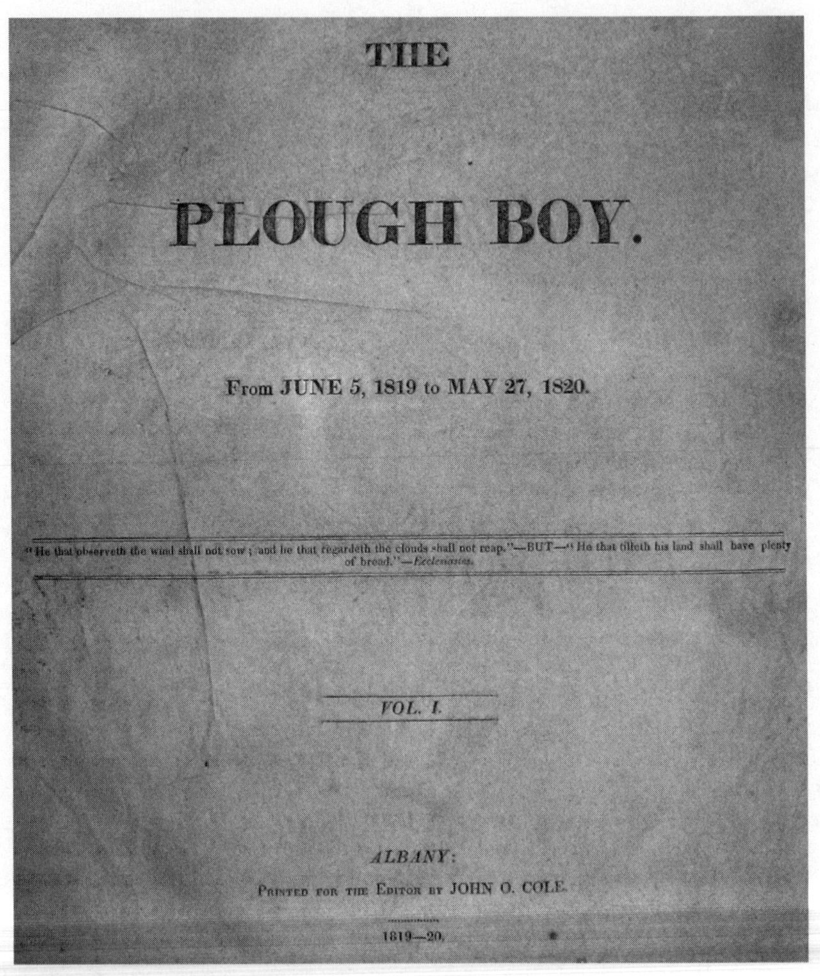

THE

PLOUGH BOY.

From JUNE 5, 1819 to MAY 27, 1820.

"He that observeth the wind shall not sow; and he that regardeth the clouds shall not reap."—BUT—"He that tilleth his land shall have plenty of bread."—*Ecclesiastes.*

VOL. I.

ALBANY:

PRINTED FOR THE EDITOR BY JOHN O. COLE.

1819—20.

FIGURE 2.4. Title page of *The Plough Boy*'s first issue, published in Albany, New York, in 1819. Courtesy of Special Collections, University of Virginia Library.

awareness of the reasons for resistance among the practical, everyday farmers. The forum of the rural press also made earlier appeals to science more pressing and more frequently voiced and opposed, especially since the press itself aimed at the same goals of improvement.

Solomon Southwick offered a similar view in Albany's *Plough Boy* during its brief four-year run. With an overtly moralizing tone, he tied his purpose of communicating agricultural knowledge to the promotion of a specific lifestyle.[27] From this publishing platform, he offered an ethical framework within which farmers could separate speculation from fact. "Henry Homespun," Southwick's pseudonym, conveyed an entire homespun ethic for the periodical, wrapping reprints and commentaries in a cloth of right living that valued industriousness and common sense while disparaging idleness and insolence. A writer using the common pseudonym "Arator" wrote to *The Plough Boy* about "the Science of Agriculture" suggesting that prudence and "industry in all our laudable undertakings" were within the "scope of moral possibility" entailed by improved agriculture.[28] Southwick agreed and used his paper to push the moral superiority of industriously practiced science. He wrote, like Fessenden, Taylor, and Peters before him and like "Arator" and other selected contributors, that those who were "careless and slothful" with "the sin of idleness" were always cast as the negative shadow to the positive frame of the farmer's virtuous lifestyle.[29]

"The Moral Plough Boy" explained his purpose as aiming "at improving the moral, political, and economical condition of the people at large." The homespun rhetoric enabling this condition was defined most clearly in opposition to dandyism. "The Homespuns and the Dandies are antipodes," Southwick wrote. Where Dandies were concerned with frivolous amusements, Homespuns were hard-working. "The Dandies indeed would be harmless, were it not for their idleness, which is always infectious," and so Southwick saw his mission as denigrating the one and promoting the other.[30]

Southwick was leveraging a tradition in American culture defined by its contrast first to British dandyism and foppery and later to the element of impracticality found throughout the new nation. Fessenden was writing in much the same spirit, if not the same terms. A good deal of Revolutionary-era rhetoric had based its appeal on promoting independent, self-sufficient textile making—textiles spun from home-grown agricultural products, of course—not only as an economic plea, but for ideological reasons, giving deeper meaning to the homespun identity. As the historian Michael Zakim has observed of Timothy Dwight's patriotic verse *Greenfield Hill,* one finds a standard comparison between "American simplicity and European pretense, locating the former in that 'Farmer plain / Intent to gather honest gain . . . / In solid homespun clad, and tidy.'" Zakim explains that the homespun tied "the productive efforts of the household to those of the nation—thus also becoming a most tangible

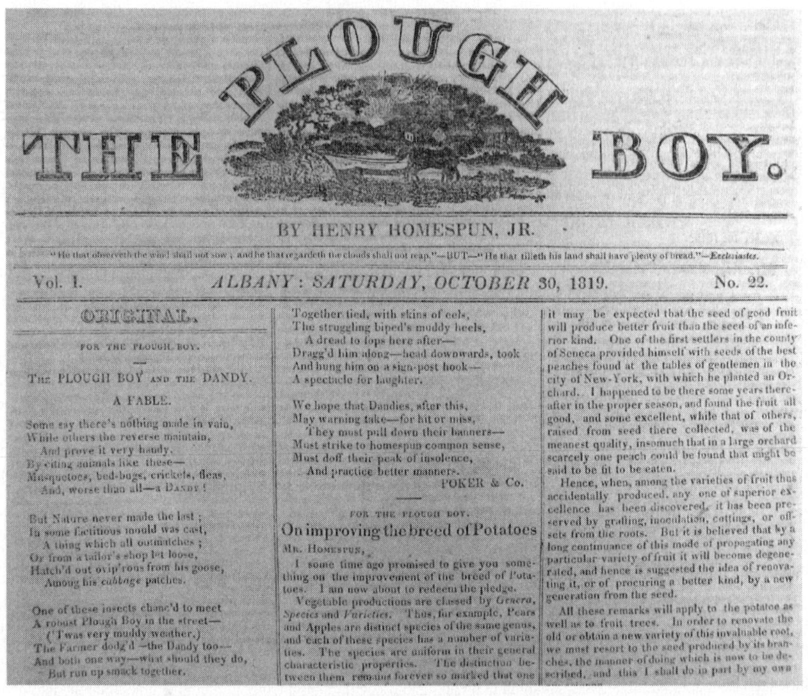

THE PLOUGH BOY.

BY HENRY HOMESPUN, JR.

"He that observeth the wind shall not sow; and he that regardeth the clouds shall not reap."—BUT—"He that tilleth his land shall have plenty of bread."—*Ecclesiastes.*

Vol. I. *ALBANY: SATURDAY, OCTOBER* 30, 1819. No. 22.

ORIGINAL.

FOR THE PLOUGH BOY.

THE PLOUGH BOY AND THE DANDY.

A FABLE.

Some say there's nothing made in vain,
While others the reverse maintain,
And prove it very handy.
By citing animals like these—
Musquetoes, bed-bugs, crickets, fleas,
And, worse than all—a DANDY!

But Nature never made the last;
In some fictitious mould was cast,
A thing which all outmatches;
Or from a tailor's shop let loose,
Hatch'd out ovip'rous from his goose,
Among his *cabbage* patches.

One of these insects chanc'd to meet
A robust Plough Boy in the street—
('Twas very muddy weather,)
The Farmer dodg'd—the Dandy too—
And both one way—what should they do,
But run up smack together.

Together tied, with skins of eels,
The struggling biped's muddy heels,
A dread to fops here after—
Dragg'd him along—head downwards, took
And hung him on a sign-post hook—
A spectacle for laughter.

We hope that Dandies, after this,
May warning take—for hit or miss,
They must pull down their banners—
Must strike to homespun common sense,
Must doff their peak of insolence,
And practice better manners.
POKER & Co.

FOR THE PLOUGH BOY.

On improving the breed of Potatoes

MR. HOMESPUN,

I some time ago promised to give you something on the improvement of the breed of Potatoes. I am now about to redeem the pledge.

Vegetable productions are classed by *Genera, Species* and *Varieties.* Thus, for example, Pears and Apples are distinct species of the same genus, and each of these species has a number of varieties. The species are uniform in their general characteristic properties. The distinction between them remains forever so marked that one it may be expected that the seed of good fruit will produce better fruit than the seed of an inferior kind. One of the first settlers in the county of Seneca provided himself with seeds of the best peaches found at the tables of gentlemen in the city of New-York, with which he planted an Orchard. I happened to be there some years thereafter in the proper season, and found the fruit all good, and some excellent, while that of others, raised from seed there collected, was of the meanest quality, insomuch that in a large orchard scarcely one peach could be found that might be said to be fit to be eaten.

Hence, when, among the varieties of fruit thus accidentally produced, any one of superior excellence has been discovered, it has been preserved by grafting, inoculation, cuttings, or offsets from the roots. But it is believed that by a long continuance of this mode of propagating any particular variety of fruit it will become degenerated, and hence is suggested the idea of renovating it, or of procuring a better kind, by a new generation from the seed.

All these remarks will apply to the potatoe as well as to fruit trees. In order to renovate the old or obtain a new variety of this invaluable root, we must resort to the seed produced by its branches, the manner of doing which is now to be described, and this I shall do in part by my own

FIGURE 2.5. Excerpt from *The Plough Boy.* This issue, of October 30, 1819, carried the fable of the Plough Boy and the Dandy. Courtesy of Special Collections, University of Virginia Library.

expression of the citizen's attachment to the public's happiness."[31] In the rural press, and with attention to land and improvement, that georgic-like expression of seeking moral and material goals was reinforced in the place-based requirements of science.

For example, in Southwick's "Fable of the Plough Boy and the Dandy," a generic Dandy and a Plough Boy run into each other on the street. The Dandy gets knocked down, dirty and muddy, then tries to knock down the Plough Boy with a stick. But the farmer wards off the blows and drags the Dandy to a sign post to hang him up as a spectacle for the public. The warning was that Dandys should "strike to homespun common sense . . . doff their peak of insolence, and practice better manners." Out of context, the caricature of the idle, disrespectful dandy stood only in juxtaposition to the esteemed, hard-working farmer. Within the rural press, however, the dandy was a recognizable contrast to the farmer's ethic of improvement and virtue.[32]

When Henry Homespun, then, advocated the tenets of agricultural science, he was siding with a specific kind of systematic work, not a universal sense of "science." When he took an implicit stand in favor of book farming, he aligned himself not with science writ large, but science for the sake of the agricultural life. Southwick skillfully distinguished between a general sense of philosophy and a specific thing like homespun science, a practical pursuit that was good for farms, good for improvement, and right for a moral society. As we see with Southwick, Fessenden, and others, the criteria for defining science in the antebellum context were not internal to its practice. The direction of credibility-gaining moved the other way. The criteria for defining science were external, being granted *by* farmers *toward* a science that was aimed at a moral understanding of right farming. Southwick's Albany paper was much like Skinner's in the upper South and Fessenden's in Boston.[33] When those editors deployed the phrase "science" so loosely and often, they had in mind the use of systematic, diligent, observational methods of practice for the benefit of the community.

Scholars writing about book farming have struggled to separate the uses of agricultural science from dominant narratives of progress, improvement, and rationality, falling prey to the same rhetorical constructs the calculating editors used. Their interpretation of book farming contrasts modernity with tradition and ignorance with innovation—those opposing book farming were ignorant and backward looking; those in favor were modernist and "scientific." Such a view, however, uses the later credibility of science as its own explanation while concealing a far deeper debate about how and why tools like agricultural chemistry became part of American agriculture. More broadly, for those seeking to wed the science of agriculture to the nineteenth-century goal of improvement, book farming stood as a precise example of the difficulties inherent in the contrasts of tradition and progress, rational and folk wisdom, almanac-based and scientific text-based information, locality and distance, and amateur and professional. Those pro-book farmers, of course, claimed that the future lay with quantified and systematic means to scientifically improved ends, that the ubiquitous call to improve could be achieved with science. Unfortunately, the historiographical focus on rural science has generally treated the science-improvement bond uncritically, assuming that it was inevitable and unproblematically achieved. But in the fields of the early American Republic, it was not clear that the science of agriculture and agricultural improvement were synonymous.[34]

The ubiquitous claims for industriousness over idleness, to "walk abroad in the majesty of virtue" and remain "strangers to vice," were not difficult propositions to accept, then anymore than now. They were basic restatements of classic Protestant values: work was virtuous. Fessenden's editorial essay on book farming was meant to place the practice of rational agriculture in a framework the practical husbandman already understood, that of the cautious and informed use of new ideas. Science, then, had to fit the farmer's ethic to be promoted, as it was one tool among many selected by the craftsmen.[35] Furthermore, and perhaps more consequentially, the agrarian community was starting to interpret nature (in the form of soil, crops, and landscape) with the instrument of science. As a new instantiation of georgic science, it was a homespun science.

References in the agricultural papers were often prescriptive, saying how things should be done, but not ensuring that they were actually practiced as such. They show us the world the improvers wanted to create. These prescriptions even came about surreptitiously. Edmund Ruffin, the editor of *The Farmer's Register* in Virginia, was a notable and unapologetically self-promoting book farmer. In the second volume of his paper, a contributor signing his name "A Book Farmer" wrote "An Apology for 'Book Farmers.'" The author invoked the writings of Arthur Young, William Marshall, Humphry Davy, Richard Peters, and even Ruffin to claim that poor *practice* caused bad farming, not the book farming advice itself. He—for we can assume it was a he, as nearly all the contributors were—also cleverly took the time to compliment Ruffin and his "commendable" example of practice. The fault of bad farming lay with the "indolence and carelessness" of the practitioner, not the theory itself.

The author's claims fit precisely into the georgic and homespun ethic that favored diligence and disparaged indolence. Considering the still unstable place of science in farmers' lives, the author understood that winning the argument for book farming meant casting it as morally superior, not deferring to the still-untested principles of accuracy or prediction (values, that is, defined by chemists and philosophers). In other words, asking farmers to use chemistry to prescribe action on their lands had to follow from placing the values of that pursuit within the farmer's perhaps idealized homespun life. The author was Ruffin himself, masquerading under a pseudonym—a practice not uncommon in the rural press—to further his own views about the merits of the science of agriculture and book farming.[36]

Despite such wily propaganda, though, the case for introducing and accepting science or agricultural chemistry could not be made from editorial perches alone; the story also involved the testimony of farmers and planters dealing with the complexities of written advice and experience on a daily basis, those not just speaking the georgic ideals, but crafting them in their work. Through account books, diaries, and letters, those farmers and planters offer more insight into matters of resisting and promoting book farming. Their records originating in rural community settings and local and familial networks are often difficult to follow and mostly avoid direct commentary on the value of science. The most telling of them, however, provide indications of perceptions of quantification, experimentation, and agricultural chemistry.[37]

To the list of presidents, wealthy planters (sometimes also presidents), regional editors, and rural press commentators presented above as actively involved in systematic studies of the farm, I add one more wealthy planter, William Fanning Wickham (1793–1880). Wickham was a Virginia book farmer. He experimented with different fertilizers, mined marl and other manures from his property, suggested methods and systems of analysis for the cause of improvement, considered the validity of new scientific or philosophical principles about agriculture, read and contributed to several rural periodicals, and even translated foreign articles for *The Farmer's Register.*[38] These activities were part of the farming practice to which he was devoted for decades. He was a well-educated and wealthy landowner of over 3,000 acres at Hickory Hill in Hanover County, situated to the north of Richmond and east of Charlottesville. He was also a trained lawyer, owner of as many as 275 slaves by 1860, and son to a century-old family of southern gentry.[39] His social stature enabled him not only to examine chemically nuanced systems of improved agriculture (by virtue of his education), but to write about his experiences with those methods with some degree of social credibility; his slave ownership and ability to conduct experiments also mark Wickham as distinct from northern contemporaries. For him, book farming represented a tension between the activity of farming and the contemplation of theorizing, between observation and speculation.

Wickham grew a diverse set of crops including wheat, corn, and clover on a series of fields. He was fertilizing wheat with marl in systematic fashion by the 1820s. As early as 1828, as he recorded in his diary, he "began to haul marle into the low field from the old bank" using "3 to 400 bushels" to the acre. He later opened "a new pit in the hill side near the river on the low grounds" and from there planned "to cover all the low grounds in the barn field with marle and to fallow it in the autumn for wheat." The corn that year was his best ever, though he did not indicate whether the fertilizer deserved the credit. "The season could not have been more favorable" was his only comment.[40] The next season Wickham concluded that "the effect of the marl . . . in the long field at the Lane is astonishing. " Using his slaves to open up new marl pits, devoting more and more labor to the projects (from one horse cart per day to three), and ever increasing the bushels per acre, he was developing a kind of protoindustrial agricultural fertilization process. His experimental success was literally visible from afar—"The young clover in the wheat field looks well especially on the marled land which can be distinguished at a considerable distance." By the late 1830s, Wickham's fertilizer use was regular and predictable; he had plaster experiments underway as well.[41]

With fertilizing techniques, crop rotation strategies, advanced mechanical implements, and even a steam sawmill by 1848, Wickham would be considered an "advanced" farmer. He was practicing and experimenting with the most up-to-date farm management and agricultural methods.[42] By then, he had extensive experience and trust in methods of fertilization and what he considered agricultural science. Still, Wickham did not always trust the dictates of scientific methods gleaned from treatises, preferring the experience gained from his own land. He was slowly complementing his extant means of learning how to farm—from his father, from his neighbors, through the experience of his early years—with the science of agriculture. His interaction with the land was thus also changing.

Wickham was considered a "methodical and observant man," as another analysis of his diaries showed. As such, he was "an exemplary farmer." He had likely shed his lawyer's identity by the 1830s, having not practiced for a decade by that time. As a lawyer, he would have been castigated as a "vice-ridden" drone. As a methodical and observant man, he was capable of promoting the qualities of "virtue, serenity, and good health."[43] Wickham fit the value model by which book farming was being judged, answering questions about virtue and vice through his attention to detail and patience for experiment. His lived experience guided his ideas about the land and suggested how to experiment for improvement. It helped that he was wealthy and had hundreds of slaves mining his marl, digging his ditches, and preparing his crops for market (see chapter 4). It helped too that he was the son of an improvement-minded planter; this erased the supposed contrast of tradition (or inheritance) versus progress. He thus offers a good example of a book farmer taking advantage of his escape from physical labor to concentrate on mental labor.

Wickham exercised his authority as a diligent practitioner in his community and through his self-representation in diaries and through the rural press. A debate in *The Southern Planter,* published in Richmond, highlights the point. In 1841, he wrote to correct misperceptions he saw in print about "ploughing" techniques that encouraged farmers to till their manure underground in the fall instead of the spring. The editor, Wickham, and at least two correspondents were advocating different techniques. First, an article from *The Genesee Farmer* in upstate New York had commented on the "scientific opinion" of a popular treatise of the 1840s, *The Practical Farmer,* noting that its method of fall ploughing was consistent with "established principles of philosophical agriculture." The article was reprinted in *The Southern Planter* and, in response and dispute, Wickham wrote in favor of spring ploughing. He there questioned the wisdom of the "scientific" opinion because it conflicted with his experience, not because it was scientific. Then a third participant entered the debate in a further reply. "A Hanoverian," as he signed the letter, wrote to *The Southern Planter* to question Wickham's contribution and the "increased improvement which that mode of using effects." The "Hanoverian"—writing from Wickham's own Hanover County—wanted to understand the difference between the uneven sets of advice.[44]

But whom should readers believe? All parties agreed on the goals: improved crop yield resulting from legitimately conveyed advice. Tilting the scales, the editor then realigned himself, siding with Wickham. To do so, he explained Wickham's credentials—"not a man likely to make a mistake"; one who "has no theory to advance [but] only repeats the facts"; "we hardly know one upon whose judgment we would sooner rely."[45] Wickham's authority in the debate was understood initially to lie in his reputation as a practical farmer. He then benefited from the sponsorship of *The Southern Planter*'s editor, who vouched for Wickham's character. The debate exemplified, first, that the *kind* of observation mattered (not whether there *were* reported observations or not) and, second, that the accepted authority of the observer ultimately settled the matter.

Wickham successfully cultivated the image of the diligent, georgic farmer. His descriptions of experimental practice did not rely on technical chemical language, though they did stand behind a planter of accepted social authority offering reports of his own positive experiences with systematic fertilizing experiments. He had tried marl, plaster, crop rotation, and clover in addition to the animal manure long used on the plantation; his later, 1850s use of guano as a fertilizer followed this mode of operation.[46] All these efforts presupposed the goal of improvement and the value of book farming. On one reading, the ploughing debate could be interpreted as a matter of competing claims between the scientific opinion of *The Practical Farmer* and the local, nonscientific opinion of *a* practical farmer. But that would be to misread the subtleties of value and virtue in the fields of the early Republic. More centrally, the debate hinged on *what kind* of scientific opinion was being wielded, where fact-based testimony had merit based on the valued source of authority, where the place of the observation gave it credibility. Homespun virtue won out.

At the same time, another Virginian, John Walker (1785–1867), was a book farmer in a different sense. He wrestled more with the authority of agricultural advice at a personal, rather than philosophical, level. As the historian Claudia Bushman has explained while detailing Walker's antebellum diary, he was also less wealthy than Wickham and less educated, had far fewer slaves (though he did have some), and was more ambivalent about taking farming advice from books. Walker lived and farmed in King and Queen County in the eastern peninsula between the York and Rappahannock rivers, called the middle peninsula of Virginia. He owned under 1,000 acres and just over 20 slaves.[47] He was a devout Methodist, having converted in 1818 while living in Kentucky. After his conversion, and after his brief foray as the model of a Virginian emigré in that earlier move away to the west, he returned to his family's land near Walkerton.

His engagement with philosophical principles of agriculture, the validity of which was the bone of contention in his view of book farming, was more troublesome than Wickham's. Walker represents yet another tension in the concept of book farming: for him, the debate was between believing in an almanac or an agricultural journal, between traditional approaches to farming and so-called modern ones.[48]

In her comprehensive reading of John Walker's journals, Bushman implicitly takes as her theme the recurring issues of trust, authority, and belief. While Walker was reading "progressive materials," she finds that he was also seeking advice from other sources of epistemic authority like the almanac. He often tested his moon-farming methods against the recommendations of periodicals like *The Southern Planter,* indicating his "continuing concern about authority and trustworthy power." The "problem of whom to believe, [of] what was the best source of authoritative advice," dogged Walker as he weighed the suggestions from written articles against his own history.[49] He also fought with his neighbors, was at ease expressing opinions on county politics between Whigs and Democrats (especially as concerned internal improvements), lived in the Methodist minority, and experimented with Thomsonian medicine. Each of those traits placed him in tension with some form of authority, be it political, religious, medical, or a combination of these. All of those traits, taken together, characterized him as a farmer struggling with multiple questions about authority and belief.

Like Wickham, Walker was using marl by the early 1830s. His diaries indicate that he read John Taylor's *Arator* and had at least a passing familiarity with issues of soil exhaustion. For him, being a book farmer meant reading the rural press and comparing moon lore against new methods of planting. In 1825, vexed by issues of authority and by the promise of new techniques, Walker constructed an experiment to compare almanac advice based on the stars against his own observations based in his fields. "Who should he believe?" Bushman asks. After all, "the competition between modern and traditional could be seen in his wavering but stubborn loyalty to planting seeds according to phases of the moon." It seems that Walker used an array of available approaches. "Walker incongruously mixed the superstitions of the past. At the same time he was reading progressive materials, he consulted almanacs to monitor the progress of heavenly bodies." But this is incongruous only if one takes the inevitability of scientific success as the benchmark against which to judge Walker. In his own context and as part of his homespun life, his dual use of almanacs and the rural press was consistent. He utilized the methods available to him and incorporated different techniques without recognizing the later clarity of differences in those methods.[50] Walker was like Wickham in the sense that he was in the midst of introducing newer methods to his agricultural practice, but differed in the scope and sophistication of his experiments.

In some cases, Walker saw a nice convergence between printed advice and his personal observations. The press was widely promoting lime use, for example. Walker's own experience with that fertilizer had been positive. He even bought a lime scatterer after seeing an advertisement in *The Southern Planter.* Walker "likely did not understand its effect on the soil," though that lack of understanding would have been typical, Bushman notes, and an artifact of a different *kind* of knowledge—georgic, hands-on, or "practical"—not a complete lack of knowledge.[51] Influenced by the press, he used a mixture of ashes, lime, and plaster to prepare the soil and his seeds for planting. He also paid attention to more precise debates such as those turning on how best to roll seeds in fertilizer before planting them, not just how to modify the soil conditions directly with the fertilizer. Consistent with his attention to the rural periodicals, to communication of agricultural practices, and to extant debates about methods, he took an active interest in the founding of the King William Working Agricultural Society in 1842. All told, Walker was a working farmer struggling to gauge the ideals of science and improvement against a long tradition of capable and successful planting.

Smaller-scale farmers also grappled with the issues of the authority of advice. John Lewis (1784–1858), a Spotsylvania County, Virginia farmer of 85 acres, kept a diary throughout the 1810s. Living in the county north of Wickham, he noted his planting patterns, his seed treatment, and the prominent theories he chose to follow while simultaneously questioning the validity of philosophy and science. Lewis based his doubts about philosophy on justifiable points of confusion over whom to believe. He was literate enough to follow the papers and interested enough to entertain theories like those of James Garnett, the popular Virginian agricultural speaker and president of the Fredericksburg Agricultural Society.[52] But Lewis was also wise enough to wonder where the credit should be given for successful improvement. Growing potatoes, celery, turnips, radishes, cucumbers, lettuce, asparagus, tomato, corn, cabbage, and wheat, Lewis bought and used plaster and gypsum, in which he rolled his seeds. Wondering when to manure his fields, he followed a method of coverage in the fall and tilling during the spring, "in conformity with Mr. Garnett's theory" (though not with Wickham's later advice, apparently). He did not elaborate specific details, but did register doubt about the methods he had been advised to use. While the "plaister" seemed to work—"The red clover 9 inches high (plaistered); 4 inch where it was not plaistered"—he elsewhere questioned the cause of his success or failure. Was it the gypsum-rolled seeds or the unusually warm spring? Was it Garnett's theory or a favorable rain? Lewis tells us little more, though, leaving his questions open-ended. One February day in 1820 it was 70 degrees at 7 a.m. Lewis noted: "Yet philosophers tell us that the greater heat of summer is produced by the greater perpendicularity of the sun's rays!" and then wondered what advantage the philosopher has when the farmer could use the same visual evidence to draw conclusions.[53]

Walker and Lewis dealt more overtly with questions of authority, trust, and belief than did Wickham. This is unsurprising given that confidence is another value often left unelaborated in discussions of social mores on the farm; questions about authority and belief will have different answers for the self-assured, confident farmer than for the self-doubter. Wickham's social status certainly brought forth the confidence of a wealthy man. His wealth enabled him not just to try new experiments but to not question his approach and techniques.[54] There is no hint of deprecation or contingency in either his diary accounts or his published columns.

Rural press advocates had long had trouble convincing average farmers that those farmers' writing held authority. Wickham was not part of that problem, but his neighbors were. The older generation of editors and agricultural society leaders made it clear that the timidity of small farmers was something to be overcome or assuaged with appeals to a greater good. Richard Peters, speaking to the readership of the PSPA in 1818, had encouraged his audience to "fear not to attempt an improvement or discovery." John Skinner editorialized in 1819 that "one of the greatest difficulties . . . in the execution of our humble undertaking [on] behalf of the farming interest, was the fear, that we should find it impossible to overcome the mistaken diffidence of agricultural gentlemen." Wilson Cary Nicholas said much the same the year before when speaking to the Albemarle Agricultural Society.[55] Walker, although not stating it so explicitly, was in a far more fragile, possibly diffidence-causing position. To be sure, psychological factors do not stand alone here, but it is probable, based on his diary and his pattern of experiment and doubt, that Walker was more likely to question his own approach. On the one hand, this led him to skepticism about book farming itself and, on the other hand, to skepticism about his own ability to refute book farming.

Lewis's confusion over the credit for his field's productivity was the bread and butter of rural uncertainty about book farming and the science of agriculture. Walker's skepticism about book farming and Wickham's continued need to clarify and promote the subject were part of the same milieu of acceptance or resistance. Again, the core of the matter was authority and belief; Lewis and Walker stood unsure about whom to believe, while Wickham affected the character of one who was believable.

What of resistance? How could it be overcome? It was hardly clear to the practicing farmer how new ideas would work or where they would lead—that is, if they *were* progressive or destructive, if they were improvements or novelties, if they were fact or speculation. Indeed, there had been ample evidence to show that not all new ideas were good ones. As Fessenden had quipped, "every novelty may not be an improvement, altho' every improvement was once a novelty."[56]

With respect to the role of science on the farm, those resisting book farming perceived a breach between their goals of improvement and the value of science. Southwick, Skinner, and Fessenden tried to redress this perception by connecting virtue and the political economy of agriculture to science. They worked hard to convince real, genuine, practical farmers that their concerns may have been justifiable against the dandies, but they were not justifiable against book farming.[57] They did this by proposing that book farming, when right, was georgic or homespun science. They had another dominant concern, which was that those already distrusted—the dandy, the insolent, the man of "whim-whams" who was "full of notions"—seemed to have undue influence on practicing farmers. So how could a practicing farmer tell the difference between valid improvements and dandyesque novelties, between legitimate observations and contrived speculation?

Given the prevailing lens of diligent and valued labor, these were questions not just about method and system, but about whom to believe and why. They were about trust and reliability. Discussions of book farming took place at a level of discourse filled with platitudes. Don't be a dandy, be homespun; be a stranger to vice, a friend to virtue. Within these terms, and lodging the authority in the hands of the farmer, whether one accepted the value of science was a matter of whether one saw the system or method as virtuously pursued and reported.

There is more to this, of course. I have been considering the ways agricultural science and chemistry gained the credibility to improve agriculture and could become a valid practice for working the land, questions whose answers cannot be deconstructed with only an evaluation of prevailing value claims. Broader philosophical, organizational, publishing, and cultural factors play important roles as well. For example, the more widely considered context of professional activity in chemical circles and the rise of organic chemistry and a new scientific profession helped give foundation to the circulation of so-called scientific activities. Conversations in the rural press taking place during the years between 1820 and 1850 occurred during a period of significant "clarification and consolidation" with respect to "improved analytical procedures . . . the growth of journals . . . [and] the rapid dissemination and criticism of results," as the noted historian of chemistry Aaron Ihde has written.[58] This too helps explain the cultural availability of chemical methods. More specifically for the science of agriculture, Davy and his *Agricultural Chemistry* (1813) and, soon, Justus Liebig and his *Organic Chemistry and its Relations to Agriculture and Physiology* (1840) were making their texts increasingly available, thus holding a place in the story about science and the farm. In fact, disseminated texts from professional chemists have been the basis for most stories about science and agriculture in American history.

Yet, those were all developments occurring *within* a social sphere of chemists and men of science. As such, they do not tell us much about how those developments were perceived from outside that sphere. When published and diffused into the hinterlands of rural America, those contributions would be perceived as *external* to the agrarian community, and sometimes violently so. Because of that outsider status, the role the theories of Davy and Liebig played in defining their own authority—in establishing their cultural credibility—would have to follow from a consideration of the farming class's values. To explain the diffusion of Davy or the acceptance of Liebig, then, one would have to consider the factors that existed in the consuming or accepting community, a story that would show it was not in fact *consumption* of knowledge and practice but *reproduction* of knowledge within a preexisting set of conditions that explains the circulation of agricultural chemistry.

The more basic point is that Southwick's "moral, political, and economical" goals were of a piece with the value of science, in just the same way that Americans like Washington and Jefferson in the eighteenth century had promoted a multipurposed georgic science. What began as an association of science with dandies, in the sense that science was speculative and theoretical, became bifurcated into dandy science and homespun science. The former was still speculative, but not definitive of all science; the latter was fact-based and derived from agricultural experience. The place of the science of agriculture in rural antebellum America was one that was understood through a georgic or homespun prism. It was alive and well, circulating indeed, but not only because of the contributions of "chimical men." This framing of the book farming debate reveals that the salient issue in the rural context was credibility, not novelty.

PLACES CULTURAL AND GEOGRAPHIC

Samuel Swartwood, a Maryland farmer, boasted gleefully in 1819 that "I desire, most ardently desire, that my favorite theory should obtain proselytes."[59] Apparently, there were a lot of Swartwoods. With the agricultural press growing in number and diversity, theories of soil fertility were seemingly endless and easy to propose. *The Southern Planter* included a passage in 1842 announcing that "agricultural theories . . . or *guesses,* for they are little better, are as plenty as black berries." Charles Botts, *The Southern Planter*'s editor, summarized that precise issue by way of follow-up to an address by James Garnett to the Agricultural Society of New Castle County, Delaware: "Certain great theories and systems, promulgated from high places, like other humbugs, have their day, until some plain farmer declares, and proves too, that the author has been misled by his ignorance of the facts upon which he has attempted to reason." Another contributor perhaps put it most boldly, noting that "Mr. Justus Liebig is no doubt a very clever gentleman and a most profound chemist, but in our opinion he knows about as much of agriculture as the horse that ploughs the ground, and there is not an old man that stands between stilts of a plough in Virginia, that cannot tell him of facts totally at variance with his finest spun theories." Liebig's ill-formed farming credibility pointed to the same contrast between fine spun theories and farm-based facts.[60]

The tenor of such remarks was reminiscent of late eighteenth-century georgic tourist rhetoric about locating improvement in the hands of the working farmer. Concerns of the mid-1800s, however, were part of a milieu somewhat different than that of earlier decades. Those earlier arguments for the place of science on the farm—that science is good—had moved into new claims for the place of properly collected "facts" on the farm—that science is good if it is sanctioned by epistemic authority, a kind of legitimacy granted only to land-based facts, not disengaged speculation.[61] Put another way, rural advocates had clarified the generality of claims for scientific studies of the land by casting them inside the deeper cultural issues of trust and credibility. All those issues were approachable through the guise of virtue and, ultimately, the virtue of the fact.

A fact could be enlightened, strong, sound, pure, and well ascertained. "Unvarnished facts are very scarce," *The Southern Planter*'s commentary on "Theories" said, "and yet, they are the only foundation upon which sound theories can rest." A report from the Hole and Corner Club of Albemarle on "EXPERIMENTS" proclaimed, "Science calls loudly now for well ascertained facts, from which she may deduce the laws of agriculture."[62] Even when reviews of prominent European chemists appeared in the American rural press, they were delivered in a context of practical farming that deferred to field experience. Negative reviews of Liebig, for example, again called attention to deficiencies in his fact-gathering skills. In this sense, when the average, nonspecialized farmer wrote to the rural press, he was saying nothing too different with respect to the rhetoric of facts than the reports coming from scientific journals.[63]

The literature of the late antebellum period shows widespread use of the laboratory metaphor for the field that served only to strengthen the importance of place-based facts. Scholars are more accustomed to discussing field-based work and laboratory-based work in a later nineteenth-century context, after the so-called laboratory revolution of chemistry.[64]

But those metaphors had already been developed in the decades before the Civil War. "The soil," said the president of the Maryland Agricultural Society, Robert Smith, "is the great laboratory in which the food of plants is prepared." The open field and the farmland were the places for agricultural chemistry development. By the late 1850s, Richard Eppes of Virginia could look upon his land and take the farm-as-laboratory observation further to claim, "A farm is another name for a chemical laboratory. It is only another way of manufacturing."[65] Liebig too proposed such metaphors of farm and lab, in the process fitting into a broader pattern of rethinking what it meant to know the farm.

Perhaps the most comprehensive way to look at the issue of book farming is by recognizing the georgic philosophy of praxis embedded within it: how those who labor in the earth know the land versus those who write in closets, speaking from disengaged speculation; the active versus the contemplative. A common value set of utility, diligence, and labor connects this praxis-oriented approach to agricultural knowledge (that to work the land is to know it) and a practical philosophy of science (that science should be based on providing practical, and practicable, results). It is not enough to say that Americans were practical, or that they promoted a Baconian fact-gathering philosophy of science. Despite the resistance to general theories in agricultural chemistry, opponents of book farming were concerned most clearly about who provided those theories and from where the facts were found. There were theories, and they were everywhere. What mattered was whether or not they were wedded to the belief that "working is knowing." James Campbell, a Pennsylvania Whig speaking with a neo-Jeffersonian voice before Congress in 1856, assumed and extended the argument that science was a tool to be applied to the farm, a set of practices and methods, by conjoining the nobility of agriculturalists with science. Let "science, applied to the culture of the earth, go hand in hand with practical labor," he would say. In the process, he subsumed the earlier material and moral discourse of georgic virtues within his understanding of the role of science in society.[66] Campbell's argument rested on the virtue of the farmer, the "noblest race."

Studying book farming in its contemporary context brings with it contradictions, inconsistencies, and vagaries, as the cases of Wickham, Walker, and Lewis highlight. Tensions abound in historical work as well: traditional versus progressive or modern, practical versus speculative, active versus contemplative, field- versus laboratory-based, market-oriented versus subsistence, or gentleman farmer versus practical farmer. Some of these, like market-oriented versus subsistence farming, seem to be historiographical constructs that have little to do with the actual nineteenth-century discussion of book farming. Others, while not constructed by historians, were politicized rhetorical maneuvers in their original use. To label an opponent of book farming "traditional" was to place him in an inescapable binary of forward or reverse. In an age of improvement, looking to the past ran against dominant social views. Similarly, but from the other side, to say that book farming was speculative and impractical set the book farmer on the wrong side of a common value set that cherished practical means and admonished "fine spun lubrications."

The book farming debate reveals the nuances of science, improvement, and the land. It shows that rural improvement advocates were producing new means for understanding the land—new forms of mediation between human and nature—with new methods for working on it. It shows too, in an era of improvement advocacy, that science had not been clearly or inevitably linked to agricultural practice. Rather, the terms of that new science of agriculture had to be debated and clarified not just by professional chemists such as Davy or Liebig, but within the communities that would grant authority to the new sciences.

Consistent across the decades was the issue of credibility and virtue. Book farming debates in the 1820s sought to distinguish between the unfettered pursuit of improvement and the correct, well-considered means of effecting it. The rural press editors had to argue for the value of science, to convince their readers that science had a place on the farm. But the meanings within book farming arguments shifted over the decades, so that a mid-century appeal to book farming—synonymous with the science of agriculture by that point—was one that argued that the virtuous pursuit of science was akin to a virtuous pursuit of farming. Culturally, social and epistemic authorities were also inseparable—the value of the fact was related to the virtue of the fact-gatherer. By the mid-nineteenth century, those promoting the acceptance of agricultural science were arguing not about *whether* science had a place on the farm, but *in what way* it had a place.

In this sense, questions about book farming tie together several issues of the place of scientific authority. Why improve? Because improvement was not only a program of economic and social progress, but also a plan for cultural stability, for utilizing new methods and practices to maintain a stronger society. Why write? Because communication was the staple of improvement, the legible means by which agricultural improvement could take place. Why resist? Not just because of mindless ignorance or feet stuck in tradition, but for valid and rational reasons that hinged on the authority of those who prescribed change and the system of belief within which the acceptance or rejection of new practices were based. What mattered? Given the system of belief and authority, the important factor in debates about book farming was *who* as much as *what.*

CHAPTER 3

Knowing Nature, Dabbling with Davy

Behold another volume on husbandry! exclaims a peevish man on seeing the title page: how long shall we be pestered with such trite stuff?

—Henry Home, *The Gentleman Farmer,* 1776

AGRICULTURAL IMPROVERS WERE ADEPT at chemical philosophy and practice on their respective lands, formulating their own ideas in relation to the leading chemical theories of the day. For them, chemistry was understood as an active prism through which to see and with which to act upon the soil. But how was agricultural chemistry used to inform perceptions of cultivated or cultivatable nature? What did it take to "know" the soil and land this way? And how did these questions fit within the tension between vitalist and materialist philosophies of nature then current in the broader post-Enlightenment, or Romantic, era?

Henry Home's peevish man would have been pestered still fifty years later, when antebellum volumes on husbandry offered insight into those questions. He could not, though, have so easily cast them aside as trite. Three agricultural treatises of the early American Republic offer insight into questions of nature and science and, at the same time, into the position of chemistry between farmer and soil. The first treatise is Edmund Ruffin's book-length *Essay on Calcareous Manures* (1832), published in Petersburg, Virginia. The second is John Lorain's *Nature and Reason Harmonized in the Practice of Husbandry* (1825), published posthumously by his wife in Philadelphia. The third text is Daniel Adams's *Agricultural Reader, Designed for the Use of Schools* (1824). Adams's work was published in Boston for distribution around New England. The treatises elaborate the place of chemistry in rural working philosophies of nature. Each was written by a book farmer for the purpose of agricultural improvement; each author used chemistry to some degree to give force to his strategies; and each work shows how chemistry was introduced as a prism through which to see the land. Like a new pair of glasses, chemistry was still very much visible. Chemistry's instrumental opacity led the users—the farmers and book farming proponents—to comment on its presence and to explain its role in their concept of improvement. It had yet to gain the transparency that unquestioned legitimacy would later bring.

The agricultural chemistry that Ruffin, Lorain, and Adams put under consideration drew explicit connections between scientific analysis and nature. It was an early example of environmental science. Writing before 1840, the authors dabbled to various degrees in the work of the foremost international authority on agricultural chemistry, Sir Humphry Davy. On that topic, Davy's name had been ever-present from the time of his *Elements of Agricultural Chemistry* (1813) and its first American reprint (1815), though soon—at least by 1840—the German chemist Justus von Liebig would become the primary reference point for agriculturalists aligning science with improvement. The two chemists—Davy in Britain, and under the auspices of the British Board of Agriculture; Liebig in Giessen, and rebutting Davy—promoted theories of the complex phenomenon of plant growth. Food sources, planting techniques, issues of solubility, and climatic conditions were each important. Both men placed chemistry at the exact center of agrarian nature, estimating that its precision and progressive characteristics would turn agricultural lands into thoroughly scientific places.

In terms of the development of soil science theory, that transition from the 1810s through the 1840s now appears as a shift from the organic, humus theory of plant growth and soil fertility represented by Davy to the inorganic, mineral theory of Liebig. Where Davy's humus model suggested that the process of fertility was living and organic, Liebig proposed that fertility was a simple matter of replenishing lost material. "A time will come," Liebig wrote in 1840, "when fields will be manured with a solution of glass (silicate of potash), with the ashes of burnt straw, and with the salts of phosphoric acid prepared in chemical manufactories, exactly as at present medicines are given for fever and goiter."[1] With respect to philosophies of nature, then, the bounded points of the 1810s-to-1840s transition correspond to a change from viewing the land as an interactive, interrelated economy of vital components to one of individual, inorganic elements, each identifiable through analysis and replaceable in quantified measure. The era helped produce and was witness to a shift from animistic and organic visions of agricultural land to the mechanistic philosophies of nature that underlay later industrial agriculture and an industrialized worldview of nature.[2]

Romantics bemoaned these developments. Samuel Taylor Coleridge foresaw them in the 1820s when he observed, "All Science had become mechanical." It had been "given up to Atheism and Materialism."[3] Environmental historians and science studies scholars of the late twentieth century have made similar observations. Carolyn Merchant traced the change as the capitalist ecological revolution that brought a mechanized nature to the fore of agricultural consciousness. She writes of the replacement of those who *imitated* nature, the animistic, by those who *analyzed* it, the mechanistic.[4] The instrumental character of sciences like chemistry, to be sure, played an influential role in establishing nature as that analyzable entity. Davy's theories were consistent with and also easily accepted into the more imitative organic worldview when he wrote in the early nineteenth century.[5] Liebig's theories of mid-century, in their mineral, inorganic framing, appeared to oppose vitalist philosophies of nature.[6] The shifts between the two men are thus more than mere chemical stories; they are also commentaries on philosophies of nature, as Romantics like Coleridge understood.

Contemporary and historical summaries of a neat transition in crisp before/after terms, however, glide over a morass of uncertainty and variety. That unclear, contested morass defined the placement of science in the daily lives of those who actually exhibited agricultural consciousness. Along with Donald Worster, whose reference to the change from Arcadian (Pastoral) to Imperial views of nature outlined a similar transformation, Merchant has mapped out the broad shifts in perceptions of and modes of interaction with nature that defined an analyzable, mechanical world. Yet the examples of Ruffin, Lorain, and Adams fit into the middle of the now-visible transition period between Davy and Liebig. Their work thus does not present a picture of stark change over time, nor does it resolve these tensions of organic and inorganic, or vital and materialist.

With respect to chemistry, Ruffin, Lorain, and Adams were familiar with Davy's work and impressed by the contributions the Englishman had made to agricultural life. On the philosophy of nature side of things—only briefly here treating philosophy and chemistry as opposite "sides," which it is my point to avoid—the authors represented nature as a living and organic system. Ruffin in particular blended vitalist and materialist concepts, while Lorain staked out a position firmly on the side of vitalism, consistently demanding attention to "the vital economy of nature." Adams was similar to Ruffin and Lorain at different points, but was more forceful than either in conceptualizing the land as an improvable entity and in forwarding Davy as the preeminent authority on all matters of agricultural chemistry. Adams was an appropriator, taking Davy's word at all times and without criticism. Lorain was a refuter, rejecting Davy's testimony as much as Adams had accepted it. Ruffin split the difference, tempering the import of Davy's theories by reference to georgic practice while acknowledging the value of his laboratory-tested ideas. Each author's view of chemistry was consistent with his philosophy of nature.

Rural knowledge about the soil in the early nineteenth century was embodied as much in mental labor as in the physical toil that dominates images of agriculture.[7] Ruffin was born in 1794 into this context of agricultural knowledge-production, but with particular geographical and cultural attention to the upper South. Lorain was more than a generation older than Ruffin, born in 1753 in the colony of Maryland though living much of his life on the backwoods frontier of Pennsylvania. His insights are thus derived from a more senior analysis in the young Republic, though with the same goal of improvement. Adams, born in 1778, split the age difference but edged geographically north, having been raised in New England. He was devoted to improvement and agriculture inside the legacy of a Puritan social structure but outside the concerns of Ruffin's slave-labor issues and Lorain's backwoods expansion. All three informed their works by appealing to the experience of the working farmer and tied the validity of their proposals to the authority of that experience.

A word or two ought to be said about my choice of these authors and sources. I focus on them not because they are definitive of the period, as any single author or even set of three could not possibly fit that requirement, but because they are illustrative of views about nature, representations of chemistry, and the motivations for associating the two. Other writers on agriculture and science might have been used.[8] The three I discuss here hail from different cultural and economic regions of the antebellum era, making it difficult to understand their uses and views of chemistry without noting the distinct political atmospheres into which those uses and views were introduced. Thus, Ruffin, Lorain, and Adams differed in their stated purposes for writing (though not in their overarching themes of "improvement"), the approaches they took to utilizing science for the benefit of the farm, and their social outlooks. They lived, that is, within different cultural architectures. Each writer was more socioeconomically privileged than the average working farmer, and so it would also be incorrect to consider them representatives of the common social experiences of mass culture. But the three shared important features and, as such, I group them together for their overtures to georgic expression, the similar formats they employed, their interest in utilizing chemistry, their concern for exploring the value of science for improvement (they are all book farming advocates), and their status as historical figures working outside the realm of a cohesive chemical discipline. They provide an opportunity to ask not exactly what the "common" person thought, but at least what the nonscientific agent thought.[9] They offer a useful blend of diversity and unity in expression.

THE NATURE OF THE AGRICULTURAL TREATISE

Edmund Ruffin had several identities. He was a Southern rural editor, a fertilizing experimenter, an agricultural treatise author, an amateur chemist, and a rabid pro-slavery secessionist. Each identity, in common, was based in rural economy and part of a worldview related to perceptions of the land. Ruffin saw that land as alive, filled with powers, subject to injury and healing, and analyzable for its discrete parts. The soil, to him, was alive and dissectible, and chemistry was the means by which that dissection could be performed. In his twenties, he was inspired by reading the fourth book of Davy's *Elements of Agricultural Chemistry.* On the basis of some of Davy's comments, Ruffin surmised that exhausted soil had lost its fertility —this, along the common antebellum fertile-to-exhausted axis of soil status—because of increased acidity. He proposed, then, to alter the acidity of his land by adding calcareous manures, specifically marl (fossilized shells) found on the riverbanks of his eastern Virginia land. On Ruffin's reading, chemistry had proven valuable as a diagnostic tool. He found the use of that tool consistent with his vitalist perceptions of the land. That is, chemistry became a prism through which Ruffin could see nature because it too interpreted the land as alive and analyzable. For him, and as a complement to his various personas, both nature and chemistry fit within what appears later as a strain between vital (alive) and material (inanimate) concepts.

Ruffin has been the subject of many studies, his life's work being variously cast as remarkable, tragic, highly influential, and deeply problematic.[10] He founded, edited, and published *The Farmer's Register,* which from 1833 ran to ten volumes before, as Margaret Rossiter has written, it "collapsed in 1842 due to an unpopular stand on a bank issue."[11] In addition to writing the *Essay on Calcareous Manures,* he served briefly as a state senator, delivered countless addresses to local agricultural societies, was the state agricultural and geological surveyor of South Carolina, and, in the 1850s, published at least four books promoting slavery.[12] Even with the fame from these efforts, two events of the 1860s were Ruffin's most notorious: his reputed firing of the first shot of the Civil War, at Fort Sumter, and his life's suicidal end with a shotgun following the defeat of the Confederacy in 1865.

As an agricultural reformer, Ruffin contributed to a tradition of rural authorship defined in part by John Taylor earlier in the century. His georgic work made frequent reference to the eighteenth-century improvers common to Scottish, English, and colonial American planters. The *Essay* he wrote as part of this tradition had a long publishing cycle. It began as a brief presentation of soil fertilizing experiments in *The American Farmer* in 1821, was expanded with unpublished revisions throughout the 1820s, and reached its first book-length publication in 1832. That 242-page text was reprinted and revised in supplements to Ruffin's own *Farmer's Register* in April 1835 (second edition) and December 1842 (third edition), then published in a fourth edition in 1844 and, as a 493-page book, in the fifth and final edition, in 1852. The 1832 edition I refer to below met with popular approval in both domestic and foreign reviews. It has since been hailed as a landmark in nineteenth-century soil science.[13]

Ruffin offers a threefold argument in his treatise. He makes the case that vegetable acid in soil causes "natural sterility," thus rendering it exhausted and unproductive. At the time, soil exhaustion was widely recognized as a problem, but explaining its cause by reference to acidic levels was novel. Readers thus considered this first premise a contribution to the chemical understanding of soil long promised through the older georgic

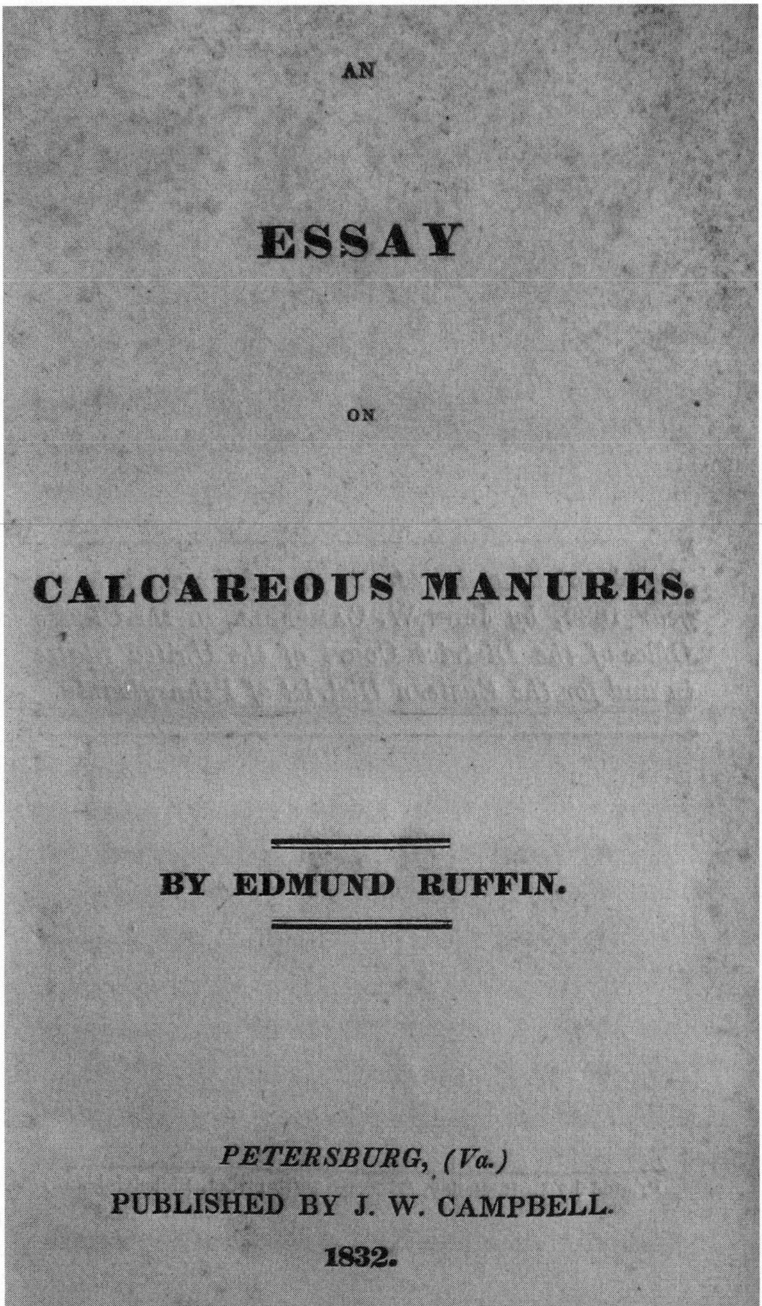

AN

ESSAY

ON

CALCAREOUS MANURES.

BY EDMUND RUFFIN.

PETERSBURG, (Va.)

PUBLISHED BY J. W. CAMPBELL.

1832.

FIGURE 3.1. Title page of Edmund Ruffin's *Essay on Calcareous Manures.* This is the first edition of five that would be produced by the 1850s. Courtesy of Special Collections, University of Virginia Library.

tradition of connecting agriculture with chemistry. Ruffin then shows that "the fertilizing effects of calcareous earth are chiefly produced by its power of neutralizing acids" (21).[14] That is, the application of calcareous earths like marl to exhausted soil can neutralize such acidity. Lastly, he explains that the then-neutralized soil can be improved with standard manures, such as putrescent animal and vegetable matter. His premise is not that the simple addition of marl to exhausted soil will restore its productive abilities. It is, rather, that marl is a useful means to another end, a necessary additive that aids the ultimate goal of soil improvement but does not directly cause it.

Ruffin presents a seventeen-chapter exposition to lay out his argument, with chapters describing the types of soils, their different capacities for improvement, and his proof for the assertions about soil properties and causes of exhaustion—proof garnered by "chemical examination," conclusions drawn from field observations, and logically constructed counter-arguments set against opposing views. Five propositions, each then expanded and confirmed, structure the bulk of the *Essay* (21–22):

1. Soils naturally poor . . . are essentially different in their powers of retaining putrescent manures.
2. The natural sterility of the soils of Lower Virginia is caused by such soils being destitute of calcareous earth, and their being injured by the presence and effects of vegetable acids.
3. The fertilizing effects of calcareous earth are chiefly produced by its power of neutralizing acids. . . .
4. Poor and acid soils cannot be improved . . . by putrescent manures without previously . . . correcting the defect in their constitution [by using calcareous manures].
5. Calcareous manures will give to our worst soils a power of retaining putrescent manures equal to the best.

He follows the proof of these propositions with direct instructions and commentary on the expense, profits, and requirements for digging and carting marl. His goal was to sustain the Virginian political economy by achieving the material gains in crop yield and productivity afforded by marling.[15]

Ruffin's view was a novel contribution to antebellum conceptions of fertilization, especially in its clarification of *how* and *why* marl worked. To be sure, marl, lime, quicklime, and other calcareous agents had long been used as manures, so the specific mention of marl alone was not the notable aspect of Ruffin's thesis. His thoughts on exhaustion made sense in part because they were tied to extant notions of a living earth available for dissection. His views offered a fluid combination of organic and mechanistic elements, as he conceptualized the cultivatable land of Virginia as having powers, capable of injury, subject to analysis, and available for healing.[16]

Characterizing the *Essay on Calcareous Manures* in any singular fashion will slight its overall ambition and appeal, but my reading is specific to the concepts of nature it offers (discussed next) and the role chemistry plays in defining those concepts (discussed later in this chapter).[17] What makes his work most relevant is that Ruffin believed the problems of soil, society, and crop growth could be solved by an analysis of the chemical and mechanical modes of plant life. We might also note in retrospect Ruffin's view relative to Davy before him and Liebig after, for he seems to fit perfectly in between, holding Davy-like vitalist and Liebig-like materialist commitments simultaneously and unproblematically.

The *Essay* offers a representation of the land. Ruffin, for his part, portrays a landscape understood for its ability to provide marketable products. In line with decades of encouragement for the systematic study of the soil, he sees that studying the land more thoroughly can make it more productive. "It is a remarkable fact," he argues, "that the difference in the capacities of soils for receiving improvement, has not attracted the attention of scientific farmers" (24). To establish that he is just such a scientific farmer, Ruffin embeds his argument for fertilization within a view of nature that is improvable and recognizable to his readers. In other words, he appeals to his audience. To the planters of the South reading his work, Ruffin argues for a specific course of action to affect improvement, something narrower than a general farming method and more specific than the call to "fertilize." His representation of the land is thus policy-driven and politically motivated.

The policy of progress-by-systematic-investigation (the call of the book farmer) is most basic in Ruffin's appeal. Improvement, Ruffin believed, will come about by "correcting the defect in [the exhausted soil's] constitution" (22). The order of nature that enables this view is one of discrete parts and interchangeability. And while the *laws* of nature governing the combinations of soil components are immutable—such as the universal need to neutralize acidic soil with calcareous manures—nature *itself* is mutable. How best to cultivate? he wonders, because while his nature is complex and interrelated, it is also elemental and extractable. Although nature is governed by an order, Ruffin believes the elements within that order can be manipulated and rearranged. It is through that perspective that his integration of vitalist and materialist concepts is plausible. To get a sense of how Ruffin views the land and expresses his policy directives, we can read him as answering three questions: What does he see as he looks around his farmland? At a single point in time, how does he envision differences across place (the state of Virginia)? And, at a single location, how does he relate human-caused differences across time?

When establishing the status of the "soils and state of agriculture" in his native Tidewater region (14–20), and then later when touring Virginian lands to the west (37–52), Ruffin offers a georgic-like topographical and geographical overview of the state defined through his own travels.[18] His panoramic views of the land are sometimes described with elegance, sometimes with blandness. To wit, the Tidewater region in eastern Virginia is "generally level, poor, and free from any fixed rock, or any other than stones apparently rounded by the attrition of water" (15–16). On the ridges of the landscape that "separate the slopes of different streams," different kinds of pines cover the silicious soils, while oaks are mixed throughout, also covering clay land. Young pines thrive on these soils when the latter are exhausted by cultivation, growing "with vigor and luxuriance" (16). Ponds of rain water fill shallow basins where whortleberry bushes grow. And while Virginians have not cleared and cultivated the lands of this area very often, those that they have were worked unsuccessfully; the land's "worthlessness, under common management" is evident from even the few examples Ruffin sees. He also highlights zones of fertility within that panorama—primarily the slopes between the ridges and streams—commenting on their once prized but now depleted status.

Ruffin provides an overview of the differences across the state of Virginia in his chapter "Chemical Examination of Rich Soils Containing No Calcareous Earth." There he describes in detail sometimes evocative, sometimes tedious, the different trees, soil types, ridge lines, plant varieties, and patterns of successive plant growth (though he does not use the term "succession") on his way to summarizing differences in soil composi-

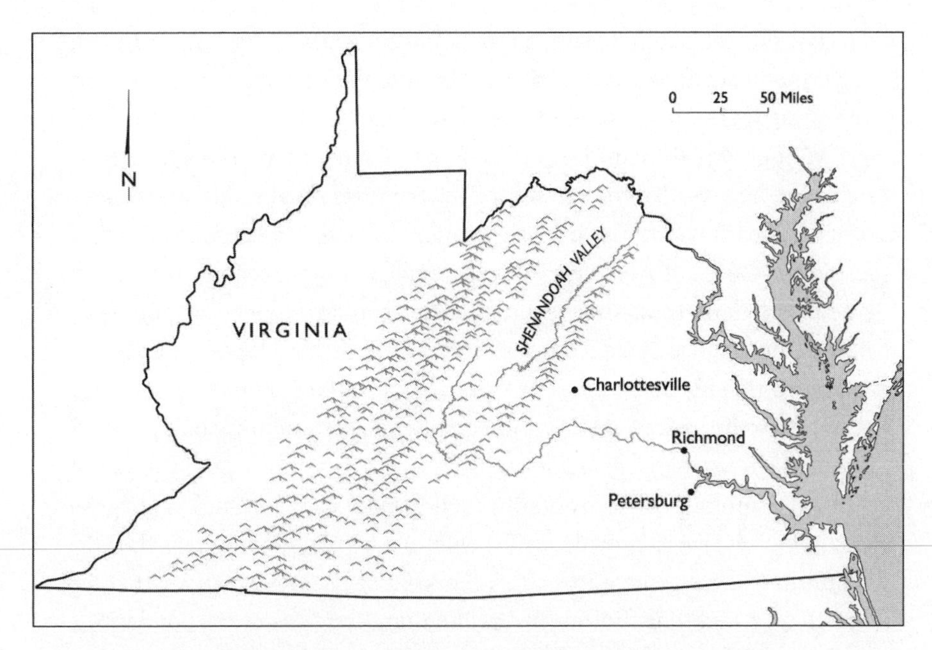

FIGURE 3.2. Map of antebellum Virginia. This shows the span of Ruffin's study, from the Tidewater area to the east of his home base in Petersburg to the Shenandoah Valley in the west. Ruffin had until the mid-1830s lived and worked at Shellbanks, Virginia, in Prince George County to the east of Petersburg. The map shows the general contours of the Appalachian range crossing the state from southwest to north central. Image produced by Bill Nelson.

tion at one time (46–50; also 37–52, passim). His tour from the fall line of the James River (near Richmond) to the Shenandoah Valley in the west—in style not unlike the Georgic Tours of the prior century—is presented by reference to a series of soil samples he had analyzed for calcareous earth (or lack thereof). The land in the Shenandoah Valley, between the Blue Ridge and Alleghany Mountains, was covered with mineral waters from active springs and was "remarkable for its productiveness and durability" (47). Other samples in the Valley were variously black, yellowish, brown, loamy, and "first rate." The "soil of first rate fertility" of Fluvanna County, splitting the geographical distance between Ruffin's home in the east and the Valley to the west, was a dark brown clay loam, both "valuable and extensive" (47). Sample "9," from present-day West Virginia, offered more variety: "High land in wood, west of Union, Monroe County. Soil a black clay loam, lying on, but not intermixed at the surface with limestone rock. Subsoil yellowish clay. The rock at this place, a foot below the surface. Principal growth, sugar maple, white walnut, and oak. This and the next specimen are from one of the richest tracts of highland I have seen" (48).

Though it is difficult for his readers to form a mental image of Ruffin's tour, he intended the numbered descriptions, totaling nineteen samples, to show variety. The differing color, richness, and compositions of fertile soil were coupled with various degrees of manageable and tillable land. Complementing his technical attention to cultivation potential, Ruffin the naturalist offered descriptions of gentle slopes, "excellent" soils, and rich black loam full of "nutmeg"-sized limestone (49). He was elaborating the variety of soil conditions, while commenting obliquely on the associated geological factors, with hills, ridges, stones, and valleys not quite matching up with any distinct soil fertility pattern. But his purpose was not to resolve state-wide issues with one walking tour across the Old Dominion, nor was he engaged in the kind of geological survey that he would later undertake in South Carolina (and that his correspondent in Virginia, Professor William Barton Rogers, would undertake in Virginia beginning in 1835 [see chapter 5]). Rather, he was more basically introducing the subject of chemical investigation and exemplifying it by comparing the non-calcareous lands across the state with the calcareous ones examined in the later parts of his *Essay*.

Ruffin then provides a review of the changes on one experimental plot of over more than a dozen years, in the process leading to his assessment of the value—through an understanding of chemical identity—of those calcareous lands. His lifelong devotion to experimentation was impressive and always geared toward redefining the land, so that readers are left with a fine-grained analysis of the changes in soil properties under his supervision. His naturalist's eye for detail lent itself to an experimentalist's view—as where he describes "specimens," not natural examples, of soil, "sugar maple, white walnut, and oak." It was also evident in his more "scientific" reports of fertilizer trials.

In his quest to convince readers that marl was the panacea farmers were looking for, Ruffin reported the success of sixteen fertilization experiments performed over thirteen years (1818–1831). In the three chapters describing them, he provides a view of changes in place over time. "Experiment 15," for example, took eight acres under second growth and produced a success story of calcareous manuring (108–109). Over five seasons, a thick pine stand was converted from sandy loam covered in "dropped and unrotted leaves" through a crop of wheat and then a planting of corn, which "excited the admiration of all who saw it." Logs, boughs, and bushes were heaped, and a wooden-toothed harrow was used to pull down furrows. The "improvement [had been] so remarkable, as to induce belief that the old fields . . . on every farm" could be made profitable. The experiment's success, Ruffin was quick to point out, could be credited to the 500–600 bushels of marl spread on each acre. As another example, "Experiment 16" took a tract where a 39-year-old stand of pines was cut down for fence rails in 1824 and marled it with 600 bushels per acre in 1825. By 1831, the wheat was "so heavy a crop" that Ruffin was in near disbelief (110). For him, the ability to transform the landscape was exciting and positive, while the methods for doing so were put under increasing degrees of technical and quantitative specificity. On this reading—Ruffin's own—this was a story of transforming the earth with scientific principles, making the unproductive productive by new modes of georgic improvement. On another reading—his audience's—this was a textbook example of cultivating the land, a how-to manual for mid-Atlantic farmers. The Southern improver had taken wild earth and made cultivatable and profitable land.

Ruffin's goal is to change the land; he develops a working relationship with that land, where the cultural and agricultural are melded through the practice of cultivation, in tandem with this view. He consistently offers views of a mutable landscape that must continue to change through the guidance of "experiments" and further "agricultural research" (3). Ruffin's reliance on and belief in experimentation and chemical analysis set him apart from earlier generations of like-minded and like-acting rural economists. In this he offers a clear example of the value of science as a new contributor to that working relationship with the land. In those practices, he has a kind of physician's view, recognizing the organic whole of the living earth, but seeking to alter it with the aid of medicines like marl and diagnostic tools like chemistry. Indeed, Ruffin "argued for a notion of expanded human control over soil fertility" throughout his career, as one of his biographers has noted.[19] This observation is both telling and underexplored. The issue for Ruffin was undeniably one of control, but his form of control promoted the human aspects of a natural order by attempting to understand, manipulate, and thus improve the nonhuman aspects of it.[20]

John Lorain also wrote glowingly about the use of chemistry for knowing the land and making it productive. He was a Pennsylvania gentleman who lived half his life in the backwoods, clearing and cultivating the earth while writing about his techniques for doing so. Lorain, like Ruffin, saw nature as a vital economy, a living, organic system of interconnected parts. In fact, he was far more forceful on this point than his younger neighbor to the south. And Lorain too had read Davy's work. But he was much less impressed than Ruffin by Davy's qualifications for dictating chemical theory to agriculturalists, concerned that Davy had not based his theories on the daily experience of farming life. Lorain provides another example of an individual who promoted the active and instrumental value of agricultural chemistry. He, however, demanded that such a chemical lens be placed within the dynamics of a harmonious organic earth, eschewing any basis in materialism.

Details about John Lorain's life and work are scant. Most of the information we have comes by inference from his published books and through scattered remarks from local archives in Pennsylvania. He was born in Maryland, of English farming immigrants, and moved to Philadelphia later in life.[21] He there joined the Philadelphia Society for Promoting Agriculture in the early 1810s, putting himself inside contentious debates about promoting agriculture in his contributions to the PSPA's *Memoirs.*[22] He then moved to the front border of the backwoods of Pennsylvania at Philipsburg, almost precisely in the geographic center of the state. From his backwoods home, Lorain wrote and published his first book, *Hints to Emigrants,* in 1819, indicating even in the title an expansionist and improvement-minded focus, that the population was shifting west, that land would need to be cultivated, and that, though it was not evident in the title, there was a right way to do this. After his death in 1823, his wife Martha had *Nature and Reason Harmonized in the Practice of Husbandry* published from Philadelphia. John Skinner saw fit to republish two chapters in *The American Farmer,* accompanied by a ringing endorsement. One scholar, evaluating Lorain's work as related to developing market conditions, puts *Nature and Reason* in a context of the class stratification being clarified with westward expansion. Another, the environmental historian Steven Stoll, places Lorain's contributions in a culture of commentary on the dichotomy between gentleman farmers and backwoodsmen and characterizes *Nature and Reason* as "a major treatise on American land use in the 1820s." Lorain's mere suggestion to harmonize nature and reason on the farm goes some way in setting the stage for the georgic-like combination of science, improvement, and nature, especially when "reason" is taken to be used synonymously with the scientific practices being promoted at the time.[23]

Summarizing Lorain's "main" argument in *Nature and Reason* is neither possible nor practical, since the work has a series of points to make and popular theories to refute. The text is over 550 pages long, broken into fifty-two chapters and four different books. "On Manures and Vegetation" (Book I) is followed by "On Cultivation" (II), "On Various Subjects" (III), and "On Gentleman Farming" (IV). While the first book treats manures and vegetation for their chemical attributes, the second book focuses more on the mechanical aspects of farming, such as crop options, irrigation, and mounding techniques. In each case, Lorain presents the "economy of vegetation" within a system of "vitality or animation" (1–34).[24] If he has one line of consistent attention throughout his work with which we can understand his views of nature, like Ruffin's strategies of calcareous manuring, it is here in his commentary on the vital force that guides the growth of crops and requires reasoned observation and evaluation.[25] And while the third book's "various subjects" range from hedges, cats, and sugar trees to a comparison of Pennsylvania backwoods farmers and Yankee cultivators, the final book is a more straightforward social historical commentary. In that last book, Lorain offers his motivation for improvement in its most georgic light, connecting the rational pursuit of farming with moral dictates of labor and "genuine principles of rural economy" (412). Throughout the work, Lorain cautions against book farming if it yields advice "opposed to reason," but approves of it if "derived from observation, reflection, and calculation" (405). He consistently argues for a subtle appreciation of the economy of nature, one that would aid the cause of cultivation by approaching fertilization and improvement within an over-arching natural order. For him, that order was vitalist, organic, and interconnected.

Lorain wrote as part of a tradition of market-oriented farmer-merchants, those who formed the core membership of the PSPA and traced their

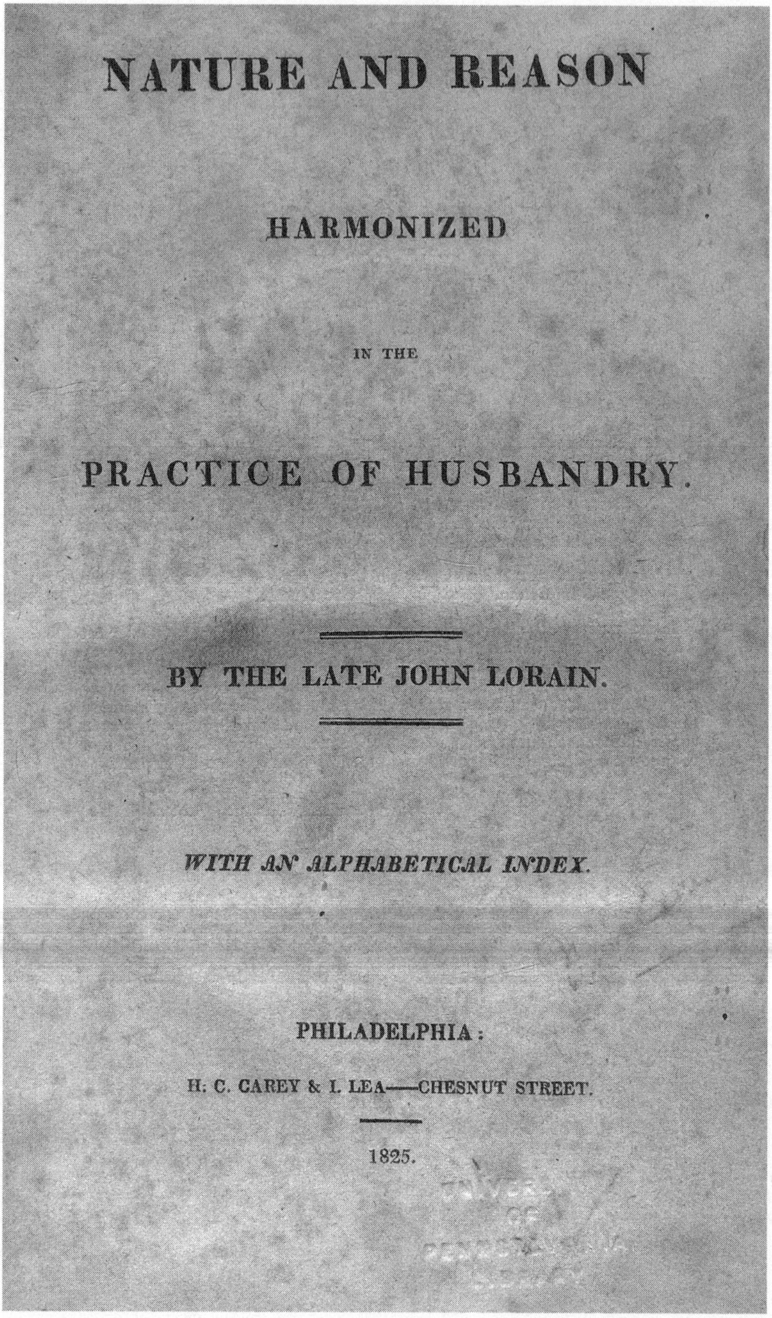

NATURE AND REASON

HARMONIZED

IN THE

PRACTICE OF HUSBANDRY.

BY THE LATE JOHN LORAIN.

WITH AN ALPHABETICAL INDEX.

PHILADELPHIA:

H. C. CAREY & I. LEA——CHESNUT STREET.

1825.

FIGURE 3.3. Title page of John Lorain's *Nature and Reason*, published in 1825. Courtesy of the Annenberg Rare Books and Manuscript Library, University of Pennsylvania.

economic and political heritage through Philadelphia and its hub-like status for goods, services, and ideas within colonial America. That Lorain wrote about the expanding settlement of the state was not unusual. His actual residence on the borders of wild Pennsylvania, however, provided an extra degree of credibility over the more urban-bound members of his social sphere. In his comments on the practice and viability of turning uncultivated wilderness into viable agricultural land, he emphasized the need to recognize and follow the vitalist economy of nature as an argument against those urban-bound associates, gentleman farmers who were proposing measures insufficiently attentive to nature's order. Judge Richard Peters was prominent within that social sphere. His 1797 *Agricultural Enquiries into Plaister of Paris,* written as part of the georgic tradition, was still being widely circulated and praised in the early decades of the nineteenth-century. Although Lorain found himself combating Peters a number of times, the problems that he and Peters met turned equally on transforming untended soil, be it clear meadows or forested stands, into productive, manageable agricultural land.

Lorain's concerns from the start, then, had a different social alignment from his southern counterpart Ruffin's and, at an important level, the over-riding cultural context of Pennsylvania agriculture placed his work in a different social setting. Aimed at this point, Book IV of his lengthy text concluded that an "improved system of management" would *eliminate* the need for slavery (525–526). New systematic practices would make the labor issues justifying the slave system in the South unnecessary; if the cultivation and management of the land followed from principles set forth by natural animated patterns, the work that had otherwise been used to overcome the constraints of those natural patterns would thus be minimized. "The time will arrive," he wrote, after having just read John Taylor's *Arator,* "when Virginians (who are certainly as just, humane, and hospitable as are the citizens of the States where slavery has been abolished) may . . . set their negroes free, if this [plan of improved husbandry] be gradually and judiciously done" (525). In this fashion Lorain's arguments were built with attention to moral (labor) and material (land) elements of the agrarian life and for an agricultural economy.

His point stood in direct opposition to Ruffin's conclusion several years later that systematically practiced agricultural improvement would *preserve* the slave system. Ruffin's pro-slavery positions suffused his life's

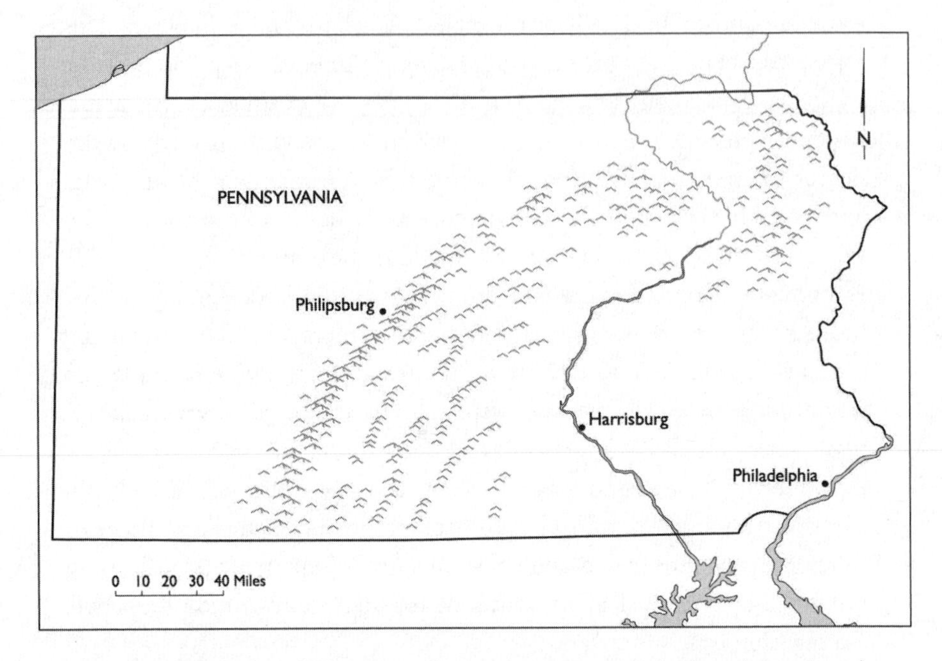

FIGURE 3.4. Map of antebellum Pennsylvania, showing Lorain's backwoods location relative to Philadelphia. The mountains indicated here represent the general topography crossing the state from south-central to northeast. Image produced by Bill Nelson.

work. When he proposed agricultural reforms he was proposing cultural reforms. His system of fertilizing and land management would thus enable the institution of slavery to thrive. In Ruffin's world, the pressing agricultural goal was to promote cultural viability by recovering agronomic strength. In Lorain's world, the pressing agricultural goal was to identify better means for settling the countryside and improving agronomic possibilities. The contrast in aims is instructive for historians seeking further evidence of connections, or lack thereof, between scientific development and slave-holding systems, yet Lorain's point in *Nature and Reason* was not arrived at after seeking to warn Ruffin and the slave-holding states. Instead, it came as his logical conclusion to an assessment of labor requirements and possibilities for more efficient cultivation on the borders of western expansion. Even though Lorain's life and writings were part of a world that differed from Ruffin's, his interest in and study and promotion of chemistry for the sake of improvement kept him conceptually linked to Ruffin. The two men asked different questions, but found the same answer. Chemistry and science were both elements of broader arguments, be they for or against slavery.

Lorain's work offers us another example of a representation of the land. His *Nature and Reason* is, very basically, about the correct way to promote the progress of cultivation. In this sense, his view is quite similar to Ruffin's, and we might expect his representations to be similar in kind. But Lorain argues that the combination of "Nature, reason, and observation . . . should be the farmer's guide" to cultivating his fields (196), as such tempering and then differing from Ruffin's appeal to put nature under human control. Lorain strikes a more harmonious, organic tone. To be sure, the differences are subtle, since Lorain does eventually want to control nature and Ruffin does claim to be controlling it by deciphering the guide provided by its laws. But the order of understanding how to interact with nature, with Lorain's entry point being participation and Ruffin's being manipulation, sets their representations of nature apart. Overall, to assess Lorain's sense of the land—before assessing the place of chemistry in his work—we can read him as answering these questions: how are fields and farmland governed by a vital economy of nature, including human and nonhuman nature alike? And in what way is the interference of human art a legitimate aid to natural, that is, nonhuman, processes of growth and change?

Lorain first explains how fields and farmland are, and should continue to be, governed by a vital economy of nature. For him, "nature" is the nonhuman component and "reason" the human contribution to the improvement that would be possible from attention to that vital economy. Lorain's purpose is to observe and promote their combination. In this, the forest provides him with a kind of experimental control, "where neither art nor ignorance has materially interfered, with the simple but wise economy of nature" (24). He watches the changes in his Pennsylvania forest as "age, disease, tornadoes, or any other cause" bring it into a new phase of existence so that he can draw lessons and apply them to his own practice of husbandry (25). Lorain notices that wherever vegetation can exist, it does. Trees fall, their wood decays, plants grow around them, and they eventually subside into the soil below. The spacing of those trees defines the spread of smaller plants and the population of animals large and small. "The fermentation and decomposition that take place within this thick body of manure furnish nutriment for the plants, and enrich the grounds." Here, Lorain points out, the new generation of plants and trees occurs without damage to roots or without forming "injurious ridges or mounds . . . or furrows." Larger animals as well as "reptiles . . . together with feathered tribes" find shelter in those trees. Leaves and crevices are also thick with insect life. Decaying animal and vegetable matter "affords living" for "incalculable tribes of animalcula." The smaller animals make do by eating the plants other animals ignore; the "manure dropped by smaller animals" fertilizes soil, returning nutrients to their source. There are so many kinds of animals, plants, and forms of nutriment that this variety "seldom seriously infringe[s] on the sustenance of each other" (26). The interchange among all of these members of the forest points out a system of necessity and cooperation within which Lorain, the observant farmer, can operate. To the modern reader, it's all either a brochure for a zoo or an expression of ecology before the term actually came into use.[26]

Beyond defining a vision of symbiotic nature, the general pattern of Lorain's example is to show how the natural action of plants and animals could change a forest. Together, those plants and animals take healthy stands of trees and turn them into populated and thickly vegetated patches of earth. It is with an eye to the organic, inter-related functioning of the forest that Lorain can answer his second question: how can the interference of human art be a legitimate aid to natural, that is, nonhuman, processes of growth and change? His answer: by understanding and working within a vital economy. Chemistry's role in this process is, first, to convey that understanding and, second, to act as a tool to promote the land's organic functioning.

When "the agriculturalist" enters the picture, that farmer can aid "nature" with art and increase fertility by continuing the general pattern of nutrient recycling already evident with natural patterns (25–26). Elsewhere in his treatise, Lorain shows that his cherished vital economy accounts for patterns of succession and development. He adduces those long-term shifts from the observations he makes on local-scale and seasonal activities. The details are often vague, and Lorain's full picture of transformation is drastically underdeveloped. It is far less dramatic than Ruffin's woodlands clearing, coming across as gentle and peaceful. But the point Lorain draws from his survey is that the natural order perpetuates itself for the benefit of its constituents and that the role of the agriculturalist is to aid this greater-than-human cause.

What matters more broadly for Lorain is that the effect of "this perfect system of economy" is equally evident in "glades and prairies, as in our forests, where nature is suffered to pursue her own course." When humans interfere with this well-tuned economy, "a new order of things takes place: the living as well as the dead vegetation found in this way is destroyed and the grounds are cultivated" (26). Lorain wants farmers to allow nature to be "assisted by art" *within* its perfect economy, not transgressed by it. Backwoods farmers generally ignore this credo, he argues, since they clear the necessary grasses and remove required root systems and decaying wood. Even the clearance of weeds could be counter-productive, as the "perpetual war against them" by humans counteracts some of the natural advantages those pests provide (28). Two decades later, Charles Dickens would tour the United States and write of his "quite sad and oppressive" view of central Pennsylvania, where the "eye was pained to see the stumps of great trees thickly strewn in every field of wheat." This was the destruction Lorain had forewarned against. Examples like this make the purpose of Lorain's social commentary clear—to denounce common backwoods practice and suggest more enlightened methods.[27]

Glades and prairies, like forests, undergo changes both natural and assisted by the art of humans. This classic pastoral assessment of the contrast between human and nature, between cultivated and wild, is enhanced into the georgic as Lorain places the labor of the farmer at the nexus of those contrasts. Change in the land itself is not opposed to the economy of nature, of course, but that which is not pursued with reason, common sense, and observation—by "cultivation," in Lorain's examples—indeed opposes nature. Despite the detail of four long books and the often rambling and repetitive nature of his writing, Lorain manages to convey a consistent point. He brings together the social concerns of expansion, its material requirements, and the vital economy of nature that structures both.

Daniel Adams was a 1797 graduate of Dartmouth College, author of numerous school books, "an eminent physician," teacher, lifelong farmer, and community leader. He was born in 1778 and died in 1864, living between the birth and the near dissolution of the nation. Except for a brief stint in Massachusetts during his younger years, he lived and farmed in New Hampshire throughout his life. *Appleton's Cyclopedia* found him sufficiently famous to include him in its 1888 compendium, highlighting his civic duties as president of the New Hampshire Bible Society and New Hampshire Medical Society. He also edited the short-lived *Medical and Agricultural Register* in 1806–1807, was the president of the Hillsborough Agricultural Society in the 1820s, and served as a member of the New Hampshire Senate in the early 1840s. Among Adams's fifteen entries in the current Library of Congress catalog are his frequently re-issued *Adams New Arithmetic* (1827), the *School Atlas to Adams Geography* (1823), and his *Agricultural Reader, Designed for the Use of Schools* (1824).[28]

This last work, a treatise of georgic orientation, commands our interest here. In the first section of *The Agricultural Reader,* the "Explanation of Terms," Adams suggests that readers memorize the seventy-nine terms he defines, as one would do with geography and grammar. A didactic tone is thus set, and the reader's interest in the work is in instruction. After that long introductory exercise, the book proceeds roughly in thirds. In the first third, Adams tackles issues of political economy and what were then standard topics for an agricultural treatise—soil, nutrition, plant growth, manure, and crop varieties. In fact, there are eight separate chapters on various kinds and uses of fertilizing manure. In the next third, he offers commentary on weeds, insects, livestock, fruit trees, and horticulture. He suggests no common lesson to be learned from each example and provides no clear connection to his earlier comments on political economy. It is worth noting too that, regardless of his goal of clarity, Adams seems to lose focus after leaving behind the more common topics of the first third of the reader. While he begins in a sustained, instructive, didactic, and interrogatory style, the book soon devolves into an assemblage of extracts, losing pedagogical focus as it goes. In the last third, he presents overviews on topics as wide-ranging as the purpose of Agricultural Societies and the parable of "the two apple trees" (where one son treated his gift of a tree with care and attention, and it blossomed; another son neglected his tree, which bore no fruit). Five rural poems end the work, and this after a series of excerpts from Benjamin Franklin. Adams quotes (without citation) the British improver Henry Home's optimistic and propagandist statement on the moral and material value of agrarian political economy, that agriculture "is perhaps the only firm and stable foundation of national great-

THE

AGRICULTURAL READER,

DESIGNED

FOR THE USE OF SCHOOLS.

"Next in importance to the great business of preparing for a better world, is to know how to live comfortably in this."
Address of Josiah Adams, Esq. Concord, Mass.

BY DANIEL ADAMS, M. D.
Author of the Scholar's Arithmetic, School Geography, &c.

BOSTON:
PUBLISHED BY RICHARDSON & LORD.
....
PRINTED BY J. H. A. FROST.
1824.

FIGURE 3.5. Title page of Daniel Adams's *Agricultural Reader.* Courtesy of University Libraries, Virginia Polytechnic Institute and State University.

ness." That tone of georgic and civic improvement pervades the *Reader,* giving it the political justification and moral foundation necessary for general use.[29]

Adams wrote *The Agricultural Reader* to utilize his educational skills for the sake of teaching agriculture. Although circulation statistics are not available, we know that his textbook-writing success already preceded *The Agricultural Reader* as all of his works were reissued frequently throughout the antebellum period. Ostensibly, and as evidenced by the very subtitle, his readership consisted of schoolchildren. But that audience ranged far wider, Adams knew, as he detailed matters in his book too "difficult for school boys to understand" though "important . . . to the farmer" (46). His grander appeal to farmers, rural press readers, and members of the county agricultural societies of New England means that he sought the same audience as that already targeted by *The New England Farmer.* His georgic context is implicitly set by tone, argument, and virtue-laden commentary and explicitly defined by drawing heavily on Samuel Deane's *Georgical Dictionary* (see 112, 114, 119, 131, 138, 188, and 189).[30]

The text's format is deliberate and orderly. Each of the sixty-two chapters, ranging from one page to sixteen, is written with numbered paragraphs to facilitate ease of reference and clarity of expression. This style was not uncommon for texts of the era, though Adams was forgoing the popular catechism model of the question-and-answer format.[31] His text, rather than falling into the more general category of a schoolbook, was agriculturally specialized. It was explicitly meant to address the science of agriculture by detailing facts, opinions, and definitions. Inside this presentation, his work is ordinary, advocating neither radical chemical notions nor controversial advice about land, soil, and improvement. Adams simply reports and then advocates his perception of the current issues concerning agricultural progress. He has neither an explicit argument to advance, nor a thoroughly explained philosophy of nature to describe. By standardizing the basics of agriculture and codifying them in his schoolbook, as with the disciplining tendencies of textbooks in general, he is in a sense "naturalizing" the idea of improvement and his representations of nature.[32]

Three aspects of Adams's view of nature are of particular interest: his perception of nature as distinct from artifice, his portrayal of nature as an economy of interrelated parts, and his understanding of nature as alter-

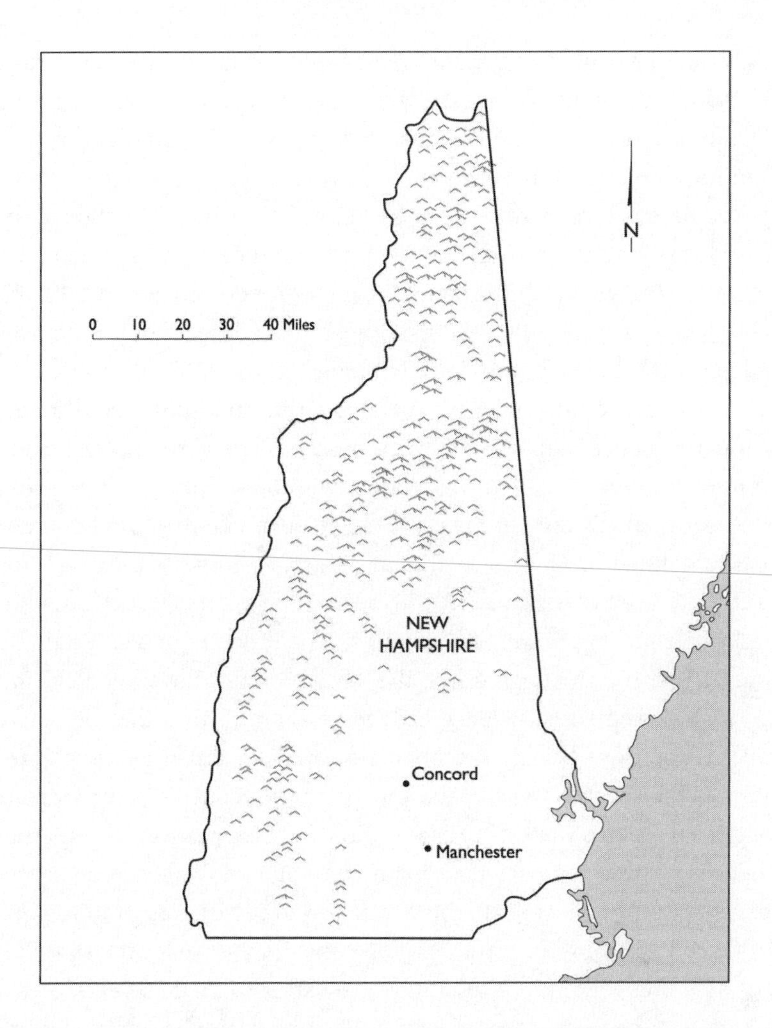

FIGURE 3.6. Map of New Hampshire. This shows Adams's location in the southern part of the state, which kept him in proximity to publishing circles in Massachusetts and other agricultural societies in New England. Hillsborough County is on the south-central border of the state. Manchester is the county seat. Image produced by Bill Nelson.

able through the action of agriculture for the betterment of humanity. These views at times coincide with Lorain's or Ruffin's and at times differ in their emphasis and purpose. With respect to chemistry, Adams was an appropriator, taking on Davy's chemistry without discussion or even a modicum of skepticism. The same is true, though less starkly, of the ways in which Adams represents nature. His views on the role of agriculture in society and the ways that agriculture changes the landscape were drawn from popular sources. They were typical of the era. He merely appropriated the less controversial notions of vitalism while introducing non-offending elements of materialism.

Adams perceives nature as something defined in distinction to artifice. He presents the natural, "artless" value of the land as understood in contrast to the civilized, human-built environment of the city. In this traditionally pastoral outlook, "happiness," he writes, "seems to have fixed her seat in rural scenes" (28). Humans are present in both locations, the city and the country, but guaranteeing virtue and fulfillment through "divine providence" depends on the rural scene. The same contrast of idle and industrious virtues that was evident with homespun rhetoric is embedded in Adams's text. In his case, the positive virtues could be understood as defined by God, as visible in nature, as cultivated in the countryside, and thus as exemplary. "The spacious hall, the lighted assembly, and the splendid equipage," he contends, "do not soothe and entertain the mind of man in any degree like the verdant plain, the wavy field, the artless stream, the enameled mead, the fragrant grove, the melodious birds, the sportive beasts, the open sky, and the starry heavens" (28). Nature is the source of virtue and prosperity; with such a commonplace perspective, Adams was on familiar ground in his appeals. As the opening to his work, these passages both justify the value of the book and set the moral stage for further arguments about how to conduct agriculture properly. Of course, other "arts and employments may serve for the *embellishment* of human life, [but agriculture] is requisite for its *support*" (28). Those other arts rely on the understanding and promotion of the natural order evident in the countryside. In this construction, Adams provides the well-rehearsed distinction between artiface and nature as one between the human-made and nonhuman-made properties of the world.

Adams also presents nature as an economy of interrelated parts. His view comes across as very much consistent with natural theology—the idea that one could know God by studying nature as opposed to knowing God from supernatural revelation—though he makes no explicit reference to the doctrine in his work.[33] His references to "Nature" pertain to that which is designed by God. This use of the term differs from that deployed by Lorain and Ruffin, who both treat "Nature" as personified shorthand for "the order of Nature." Adams's penchant for natural theological constructions was not, he suggested, intended to say that humans should avoid improving upon the natural order, that tampering with God's design, for instance, was sacrilegious. It was, rather, aimed to suggest that improvement of the land would come about by closer attention to the processes of an active nature, a form that could be understood and followed. "Nature," he writes, "is continually holding out intimations of her designs would men but regard them. Her operations are always correct, and never directed to deceive; and he who follows nature, watches her intimations, seconds her efforts, and studies her designs, will unquestionably become the most successful cultivator" (149). Like Lorain, Adams instructs his readers that this process is defined through interactions among existing parts of the soil and atmosphere. "The economy of vegetation, the nature and application of manures, and the rotation of crops" would provide the basis for a systematic agriculture that demands attention to the interrelations of nature's components (35). Putrescent manures, for example, exemplify the actions of the economy of nature by returning to the earth "something . . . which makes it richer." "Everything which has possessed life, whether animal or vegetable, having undergone putrefaction and being returned again to dust, becomes food for the support of vegetable life" (32). Matter is "entering continually into new forms and new combinations"—it is always being recycled as it finds its way into new life forms.[34]

Adams presents the active dimension, the agency of nature, time and time again by representing nature-as-process. For him, the constituent parts of plants, soil, and all of rural economy are merely reshuffled from one part of nature to another. His economy of nature asserts that the basic particles of plants like carbon, hydrogen, oxygen, and nitrogen simply move from one living being to another. Action comes about through the putrefaction of dead animals and decaying vegetable matter. Adams suggests that this active basis is another way to interpret vitalism. However, with his emphasis on the process and changes of rural economy instead of the simple insistence that it is alive, his work bears a slightly different character from Lorain's.

This goal of improving and changing nature in Adams's work is consistent with that of Lorain and Ruffin. Adams's metaphors of process and activity for nature lend themselves to the actions of humans for changing the environment in a clear manner. "Not one particle," he writes, "probably has ever been added to or taken away from the great mass of matter since the creation of the world" (47). Nature "herself" changes, not by addition, but by redistributing the existing components of the world; humans are thus justified in altering the landscape too. In fact, though they can change nature, they cannot do so destructively. Adams tells his readers to study the design of nature first, but then to apply systematic studies of plant growth and nutrition to the land. The role of those systematic studies, the work and goal of "gentlemen of science," was to inform tillage, rotation, and manuring. Adams maintains that these are standard approaches, but insists that their capability as agents of improvement depends on alliance with natural processes. The strictly human ability to conduct experiments, analyze soil, and manipulate the texture and quality of that soil guaranteed that the landscape could be changed for the better (37–44).

Neither Ruffin, Lorain, nor Adams explicitly sought to provide descriptions of their landscapes for pastoral appreciation even though many passages portray images of verdant fields, lush forests, gentle slopes, fertile ridges, and idyllic streams. Each of the works was underpinned with an active, georgic ethic and each proposed variants of book farming as positive contributions to agricultural society. What is more, each text was pervaded by an underlying assumption of the cultural validity of agrarian identity.

Adams identified himself as a farmer in the sense that most early Republic Americans did—farming as a state of mind, a notion of identity which included virtually all Americans. He did work the land, though as a physician and publisher he did not do so for subsistence; he shared and promoted his fellow New Englander John Adams's view that "the Enthusiasm for Agriculture like Virtue will be its own reward. May it run and be glorified."[35] Lorain was a farmer in a more practice-based sense, living the later years of his life on the land for the purpose of cultivating both his food and his lifestyle, even though he was also a merchant. Ruffin, like Lorain, matched the description of a farmer in the sense we commonly think of today. He too made his living from the land and based his life and viewpoints on the site of the cultivated field. But he too was more than a subsistence farmer, as the second half of his life kept him to publication schedules, geological surveys, and a pattern of public speaking and civic involvement. Furthermore, Ruffin's place in the upper South cast him as a member of a different sort of agrarian culture, one reliant upon and dedicated to the promotion of a slave-labor system.

The views of cultivated and cultivatable land in the three treatises offer a map of the 1820s and 1830s that is at once unified—all three were rural economists, concerned with agricultural improvement, and commenting on the tenets of agricultural chemistry—and diverse—their writings presented differing degrees of specificity, assumed different emphases on vital or material concepts, reviewed different characteristics of nature, and were motivated from different cultural architectures. Each writer saw nature as an "it," as something governed by laws that were immutable but identifiable, and through a georgic sense of improvable material and moral sustenance. But where Lorain saw the land as participating in a wider economy of nature, Ruffin saw an extractable and analyzable material substance. Lorain's plant nutrients were exchanged with other animals and through interaction with the atmosphere and the vegetable manure that came from leaves and compost; Ruffin discussed the individual, isolatable properties of soil and the particular means to act on those properties. This was a contrast between the seemingly organic and holistic and the elemental and reductionist, but it should not be overdrawn. To be sure, Ruffin did not deny the wider natural context of animals, vegetables, and minerals, but instead saw *improvement* as deriving from specific attention to particular components of the soil, whereas Lorain saw improvement as following the use of human aid as part of, not opposed to, the economy of nature. Both men were at once grappling with Romantic philosophies of nature consistent with a humus-based *organic* theory of plant nutrition and aware of the *inorganic* basis for recipe-like fertilizer strategies like Liebig's dream of manures "prepared in chemical manufactories, exactly as at present medicines are given for fever and goiter." For the chemist Liebig, the procedure was to add a quantity of X, then some of Y to the soil, for the plants to grow more ably.[36] For Ruffin, Lorain, and Adams, the matter was not as straightforward.

Adams balanced the complexity and reducibility of natural components, but in his presentation he advocated the pursuit of rational principles of agriculture. It is possible that we can learn more about antebellum attitudes toward cultivated land from Adams's standard-fare presentation than from Lorain's disputatious or Ruffin's theoretical contributions. He was less interested in confronting authority figures or proposing novel practices and theories, and as such he was more apt to present a conventional perspective. Given that, we find that the agricultural improvers of the era truly were comfortable mixing vital and material, organic and mineral concepts in their daily efforts toward improvement.

In working out their respective conceptions of the land, the authors were establishing a criterion for knowing nature. That is, with Lorain, vitalist predilections were the marker for a kind of experience; with Ruffin, an acknowledgment of soil properties served the same role; for Adams, recognizing the exchange of existing components within nature's design signified the right knowledge. But what is "knowing nature" in this context —how does one know nature? What kind of nature is there to know? Only farmers, people who knew the land from daily interaction, could come to this kind of knowledge with their kind of experience. The authors thus define nature in such a way that agriculturalists *in* the field are the authorized agents for knowledge *of* the field.[37] Hence, as we will see, their naturalist-like descriptions lead to discussions of chemistry as, most effectively, a tool in the right hands. In the texts, this comes across as a sort of expert's regress. The only right way to use chemistry is one that operates in concert with how nature really is; the only way to know how nature really is requires intimate experience with it, vitally, in the field. Therefore, chemistry works as a valid tool if the right people use it.

REPRESENTING CHEMISTRY

Edmund Ruffin tempered his acceptance of Davy's chemistry. He referred to Davy often and claimed that his chemical inspiration came from reading Davy's work. But he was not uncritical in that reception. He questioned many of Davy's insights and sought to improve on his work rather than accepting it wholesale. John Lorain, by distinction, was a refuter of Davy's work. He was at pains to devalue Davy's theoretical acumen and show that Davy's experimental reasoning was flawed and misleading. On the opposite side of this spectrum was Daniel Adams, the appropriator. This identity held true in a strict literary sense, since Adams's textbook was comprised of scores of excerpts from other works meant to represent his views (i.e., major portions of his book were cribbed from a great number of other sources). It also held true in a more interpretive sense, since he used those references without critical filtering, choosing to import the view of his source without questioning its soundness. In line with this, Adams merely appropriated Davy, taking at face value the definitions of soil, theories of plant nutrition, and expressions of vitalism that came from his work.

The authors used chemistry in various ways to mediate their views of nature—Ruffin the Temperer of Davy, Lorain the Refuter, and Adams the Appropriator. Davy stands in as the main figure of reference, but my point is to generalize from his example. What did the authors think of his work, what did they assume chemistry was and what it was useful for, and how did chemistry act as a prism for their perceptions of the land?

Chemistry, Ruffin writes, is the "true philosophical mode of examining questions of agriculture" (40).[38] In pursuit of its utility, he engages an eclectic array of chemists and agricultural improvers both foreign and domestic. His work refers not just to Davy, but to the European chemists Richard Kirwan, Torbern Bergman, and Frederick Accum; the British agricultural improvers Arthur Young, John Sinclair, and Archibald Cochrane; the American geologist and chemist Parker Cleaveland; and various mid-Atlantic georgic figures like John Skinner, John Taylor, John Bordley, and the wealthy Virginia statesman John Hartwell Cocke. The *Essay on Calcareous Manures* is a chemical work in that it is based on Ruffin's diagnosis of the chemically determined properties of soils, his theorization of the action of calcareous matter to neutralize acidic earth, and his enrollment of chemical theory and experimental reasoning to carry his argument. In fact, unlike Adams, though like Lorain, Ruffin is explicitly proposing an argument in his work that he presents as chemical.

Ruffin directs and orders his chapters to define what actions farmers can take to improve the quality of their soil. His guiding premise is that such quality can be assessed, compared, and then acted upon. He distinguishes in a number of passages between mechanical and chemical means for soil improvement, parsing the two approaches to show that his solution is chemical. Though he was without extensive formal training in chemistry, he read widely and was confident enough to apply his reading in practice. His wide range of technical references thus lends support to the tenor of his work which suggests, in rhetorical style and with the posture of a working farmer, that his is a work of chemistry for the "practical" farmer.

Ruffin speaks more about chemistry as a form of inquiry, a process to be pursued, than as a set of theories. Not surprisingly, he offers no simple definition of what he means by chemistry. Rather, he indicates his understanding of the field by reference to its analytical and investigative role. The constituent parts of soil have been "obtained by chemical analysis"; "chemical tests" indicate calcareous presence (34, 35). The analytical results of "The Chemical Examination of Various Soils"—the basis for an entire chapter—"completely establish [the] general rules" of the primacy of calcareous soil as the basis for soil fertility (44). Thus, he presents chemistry as an instrument in two senses: first, it is an investigative way to understand agriculture, it is the "true philosophical mode of examining questions of agriculture"; second, in a more direct material sense, the instruments of analysis are "another means for knowing" the properties of the soil and thus the key to solving problems of soil exhaustion (40).[39]

In another way, what Ruffin means by chemistry is the work of Humphry Davy. Although he refers to a wide range of chemical treatises, his main influence and reference point is Davy, whom he considers "the highest authority."[40] For one thing, picking up Davy's *Agricultural Chemistry* in the late 1810s gave Ruffin his original inspiration to investigate calcareous manures.[41] For another, Davy's influence was consistent with Ruffin's view of the instrumental role of chemistry as a tool useful for agricultural, and thus cultural, ends. On theorizing the action of lime, for example, Ruffin accepts Davy's view because it directs him to practice manuring in a specific way. He understands the explanatory power of Davy's theory of lime in this case less for its academic value than for its power to "deduce proper practical use" (182). The pneumatic apparatus Davy constructs in the fourth lecture of *Agricultural Chemistry,* whose "accuracy is almost perfect," also appeals to Ruffin's sense of the role of chemistry for the farm. The instrument relies on "well established facts in Chemistry" (41). In the context of his *Essay,* Ruffin's concept of "well established" is guaranteed through "the practical value of an analysis" (40; also see 36).

But Ruffin does not accept Davy uncritically. For Ruffin, the chemist's taxonomy of soils and definitions of soil varieties are particularly problematic, a flaw not unreasonable given Davy's occupational distance from actual field conditions.[42] Another disparity in views is the men's respective attributions of mechanical or chemical explanations to the action of marl. The prevailing view of the chemistry community at large was that "chalk and marl and carbonate of lime *only improve the texture of a soil,*" which was a mechanical explanation for its utility (40). But the differences between mechanical and chemical action were central to Ruffin's own thesis of the action of calcareous matter (11; 69–77). Ruffin proposes that lime and acids are attracted to one another through laws of affinity and, thus, by processes he believes are only describable as chemical (8–13, 39). Therefore, despite his acceptance without question of analytical results, he rejects Davy's mechanical grounds. The georgic context helps frame the Virginian's position in this regard: Ruffin was on firm conceptual ground, in his own mind at least, when questioning the "men of science" who considered marl's value mechanical because he understood its use through experience, not speculation. Adding grist to the concurrent book farming debate, Ruffin was casting the legitimacy of his views through the lens of a georgic farmer, not a dandyesque philosopher.

Ruffin's concepts of soil and nature as seen through a chemical lens are clearly expressed in his experimental reports. "Experiment 10," in particular, stands out for its combination of health metaphors, vitalist views, and materialist diagnoses. The story of Experiment 10 is cautionary since it answers concerns that marling can destroy crops, rather than improve them. Marl can be "over-dosed," causing "injury" and "disease" even though the "recurrence of evil" that can be "inflicted" by using too much marl is avoidable (100–102). Like medicine, it must be applied lightly at first, and then increased if the symptoms of unproductive soil continue. Ruffin describes corn ravaged by forces as destructive as insects. The "gloomy prospect" of the crop's yield sent a pall over the fields. Leaves shriveled and died from "injury." "Remedial measures" were necessary for the corn to be "relieved of the infliction." The "ill effects" of calcareous manures were everywhere apparent, but they were fortunately confined (120–121). For Ruffin, chemistry was the diagnostic tool used to identify that soil's poverty and ill health.

Ruffin's underlying assumptions cast the soil both as an active agent, working either to turn against improvers or to help promote the improvement "it" should by all rights want, and as a passive vessel into which measured quantities of fertilizing agents could be poured. To avoid "the recurrence of evil," Ruffin suggests moderating quantity. In a tradition common to domestic labor and medical chemistry alike, he invokes a recipe view of fertilizing, one that follows partially quantified instructions in a step-by-step manner. He also infers that the field is a kind of laboratory where soil serves as the beaker, or the body, to be filled or emptied per instructions.

Ruffin's belief in universal and timeless laws of nature was applied not just to nature in the abstract, but to actual physical soil. "The component parts of calcareous earth *always* bear the same proportion to each other," namely, 43 parts carbonic acid to 57 parts lime (italics added); one ounce of water was *always* equal to the amount of carbonic acid in two grains of calcareous earth (41). Thus, chemistry was also useful for revealing the immutable conditions of agriculture, even if a primary difficulty for the science of agriculture had been reconciling the always local factors of climate, labor, and soil with such universality.

Ruffin positions his work as both theoretical and practical, though he avoids developing a full theoretical edifice and eventually moves away from the practice of farming in his later years. He was part of the social space between the farmer who required testimony from other practicing farmers and the chemists who lacked the practical experience to recognize the action of manures in the soil. Tempering Davy's full import, Ruffin's work bridged two worlds, conceptual and practical. As a sort of boundary object, chemistry could be invoked in the farming world for its practical value and in the world of the chemists for its commentary on nonmechanical aspects of manuring strategies.[43]

Lorain had a different view. He was not satisfied with the prevailing assumptions of agricultural chemistry he encountered, mainly because of their lack of attention to knowledge gained through practical engagement. He refers to the standards from Britain—Archibald Cochrane on connecting chemistry to agriculture, Henry Home's then seventy-year-old work on English gentleman farming, and Arthur Young—as well as the American John Bordley. Lorain also writes an extended review and comment on John Taylor's *Arator* and draws heavily on Erasmus Darwin's (1800) *Phytologia*. His tone is acerbic. Finding fault, Lorain considers Darwin's observations well summarized but sharply rebukes him for being insufficiently attentive to actual agricultural practices. Davy, the most popular chemist of the day, is subjected to scrutiny and ridicule more than any other source. His work is frequently introduced only to be quickly refuted. The "facts," he states, "are very contrary to [Davy's views] in the backwoods" (530).[44] The vitalist science that Lorain promotes not only coincides with broader Romantic conceptions of nature, but indicates how he considers the twin themes of nature and science: he is opposed not to science, but more specifically to non-vitalist representations. This position places chemistry in a more diffracted light, containing and representing multiple meanings and uses that defy simple categorization.

Lorain's overall dissatisfaction with connections of chemistry and agriculture is methodological, claiming that those like Davy and Darwin have used inadequate *methods* (practices) when developing their ideas. He devotes nearly the entirety of the first book of *Nature and Reason* to refuting Davy's work—theories about the value and action of marl (4–6), the effects of gypsum on vegetable decomposition (15–16), the source of plant nutrition (23), the source of moss (29–30), the action of sap (41–42, 530), and the living components of trees (61–69). But Davy is in for much more. Better that he were only methodologically unsound. Lorain finds him contradictory and obfuscating as well. By refuting Davy, Lorain takes the tack of recognizing and promoting the value of chemical investigations while remaining skeptical about the value of the chemists. Davy is the foil against which Lorain distinguishes between chemistry as rational observation and chemists themselves as credible observers.

In true georgic fashion, Lorain suggests that Davy's distance from actual agricultural conditions accounts for his errors. The Englishman is wrong because "some of his theories are not in unison with the economy of nature" (526). While "Sir H. ought to have recollected" what "every schoolboy" knows about the life of trees, for example, "it would seem [he] suffered his imagination to soar too high" (43). Poor, ill-conceiving Davy goes beyond the facts of the farm to formulate his ideas. Lorain's quip then further denigrates Davy since "experiments prove nothing more than every man who has been conversant with trees, has known for time immemorial" (69). The value of chemistry in this instance is to further what every schoolboy and practicing farmer already knows, not to contradict it. Chemists who negate common knowledge—common sense observations, in Lorain's terms—are putting forth proposals opposing the economy of nature.[45]

The purpose of Lorain's work was to offer rational insights without the detriment of distance from the farm and to do so by tying his perception of chemistry (as a form of reason) to his view of the land (as a vital economy). He seeks to "make nature, reason, common sense, and observation alone his guides" in agricultural improvement. This has him using experience-based insights to formulate a strategy for improvement. That strategy combines the identification of natural laws with the unique human capacity to learn from reason and observation to make nature better. Lorain respects chemistry, the example of that human capacity, as the method by which the combination of nature, reason, and observation could be achieved.[46]

Lorain's representations of soil, like Ruffin's, are based on metaphors of life. With those, we get another example of how his views of the land and representations of chemistry found common ground. Plants grown in "rich soils" resist more easily "the various injuries to which vegetation is subjected" because those plants and their soil have been enriched by natural animal or vegetable matter (17). "Life gives a peculiar character to all [nature's] productions: the power of attraction and repulsion, combination and decomposition, are subservient to it" (13). Therefore, farmers should first recognize "the living principle" of the soil and the various members of the economy of nature before attempting to improve their lands (50). Vital chemistry is the only worthy kind; any practicing farmer would know this.

The experience-based structure within which Lorain's improvement proposals are based may say more about his differences with Ruffin than larger socioeconomic and political factors (such as slave labor) can. Because Lorain understands soil as alive and participating in nature's design, he ultimately proposes schemes for improvement that differ from his southern neighbor. For example, Ruffin writes to explain how calcareous manures can neutralize existing soil properties to allow for the successful addition of further fertilizing agents. Lorain, however, believes that using what he calls stimulating manures (like marl) is mostly unnecessary if farmers allow naturally produced vegetable and animal matter to do its work. Nature knows what it is doing. The role of human interference is to promote, not circumvent, knowledgeable nature.

Lorain and Ruffin show their common appreciation of chemistry, but with different senses of its place in the scheme of improvement. The two emphasize chemistry as a form of inquiry—a methodological example of how to harmonize nature and reason—rather than as a body of theoretical dictates alone. But whereas Ruffin's instrumental use proposes chemistry as a system of analysis, Lorain's instrumental use proposes chemistry as a form of argument. Thus, the reason and common sense he seeks is that which is aided by the georgic art of chemical inquiry.

Lorain's text gave attention to the role art should play in altering nature. By elucidating his views of science, he also shows that chemistry acts as that art with which humans know nature. Additionally, *Nature and Reason* shows that, no matter the difficulty of sustaining a consistent theory of agricultural chemistry, the georgic goals of improvement—social, economic, material—were being entrusted to some combination of nature and reason. Lorain's vitalist chemistry coincides with broader Romantic conceptions of nature, while underscoring his consideration of nature and science. "Science" was not the target. Non-vitalist representations were. His criterion of knowing the land was defined not only through the practice of farming but by recognizing the vital economy of nature.

Adams engages a narrower range of authors on the science of agriculture than either Ruffin or Lorain. What is more, many of those references come about secondhand. He frequently quotes and excerpts lengthy passages from other treatises—in many cases without citation or quotation marks—accepting and importing the views of those authors without any degree of translation, critical filtering, or analysis. In addition to Davy, the *Reader* also references Richard Peters, Arthur Young, and Samuel Deane (and the *Georgical Dictionary*). In several instances Adams copies material from the similarly uncritical lens of Thomas Fessenden at *The New England Farmer.* Overall, while his range of scientific references is narrower and his discussions of the value of chemistry for the farm are vaguer than either Lorain's or Ruffin's, he is more optimistic about the promise of chemistry for agrarian culture.

The New Hampshire improver provides a basic and general definition of chemistry in the list of terms that begins his text, calling it "the science which enables us to discover the nature and the properties of all natural bodies" (15).[47] Beyond the semantic definition, he views chemistry technically in three ways. It is a specific example of the general approach of science, that type most applicable to the farming classes. More precisely, chemistry is that kind of science which can inform farmers about the composition of their soil with its analytical techniques. And finally, chemistry, which "[throws] light on the subject" of agriculture, is the resource of the most enlightened and improvement-minded farmers, those he calls practical scientific farmers (103). Tying together local material gains and broader cultural aims, Adams restates the grand hope of Enlightenment progress by quoting the respected antebellum geologist and naturalist Samuel Mitchill: "as the ingenuity and invention of man may increase to an unknown and inconceivable degree, so may the improvements and arrangements of husbandry keep pace therewith, until the most fruitful spot that now exists, may produce a tenfold quantity, and the land which now supports a hundred men, give equal enjoyment to a thousand" (33). The case provided in the *Reader* is unique in that such improvement and progress are enabled by chemistry.

According to Adams, all writers on chemical method, and not just the most popular professional chemists, are worth the attention of aspiring farmers. What matters is the cultural status of the proponent, not the mere fact that there are chemical ideas being proposed. In this style, Peters, Fessenden, and Young are as valid as Davy and Priestley. To be sure, Davy, the "celebrated chemist in England," is Adams's most common reference point, providing the basis for his views on the number of elements (Adams writes that there were 47 at the time, according to Davy), analysis of the abilities of soil to retain moisture, and details about the fermentation of manure and what this means for fertilizing fields (16, 43, 59). Adams provides no indication that he disagrees with Davy, or any evidence of dispute with the practical payout of chemistry as an agricultural science. But all told, Adams writes outside the burgeoning tradition of connecting chemistry to agriculture, co-opting mainstream arguments rather than contributing to them.[48] He thus is less concerned with Davy's scientific credentials as with his reputation for approaching agricultural problems with chemical methods. He is also far less critical than Lorain when interpreting the theoretical basis for Davy's views.

Like a craftsmen selecting a tool, Adams places chemistry within his work for its use value, focusing on how it can best be used to achieve the improvement he seeks. As with Ruffin, analysis, as the best example of aiding practice with method, is the way to gauge use value. (Thus, where Lorain finds fault with the methods of Davy, Adams appropriates Davy and chemistry because at least they offer *a* method.) Adams's assessment of analysis, in this instance, is meant to demarcate not who can do it, but that it can be done. Consistent with this view, he allows his authorities to speak for him, leveraging their social status along the way—Peters the Pennsylvanian gentleman, Fessenden the New England editor, Young the respected British agriculturist and improver, and Davy the famed chemist of England.

Adams's views of improvable nature and the role of the farmer in that process are further underscored in an address he delivered to the Hillsborough Agricultural Society as its president in 1825. He reemphasizes there that to become a practical scientific agriculturalist "it will be necessary [that the farmer] should have some acquaintance with the principles of natural philosophy, and especially of *agricultural chemistry*."[49] He delineates the identity of theoretical farmers as "the worst of all farmers," distinguished by their promotion of theory *without* experience (148). In contradistinction to these social outcasts, Adams—and by extension the most authoritative figures of the age—advocates practical scientific farmers because they most properly practice agriculture "in the light of science" (149). Chemistry fits in precisely at that conjunction of the rhetorical (as with his speech and text) and cultural (by deference to legitimate practitioners).

Adams believes in a regulated and God-given economy of nature. As a tool, chemistry is the means by which to identify the laws of that economy. In one example, this leads him to conclude that the earth "not only retains moisture in itself, but has the power of attracting it from the atmosphere; and, what is still more wonderful, *attracts it in exact proportion to its fertility*" (43). His message is that "Nature" knows what it is doing. In this case, the duty of the farmer is to avoid contradicting nature's laws or overcompensating with fertilization (often understood as the means to promote the retention of moisture), and to allow for the amount of moisture in the soil that the earth already knows it needs. But how can he be so sure? Well, a study by Professor Davy confirmed it (43 and 51). Chemistry, through this channel, is meant only to assist farmers in helping the soil achieve its natural fertility.

Adams's descriptions of soil in the *Reader* fit into his notion of the agency of nature and humans to alter the landscape summarized above. Through chemical investigation or investigation by proxy he has come to understand that the soil is active, just as nature in general is. It "attracts" moisture, has a "power" to decompose matter, and actively "promotes" putrefaction (37–43). Since he begins with the view that different soils have different "powers" to retain moisture and since he conveys the view that moisture and putrescent matter are the staples of plant growth, his plan for "improvement" is to promote the moisture- and putrescent matter–retention abilities of the soil. Borrowing more from Davy, Adams categorizes the soil in four types: clay, sand, lime, and magnesia (36–38). This matches Lorain's taxonomy of soil types and duplicates the categorization that Ruffin sought to modify.[50] (Ruffin did not consider magnesia a type of soil despite Davy's definition of it as such.) Adams spoke of the life and death of soil, of how one could "impregnate the soil" with manure, but beyond that did not force the vitalist position as often or as clearly as Lorain had (63).

Adams at times obscures a pastoral-georgic distinction, but otherwise mostly aligns himself with georgic inclinations by coming to know the abstract economy of nature, and the even more abstract natural laws governing it, through the direct and concrete practice of agriculture. Like Lorain, he understands and highlights the role of human intervention in the processes of nature. But whereas Lorain enforces a distinction between those who practiced chemistry and those who practiced agriculture, Adams introduces the element of chemistry to make a precise distinction between two *types* of agriculturists: the practical farmer ("who practices agriculture as an art . . . but who has never studied it as a science") and the practical scientific farmer ("who, with the *practice,* unites the study of agriculture as a science").[51] He glosses over Lorain's precise clarification between a good method—chemistry—and the one who offered the method—the chemist—since he was engaging with the tenets of scientific improvement in a far more general way.

With chemical analysis, farmers could identify ingredients and from that calculate and measure the missing components of their soil. By encouraging the replacement of missing ingredients, Adams tends toward an inorganic, reductionist view of a soil made of individual and discrete components. Since improvement was achievable by attention to soil conditions, the fertility of soil, it follows, could be understood by regulating its components.[52] As it happens, Adams at once offers conflicting views of the land, both vital and material. But this too, rather than representing a contradiction, is another comfortable mixture of the vitalist and materialist perceptions of nature as evidenced with the soil.

Adams also recognizes the distinctly human capacity for such scientific investigations, placing that ability in a context of georgic science by associating the labor of experiment and experience with enlightened study. The best scientific agriculturists, he writes, "have not only studied agriculture in the closet, but have contemplated it in the field" (156). As he later clarifies, "It is not the circumstance simply of being *employed* in the labors of the field, as some seem to suppose, which makes the accomplished farmer, for then the ox might aspire to this character as well as his master." Rather, "it is *viewing those operations in the light of science*" that matters.[53] While his criterion of knowing nature rests on georgic measures just as Lorain's and Ruffin's did, his concern is to nudge the farming class away from rejecting chemistry as simply a practice of others. Dabbling with Davy by appropriating him without doubt, his uncritical citation serves to represent chemistry as a legitimate aid to the practice of agriculture, not a replacement for it.

In early American rural culture, agricultural chemistry did not yet represent experience as it later would. Instead, it was a tool *for* the agriculturally experienced. Ruffin, Lorain, and Adams were all members of that robust culture, no matter that the age of industrialization would soon be upon them or that they spanned the geographically and culturally diverse eastern seaboard from New Hampshire to Virginia.

The first half of the nineteenth century found a host of improvement advocates ushering in stark changes in views of nature that were in part shaped with the practice of chemistry. The front half of that shift from an earlier, uncodified promotion of systematic means for agricultural ends was set in the late Enlightenment and early Romantic period. That world was still managing animistic and organic theories of nature. Improvement advocates were slowly displacing those theories by materialist doctrines that aided an increasingly commodified environment. The back half of that shift to more mechanistic views of agriculture was set in a nascent industrialized society that treated nature as a mine-able resource and improvement as a fundamental value. In the domain of agricultural chemistry, differences between Humphry Davy and Justus von Liebig map onto the shifts between those two historical worlds, the former vitalist and organic, the latter materialist and mechanical.

Temporally, the early American Republic covers the same spectrum crossed by Davy and Liebig. In political terms, it begins with a world of uncertain national origins and moves to at least a modicum of stable cultural identity. In that time, aligning the virtue of the farming classes with the moral authority of the national psyche dominated social discourse, rural and urban alike. Improvement had cultural, moral, epistemic, and environmental connotations; the georgic spirit combined them.

The early Republic nature John Lorain represented was all-encompassing, less defined by opposition to terms such as "human" or "artifice" than as facilitating them. "Nature understands [how to operate] where art has not interfered with her simple but perfect system of management," he wrote. He also emphasized practical concerns as distinct from pastoral leisure. "Poets," he went on to say, "attribute to rural pursuits, all the rational pleasures which constitute the chief happiness of man. In doing this, they, however, appear to have forgotten that these beautiful scenes which they so elegantly describe, are the effect of immense labour and fatigue."[54]

Edmund Ruffin saw his era as one "driven by necessity [where] a spirit of inquiry and enterprise has been awakened, which before had no existence." He wanted to return to a prior state of nature, but not for Romantic, primitivist ideals; instead, he sought a return to a prior state of natural fertility only for the ability to improve and progress.[55] Improvement was the ability to make unproductive land alive again. The *Essay on Calcareous Manures* described a Virginian landscape where soil composition and properties defined differences between eastern Tidewater lands and the Shenandoah Valley to the west. Slopes, ridges, tree stands, streams, mineral springs, marl pits, limestone hill sides, and the falls of the James River define only the contours of his landscape; silicious, aluminous, calcareous earths and various vegetable acids define the actual land.

Daniel Adams promoted a "naturalized" vision of nature, wording that I recognize is odd but which captures well his uncontroversial and uncritical stance. By representing nature through the re-presentation of others' views of nature, he furthered prevailing perceptions of the land. Chemistry in his text offered an example of the means to align scientific investigation with agricultural improvement which, he might have said, was the natural course of action.

By deliberating on the value of experiment, systematic study, rationality, and method for the improvement of their lands, these improvement advocates were developing their own sense of agricultural science and their own views of progress and science. They defined the moral and material place of science in agrarian America by its circulation amidst the values of practical farmers and their practices with the valued tool of georgic science. Early national leaders such as Washington, Jefferson, Peters, and Taylor had participated in a georgic conversation with practical farmers like John Spurrier, John Binns, and a host of others, deliberating along the way on the place of systematic attention to observable agricultural conditions (chapter 1); the nuances of the book farming debate showed science gaining credibility as a farming practice—earning a place in rural America—by fitting into dominant and preexisting rural virtues (chapter 2); and the treatises by Ruffin, Lorain, and Adams helped elaborate the place of chemistry in rural working philosophies of nature. The place of science in the fields of the early American Republic may not have been presented in a consistent manner by its advocates, but it was a consistent point of conversation among a farming class that sought to align the dual moral and material mandates of American improvement.

PART II THE SCIENCE OF PLACE

CHAPTER 4

The Agricultural Society, the Planter, and the Slave

PRODUCING SCIENTIFIC VIEWS OF VIRGINIA COUNTY LANDS

The first main step towards [the work of recovery], is to make the thieves restore as much as possible of the stolen fertility.

—James Madison, "Address to the Agriculture Society of Albemarle," 1818

IN THE FIELDS OF THE early Republic, science sat at the nexus of soil, fertility, and improvement. This chapter and the next put a spotlight on that nexus. They bring a geographically circumscribed focus on Virginia alone to explore how improvers there used field-based experimentation, a particular kind of science, to re-envision those fields. Such a process of conceptualizing soil was implicit in new ideas about fertility and explicit in new practices of fertilizing and land management. In those practices, members of Virginia's agricultural societies (this chapter) and the contributors to the Old Dominion's first statewide geological and agricultural survey (chapter 5) were moving beyond more rudimentary means of physical soil manipulation into what they considered more sophisticated chemical and geological means. By conducting experiments, pursuing systematic fertilization plans, and reporting and debating the validity of each, the Virginians were using science to interpret the agrarian nature that defined their lives. Along with this geographical and cultural focus on Virginia in Part II comes a shift in this book's basic questions: where the first half elaborated the cultural standing of the sciences of agriculture, this second half emphasizes how scientific measures were used to define land and landscape; while Part I addressed what people thought of science, Part II will consider how they used it to mediate their interactions with the land.

Soil identity was the primary issue in Virginians' efforts to use science for land improvement. Thus, the perceptions of those doing the identifying are central to an understanding of shifts toward scientific views of tillable land. In this vein, James Madison embedded a set of assumptions when he gave a presidential address in 1818 that would necessitate thinking about soil and managing land differently. By seeking to recover "stolen fertility," Madison provided a view of a replaceable store of fertile elements in that soil, a view that assumed the earth was material and depletable in quantified measure. He was an adept farmer, so these views were not developed from nothing. Madison was also an interested party in pursuing the best possible strategies for agricultural improvement in all the georgic senses brought out in Part I of this book. By referring in his speech to science and "the latest chemical examinations of the subject," he indicated to his audience how the twin problems of soil fertility and agricultural improvement could be approached. He represented and promoted the georgic ethic's dual focus on moral and material improvement.[1]

It was not as president of the United States that Madison argued at length for increased attention to the causes of sterility, urging "improved" processes of land management. Madison had recently been chosen president of the Agriculture Society of Albemarle (ASA) near his native central Virginia county, one of the very many civic organizations directed at making local improvement that sprouted in the early Republic.[2] As an indication of his agricultural identity, he cherished that election as much as any previous one. Thomas Jefferson, another visible figure during the ASA's early days, considered Madison the "person who unified with other science the greatest agricultural knowledge of any man he knew. . . . He was the best farmer in the world." A historian of Madison's home county has noted along the same lines that the ex-president was one of "the most active and best known local reformers."[3]

In his first address, Madison the famed statesman and planter told a receptive audience that "the study and practice of [agriculture's] true principles have hitherto been too generally neglected." This was not entirely true, as a century of treatises, tours, and addresses attested. Surely Madison knew that. But if the problem was a loss of fertility, the solution was to restore it. And if restoration of fertility or reversal of exhaustion was possible, then identifying the true principles of its cause was necessary. To achieve this, Madison and his ASA believed they needed to detail "a mode of conducting agricultural experiments with more precision and accuracy."[4] The identity of the soil lay at the root of the problem—soil as something improvable; as an entity governed by principles; as matter that could be studied, analyzed, and experimented upon; and thus as something that could be made more productive.

Virginians in all areas of the Old Dominion used systematic and experimental measures to define rural lands, looking to agricultural societies like Madison's ASA as loci for directives about such activities. Through such organizations, improvers were developing new means to interpret their lands in what would later be called a scientific way. In the process, how did county-based agricultural societies organize their activities to produce a generalized, instead of merely localized, interpretation of their soil and land? What did individual planters do to introduce quantified and systematic studies of those lands? And what labor was actually involved in performing those field-based fertilization experiments?

Virginia in particular provides an excellent site for posing such questions for reasons of both environmental and scientific historical significance. Although many citizens thought the health of the state's environment was declining, as measured and understood by agricultural parameters, they considered science a possible solution for that negative change.[5] In 1816, Virginia's House of Delegates established a Board of Public Works to provide measures for reversing, they said, "familiar scenes of poverty and decline." The Board was authorized and funded to employ an engineer, to conduct surveys, and to patronize organizations deemed worthy of support—the State Geological Survey, begun in the mid-1830s with a strong agricultural strain, was one of its most visible scientific endeavors. In the 1820s, the respected improver James Garnett did not mince words when writing to his fellow Virginian John Randolph that "Virginia—poor Virginia furnishes a spectacle at present . . . her Agriculture nearly gone to ruin from a course of policy which could not well have been worse destructive if destruction had been its sole objective." He then suggested that if practiced correctly the science of agriculture could help stave off further decline, along the way improving Virginia's lot and avoiding the embarrassment of spectacle. Yet at the state's 1830 constitutional convention, Benjamin Watkins Leigh, later one of Virginia's senators in the United States Congress, noted a still present "tripartite decline" in his state: "a loss of 'Genius,' a loss of wealth, and a loss in population."[6] While indicating skepticism about the cultural and agricultural health of the state, such proclamations made clear the state's eagerness to identify new means for improvement.

Virginia's historians have long articulated the commonwealth's unique identity as a member of the Union and the unique circumstances from which the issues described above were raised. Geographical, governmental, agronomic, and political economic contours marked those conditions. Territorially, the Old Dominion was the largest of the original states. In his 1780s-era *Notes on the State of Virginia,* Jefferson was commenting on a territory that had until recently ranged from the Chesapeake Bay to the Great Lakes. Although Virginia would soon cede present-day Kentucky (it gained statehood in 1792), the commonwealth still remained the most geographically expansive of the former colonies. After the Revolution, Virginia's stature was further enhanced by its position as the most populous and most politically powerful. The Virginia Dynasty of Jefferson, Madison, and Monroe held the United States presidency for the first quarter of the century. Except for John Adams's single term, by the 1820s the nation had been guided by Virginians for its entire existence.

The dynamics of local political power structures and governance also differed relative to states like Pennsylvania and New York. State-funded support had already come to agricultural reform societies in those states by the 1820s, where the flow of capital expenditures often followed from the tap of urban capitalist politicians. In Virginia, however, most of the agitation for reform developed locally and county-wide with the support and impetus of powerful planter elites. This localized attention was in keeping with Jeffersonian ideals, where the drive for reform began from specific sites in the commonwealth and moved up to eventual statewide organization.[7]

The diversity of Virginia's agronomy, unique among the states, was in part shaped by geographical and geological variety. Eastern counties of the Tidewater and counties along the southside border with North Carolina were prime tobacco country. Virginia was the leading tobacco exporter in the nation in the 1830s, although only one-quarter of the counties were involved in its production. Even in the late 1830s, cotton was still grown in the lowland southeastern counties. Corn production placed Virginia third in the nation; it was fourth in wheat production. Farmers across the state kept animals for labor and for food—cows, pigs, sheep—and grew a variety of crops for both subsistence and sale. The Shenandoah Valley was well known for its grains, among them rye, oats, and barley, along with its famed wheat production that outpaced the Piedmont, Tidewater, and western coal field regions. Potatoes too were grown in every county. The Alleghany counties to the west—mostly present-day West Virginia—were of higher elevation but, especially near the Ohio River, still impressive in their corn production. Thus, not only was Virginia agronomy distinct from that of other members of the union, but various crops and pasture usage differed within the state itself.[8]

Then there is the slavery question. Since, and even before, the founding of the nation, Virginia's native sons had debated the meaning and relevance of the peculiar institution for its identity, culture, and agrarian practices. Historians have argued about it ever since, unable to find that elusive grand narrative that could speak to the ambiguities and contradictions of the institution and its political economy. But rather than slavery as signifier of southern culture and identity, the relevant subset of that vast category in the present discussion is slavery and agricultural improvement. Furthermore, when one examines not just slavery and agriculture, but the place of science in a slave-labor agricultural system, the picture is yet more unclear, clouded by the prior ideological commitments of contemporary political observers and later scholars alike. Some years ago, the historian Drew Gilpin Faust provided an entry into the general topic that avoided such ideological directives in her elucidation of the problems of the intellectual in the South. Her study followed a "sacred circle" of five men, among them Edmund Ruffin, who represented the alienation and disappointment of a southern elite. Although Faust positioned science as a distinct and pre-given entity—something standing autonomously outside society rather than the culturally embedded practice-in-the-making shown in this book—her study confirmed the point that any examination of ideas in the South was also one of cultural and moral commitments. Virginia's transition to scientific agriculture would be problematic due to slavery; but rather than looking in retrospect to mark that eventuality, the discussion here positions science as a subset of the practices of everyday life in the fields of the Old Dominion, as Virginians did, to show that the science was not just about ideas, but cultural fabric too.[9] In the end, the story of how those in this part of the early Republic used science to understand their soil and their lands was set against the backdrop of, but not precluded by, a slave-holding system.

Hardly naive about these issues, the slave-holding Jefferson was concerned from the start about agricultural viability in his home state. He had argued in *Notes on the State of Virginia* for gradual emancipation and considered that the best solution might be the establishment of colonies for freed slaves. The burden of those who held this view was to develop the most feasible way to accomplish such resettlement at a plausible cost. Into the 1830s, James Madison was vocal along with regional state spokesmen like Garnett and Leigh that Virginia's decline was evident; a member and one-time president of the American Colonization Society, he too considered the agroeconomic decline of his home state the effect, slavery the cause. Although all southern states wrestled with similar questions, Virginia's cultural and political status in the nation led the matters to be more pressing and more basically articulated. The historian William Shade notes that by the later decades of the antebellum period, some advocates in the state were proposing solutions to the slavery question that shifted away from "the impracticality of emancipation" by claiming that such a policy "would be both unnecessary and unwise." Making the slave question unnecessary would require a stronger agricultural reform effort, one in which the place of science was still undetermined.[10]

Critiques of the Southern system were thus abundant. In one form, as Madison, Garnett, and Leigh show, they arose from within the South itself. The gentlemen had disparaged dominant slave practices in Virginia, speaking for their organizations and for the perceived fear of cultural decline and actual material decline evident with soil exhaustion. Just as pessimistic, John Hartwell Cocke would write in 1835 that Virginia could not be redeemed from "the *curses* (for there are many) which in the wisdom and justice of Providence he has imposed on us." Coming from within as they did, these views provided the motivation for the improvement practices necessary to maintain the slave-based economy.[11]

More pointedly, northern observers and foreigners reported that the South's economic woes were part and parcel of the moral woes of slave-holding systems. Without the cultural attachments and with the possibility of an outsider's perspective, they used observations of exhausted soil and declining cultural viability to illustrate slavery's moral and material illegitimacy. Along with tourists like Alexis de Tocqueville and Charles Dickens, Frederick Law Olmsted, whose travel journals were published in the years before his plan for Central Park took root, reported on the weakness of reform efforts he found on his own tours across the South in the 1850s. "The improvement, or even the sustentation of the value of [Southern] lands," he wrote, "became a matter of minor importance." Given his goal of understanding what made the South distinct, and finding the slave economy a most visible and explanatory means to do so, the now-famed urban planner brought assumptions about the character of the South with him to interpret events on the ground.[12]

With his ideological baggage in tow, Olmsted believed it was precisely the failings of slavery that explained the failings of agricultural improvement. More precisely, for Olmsted and many another Northerner the fact of slavery offered an interpretive lens through which to understand the prominence, or lack thereof, of scientific agriculture. The observers then suggested that slavery prevented scientific experimentation on the land, a point that the evidence from Cocke, Wickham, Walker, and Lewis calls into question. Virginians made the same observations as Olmsted—and carried as heavy a burden of ideological baggage, if not more—but, rather than using them to explain away their lack of systematic reform, took them as a reason for more organized and experimental efforts. Science as a means for agricultural improvement was still flexible and indeterminate, part of the debate about cultural reform rather than standing outside it. In the end, then, the dynamics of slave-based political economy went along with Virginia's other distinguishing geographical, governmental, and agronomic features, allowing it to claim a distinct character relative to other states.

Even so, although many Virginians and Virginia's historians thereafter have characterized this uniqueness of the commonwealth for better (as a source of pride) or for worse (as a site of decline), the Old Dominion was quite similar in many ways to other states of antebellum America. For instance, Virginia's pervasive concerns for promoting population, stability, and agricultural economy were common to all the eastern states. From Boston, Thomas Fessenden had argued for agricultural improvement to stave off emigration to Ohio; in Virginia the improvers likewise sought to hold back the tide of emigrés to western states. *The Richmond Enquirer*, for one, reported in 1837 that improvement efforts could "arrest the flood of Emigration [to western states by] unfold[ing] our natural resources." The state's patterns of political development, as with early Whig and Democratic party systems, were also consistent with those in other regions, growing up and finding challenges there just as elsewhere. What is more, "the same dynamic economic and social development that characterized the country as a whole" after the War of 1812, as William Shade has shown, was evident in Virginia. The state, he argues, was "less exceptional and more commonplace than either its hagiographers or critics have conceded." From the earliest days of settlement, from the tobacco of Jamestown to the fields of George Washington's Mount Vernon estate, Virginians had looked across the countryside to understand their landscape as a site of work, as governed by God, and as comprehensible through the guidance of paternal wisdom, local lore, and the direction of an almanac. This too was a common view of the land (but for the tobacco) and a shared experience among the states. Yet, as Madison, Garnett, and a raft of others so ably put it and as many an observer from afar could see, those lands were clearly declining by the early nineteenth century, leading to widespread efforts in recovery.[13]

IMPROVEMENT AND THE COUNTY AGRICULTURAL SOCIETY

When Alexis de Tocqueville toured the country, he found among other things that "Americans . . . constantly form associations . . . religious, moral, serious. If it be proposed to advance some truth, or to foster some feeling . . . they form a society."[14] County-based agricultural societies were one such association. Part of what Tocqueville saw as an ever-present reform movement, they offered community members of the time a site for organizing, discussing, and debating the merits of field experiments while structuring activities to impose a more focused gaze on local lands. They offer historians a vibrant forum in which to examine how improvers used science to redefine their environment. They also suggest how Virginians addressed the concerns expressed by neighbors and critics, in the process helping to build the conditions that made possible the later legitimacy of scientific agriculture.

In Virginia, planter elites took a leading role in organizing agricultural societies that brought issues of soil property identification to the fore. The societies provided a forum for organizing and directing increased attention to principles of land improvement. They also advocated a series of measures that defined specific land management practices, making systematic and experimental identification of measures of fertility the centerpiece of such efforts. "Fertility" was code for "soil status," and the common plea to restore fertility and alleviate exhaustion was, in subtle ways, a call to identify soil properties with more scientific attention. It is not clear that the societies themselves changed concepts of soil either instantly or directly. However, through direct action such as surveying soil conditions, and indirect action such as promoting systematization and encouraging communication to others, they put a significant focus on the importance of defining the land anew.

Madison's ASA was in fact one of two groups of reform-minded planters forming in Virginia nearly concurrently, each exemplifying the collective spirit Tocqueville noted in his travels. In addition to the ASA was the more ambitiously named Society of Virginia for the Promotion of Agriculture or, as it was more colloquially known, the Agricultural Society of Virginia (hereafter, ASV). The ASA, founded in the spring of 1817 and the ASV, founded the next year, had only a small overlap in membership, though a large overlap in geographical focus. While the original thirty members of the ASA hailed from five counties in central Virginia (not just the eponymous Albemarle County) and met in Charlottesville, the more than two hundred in the ASV came from as many as fifteen counties to meet in Richmond.[15]

The membership of these groups was impressive, to say the least, lending the necessary cultural credibility to their efforts in refashioning mid-Atlantic farming practice. Both the ASA and ASV boasted community leaders, state senators, governors past and future, United States senators, and even ex-presidents as members. The ASA chose James Madison as its first president in its inaugural meeting. Jefferson, a seemingly ubiquitous presence during periods of agricultural reform, was also on hand for the first ASA meeting. In 1825, the ASA's second president, James Barbour, was then the secretary of war in John Quincy Adams's administration. In Richmond, the larger ASV elected John Taylor—former U.S. senator, georgic author of *Arator,* and respected state-wide leader—as its first president. The ASV's vice-president, Wilson Cary Nicholas, would soon be Virginia's governor.

The ASA and ASV understood the pressing problems of unproductive soil. By 1818, their members had been fighting those problems for decades

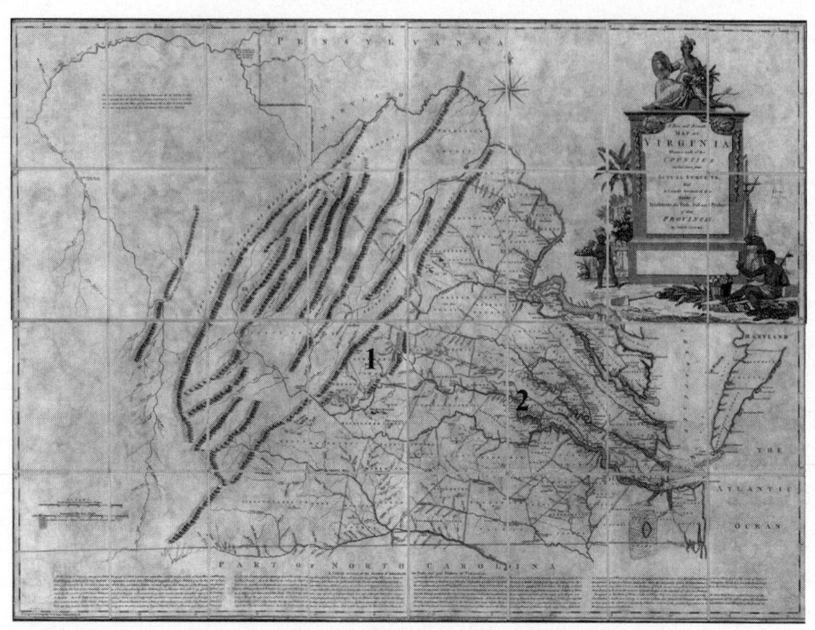

FIGURE 4.1. A county map of Virginia from the late eighteenth century. Albemarle County, near the center of the full map, with the first line of ridges bordering its western boundary, is designated as "1." The ASV was based in Richmond, Henrico County, designated as "2." This map gives little indication of the settlement patterns in the Shenandoah Valley, just over the ridges that mark the western border of the designated counties, or in the region even farther west that later became West Virginia. It does, however, indicate the breadth of the Piedmont and the dominant populated regions of Old Virginia. Library of Congress.

by altering the physical demands on the land. In terms of physical techniques, simple practices like crop diversification, field rotation, and the use of newer mechanical plowing implements helped in this regard. Each of those measures, the planters knew, had been a staple of eighteenth-century agricultural treatises. They were, as Carolyn Merchant has described them, core elements of the nation's agroecological system.[16]

The planters were aware that monoculture was detrimental to their land, having come to understand through decades of experience that diversity in planting was the key to sustainability. Of course tobacco had been the staple crop of the state, the earlier purpose for Virginia's existence. But by the early nineteenth century the monoculture tobacco paradigm that had been recognized as harsh on the soil by all who worked it had already shifted into diversified planting. Madison himself noted that "tobacco . . . though of great value, covers but a small portion of our grounds." Michel Chevalier, the 1830s French tourist whose historical significance has been obscured by his more famous countryman Tocqueville, found Richmond's "flour markets" impressive, not its tobacco.[17] The variety of grain crops noted above, the oats, rye, wheat, corn, and other cereals of the area, were on the rise during the antebellum period, while livestock and poultry flourished. Based on the planters' view of their land, antebellum Virginians were gaining significant experience in changing soil productivity by increasing their crop diversity. With tobacco acting as neither the sole nor economically dominant crop it had been, planters were learning about the opportunities to manipulate soil status with mechanical changes.

The planters' experience with physical measures of manipulating ecological conditions—with crop rotation, plowing, new plant varieties—underscored their know-how as they gathered to discuss further improvement efforts. They systematically augmented their view of the land with more direct attention to soil conditions. They shared a common view of their landscape, one visible as a patchwork of crops, defined through experience, and measured qualitatively and mostly in an ad hoc fashion. As evidenced by the continuing vitality of the agricultural organizations, the georgic cultural credibility of those community leaders, furthermore, lent local legitimacy to the proposals they put forth.

The ASV convened their inaugural meeting in Richmond in March 1818 to direct increased attention to principles of land improvement. This, the simple fact of organization, would be the first of several means for bringing issues of soil property identification to the fore. At the Richmond meeting, they put their static views of the land into a dynamic historical trajectory. The members did so by offering an origin story that traced the history of Virginia, in a markedly abbreviated form, from English settlement through the recent stage of exhausted soil. Upon settlement in the seventeenth century, they observed, all the land was "wilderness covered with thick forests." With increasing trade opportunities and the concomitant growth in population, though, "cultivation . . . produced by necessity" soon became the hallmark of the colony. Finally, the legacy of cultivation led directly to "exhaustion," an "unavoidable consequence." The situation, the members contended, was understandable: "New lands invited and rewarded the labourers; and cutting down and wearing out became habitual." Yet although the pattern had been inevitable, the ASV officers believed the trend was not irrevocable. "Intelligent . . . citizens have appeared in different parts of our country whose judicious exertions have demonstrated that our lands are capable of recovering their original productiveness." It was even possible, they dared suggest, to go beyond the simple restoration of past conditions to actually achieve "increased fertility."[18]

In that first meeting, ASV members understood their current state by narrating the action that led them there. They had written a soil exhaustion origin story. "The land of our ancestors which nourished our infancy, and contains the bodies of our fathers," the officers noted, "must be improved or abandoned." The options were simple, either leaving the Old Dominion behind—a common decision as the expansionism in Monroe's Doctrine signified—or finding the means to stick it out. The ASV planters, however, were already committed to local improvement, their reference to abandoning the state merely a rhetorical flourish. To this end, those at the ASV's meeting would of course be an enthusiastic audience for Madison's view at the ASA on the value of "chemical examinations," a view that moved beyond older mechanical practices that already underlay land management across the state. The opportunity afforded by the mere fact of organization would bring better techniques to the fore, combining the older mechanical approaches with newer, more methodical chemical and scientific ones. For ASA and ASV members, more systematic attention to practices would make their soil more fertile where fertility was the very metric of improvement.

Once the ASA and ASV established their organizational constitutions, complete with origin stories, patriotic charters, and the georgic credibility to hold it all up, the planters moved on to other means for bringing issues of soil identity to the fore—advocating specific measures that defined explicit land management practices. The officers stated their premise clearly: "In every science, and none more than in agriculture, theories should be tested by experiment and systems founded on facts." The reformist improvers recognized that "habits of attention to, and observation on, all the daily operations of a farm" were essential to their goals.[19] Thus, they proposed a series of specific activities, ranging from the conceptual and theoretical to the practical and political. They would not speculate, but rather report empirical observations from actual farming practice.

II. SCIENCE OF AGRICULTURE.

1. Soils.
   a. Formation; b. Classification; c. Topography; d. Elements of fertility; e. Renovation.
2. Causes affecting vegetation, independent of soils
   a. Climate; b. Heat; c. Light; d. Electricity; e. Moisture; f. Atmosphere.
      1. Hygrometry; 2. Meteorology; 3. Composition of Atmosphere; 4. Elements supplied by it to animal and vegetable life.
3. Mechanics of Agriculture.
   a. Law of mechanical forces and powers; b. Hydrodynamics; c. Principles of draught; d. strength and durability of timbers; e. Levelling; f. Draining; g. Irrigation; h. Construction of roads and bridges; i. do of farm implements; j. House-building.
      1. Residences; 2. Barns; 3. Negro quarters.
4. Rotation of crops, theory and practice.
5. Manures.
   a. Classification; b. Composition; c. Comparative value.
6. Botany of Agriculture.
   a. Principles of botanical classification; b. Agricultural plants; c. Edible vegetables; d. fruits; e. Pestiferous weeds; f. Timber trees.
7. Stock breeding.
   a. History of breeds.
      1. Horses; 2. Cows; 3. Hogs; 4. Sheep; 5. Poultry.
   b. Principles of cross breeding.
8. Labour-saving machines.
9. Insects injurious to agriculture.
10. Mineralogy and geology.
11. Comparative anatomy and vegetable physiology.
12. The veterinary art.
13. Chemistry, analytical and agricultural.

FIGURE 4.2. "Synopsis of Subjects to be Embraced in a Course of Agricultural Lectures (Agricultural Society of Albemarle) 1822." The ASA's efforts to improve soil conditions were evident in the educational plans the members proposed. This "Synopsis" provides the basic course of study necessary, in their view, to align science with agricultural improvement. Courtesy of Special Collections, University of Virginia Library.

At a more practical level, the societies' proposals included several avenues of written communication. One example was writing for the press, as evident with the ASA's publishing agreement with Skinner's *American Farmer* and the ASV's agreement with Thomas Ritchie's *Richmond Enquirer.* The ASA noted that "former attempts to establish Agricultural Societies" failed in Virginia not for lack of valuable information, but because they had not followed through on timely "communication to the public."[20] The new opportunity for collaborative improvement afforded by organization was meant to correct those prior flaws.

Reports of members' land-management practices were another form of such communication. These reports in themselves were another way to bring issues of soil identity to the fore by promoting the systematic and experimental identification of measures of fertility. At this most direct level, the societies' proposals included typical organizational activities. In addition to the commissioned reports, these included awarding prizes for demonstrations of best farming practices, holding annual fairs to exhibit examples of good cultivation, keeping extensive and quantitative accounts of practice in personal diaries, promoting the trading of seed varieties to enhance diversity in crops, and acting as special interest groups before Congress. Each example was an instance of encouragement in the pursuit of improvement; all of the activities together were intended to introduce measures of soil property identification to farm practice for the sake of agricultural progress.

The ASA brought together the general and practical when it outlined ten "Objects for the Attention and Enquiry of the Society" in October 1817. The list ranged from animal care, crop rotation strategies, fertilizing options, and mechanical implements to calendars of work, building structures, and a "succinct report" of rural economic practices by members. The members saw this last as the most appealing of the ten statutes, promising that "a judicious execution of this article alone might nearly supercede [*sic*] every other duty in the society."[21] It was a clear example of going beyond the traditional pillars of their agroecological system and into new realms of codified and systematic studies. The members followed up on the idea of the report in their subsequent meeting, recommending that "each member of the Society be required to make a report of his own practices in Agricultural and Rural Economy." They asked questions about rotations of crops, number of acres under the course of cropping, quantity and description of manure carried out yearly, quantity of plaster used, general description of the soil, and the labor-saving implements used. They understood chemistry, as evident through the case of fertilization experiments, as a practice on the farm, making observations of the micro-identity of the soil as important as reports on its macro-manipulation. On the face of it, the reports would give a synopsis of cultivation strategies. They indicated how members treated their property, what they expected from it, and how their practices were either exhausting or restoring fertility. More deeply, the ASA was evaluating soil content and

Heads	Answers.
Rotation of crops.	
Average produce of each crop p acre.	
Number of acres under the course of cropping.	
Quantity of Land cleared yearly.	
If any, what proportion of worn out Land.	
Number of Hands, Horses and Oxen employed.	
Quantity and description of Manure carried out yearly.	
Quantity of Plaister used—at what rate—and with what effect.	
General description of the Soil of the Farm.	
Number and description of Labour saving machines	
Number and description of wheel carriages used in the operation of Husbandry.	

FIGURE 4.3. The Survey—a questionnaire—issued by the Albemarle Society of Agriculture at its November 4, 1817, meeting. By this "subjoined Formula," the officers wrote, "each member of the Society [will] be required to make a report of his own practices in Agricultural and Rural Economy." "Minute Book of the Albemarle Agricultural Society," 273–274. Courtesy of Special Collections, University of Virginia Library.

identifying the degree of local exhaustion, in the process promoting the view that the earth was subject to codifiable scrutiny and analysis.

The scientifically literate men of the ASA also offered prizes for outstanding essays and experiments. In addition, they hosted fairs to showcase successes in animal breeding and plant growth. Local farmer John Craven won the 1828 prize for his exemplary land management skills. William Meriwether came in second because his report, though adequate, was not quite as indicative of a broad fertilization plan.[22] The society was creating standards of writing and experimenting through the formats it encouraged and the prizes it awarded for meeting those standards. Craven excelled at meeting those new standards; Meriwether was at the time a tad less capable.

In Richmond, the ASV also encouraged members to become diarists, tabulating the results of their practices for personal advance and for the organization's collective benefit. Many of those members, such as Dabney Minor, a planter from Orange County (bordering Albemarle County to the northeast) followed suit by recommending that readers of the rural press keep an agricultural diary too. Minor noted in an article in *The American Farmer* that by maintaining records and observing soil properties "we shall derive both satisfaction and improvement. We can mark distinctly the results of various improvements and experiments; [it will] banish those loose, haphazard, careless, and guess work habits, but too prevalent among us, and unquestionably the bane of all good husbandry."[23] Joseph Cabell, a planter, member of the ASA, and longtime friend of Jefferson, was similarly inspired to keep detailed records. The papers of scores of other planters repeat the same example. Planters like Minor, Cabell, and their neighbors were empirics, though attentive to theories of plant nutrition and soil composition. The societies' ideas for action were moving beyond the meeting halls of their Charlottesville and Richmond bases.[24]

Seed trading was still another example of the societies' attention to altering soil. With new seeds, the members were changing the demands of the soil's food supply to the new plants. That is, introducing new seeds to the fields was not only about bringing in new products; it was simultaneously about more efficiently extracting the nutrients of existing soil by leveraging the different nutritional needs of foreign seeds. In that simplest economic sense, introducing seeds from different lands could increase the variety of crops available for market and, thus, create new markets and specialty products. In its first seven years, the ASA exchanged seeds of rare crop varieties with China, France, Portugal, and several South American countries. The trade was not only international and inter-regional, but also intrastate.[25]

John Hartwell Cocke was a dealer of sorts for the Fluvanna County region (bordering southeast Albemarle County, to the west of Richmond), supplying White May and Mexican Wheat to interested parties.[26] He used his influence as second vice-president of the ASA, as well as a former military and current political leader, to manage the traffic of seeds and crops. The seeds also provided disease-resistant strains of common crops, like wheat, which had been hit hard by insects, weather, and other destructive forces. The story of the Hessian fly in Revolutionary-era America stands as one popular episode in this vein, where planters sought disease-resistance from various wheat strains sent from afar. Cocke, in fact, offered several reports to the ASA on that very subject.[27] The seed trade was a way to investigate changes in soil productivity with precise experiments on specific plots of land.

Broadly conceived political action offers a final sense in which the societies fostered attention to altering the soil. The ASV and ASA sought to influence economic and social policy, promoting their networks of influence in the process. These were not simply groups of men who wanted better farming for farming's sake; they were powerful political entities that sought to control the sanctity and future of the Southern plantation system and, as such, were overtly political. The members were driven by the desire to maintain economic (read: slave-labor) power, while simultaneously defining their goal as one of agricultural improvement. As ever, the strong political and cultural component of the societies could not be divorced from their agricultural goals. The material ideals were of a piece with the cultural and moral ideals.

John Taylor, in his addresses to the ASV, advocated using reports of experience and the testimony of farmers as a basis for legislation in Congress. Others used the Society as a platform from which to oppose Northern attempts at industrial inroads in the South that could "lead directly to an insurrection of our Slaves." John Hartwell Cocke, as a central figure in the ASA, called a special meeting of the Society in 1819 to oppose the Society for the Improvement of Domestic Industry's attempted moves in the South. Perceiving them as a group of "Northern Abolitionists" scheming to unsettle the slave population, the ASA was positioning itself to advance its own memorials in Congress. The "Northern Projectors, are taking measures in a clandestine manner to obtain signatures to their Memorials (at least in favor of additional taxes to us), in the Southern Section of the Union," wrote James Garnett.[28] The improvement-minded men of the ASA would have none of it.

The members agitated explicitly for political reform and policy creation through the many "remonstrations" and "memorials" prepared and sent to state and federal assemblies. There were also many less hostile examples. The ASA's 1825 call to the Virginia General Assembly asked for support of river improvements that would encourage trade and market viability; later, the treasurer was authorized to put any "disposable funds" into the stock of the Rivanna River Navigation Company. An 1820 memorial outlined the argument against protectionism, with reference to Adam

Smith's *Wealth of Nations* and counter-arguments to Alexander Hamilton's perceived pro-monopoly policies.[29] The ASA also began a fund for "the Establishment of a Professorship of Agriculture" at the new University of Virginia. Members considered the professorship a necessary component of "the march of Agricultural Improvement already so happily commenced." Their call for such a fund was hardly unexpected, since many of the ASA's well-connected members, Cocke among them, were on the new school's governing board.[30] These efforts tied local attention to soil conditions and national attention to economic policy together, shaping a multifaceted notion of improvement—of soil, community, and culture—through the activities of the agricultural societies. With Virginia still standing as the most influential and populous state into the 1810s, the groups' agitations for programs of agrarian political economy were consistent with assessing the cultivated landscape of rural America more broadly.

The members of the ASA and ASV were not unique in the early national period. Both the forum provided by agricultural societies and the cause of their formation were already widespread. Pennsylvania, New York, Massachusetts, and Connecticut all had regional groups devoted to advancing agricultural causes.[31] The Philadelphia Society for Promoting Agriculture had originated in the 1780s when a group of urban elites—"men of property and education," founder John Beale Bordley called them—sought to improve agricultural productivity and thus economic power.[32] By the turn of the century the PSPA's activities were temporarily subsiding, only to be rejuvenated in the second decade of the 1800s around the same time the ASA and ASV formed in Virginia. By the time Tocqueville toured the countryside, his observation about the flurry of "society" activity was easy to make. A mere review of Tocqueville's list—"religious, moral, serious"—shows that the agricultural societies were especially potent: they pulled together so many of the ostensible reasons for association, including the diffusion of books, community education, and the promotion of a moral and near-religious spirit. They combined ideas, people, and material to produce specific actions on the land, providing the forum for reassessing issues of soil identity.

SURVEYING ESTATES AND PLANTATIONS

Merely organizing efforts to improve agriculture added only so much to specific changes in perceptions of nature. But the ASA and ASV did more. They offered the means and the forum for scientizing the land as much as forwarding the idea of it. In an era where the soil was a political, cultural, and material locus, they promoted attention to soil identity and increased the nexus of practices *on*—allowing for changing views *of*—the land.

The "succinct report" of rural economic practices put out by the societies goes farther in exemplifying what went into introducing newer views of the land. Those reports were localized agricultural surveys. They were neither strictly cartographical, natural historical, geological, nor chemical, but nearly all of these at once. Though the responses were in reality rarely "succinct," the ASA and ASV helped pioneer the questionnaire format in their reports (surveys), asking their members to provide personal assessments of their landscape, crops, and soils. Those estate surveys offer us a site from which to examine a short list of parallel and overlapping efforts in society activity—surveying, experimenting, and manuring. Each was an effort to re-identify and redefine the soil, indicating how specific individuals were following in the spirit of the agricultural societies to re-vision nature.

Two cases in particular show how land could be viewed through a systematic or scientific prism. Both cases begin within the scope of the society-influenced surveys. The first is the example of John Hartwell Cocke (1780–1866), prominent ASA member, influential Virginia statesman, dedicated improver. He was a planter moving from qualitative and systematic accounts of soil identity to quantitative and technically specific ones. The second case is that of William Fanning Wickham (1793–1880; see chapter 2), a figure more indirectly involved in society activity through correspondence and kin rather than the direct involvement in society administration Cocke represented. Wickham too stands as a planter applying increasing scientific strictures to land management over the decades of his life. The two men, both slave-holders, worked and lived within the geographical and organizational scope and intellectual ethos of the improvement societies. A retrospective re-creation of scientific agriculture's path from the 1810s to the 1850s would highlight the increasing presence of chemical analysis given credibility by scientific theory. Cocke's and Wickham's examples allow us to follow the same path as it developed in the personal context of two wealthy improvement-minded planters.

General John Hartwell Cocke had his hand in nearly every antebellum reform in the state. I can only begin to account for his role in agricultural reform. He was a founding member of the ASV and ASA and a longtime holder of the post of second vice-president in the latter organization. His voluminous correspondence records indicate the pivotal role he played in seed trading (and thus crop diversification) around the mid-Atlantic. In the 1820s, he was an original member of the Board of Rectors at the University of Virginia. He was also the head of the state Board of Public Works for many years, steering scientific projects such as the State Geological Survey and large transportation projects, including the James River and Kanawha Canal. He transmitted this spirit of civic involvement to the next generation as well. By the 1850s, his son Philip St. George was the superintendent of the Virginia Military Institute, where some of the earliest efforts at academic agricultural chemistry were taking place.[33]

Cocke's land management practices changed over time to incorporate more technically quantified measures of soil conditions and properties. His progress in this regard falls into three phases. The first is from his early adulthood to the early 1820s. His first surveys were made during the early years in his life and then encouraged by the newly organized ASA. They are general and qualitative, even if systematic and orderly. In a second phase, across the decades of the 1820s and 1830s, he adds more quantitative and experimental depth to his approach. Finally, from the 1840s to the end of his life just after the Civil War, his notions of soil identity and improvement practice become analytically determined with the combination of instrumental analysis, experimental practice, and systematic quantification. The ASA and broader population considered the differences between the two management tendencies, qualitative and quantitative, to be the differences between less scientific and more scientific. My characterization of Cocke's land management practices is thus a reproduction of the positivist-based association at the time between quantification, science, and modernity.[34] Put another way, Cocke is precisely the kind of gentleman carrying forth Jefferson's vision of a scientifically identifiable landscape, and exactly the kind of gentleman someone like Dostoevsky could have pointed to as representative of the science-is-improvement view in later years.

As chair of the first and second meetings of the ASA (before Madison was elected president), Cocke explained the "Objects for the Attention and Enquiry of the Society" before volunteering to report on his own personal survey at a subsequent meeting. In May 1818, he read that report, noting his crop rotation techniques, manuring policies, and description of the soil. This overview of the "Agricultural and Rural Economy" at Bremo, his estate in Fluvanna County, was meant to illustrate what the members sought.

Cocke explained in his report that his lands were of various quality, the better of them "composed entirely of James River bottoms," with other soil of "inferior quality" set higher up and away from the river. In one field under review, he estimated that he cleared five acres a year and worked 75. The survey asked "what proportion of worn out land" he dealt with. He estimated fifteen acres. On that particular field, he manured heavily—"about 20 loads" carted out—but used no plaster. On another, far larger area that was divided between "low ground" and "high ground," he cleared twelve to eighteen acres annually, grew crops on about 550 acres, and considered none of them "worn out." He manured that land with up to 600 loads and there used plaster liberally, "as recommended by *Arator.*" However, though he saw the direct efficacy of the manure as a fertilizer "in all those experiments" using plaster, he doubted whether he "derived any advantage from its use."[35] Animal manure was beneficial; plaster and lime were not. The same was not true on the "High Grounds" planted in wheat, where he "experienced the most wonderful benefits from [plaster's] use." He estimated that the differences in outcome were due to the different soils. "Upon the red soil" he saw that plaster helped. On the "gray soil mixt [*sic*] with gravel" it was less helpful. These assessments based on recognizing different soil types and recording the results of direct experiments were the high point of interest for fellow ASA members, who were looking for encouragement to seek even better means for soil identification and experimental acumen.

Cocke's planting techniques were rather uncontroversial and, in kind, his report was not contentious. (The minutes of the meeting record no objections to the report.) He was qualitative and general in his assessments. Yet he insisted upon the systematic attention he gave to the questions in the survey and to the practices underlying his answers. "System and order," he said, "are the grand secrets of using our little span of time to best account."[36] The ASA audience was less interested in the degree to which his practices were sophisticated or "scientifically" informed than in the fact that Cocke had acted virtuously, that he used a system, was diligent, and appeared methodical. He actively crafted this ethos, as his testimonies in speech and in *The American Farmer* reprints indicate.

His local agricultural survey was a model. It provided the form and focus sought after by the original survey request: to know your lands, to understand the soil in use, Cocke answered, required attention to detail and awareness of difference. It showed farmers how they could improve productivity by systematically investigating and manipulating soil conditions and properties. More than a decade later, the secretary of the United Agricultural Societies of Virginia (UASV) called upon him to answer a request for a similar survey of Bremo. That secretary was the ever-present Edmund Ruffin; his organization had been founded in the 1830s for collaboration between various county societies. Cocke strutted his authority as he explained to Ruffin and the UASV that wheat soils were best for fallowing and sandy soils best for deep ploughing. For the best fertility on light soils, Cocke explained, use "18–20 loads" of manure. By the time of this later survey, into the 1830s, Cocke's increased attention to the questions of soil identity led him to recognize the diversity of soil in a way more precise and investigative than he had done in earlier years. From light to dark and sandy to coarse, he delineated the character of his lands with an increased systematic attention and with greater precision so that fertilizer recommendations were more precisely tailored to soil conditions.

In his second phase of increasing attention to soil identity, Cocke moved beyond providing general commentary on the status of his land. He actively sought to introduce the latest models of scientific practice into the management of his estate. He even started to treat scientific reform as an article of faith, speaking of it as a hallmark of the age. Writing to his son, John Jr., in 1836, he claimed that to promote "scientific principles is one of the few Monuments of our day and generation."[37] As the historian John Majewski has noted, Cocke "wanted nothing less than the rejuvenation of the Old Dominion founded upon new respect for work, discipline, and science."[38] But Cocke's science was still imprecise and, while he was steadily introducing more quantitative measures, his use of it for defining the soil was still in part qualitative.

Cocke's science, in fact, was well situated within the transition period between Davy and Liebig highlighted through the examples of Ruffin, Lorain, and Adams in chapter 3. He was hardly an innovator or a leading theorist. His views of the soil at times tracked along with the rise of a materialist agricultural chemistry, at times lagged behind. For instance, in other matters of estate management he mostly put forth an older but more familiar recipe-like style of chemistry. Cocke took advice from his old copy of *The Farmer's Pocket Guide,* a French book of agriculture from his school days in 1804. The *Guide* recommended preparing seeds "for poor sandy soil" by a sequential process that included mixing portions of ingredients in a sort of fertilizing stew, taking "twelve or thirteen pounds of sheep dung" and salt petre to a boil and then, adding wheat seeds to the broth, boiling again for "eight hours a bushel," followed by a step where the treated seeds were dried and sowed as required.[39] In a process for detecting adulterated wines, as a later example, he applied a laborious and complex technique that took "equal parts of calcined oyster shells and crude Sulphur in fine powder" and put them in a crucible under heat for fifteen minutes. Then he was instructed to "take it out, let it cool, beat the ingredient to powder, and put them into a corked bottle." This was only the half of it, since his recipe listed a full eleven more steps.[40] The processes were nonspecific, forgoing precision in detail—"twelve or thirteen pounds"; drying "for about an hour"—though referring to quantified measures. Thus, while Cocke brought a refined sensibility for increased quantification and the appearance of chemically attentive studies to his soil, he also relied on tradition-laden recipe formulations that had more in common with an eighteenth-century chemical tradition than a nineteenth-century science of agriculture.

For all his social prowess and decades-long leadership capacity in central Virginia, Cocke was not a remarkable chemist or philosopher. The influence of his ideas was probably less impressive than the example of his practices.[41] Despite innumerable reports to *The American Farmer,* a long record of addresses to the ASA, and a position as a focal point for the foreign seed trade—and, thus, a leading spokesman for experimentally derived planting practices—Cocke's own proposals were sometimes ill informed. In "An Essay on Agriculture," for example, he proposed a new system of husbandry based on horizontal terracing and justified by the view that water is "the chief element of fertilization." This foray into theorizing fertilization strategies and plant nutrition fell short, and quickly, as his friend and fellow ASA officer Nathan Cabell pointed out. In a reply to the essay draft, and putting it gently, Cabell noted that "every farmer knows that water is indispensable to the growth of plants," but whether it was "the greatest of fertilizers" was unclear. "Agricultural chemists" were working on such problems, Cabell wrote, and they had advanced beyond the theory of water as the universal and basic fertilizer.[42] Cabell was being kind to a man of greater influence and public stature. Cocke was wrong, drawing on far-outdated concepts of fertility.

In his third and last phase of increasing scientific attention to the farm, Cocke demanded more systematic reporting along with more precise analysis. Both were features of his land management practice for decades, but both had come in for increasing technical specificity over that time span. By the 1840s and 1850s, his references to scientific improvement were couched in newer terms of system, communicability, and translatability. In one case, he wrote to *The Farmer's Register* to suggest that in "the whole circle of the science of agriculture I presume . . . [that there can be no subject] of more importance than the best rotation of crops."[43] While he provided only minimal quantitative evidence for the three-field system in that particular letter, he insisted that systematization was the key to improvement. In this case, though, he emphasized to the readers that quantitative measures were necessary and stated clearly that science, for him, was the practice of quantifying observations systematically. By creating a reproducible system of crop placement, and by recording the results of the benefits from such rotation, the farmer would improve his lot. The proposal was not much different from earlier ones espoused by the ASA and ASV; it brought Cocke's earlier enthusiasm for system back to the fore. But his comments were significant in that they exemplified a view of everyday practice, not abstract theory. He indicated that action on the land was always pursued best when it was grounded in scientific agriculture. His letter was a statement of standard practice, not one meant to introduce new ideas.

Cocke's example shows the increasing presence of chemical practice as part of a broader regional improvement trope. Here the planter of Virginia slowly institutes scientific principles on his land as the very idea of a scientific principle was coming into modern clarity. By the 1850s, Cocke was relying on official chemical analyses to define his soil and guide land management. He structured his plantation management around definitions of the soil that came from laboratory equipment. In the 1840s, he used the services of Robert Rogers, then a chemistry professor at the University of Virginia, to analyze his soil. By the 1850s, he was receiving analytical reports of guano to two decimal points from the new official state agricultural chemist of Maryland, David Stewart. Cocke's was a scientific agriculture.[44]

John Hartwell Cocke's example is not so much revolutionary as evolutionary. His case does not show a distinct switch from the complete absence to dominant presence of scientific agriculture. Instead, he was a planter gradually shifting land management practices over a half-century period to newer and more popular scientific techniques, from qualitative to quantitative, from mechanically systematic to chemically scientific. Cocke's initial survey report to the ASA noted estimates of acreage and fertilizer usage, suggesting his awareness of the value of fertilizers and a parallel search for more specific knowledge about how they could be applied. Noting the specificity of soil types, he soon sought to complement his lands with formulated batches of fertilizer based on soil types understood through analysis. Later, as his discussions about the Bremo estate came into increasing technical specificity, his actions on that land followed suit. The estate left no evidence that would indicate its actual chemical composition, so we cannot say to what degree and with what results his changing views of the land led him to alter the physical and chemical constitution of it. But the records of his estate reveal that over a five-decade span, Cocke was introduced to new, increasingly technical, and quantitative techniques for land management. As his concepts of the soil began to change, so too did his practices. By the 1850s, he treated his soil as a scientifically analyzable entity because he understood it as such. What is more, he offered his own sense of what a science of agriculture must be, sometimes in tune with prevailing chemical theory, sometimes not, but always based on reference to local experience.

William Fanning Wickham was also an estate surveyor and nascent scientific enthusiast. He, like Cocke, began to interpret his soil with technical precision and analytical specificity over the course of the antebellum period. He was too young to help found the first wave of agricultural societies in Virginia, but his life was suffused with their activities and the ethos of science-based improvement. His father, John, was a founding member of the ASV and frequent correspondent with members of both the ASA and ASV.[45] Edmund Wickham, an older kinsman of William's, corresponded with Cocke in the mid-1820s when the two gentlemen exchanged ideas on cultivation techniques. Edmund also requested the apparently popular Mexican wheat from Cocke's store of seeds. William's father-in-law, Robert Carter, was another prominent Piedmont planter and likewise had strong ties to the agricultural societies. Two other Carter family members, Hill Carter and Williams Carter, were also founding members of the ASV.[46] William benefited from the society activities, seed trading and estate surveying in particular, while growing up in the context of improvement his father, father-in-law, and extended family helped foster. His case, then, is indirectly related to the societies' survey efforts, but directly related to the spirit of science and agricultural improvement they advocated.

William inherited the "Hickory Hill" plantation from Robert Carter upon his marriage to Ann Carter in 1820, further cementing his place as a planter of regional prominence. Over five decades at Hickory Hill, his shift in views of the land was more distinct than Cocke's, a transition more revolutionary than evolutionary as seen in his quantified accounts of fertilizer experiments, estate surveys, and technological enthusiasm. He began his career as a lawyer, only taking to farming after marriage. His later work was suffused with references to scientific interest, quantified accounts, and technical demands. He was an especially active planter, as his contributions to *The Farmer's Register* and his public identity in debates from *The Southern Planter* show (chapter 2). He engaged in daily management by walking the fields and conducting personal surveys of experimental progress even though he employed overseers, a degree of engagement not common among the planter elites.

Wickham's quantified accounts of field management illustrate his developing character as a scientific agriculturalist. The former lawyer took to experiments and systematic planting quickly, soon preferring the precision of numbers over the uniqueness of physical signifiers. He was fertilizing with marl in systematic fashion by the 1820s, along the way leaving behind records of an increasingly industrialized marl-mining operation. By 1828, he scrutinized his fields for the bushels of marl they held; he calculated their productivity with consideration of cartloads of marl, bushels used per acre, and tons per harvested acre of wheat. Even the naming convention for his various fields reflects the mindset Wickham developed. By the later 1830s, numerical references came to replace the older place-based referential terms. In his surveys, we find that what was the Lane Field, through which Hanover Court House Road ran, became Field No. 1.

(Figure 4.4 shows the field still named The Lane, though diary and account book records refer to Lane as Field No. 1.) The "Low ground field" that abutted the Pamunkey River became Field No. 3. His earlier diary references to the Shop Branch Field, The Lane, South Wales, and the farm pen became codified in later years as Field No. 1, No. 2. No. 3, and No. 4.[47] The change from place-based to numerical naming was a distinct indication of how Wickham viewed his land. The particularity of the Shop Branch Field, named for the stream that ran by his workshop, became disembodied as a numbered place, devoid of identifying features beyond its code. Wickham's preference for the quantified and systematic treatment was consistent with such changes in viewing the land: his technical management of the fields was guided by his perception of the quantified and systematized identity of that land.

Estate survey practices also exemplify Wickham's land management style. Later records of the Hickory Hill estate show surveys conducted in the more traditional cartographical sense—a map of the property showing boundary lines, angles of intersections with roads and rivers, and field locations.[48] But the survey reveals more than cartographical details; it also demonstrates the ramifications of Wickham's decades of scientific management. For example, an 1878 survey Wickham commissioned, made by chains and measured in poles, offered a unique and telling historical retrospective on fifty years of environmental change at Hickory Hill. The surveyor, M. A. Miller, had trouble on at least a half dozen of the sixteen parcels locating the original corner markers. Where a "Spanish Oak" had been called for in 1825 there was "none now found." Red oaks, dogwoods, pines, hickory, sweet gum, and birches once stood to define the parcel corners, but were long gone by the 1870s, either felled by time and decay or by axe and miscellaneous clearance techniques.

The loss of specific trees itself is not remarkable and could be easily accounted for by the "natural" processes of landscape aging. But in Wickham's case, these changes were a hallmark of active engagement and alteration promoted from within and by specific planning. He had opened a series of marl pits over his career, a new one every few years, shifting from river banks to field pits as he went along. He had expanded fence lines frequently, reordering the clear and naturally delineated (by trees) boundaries with decades of experimentation. The result was that he eventually had to resurvey his property based on artificial, disembodied num-

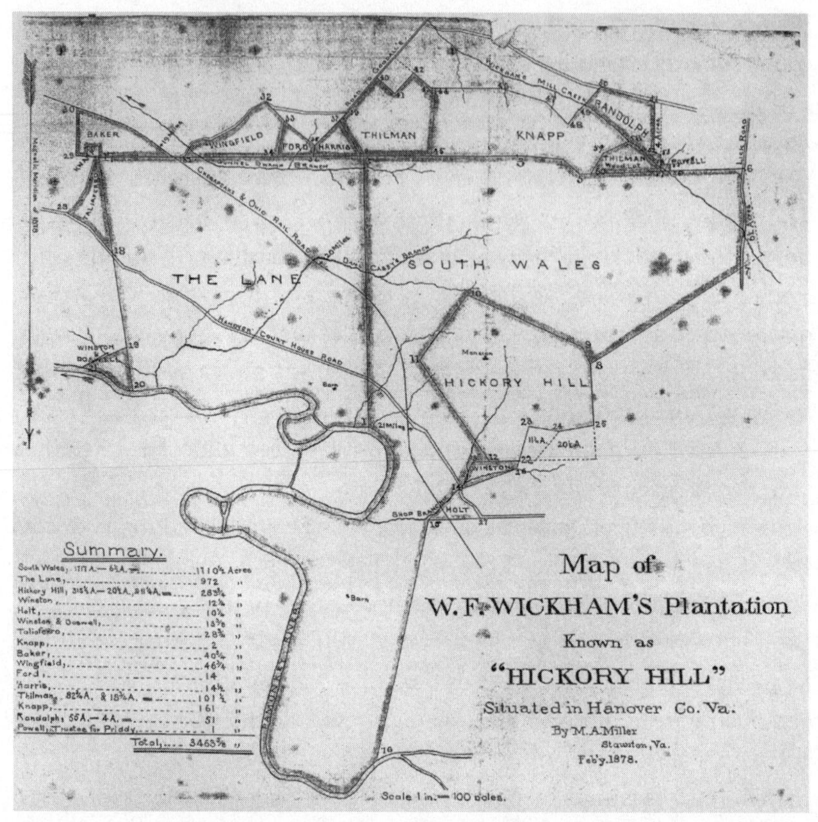

FIGURE 4.4. William Fanning Wickham's Hickory Hill Estate, including all of his fields. The platte, commissioned in 1878, shows the many expansions beyond the original Hickory Hill property of a half-century earlier. Reproduced by permission of the Virginia Historical Society, Richmond.

bers. The resurvey represents a new fashion of identifying landscape that contrasts with the earlier, uncodified one. In one final mention of lost place-names, Miller, after examining a dozen parcels before, noted that he had "found the cherry tree with the axe marks in it." However, it was "dead and fallen."[49]

Wickham's technological enthusiasm provides another example of how he treated his land. Throughout the decades before the Civil War, Wickham rushed to get the latest mechanical implements. He utilized mechanical implements as often as possible, operated a watermill on his property, and even petitioned to dam his stream further—Wickham's Mill

Creek, it was called—to enable still more power. He rooted the technological artifacts and quantified surveys in the same conceptual framework as his earlier experimental operations. His contributions to *The Southern Planter* about plowing techniques were quintessential examples of the worldview he was then bringing to his land. In Wickham's rural press contributions, it was "scientific opinion" that underlay his proposals. His own experience walking the grounds of Hickory Hill confirmed them. His use of science for agricultural improvement, then, became one exemplified through quantification, experimentation, and systemization. He changed the way he managed his land because he introduced new ways to see it.[50]

Cocke and Wickham crafted a delicate association of intellectual and physical labor conveyed within the legacy of a georgic ethic. What these planters think about the soil is reflected in what they say, of course, indicating their ideals and prescriptions. But words are tenuous and liable to insincerity. What they believe, however, is also reflected in how they act, indicating their priorities in the causes of improvement and profit. Many more planters fit the role; Cocke and Wickham offer but two clear and accessible accounts. They show how planters were taking the directives of the agricultural societies—either by creating those directives personally, as with Cocke, or by growing up in the context of their development, as with Wickham—and re-envisioning their land in kind. Fields, instead of being named for streams, were labeled numerically; soils, instead of being analyzed with qualitative measures, were subjected to specific technical instrumentation; crop productivity, instead of being improved by the simple and almost passive addition of fertilizer, was augmented through methodical, systematic, and well-documented experiments, the improvements guaranteed through direct attention, recorded quantification, and technological assistance.

THE LABOR OF EXPERIMENT

What labor went into producing a scientized land? If my theme is the production of scientific understandings of place, then what were the mechanisms by which such practices occurred? If the underpinning of the entire improvement ethos was "collecting . . . facts and experiments," to quote the ASV, then who did the collecting?[51]

In antebellum Virginia, the answer is somewhat obvious: slaves.

Farmlands from the Tidewater, across the Piedmont, and, to a lesser degree, the Shenandoah Valley, were structured by a slave-labor system. The amount of labor required to procure, transfer, and manage marl for tests of restoring soil fertility provides a subtext to the practice of science in Virginia's fields. Marl and plaster experiments were widespread throughout the state, so the use of fertilizers as agents of improvement provide a good site for the integration of labor, science, and place. The historian Steven Shapin once brought the identity of early modern technicians to the fore by referring to them as "invisible technicians"—the people behind the scenes who allowed Boyle's and Hooke's and Newton's experiments to take on meaning. With the case of nineteenth-century rural experimental practitioners, the ultimate invisible technicians were slaves, the ones who actually did the work to identify and then change the soil.

Slave labor was the assumed bedrock of land management and a defining feature of the southern agricultural landscape. In detailing the epistemic differences between the chemistry of Edmund Ruffin, John Lorain, and Daniel Adams in chapter 3, I glossed over the practical differences in each advocate's application of chemistry. Those variations are especially evident with the differing regional uses of labor to change the land. It would be impossible in the context of Virginia's agricultural societies not to take further note of the labor on the farms.

The 1817 membership survey conducted by the ASA (figure 4.3) asked planters to report the "Number of Hands, Horses, and Oxen employed." It also asked about the "Number and description of Labor saving machines" used on local plantations.[52] Both questions were meant to gauge how many laborers the planters had and how best the planter could avoid using them. In Albemarle County, "hands" and "labor" were euphemisms for slaves, terms that removed the direct human element of the work, instrumentalizing labor in the process. In Pennsylvania, John Lorain had suggested that part of the purpose of improvement was to abate justifications for slave labor. In Virginia, Edmund Ruffin spoke to the contrary, suggesting that agricultural improvement could save the slave society. Societies like the ASA and ASV were essentially Ruffin's source and outlet for those notions.

The ASA's "Show and Fair" of 1828 carried forth their earlier questionnaire system in the form of physical presentation. John Craven was the first place award-winning planter at that year's show, having demonstrated the most improvement on his lands as gauged by the twenty-two answers he prepared for the prize committee and by visual inspection. Two of the questions, numbers 10 and 22, dealt with human labor. Question 10 asked, "How many labourers do you work regularly on your farm, and of what description?" Craven noted that he worked "eleven men, six women, and one boy, with the addition of three watermen." He scarcely hired day laborers, he added, not even on heavy harvest and planting days. A mere $30 a year from his budget was devoted to overseeing the labor force. Question 22 asked, "What is your mode of managing your negroes?" The hardest part about "negro management," Craven explained, was "obtaining from them their due and necessary portion of labor." Speaking as if they were mechanical implements, Craven placed the laborers necessary for managing his land into the framework of maximal efficiency. The slaves whose labor ran his plantation were anonymous in the reporting; to the ASA they remained only signifiers of labor rather than individual humans dignified by work. Notably for Craven, that machine mentality preceded his full use of mechanical implements. His use of mechanical implements—because he was adept at using available farm machinery—and his ideas about a strictly managed labor force, in Craven's case, were part of one common view of land management.

The ASA assumed in its prize-judging format that the best means for improvement included careful application of fertilizer in controlled and systematic ways. Two other questions, numbers 17 and 18, alluded to the details of fertilization. Question 17 was twofold, asking, "What quantity of manure is annually made on your farm? [and] at what season carried out and in what manner applied?" Question 18 asked specifically about plaster. Craven, in fact, did not specify how much manure or plaster he used that year in the main body of his answers. But in his general improvement plan he made clear that his farm labor was devoted to a well-planned combination of rotating, tilling, and fertilizing. The judges were sufficiently impressed by Craven's use of his twenty-one slave laborers. His demonstration that a diligent fertilization plan was central to his land management convinced the ASA's panel of reviewers.[53] Rotation schedules, plowing techniques and times, and fertilization plans were all integrated into the practice of agricultural improvement, a system that owed its viability not just to Craven's awareness of systematic principles of soil management but also the slave labor that mediated that awareness.

William Meriwether placed second that year. He was another ASA member, one of their officers, and the owner of 900 acres and a team of slaves in Albemarle County. He used "seven men, three boys, and one woman" to manage his land, he claimed, but employed an overseer to implement the systematic fertilization plan and so could not comment directly on the details of manure and plaster usage.[54] The judges likely recognized his disengagement from daily management, using it to justify Meriwether's second-place finish.

Wickham also hired men to oversee his labor force. Unlike Meriwether, however, he still walked the land himself, even though he owned many more slaves than either Meriwether or Craven. Wickham knew exactly how much manure was being spread, in what places, at what time, and with what amount of labor. When he began mining marl in earnest during the early 1830s, he was tallying the hands per cart, carts per day, and bushels of marl per acre. In 1830, he had an "inventory" of 142 slaves, not all dedicated to the field, but at least several dozen working to plow, plant, and fertilize. When Wickham wrote to *The Farmer's Register* or traded stories with his extended family, he was observing the results of hours and days of work to dig and carry marl from the span of his more than 3,000 acres. With distances from pit to field of up to three miles, this traffic in marl manuring was both difficult and impressive.

Over the years, Wickham had his slave force vary the use of marl between 300 and 1,500 bushels per acre. He generally had it applied to 18-to-100 acre plots. A modest estimate of marl used per acre—say, of 600 bushels on each of 20 acres—would suggest that his slaves were carting and applying 12,000 bushels of the fertilizer for one application. In the earlier years of Wickham's over-enthusiastic marl application, 1828 and 1829, his usage quadrupled from "3 to 400 bushels" to "1000 to 1200 bushels." The young enthusiast claimed that his "quantity . . . varie[d] from one to two thousand bushels per acre" by 1831, but it seems he hit a peak at about 1,500 bushels per acre on "about 30 acres" that year. "All hands are now getting out manure on the marled lands on this side of the shop branch," he noted without elaboration in his diary. In 1839, it took a full month to marl twenty acres, running four carts every day. Wickham's earlier careful allocation of work duties, assigning specific slaves to work particular plots around the plantation, gave way to all-out efforts by 1834. By then, "everyone [was] scattering marle and fencing." It is still unlikely that all 137 slaves (the number on record just a few years later) were

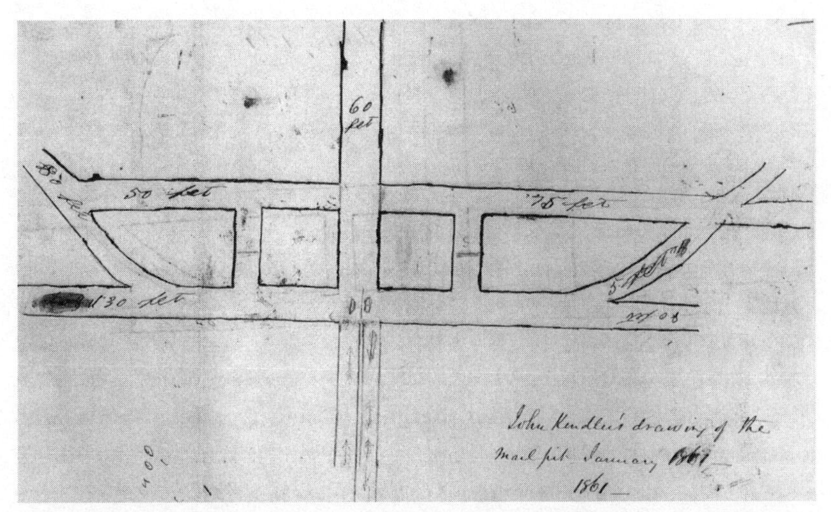

FIGURE 4.5. Wickham's marl pit schematic, as of 1861. The hand-drawn map offers a visual example of the deliberate and rationalized approach to mining marl, with perpendicular lanes and avenues of transport out of the pit and toward the fields. Wickham does not provide commentary on the map, but its visibility helps emphasize the routine and experience of marling. Reproduced by permission of the Virginia Historical Society, Richmond.

devoted entirely to marling, and that he more likely was referring to all of his field slaves, but these episodes show that Wickham clearly had a dedicated interest in this new systematic mode of improvement.

This was a tremendous amount of work, requiring a large, healthy labor force, able horses and oxen, well-constructed carts, and the roads, tools, and supervision to manage it all. What is more, even this basic summary glosses over the full degree of physical labor, since running one "cart" meant digging it out, loading it up, transporting it across the fields, and delivering it to the experimental field to be adequately spread. Four times a day was a lot. Carts were heavy. Fields were hot. When the process was put in motion and not just reflected upon, as Wickham did in his daily diary, one might appreciate better how much work went into the fertilization experiment of an antebellum plantation. Wickham explained his 1832 procedure for application: "[one] marle cart runs and a three horse plough follows." It was this process, honed by a dozen years of practice, that he advocated in the debate in *The Southern Planter* ten years later.[55] To facilitate these practices, he wrote excitedly about opening up new marl pits in 1833, 1834, 1838, and 1841. Underneath Wickham's efforts to develop Hickory Hill was his desire to redefine the soil, to bring it under dutiful observation and direct management by methodical planning. The continued use of a healthy work force guaranteed the possibility of his experiments.

Wickham's peer Ruffin had insisted in his *Essay on Calcareous Manures*, "Whoever uses marl, ought to know how to analyze it."[56] But missing from Wickham's account is a pattern of marl analysis. He reports generic assessments of the fertilizer's properties in many instances. Of a new batch in 1841, he commented, "The quality seemed good."[57] Wickham likely read Ruffin's *Essay* after its 1832 publication, but his papers show no record of having purchased the requisite analytical equipment or of hiring the analysts who were available just a decade later, as Cocke did. (If he followed Ruffin's advice, he would have used a version of Davy's equipment, including glass bottles, a stop-cock, a bladder, and muriatic acid.) His own personal experience, though, born of years of supervising digging, carting, and spreading, undoubtedly aided his successful assessment of the various qualities of marl. Wickham's slave laborers certainly became more proficient over time, identifying marl pits more readily, spreading the marl more easily, and noting its success or failure more quickly. Evidence of such is not forthcoming in Wickham's papers, though the range of his records about the centrality of fertilization experiments suggests that his invisible technicians were becoming quite skilled. The further matter of slave knowledge itself—what they brought to the land, what they developed as part of their daily lives—would add grist to the details of local experimental knowledge based on labor. Unfortunately, Wickham does not provide that evidence either, nor do we have access to those details from other local accounts.

After years of the near perfunctory use of marl to alter his soil properties, Wickham's daily comments on experimentation eventually indicate less attention to the mechanics of the process and less curiosity as to its success or failure. Marling had become routine and, with it, the experimental groundwork with which he began his career became transparent to his daily practice. What began as an effort in agricultural experimentation—in part as influenced by his father's own ASV and also with the encouragement of Ruffin—became the normal pattern of life for Wickham and its own explanation for the activities of and need for slave labor.

It is no wonder that he switched to numerical names for his fields by the later 1830s; such technical detail was consistent with his developing view of the land as a site to be measured, processed, and manipulated. Inside that detail, the soil of Hickory Hill had become identified as a site of quantified matter.

Wickham's history complements that of the elder statesmen, John Hartwell Cocke. It also fits with the view of the land Ruffin was producing. Ruffin, in fact, had confessed as early as 1821, "The Labour required for using shell marl, is now the greatest obstacle to the practice."[58] We know the two men were peers and correspondents, so this similarity is not surprising. Historian David Allmendinger has reviewed Ruffin's labor schedule and demands, reporting that Ruffin followed Smithian ideals of economy to permit "the division and specialization of labor for different categories of slaves." As Allmendinger notes, Ruffin deployed nineteen slaves in an episode in 1824 over a four-day period—nine young men, six young women, two boys, one girl, and one old man—just to clear the land so that marl *could be* dug and carted away. That digging and carting, as with Wickham, required even further labor. Marling had to become part of the daily management of the farm if it was to succeed. Just as with a scientific experiment, "the slave forces had to be devoted to specific tasks routinely or they made mistakes through inexperience and forgetfulness." Thus, these activities required a modicum of training and a great deal of supervision. The general point that scientific agriculture took work must become central to any account of the practice of scientific improvement in Virginia. The slave laborers worked without recognition but were still visible as the machinery driving fertilizer from river bank to lush field, acres and sometimes miles away. That point also serves to help distinguish patterns of land treatment in the South versus the North and to suggest the particular tensions of local and universal that came about through the use of science for Southern improvement.[59]

CHANGING PERCEPTIONS BETWEEN THE LOCAL AND THE GLOBAL

In attempting agricultural reform with scientific means, the members of Virginia's agricultural societies lived with frustration and contradiction. Politically frustrated, they advocated regional identity while working within an atmosphere of increasing sectionalism. Madison's 1818 address was overtly devoid of expressions of sectionalism, but contemporaneous memorials to Congress, the pursuit of funds for agricultural professorships, and remonstrances against the imposition of pro-manufacturing tariffs bear witness to the growing disparity between the political economies of the South and North. Intellectually frustrated, Virginians sought to introduce purportedly universal principles of soil identity into visibly local contexts of agriculture, contexts defined as much by climate and labor as by the soil underneath them. George Perkins Marsh, recognized with Thoreau as a seminal figure in environmental history, would later write: "In proportion to man's advance in natural knowledge, and his consequent superiority over outward physical forces, is his emancipation from . . . local causes." Was Marsh's view true? Would more knowledge of nature lead to "emancipation" from local constraint?[60] Virginians already thought so, though they balanced that hope with a concurrent plea to remake and strengthen a unique Virginian cultural identity. On the one hand they sought emancipation from local constraint; on the other, they cherished it.

John Taylor, speaking in 1818 as the president of the ASV, had laid out the problem of, and possible solution for, introducing science into local reform practices. "Sciences, universally the same," he said, "enjoy the great advantage of reaping harvest from every clime, and of being enriched by the contributions of every language. But, the subjection of agriculture to the climate, soil and circumstances of the position, upon which she must exert her talents, renders her unable to extract a system from foreign compositions suitable to dissimilar meridians; and exposes her to errors and disappointments, from incongruous imitations of foreign practices."[61] Planters like Cocke and Wickham, following Taylor, worked through the tensions of political and intellectual frustration by introducing systematic studies of soil. Using unnamed and uncredited slave laborers, they sought to redefine their land and reinvent their management practices. The entire brunt of the agricultural society movement, and not just the scientific reforms advocated within it, was aimed at unifying differences. Collective action for agricultural improvement was exactly the morally binding benefit that georgic advocates like Jefferson and Taylor had trumpeted in prior decades.

Taylor's observation that science was universal and agriculture always local brought out the possible inconsistencies of combining the two. Yet his was only an instance of the recognition of tensions between locality and universality. His fellow ASV members—and the very premise of the organization he justified through its local attention—were turning to the supposedly universal tool of science as a way to eliminate the constraints of physical locality. In his report to the ASV, James Garnett made sure to tie his proposals to the distinct local nature of the soil, "neither stiff nor very light," on the Rappahannock River banks where he lived. In the process he avoided generalization yet presumed that his insights would be applicable beyond just his land.[62] Nearly concurrently at the ASA's meetings, Madison was highlighting the variety of kinds of soil and plants while simultaneously holding out hope for the discovery of universally "true principles" of agriculture. This tone of ambivalence remained common throughout the era of internal improvements. Erasing local specificity for universal relevance may have been the goal, but tensions between local and global were persistent, not transcended.

A series of factors worked together to promote the common identity of soil that would allow generalizations from the local to the regional, from where they then could go, improvement advocates believed, from the regional to universal. Exchanges within a strong social and kin-based network of planters, for one thing, produced the context of local improvement in Virginia, the interdependent character of which worked to generalize approaches to reform. The agricultural societies were social clubs, gathering places where men could trade stories, complaints, and suggestions. Like the church, the courthouse, or the family dinner, the societies provided a forum for exchange. These same dimensions of kinship underpinned the contemporaneous debate over book farming, where the spread of information and advice went beyond mere individual proclamations. The agricultural societies likewise provided a nexus for disparate views and a clearinghouse for unifying them. The groups also provided an organizing force for people, practices, and knowledge in their identity as common meeting places. Meetings offered the opportunity for face-to-face interaction; in print, with the addresses published in *The American Farmer,* and in state-wide newspapers, though, the members' comments and opinions could spread beyond the meeting room. Through these ostensibly local mechanisms advocates distributed experimental reports and advice on improvement strategies from afar that they considered equally relevant everywhere. The societies also circulated books on agricultural theory and labor-saving devices that aimed similarly to advance the cause of improvement no matter what county. They acted as minor publishing houses in and of themselves and used their influence to coordinate regional publications.[63]

The dynamic forum afforded by the societies encouraged an active form of knowledge and practice. That social order began with certain virtues. As evident in the societies of central Virginia, that order was predominantly paternalistic, an outlook consistent with a similarly paternalistic concept of slave-ownership writ large. Jefferson had perceived this in the decade before the ASV and ASA's founding. "No sentiment is more acknowledged in the family of Agriculturalists," he argued, "than that the few who can afford it should incur the risk and expense of all new improvements, and give the benefit freely to the many of more restricted circumstances."[64] Wilson Cary Nicholas, as vice-president of the ASV, repeated the line, though with specific emphasis on both intellectual and financial advantages: "The establishment of agricultural societies offers . . . a cheap and permanent channel of communication to those who have skill and knowledge to impart."[65] The societies offered an opportunity for planters to exercise moral distinction while promoting material practices. The members also utilized their prevailing social standing as pillars of influence. Thus could the activity of a group of regional planters be advertised to a wide geographical readership.[66]

The issue of scale and extrapolation was not unrelated to the tension between local and universal. The ASV claimed from the start that scientific improvement had demonstrated "a most beneficial influence in neighborhoods," but that a full state-wide scale was still "necessary."[67] Their claim of state representation, even the rhetorical move of naming their society to suggest a statewide scope, was one way to assist this increase beyond the local scale. They never represented all of Virginia, the core of their members residing in the same Piedmont counties as those of the ASA. As for daily farm management, however, they wanted to generalize local techniques into state-wide practices. The ploughing and irrigating measures of Albemarle County were meant to be equally applicable to other counties. The best manuring techniques were worth reporting because they could be adopted by any reader of *The Farmer's Register* or the Memoirs of the Society. To strengthen these overtures, the members developed intellectual and conceptual arguments, improving cultivation in local settings by reference to scientifically informed and universal concepts of soil composition and nutrition. The entire trajectory of scientifically informed agricultural improvement, as envisioned by Taylor, Cocke, Wickham, Ruffin and their neighbors, was to expand from local practices to broader geographical, let alone cultural, applicability.

Historians often look to later eras for evidence of a scientized landscape, thereby taking for granted the scientific and environmental developments that the societies highlight.[68] But important and wide-ranging effects followed from the activities of reform in antebellum Virginia, even if the scale was more local, the institutionalization less concrete, and the skill and adeptness at chemistry and science less lauded than with the largely federal and institutional examples of the later nineteenth-century. Seed trading, reports of observations, questionnaires, and forums for strategizing animal and land care were all elements of an experimental approach. They acted as mechanisms for people to interpret places scientifically. Fields, no longer understood simply as plots of individual property, could become treated and managed as experimental places. The record of the press, personal papers, and account books of Virginia farmers and planters reveals Virginians using the language of "the science of agriculture" to create a sense of place beyond their own boundaries. Common terms, language, methods, and mechanisms helped situate science as an instrument for that change. At the same time, local and regional mechanisms helped produce the cultural conditions of experience, familiarity, trust, and awareness within which later developments could make sense. Later views of the environment as a place defined by science owe a conceptual debt to mechanisms developed by improvement societies like the ASA and ASV.

CHAPTER 5

The Geological Survey, the Professor, and His Assistants

PRODUCING SCIENTIFIC VIEWS OF THE STATE OF VIRGINIA

Whilst engaged in the improvement of the State . . . the great wealth which lies buried in the earth . . . only requires examination of men of science to bring before the country, and make known its value.

—Governor John Floyd, "Address to the State Senate," 1833

IN 1837, HEZEKIAH DAGGS, a farmer from the fertile Shenandoah Valley of Virginia, sent a letter to the state's foremost natural philosopher, William Barton Rogers. "I take the earliest opportunity of sending you five bottles of Sulphur Water and a specimen of lime stone," he wrote. "A specimen of what I suppose to be shale," he added, was included in the mailing. Daggs and Rogers did not know each other; their correspondence might not be expected. But the farmer was interested in increasing agricultural yield, locating coal, and profiting from the possible medicinal benefits of his mineral springs. He had heard that Rogers was involved in a state-funded project to do just those things. Daggs was participating in the same circulation of systematic soil practices that county and regional organizations along the Atlantic coast were then making visible. He wanted to improve his property with the aid of Rogers's "specimen" analysis, to make it more productive through the use of science.[1]

At the time, Rogers was not only a professor of natural philosophy and chemistry at the young University of Virginia but also the official in charge of Virginia's first Geological Survey. As the most prominent geologist and chemist in the state, he had been fielding requests from all over the Old Dominion to analyze water, soil, and mineral samples for several years. By 1835, even before the survey began, his notebook on the "Analyses of Marl, Sand, and Soils" was 43 pages long.[2] When Daggs wrote,

Rogers was preparing for his second full season of the survey. He sat at his home in Charlottesville coordinating assignments to his paid assistants, reporting to his superiors at the Board of Public Works in Richmond, and lining up unpaid contributors from around the state—people like Daggs—to assist in the collection of samples, or "specimens."

During the era of internal improvements, as the historian Steven Stoll has argued and as Governor Floyd and others in his constituency exemplified, practices of land improvement like the survey "blended ecology with ideology, practice with politics, nature with the future of the Republic."[3] Nowhere was that more evident than in Old Virginia. With its agroeconomic dimensions, the role of the scientific survey as an agent of internal improvement extended beyond the geological realm.[4]

The survey was more accurately an agricultural, chemical, mineralogical, and geological inventory of the state's natural components, despite its sparse geological moniker. In Virginia, the state legislature structured it to contribute to questions of agricultural improvement and soil identity along with mineral wealth and coal. This was no small overture, because eight out of ten Virginians were working farms in 1840—or, in many cases, overseeing slave forces that performed that labor. Wheat, corn, oats, and potatoes grew all over the commonwealth in addition to the well-known tobacco grown only in a quarter of the state's counties. Agricultural leaders, constantly attentive to the centrality of soil fertility and soil identity, filled the ranks of survey proponents. The influential agricultural reformer John Hartwell Cocke was for a time the head of the Board of Public Works that oversaw the project; Edmund Ruffin publicized the early survey reports in his *Farmer's Register* and made clear their benefit to farmers; and *The Richmond Whig* editorialized retrospectively that the primary objective of the survey was "the better development of the State's Agricultural resources." William Barton Rogers, the survey's chief architect and supervisor, argued before the survey's inception that "on the subject of a geological and chemical survey [Virginia] would behold, spread out beneath her soil, the rich earths, which [are] soon to diffuse fertility over the hills and plains." He would later emphasize that his work would garner "the attention of the agriculturalist."[5] The legislators believed making a more systematic, more scientific study of that land would provide a new method for identifying and assessing Virginia's cultivated and cultivatable areas. They wrote that, with "a view to the geological features of our territory, and the chemical composition of its soils, minerals, and mineral waters," Virginia could generate a more productive agronomy and provide a more accurate assessment of mineral availability. The state officials foresaw these benefits accruing through the process of scientific surveying.

County-level activities highlighted the growing antebellum circulation of agricultural chemistry. The Old Dominion's first state Geological Survey (1835–1842) highlights this circulation on a larger political and geographical scale. Even with differences in scale, the two kinds of surveys had much in common. Among the things they shared were a conceptualization of science as a practice useful for agricultural improvement, a reliance upon the many instruments and technologies embedded within the process of surveying, and a belief that *the surveys themselves* were tools for bettering the land through systematic study and for bettering the moral and cultural strength of the state under the label of improvement.[6] With those features in mind, by the 1830s the Virginia legislature had the available conditions before them—conceptual, material, and organizational—to warrant a survey of the commonwealth's resources. Rogers, his staff, and his array of statewide contributors provided the information resulting from their project through an always arduous, sometimes tenuous coordination of field-based examinations and lab-based processes. The survey was also indicative of the place of science in the civic context of improvement. It indicated a fruitful interplay between nonscientific agents across the state and state-authorized scientific agents and an openness to new ideas about, budding interest in, and acceptance of science as a valid means to understand the land.

Instead of the wealthy planters of the agricultural societies, though, the state survey was organized and operated by Professor Rogers, an agent whose experience was chiefly chemical and whose character was widely respected. Instead of the fairs, prizes, and meeting reports that facilitated the system of measuring and reporting for those county improvement activities, the Geological Survey detailed the state's soil composition with a system of correspondence between Rogers, his field technicians, and numerous local contributors from across the commonwealth. And instead of the local contributors of county efforts, the state survey included that roster of paid assistants and unpaid local farmers from across the Old Dominion who did much of the actual work of science, taking measurements, recording observations, and mailing soil and rock samples to Rogers.[7]

Even more substantial than just distance in that statewide scope was topographical and agroeconomic variety. Albemarle County, for example, had a generally cohesive set of agricultural products. When the officers of the Agriculture Society of Albemarle pursued surveys of local estates, they were dealing with a relatively narrow range of results; their diversity in crops and natural features was circumscribed by the local bounds of the county. The entire state, on the other hand, had varying climates, altitudes, forest cover, and other environmental features. To pursue a statewide survey conducted with the same conceptual basis as the county surveys required attention to a far greater range of results. Consequently, the tensions of local and universal introduced with county work were again brought to the fore.

Yet, though Virginians may have looked from smaller scale to larger, they were at the same time providing a bounded example relative to the broader national scene. At that level, scientific surveys were among the wide set of projects defining the era of internal improvements in American history. Like canals, railroads, and turnpikes, the more popular of that set, the scientific survey was a site of coordination between private and public individuals aimed at creating progress for state entities.[8] Virginia's decision to organize a scientific survey was thus made in kind with its neighbors. Just as northern entities—Albany's *Cultivator,* Boston's *New England Farmer,* as well as the writing of Pennsylvania's John Lorain and New Hampshire's Daniel Adams—had participated in the common cause of agricultural improvement and soil identification with southern papers like *The Farmer's Register* and planters like William Fanning Wickham, John Hartwell Cocke, and Edmund Ruffin, so too were the different states of the Union venturing into deeper geological detail with common enthusiasm. In the mid-1830s New Yorkers, for example, were simultaneously promoting a study to identify resources under the earth as they too, basking in the success of that other famous improvement project, the Erie Canal, funded a Natural History Survey for six seasons beginning in 1836.[9] In Pennsylvania, the geologist Peter Browne pitched the idea of a survey because agriculture would be "greatly enriched" by it. The period between 1820 and 1840 has even been called the era of the state survey.[10]

When Virginia voted to fund the project in 1835, it was the sixth state to do so, after North Carolina, South Carolina, Massachusetts, Tennessee, and Maryland (but before New York). Within the next few years, New Jersey, Connecticut, and Ohio followed suit. They each had their eyes on cataloging and cultivating all manner of nature's resources within an agrarian milieu; they were all still operating in a time of agricultural dominance in the political economy of the Union, fitting neatly into the climate of agricultural improvement. Their surveys forwarded the georgic ethic, in spirit if not name, by virtue of their relation to agriculture and their integration of civic goals. As with the georgic or homespun ethic, advocates conceived of economic and political aims in kind with moral and cultural directives.[11]

Yet Virginia's own political and cultural concerns also motivated its survey. James Garnett's dismal appraisal of a Virginia "nearly gone to ruin" from destructive agricultural policies brings that regional uniqueness to the fore. The dual desires to preserve the strength of an *economically* viable Old Dominion (by growing the agroeconomy) and redevelop a *cultural* concept of Old Virginia (the one being defined by the eastern planter elites with reference to productive soil from centuries before) went hand in hand. The same worries that attended county-wide inquiries on soil fertility and identity were present at the state organizing level. If agitators for improvement could increase soil fertility they would encourage the population base, and by extension the slave culture, to remain near the coast. Put another way, as with the local and regional efforts at systematic and experimental reform examined in chapter 4, Virginians were restructuring their political economy with attention to social capital and environmental health. Moreover, the survey offered a way to address not just the dual preservation ethic, but two temporal facets of improvement as well: it would strengthen the future political economy by systematically cataloging nature's resources while replenishing the past by making exhausted soil fertile again with the practice of agricultural chemistry.

Even in Virginia the thought of systematically analyzing a state's natural features was neither new nor unique in 1835. Jefferson's 1780s-era *Notes on the State of Virginia* was a state survey of sorts. There, he assessed the status of the natural features of the state in categories such as rivers, mountains, and "productions mineral, vegetable and animal." Despite the patriotic purpose and tone of the book, he was still left to bemoan in anticipation of Hezekiah Daggs fifty years later that, among other things, "none of [the medicinal springs have] a chemical analysis in skilful [*sic*] hands, nor been so far the subject of observations." An anonymous contributor to Petersburg's *Virginia Argus* would write in 1816 that his community wanted "more minute details of the soil, climate, vegetable productions, local advantages, and progressive improvement." The record of the early agricultural press only furthered the same point. Nearly a decade after the rise of the rural press, twenty years after the Petersburg contributor summarized the issue so succinctly, a half century after Jefferson's *Notes,* and with the examples of nearby states standing clearly before them, Virginians voted to fund an official, systematically organized scientific survey.[12]

ROGERS'S SURVEY AND THE DYNAMICS OF VIRGINIA'S TERRITORY

By 1835, the General Assembly's vote in favor of a survey was an effort to remake the Old Dominion economically, politically, and quantifiably. As dictated by the legislature, the first year's work would be a reconnaissance providing "a view to the geological features of our territory, and the chemical composition of its soils, minerals, and mineral waters." To accomplish this feat, the Board of Public Works would appoint an official surveyor, and an annual report authored by that surveyor would provide evidence of the value of the work.[13]

William Barton Rogers had a hand in drafting early proposals for a survey. His appointment as the official state surveyor in early 1835 was thus not surprising. Then a professor of natural philosophy at the College of William and Mary, he would move to the relatively new University of Virginia (founded in 1819) by the time the survey began in full in the spring of 1836. The mid-Virginia geography of Charlottesville was more hospitable to his recurring health problems (Rogers apparently did not fare well in Williamsburg's humidity) and to the organization of a statewide survey because of that more centralized location. He remained there until leaving for Boston in the 1850s to work on founding MIT. At the request of several counties in the northern Shenandoah Valley, Rogers had co-authored a "Report from the Select Committee of the General Assembly of Virginia . . . Praying for a Geological Survey of the State" put before the Richmond politicians in 1834. Even more directly, he acted as the survey's key advocate when he spoke directly to the General Assembly at the February 1835 legislative session.[14]

Rogers had the chemical knowledge to understand how to analyze the samples he would be collecting. His prior experience had been mainly analytical and developed through laboratory-based instruction to students. He was no farmer and made no claims to be directly acquainted with the daily travails of soil management. His ability to analyze soil samples, though, was legitimized by professional academic standing. It helped too that Ruffin was publishing Rogers's work in *The Farmer's Register,* thereby giving him a reputation among farmers as someone interested in practical pursuits and someone vouched for by a rural press editor.[15]

Rogers studied chemistry first at home under his father, Patrick Rogers, an Irish immigrant who had studied medicine and first made his name in America with a popular series of public lectures on chemistry. William then furthered his studies at the College of William and Mary, where his father had become a professor of natural philosophy and chemistry by the 1810s. (William inherited the appointment after his father's death in 1828.) His strictly scientific publishing record by the 1820s was confined to reports on chemical analyses and instrumentation. His geological experience was tied to this chemical familiarity. Because the two kinds of sciences were so interwoven at the time, credibility in one field meant credibility in the other.

Governor Littleton Tazewell, prefacing Rogers's first-year reconnaissance report of January 11, 1836, to the House of Delegates, claimed that it was Rogers's "reputation as a geologist and chemist [that] induced the board without hesitation to appoint him to make said reconnaissance." His younger brother Henry Darwin Rogers was a wayward Owenite recently returned from an immersion in geological debates in London. It was from him that William learned of the newest European theories of mountain formation, stratigraphy, geological nomenclature, and technique.[16] Given Tazewell's view of the matter, William's appointment to the post of state surveyor was neither unexpected nor contentious. With that appointment, the project gained currency from not just the economic and political quarters that funded it, but the educational and scholarly ones too.[17]

In general, Virginia politics were strongly localized, county-wide decisions and party affiliations mapping onto geographical features as well as pockets of religious denominations. It would appear from its success that the survey bill was an effective political article, including enough material to appease disparate constituencies without presenting too many measures of strictly local benefit. With a nod to agriculturalists, coal mining advocates, canal beneficiaries, and grain and tobacco market agents from east to west, the survey was acceptable financially. With support from various cultural outlets across the state, including the academic center of Rogers himself and the wealthy planters who were already invested in county surveys of similar organization, the Rogers survey had the opportunity to influence more than just the scientific readers of his work and more than just the economically minded politicians seeking a return on their investment.

Temporally, Rogers planned the survey season to run from mid-spring to mid-fall, depending on long days and the time outside of the university's academic year to coordinate and manage the process. This latter point was significant because Rogers was tied to his professorial duties throughout the duration of the project. Spatially, the project covered nearly all corners of the state at some point in its years-long duration. In terms of land use, those regions were themselves uniquely defined. Whether a region was distinguished primarily by agricultural, coal, or other market interests depended on the particular topography of that region, meaning that there was a reciprocal relationship between geography and economy.

Virginia's topographical variety reveals the interdependence of land management practices and social fabric: the dynamics of slave labor were often tied to elevation, with slave culture relatively limited above the thousand-foot line; patterns of emigration were connected to soil conditions; the problems of tobacco-caused soil exhaustion, so often used to explain the impetus for Virginians to enact programs of statewide revitalization, were less substantial in the west than in southern counties; and older geological strata were tied to the differential availability of fertilizers like marl and lime. Even the central Piedmont counties were already diversifying their economic basis by the early nineteenth century.[18] For the survey, the larger plantations of the east, central, and southside counties sought more agricultural attention and market opportunities, whereas those representing the western coal fields and unearthed coal deposits around Richmond—some of that "wealth buried in the earth" of which

Governor Floyd spoke—sought the geological knowledge of rocks, minerals, and ores.

Over the span of the five years after the reconnaissance year of 1835, Rogers focused on these different areas of the state, as is evident in the concentration and development of each year's reports. The first year emphasized the eastern Tidewater and the peninsular marl regions and coal deposits near Richmond, ending with a brief glance toward southwest counties down the Shenandoah Valley. Figure 5.1 shows these regions.[19] The second year was spent finishing up the eastern observations, and then stressing "all of the region lying between the Blue Ridge and the first escarpment of coal-bearing rocks of the Alleghany proper." The third year, 1838, bore down on the same western area as the previous year, with greater detail and a larger set of samples. The fourth year saw additional attention to the wide Piedmont between the Blue Ridge and the counties near the capital of Richmond, while the fifth and final year related "chiefly to the marl region between the Potomac and Rappahannock rivers, the northern district east of the Blue Ridge, and the great western coal region."[20]

Rogers hired as many as five assistants at a time to cover such a wide expanse. His younger brother Henry was the first assistant to join, helping out during the initial reconnaissance year of 1835. Thereafter, though, Henry would be too preoccupied running the Pennsylvania and New Jersey surveys. The rest of the team dispersed to study the geological features of the Appalachians, the fertile agricultural Valley of Virginia (the Shenandoah), the coal regions, the marl and clay regions of the eastern peninsulas, the mineral springs of the Valley, and the various features of the Piedmont.[21] At times, the survey was a Rogers family affair. Henry helped that first year; Robert, the youngest brother and a noted chemist, helped out on field trips and with analytical laboratory support; and James Rogers, the eldest brother who was also trained as a chemist, assisted on and off—paid and unpaid—over the entire span of the survey. William Aikin, George Boyd, Caleb Briggs, Charles Hayden, Thomas Ridgeway, and Israel Slade served as the other assistants over the years of the project. In prior experience they ranged from the freshly trained Briggs, who had just completed his studies at Rensselaer Polytechnic Institute, to the veteran William Aikin, a former worker on various New York surveys.

During 1836, Rogers was allotted $3,000 to fund himself and just one assistant, thereby doubling the amount authorized for the 1835 reconnais-

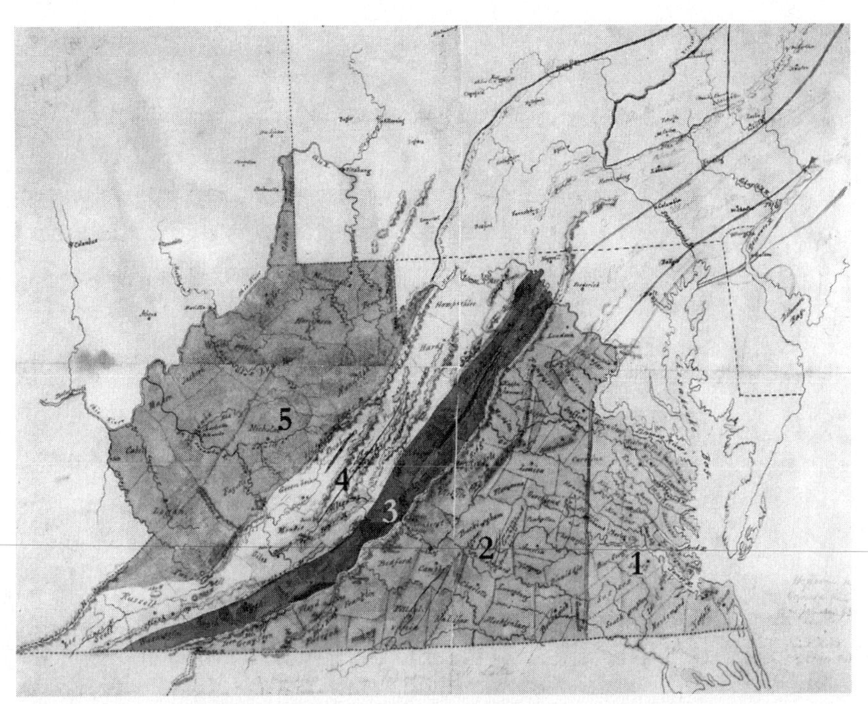

FIGURE 5.1. A partially completed geological map of Virginia, showing basic zones of the state, 1835. The eastern section represents the Tidewater and peninsular marl regions, here labeled "1." The Piedmont ("2") extends from the marl region to the Blue Ridge Mountains in the middle of the state. The Shenandoah Valley ("3"), the slim diagonal on the map, is in between the Blue Ridge to the east and the Alleghany region to the west. That Alleghany region is divided into two parts—"4" on the eastern half, and "5," the western coal fields. Modified with numerical annotations. Reprinted with permission of the Library of Virginia.

sance. Over the years, however, Rogers paid out of pocket for several of the positions because there was never just one assistant helping out. As one might expect, money was always an issue, and the tense letters between Rogers and his contact at the Board of Public Works, James Brown, Jr., speak clearly to this. Rogers did receive a modest increase in funding in 1837, but never anything to match the grand designs of some other states. New York, for instance, estimated costs for their ambitious Natural History Survey at $109,000 and employed seventeen paid scientists. By the end of the project, Rogers had spent $40,000 in just over six years. The state only authorized a total of $26,500, so Rogers's personal investments were significant.[22]

The General Assembly sought an "annual report" written by Rogers as a key outcome of the survey. At a broad level, these reports were components of a context of communication and documentation of which county surveys, agricultural treatises, and even Georgic Tours of the prior half century were also a part. At the more precise local level, Rogers wrote from information provided by assistants and from his own field travels. The assembly also wanted Rogers to "collect and catalog specimens of rocks, fossils, ores, mineral, compounds, and organic remains [to be distributed to] principal institutions of learning in the State."[23] The annual reports provided the means by which direct observations, filtered analyses of samples, and collated assessments of rock, water, and soil quality made their way to the Virginia legislature and ultimately the general public. They represented the public face of the survey. Along with the mineral cabinet that Rogers amassed, they offered visible proof that the assembly's money was well spent. Likewise, the reports were documents of value which could be transported anywhere; their printing and dissemination was predicated on a belief that the information inside was distinct, real, stable, and beneficial. The reports contained rhetorical representations of the nature of Virginia, expressed in the idiom of scientific analysis. The process of surveying the state was a transformation of the land from a site individually experienced, and uniquely conceptualized, to a series of "specimens," each identifiable in a codified, universalized way.[24]

THE LOGISTICS AND LABOR OF SURVEYING

From the start, Rogers, his assistants, and the roster of contributors to his efforts from across the state were mutually charged with the task of organizing and deploying a scientific and technological apparatus for the purpose of measuring, assessing, and defining the land. This task further cemented the model of science as an agent of improvement—blending notions of "practice" together, so that science was a practice, an activity, of practical import—in the eyes of Virginians. It also helped promote the similar goals of smaller county surveys and regional efforts by integrating technical analysis with the goals of cultural improvement. But it was not easy to coordinate the everyday logistics of the survey. Such features had to be created and then produced over the long years of the project. In this process, Rogers had the help of preceding and contemporaneous projects from other states, most directly those of New Jersey and Pennsylvania, where his brother Henry was the state surveyor. The different scale and geography of Virginia's survey, though, tempered the impact of such concurrent examples. The pioneering aspect of all the surveys, in addition, made all the work somewhat tenuous. Rogers's enterprise required attention to developing material practices, but it also required social innovation. These social and material features speak to the devil-filled details required to produce a project that could define and improve nature.

The social dimensions of the survey work included establishing effective correspondence networks with assistants, developing a reliable group of sample collectors, and accomplishing the rhetorical feat of producing a report to suit the needs of the state. In these, Rogers crafted his identity not as a farmer but as an academic, with an audience not of similarly minded neighbors but of political constituents with various reasons to support his efforts. The burden for Rogers and the survey personnel was to create a space where disorganized, uncollated information could be fashioned into codified, polished data. By establishing a cohesive social organization, Rogers was building a central component of such a space. From his home base in Charlottesville, he thus immediately had a series of organizational wrinkles to iron out.[25]

It would seem that one of those aspects, developing a system of written correspondence between Rogers and his assistants, would be unproblematic. Just write; they will get the instructions and then write back with questions and answers. Yet the record of letters from the survey shows that the assistants experienced a consistent and irritating pattern of communication. Letters arrived late, or not at all; goods were damaged in the mail; instructions were unclear to those gentlemen in the field. The very roads that would later enable a more efficient postal service were still then but a part of the agenda of improvements for intrastate infrastructure. As the transportation historian George Taylor dryly observed, the roads were "invariably poor."[26]

Assistants often awaited their instructions in the field. Sometimes they had moved on to the next town before finding out, with the next letter, that they had overlooked a key observation at the last site, or mistaken one measurement location for another. The matter of the mail was thus a daily problem. The record of one assistant, Charles Hayden, is nearly comical in his consistent appeal for more mail. Barely a letter to Rogers fails to note a package not received or instructions late in arriving. Several letters are sent that seem predicated entirely on wondering whether or not a previous letter had been received. The other assistants—Briggs, Slade, Boyd, Aikin, and even William's brother James—made similar complaints over the years, almost never failing to devote a part of every letter to the very matter of the letters and to the problems they encountered because of miscommunication.[27]

Superficially, this problem might appear to be one of reliable transportation. That is true to some extent. But more than that, it was a problem of connecting the safe haven of the tamed town with its university and laboratory, Charlottesville, to the more rustic country, where horses and letters and packages had to contend with inclement weather and mountains and thick brush across trails. Charles Hayden complained of heat and bugs, and their part in preventing his work. Israel Slade had similar experiences and obstacles, lamenting his lame horse and lost trunk. William Aikin could not pass certain roads because of mud and exhausted horses. James Rogers experienced bad weather, poor health, a lame horse, and a broken piece of equipment all at once. William Rogers addressed the problems often in his annual reports, talking about "the fatigue and privation" to which they "were frequently exposed." He sometimes observed the pastoral experience of it all, noting that their troubles were now and again "a little lightened by the animating influence of scenery, at once wild and beautiful and sublime, rich in subjects for the artist's pencil."[28]

Fatigue, privation, and roadless miseries are classic problems of field scientists. But to establish a viable survey system one had to acquire information from the field and incorporate it into a presentable report at the home base, a process that gave no attention to the work of the assistants, or their travails, that accompanied data acquisition. On the one hand, this process of backwoods work is a story of the development of field sciences in America and of how published scientific data is born of the labor of uncertainty. On the other hand, it is a story of how the knowledge of the land was augmented in an agrarian world from georgic experiential knowledge to knowledge produced through disengaged analysis. The scale of the state survey was grander, but the role it played in developing a science-improvement combination was in keeping with other smaller efforts in the early Republic.[29]

Rogers dealt with another organizational wrinkle when he had to identify reliable rock, soil, and water sample providers and then use them as partners in the survey enterprise. Unlike the county improvement societies, where membership rolls defined the range of contributors, Rogers had to develop a roster that extended across the entire state. His annual survey reports list pages of tables of analytical results for samples of marl, soil, rocks, and water from the Tidewater of the east to the mountains of the west. But where did all these sample providers come from?

In part, the official survey was leveraging a system of sample analysis acquisition that already existed throughout the commonwealth. Even before the survey began, and in part because of the activity of county improvement societies, Rogers was receiving and analyzing samples from local farmers. In 1834, he wrote to his brother Henry that "letters are coming to me every mail asking advice on the subject of marl or some other thing." He would later write in the annual report for 1837 that "the high value of these researches, manifested by the eagerness with which the chemical details in the annual reports are referred to practical objects, is still more strikingly indicated by the numerous enquiries addressed to me, and the numerous specimens transmitted for examination from various quarters of the state." Samples were thus relatively accessible as the survey in a short time gained a reputation for its expected utility.[30]

What is more, once the official survey was under way some contributors were brought to it by contact with the traveling field assistants; many others learned of it in the rural and urban presses. *The Farmer's Register* reprinted the survey's first-year report in its fourth volume and made occasional comments on it. Local papers in Richmond also published comments and editorials about the survey and its associated legislation. Rogers proudly told his brother Henry that the "daily papers of Richmond have lauded my efforts in a very complimentary style." Some contributors knew of the survey through their relationship with Rogers, the university, or the state government directly. Joseph Cabell, for one example, was a prominent state senator, Lancaster County planter, and marl sample contributor to the project. He knew of the survey because he was both a personal friend of William Rogers and an associate of the university. In addition, Rogers himself and the survey assistants gathered a small percentage of those samples firsthand during their travels. The process of locating contributors, then, was less fraught with difficulty than that of establishing a system of correspondence.[31]

Rogers's rhetorical skills also figured in his organizational and non-technical efforts. His legally mandated reports to the Board of Public Works provide but a gloss on the field workers' training, assignments, and results, a gloss that enabled field results to appear problem-free when presented in the report. One of those field assignments was to "ascertain by accurate barometrical observations the height of the principal mountains in the State."[32] Difficulties in coordinating instructions and assignments by way of mail provide the background for the challenges present in achieving this task. The assistant has been hired, the assignment given, the travel to the mountains made, the boiling-point thermometer—used to measure altitudes by comparing relative boiling points of water at different heights—in hand, and the measurements recorded and then communicated by mail back to Rogers. In the final report, and citing "sufficient evidence," Rogers reports the "accuracy of the heights computed" from altitude measurements with great satisfaction.[33] Yet the survey's correspondence records show that repeat measurements were few and usually inconsistent. Thomas Ridgeway took nearly fifty pages of boiling-point measurements on his tour of mountain peaks across the western part of the state, never once providing more than two measurements per peak, and often just one.[34] Rogers's rhetorical accomplishment was to define such data as sufficient evidence. His management ability was apparently solid enough to compose a clear report even though the avenues of communication were fraught with difficulty. At the same time, however, the transformation of raw, "natural" data from the field might not have been as unfiltered as a mere perusal of the report would lead one to believe.[35]

In producing marl sample analyses, Rogers dealt with the problems of correspondence, network creation, and rhetorical output all at once. Marl samples were mailed to him in Charlottesville; the contributors who sent them thus became involved in the process of the survey; and the annual reports distributed the analytical results sought by those contributors and other state agents.[36] Rogers's geological map of Virginia (figure 5.1) indicates five basic regions, the first two of which were heavy marling territory. Marling activities from those areas and then in the survey were set against a preexisting background of fertilizing experimentation. Edmund Ruffin, for instance, had advocated marl use and its benefits for improving soil and society in the *Essay on Calcareous Manures*, a work first published in 1832, just three years before the Rogers survey.[37] Rogers even contributed an article on marl analysis to *The Farmer's Register* in 1834. The marl analyses of the Geological Survey highlight its agricultural context; marl, lime, and other fertilizers were as much on John Floyd's mind when he spoke of the "wealth buried in the earth" as ore and coal.

Rogers believed that the task of unearthing fertilizing wealth was relatively easy. He argued in his 1837 report that from "the extent of its exposures in many places, from its great richness in carbonate of lime, and from the facility with which, without any previous preparation, it can be applied to the soil," the only thing left "for the attention of agriculturalists" was to identify the location of the marl and to evaluate its quality. After several years of survey results had been loosely compiled, Rogers proudly presented a comprehensive table of "every variety of marl met with" in the eastern farming district of the state in his 1839 report.[38] Previous volumes had included scattered tables, so the 1839 version was but the tabulated culmination of the fertilizing arm of the survey. Those summaries of the description and composition of samples included in the annual reports tell of a process that began with soil in the countryside or on a farm and ended with an analyzed scientific specimen. Figure 5.2 shows a detail of one of these from 1837.

A large number of the letters and samples came from the marl region labeled "Miocene"—region "1" on the map in figure 5.1. This area included the counties of Surry, Essex, Middlesex, King and Queen, Isle of Wight, Northumberland, Lancaster, Nansemond, and others to the east. In the cases shown in figure 5.2, the likes of Mr. Bagby and Mr. Pollard of King and Queen County and Mr. Cabell of Lancaster County—this was Joseph Cabell, friend of Rogers and associate of the university—were members of that new network of sample contributors and participants in the survey. From them, Rogers collected, analyzed, and collated the marl samples, creating a classification. There were white, blue, and yellow marls, yellow, green, and light sands, for example. While organizing these different types of fertilizer, the contributors complemented technical quantification with verbal descriptions of the samples. Some shells were "finely decomposed and partially cemented" while others were but "fragments of shells." Some were "tenacious" while others were "in nodules." Interested parties could review the output of the survey for details on marl quality as described by the quantity of carbonate of lime—the active ingredient—and as compared with samples from adjoining lands. When Rog-

MIOCENE MARLS.

Localities.	*Observations.*	*Carb. Lime.*
LANCASTER.		
Capt. Ja's. Robinson's,...	Small fragments of shell in a ferruginous sand—green sand a trace,.............	42.0
do.	ditto, ditto,	37.5
Mr. Yerley's,...........	ditto, ditto, rather compact,...	30.6
do.	Yellow—aluminous—green sand a trace,...	12.2
do.	Yellow—consisting of shelly fragments partially cemented—green sand a trace,....	21.0
Mr. Cabell's,............	Shells decomposed and partially cemented,	42.0
do.	ditto, ..	46.5
Mr. Callahan's,..........	Yellow—fragments of shell in ferruginous sand—large grains of green sand in considerable quantity,.....................	21.5
Mrs. Palmer's,...........	Yellow—small shells and fragments—green sand a trace,.........................	32.9
Benj. Walker's,..........	Blue—green sand a trace,................	18.0
Warner George's,........	Blue—shelly fragments—green sand a trace,	14.7
Col. Palmer's,...........	Light—conglomerated fragments of shell—slightly compact—green sand a trace,....	57.0
do.	Shells decomposed and partially cemented,.	37.5
Dr. Jones's,.............	ditto, porous,..........	86.8
Union mills,.............	Yellow—small shells in ferruginous sand,..	23.8
do.	Light—quite compact—shells small—green sand a trace,.........................	62.5
Col. Phil. Branam's,.....	Blue—tenacious—small shells—green sand a trace,..............................	21.5
KING & QUEEN.		
Piedmont,...............	Blue—containing fragments of shell—green sand a trace,........................	33.6
do. lower bank,.....	Blue, ditto, ditto,.............	22.2
Mr. Bagby's,...........	White—shells finely decomposed and partially cemented,......................	80.6
do.	Blue—containing fragments of shell,......	30.6
Mr. Mann's,............	Light—nodular,........................	78.4
do.	White—containing small fragments of shell,	80.6
Mr. Burton's,...........	Light—shells decomposed—occasionally cemented,...............................	85.2
Mr. Atkins's,...........	Small fragments of shell,................	76.1
Mr. Ryland's,..........	Shells decomposed,.....................	46.5
Mr. Motley's,...........	Blue—small shells and fragments—green sand a trace,.........................	14.7
Mr. Pollard's,..........	Blue—fragments of shell—green sand a trace,...............................	21.5

FIGURE 5.2. Detail of a Miocene marl table of analytical results. The first column indicates the marl contributor; the second column offers a verbal description of the sample; the third column provides the analytically determined carbonate of lime composition of the marl. Rogers, *Geology of the Virginias*, 151–153.

ers had taken geological samples across the state, checking the rock strata at various points along a traverse line, he would then generalize from those selected points to a view of the state as a whole. When he collected these marl samples he was doing much the same, generalizing from the local to the state scale to redefine place with science.

The results of his analyses served the professor well, because they helped solidify ideas about the relative ages of layers of earth examined with geological interest. Henry Rogers had already completed work in New Jersey identifying "green sand" as a constituent of the marl beds. Together, the brothers extended this work into geological considerations of the Appalachian chain. The results also served the state constituency well by proving "useful to individuals interested in knowing the value of their marls," as one annual review put it. As if to make this point more salient in the reports, Rogers reminded those authorizing the work that "further detailed information in regards to [the analyzed] rocks [was] so important in their application to agricultural and architectural purposes."[39]

In the process of analyzing marl, Rogers performed a subtle and helpful maneuver: by taking the work of marling farmers into the statewide authority of the survey, he had reconfigured the sense of "value" to the benefit of both himself and the citizens of Virginia. Marl analysis, by appealing to Rogers's own theoretical designs, the farmers' pursuit of economic benefit, and the state delegates' demand for political utility, represented an integration of geological, agricultural, and political goals. At the same time, through the process of collection, analysis, collation, and report-based reproduction Rogers refashioned the individualized, particular marl, water, or coal samples into a generalized portrait of the state. In this process he transformed something like marl, for example, from a handful of nondescript crushed shells to a precise scientific specimen. Colonel Branam's piece of marl from Lancaster County becomes a 21.5 percent carbonate of lime specimen of blue, tenacious, small shelled Miocene marl, with traces of green sand.[40]

The material and technical dimensions of the survey work contributed to the same process of environmental knowledge production characterized by the social elements discussed above. These included, first, the general use of chemistry and geology as tools for inquiry across the state; second, the deployment of field-based instruments like thermometers for examining features of the natural landscape; and third, the use of specific analytical instruments like the marl analyzer to study compositions and qualities of soil and fertilizers. Each of these aspects defined the active, mediating character of the survey. It was not just the analytical equipment that characterized the material dimensions of the survey but the basic use of scientific inquiry; the sciences themselves, in the sense of georgic science, were conceived of and then used as tools for studying the state. Given the practical bent of the scientific surveys, their various goals and audiences, and their basic object of the land of Virginia, the sciences defining these surveys represented the evolution of georgic science from earlier in the century.

In an appendix to the 1837 report, Rogers describes the multifaceted role chemistry played in his pursuit of practical benefits for the state citizenry: "The amount of chemical investigations thus bestowed upon the materials of economical value, collected in our explorations or forwarded to us from localities not visited, though not mentioned in the annual reports, forms a very important item in the yearly operations of the laboratory, furnishing useful facts and valuable suggestions in relation to the nature and appropriate application of our marls, limestones, iron ores, and other important mineral resources, and thus silently, but largely and continually, diffusing information of immediate practical utility to persons in every district of the state."[41] Rogers speaks to several issues here, reinforcing many of the broader meanings of the survey in general. He indicates that contributors are sending more material for analysis to his lab than he can report on, as more and more specimens from "localities not visited" remain unanalyzed. While treating chemical analysis as a tool useful for achieving his goals, he is also diffusing the information "of immediate practical utility" not just from his office to the state constituents, but between local areas amongst themselves—"silently" and thus invisibly. Rogers presents all of these activities as useful and directly beneficial to the citizens of Virginia.

The process for analyzing water samples offers an example of material coordination in the survey. Many of the water "specimens" came to Rogers from spring operators in the mid–Shenandoah Valley—Hot Springs and Warm Springs, White Sulphur Springs, Red Sulphur Springs, Sweet Springs, and others. The contributors, aware of Rogers's work and tapping into the possibilities of commercial advance, wanted to capitalize on the association of mineral springs and health.[42] From the Shenandoah Valley or west, John Sites of Rawley Springs, William Seymour of Howard's Lick Spring, and Samuel McCamant of Grayson Courthouse sought analytical assistance for their springs. Hezekiah Daggs, another neighbor from the Shenandoah Valley, wanted to leverage all manner of analysis from Rogers, asking for a chemical examination of not just his spring water, but also

Hot Springs, Bath Co., Va.

BOILER BATH.

Temperature, 106°.

Solid matter procured by evaporation from 100 cubic inches, weighed after being dried at 212°, 13.25 grains.

Quantity of each solid ingredient in the 100 cubic inches, estimated as perfectly free from water:

Sulphate of Lime,........................	1.750 grains.
Sulphate of Magnesia,......	0.890 "
Sulphate of Soda,........................	1.437 "
Carbonate of Lime,.......................	6.557 "
Carbonate of Magnesia,...................	1.153 "
Chloride of Sodium with traces of Chloride of Magnesium and Calcium,.............	0.570 "
Silica,..................................	0.703 "
Peroxide of Iron,........................	0.020 "
Organic Matter,......................	in small amount.

Volume of each of the gases contained in a free state in 100 cubic inches of the water:

	Cubic inches.
Carbonic Acid,...........................	11.007
Nitrogen,................................	1.790
Oxygen,..................................	0.220
Sulphuretted Hydrogen,...................	a trace.

FIGURE 5.3. Selection of analytical results from mineral springs analyses, as tabulated in the annual report. Rogers, *Geology of the Virginias*, 555.

limestone and other field rocks. At least one gentleman, from the Black Sulphur Springs Company, sought mineral water analysis in what he thought was compliance with a law that required official measurement of mineral content for spring operators. It may be that the businessman was misled or misunderstood state law (or that I cannot find the legislation to back it up), but either way water samples made it to Rogers from a variety of sources.[43] The traffic in mineral water samples was heavy.

To analyze these samples, Rogers combined field work with lab work. At the springs his assistants measured temperatures while he ran evaporations on boiler plates in Charlottesville. The purpose was to define the chemical proportions of the resulting precipitate: sulphate of lime, sulphate of magnesia, carbonate of lime, chloride of sodium, and so on. He also calculated the volume of the gases let off in the process: nitrogen, carbonic acid, oxygen, and sulphuretted hydrogen, for the most part. Figure 5.3 is one example.

The details of analysis are many, as Rogers includes the analytical results of some thirty springs, each going to show that he and his political supporters were actively seeking to redefine the state's environs scientifically. His reports include an eighteen-page section devoted just to the mineral water analyses. The details are just as great for analyses of soil and rocks, samples of which were examined for reasons similar to water, even if unrelated to health. Rogers used a straightforward analytical procedure on rock and soil, generally treating the soil with hydrochloric acid, then with ammonia, and then drying it to leave a mineral precipitate.[44] As evident in dozens of pages of results for iron, lead, soil in general, limestone, marl, and the spring water analyses, Rogers had taken samples from their sources, collected them in a laboratory, placed them under analysis, and reissued them as lists of chemical constituents.

Two brief examples, the thermometer and marl analyzing apparatus, further delineate the ways survey personnel used material elements to produce new knowledge of the land. These survey instruments were less than ideal or pristine, as plates, glassware, heating elements, receptacles, and soap dishes made difficult travel companions. Thermometers were well used and often broke. Assistants employed them to measure altitude (by relative comparisons of the boiling point of water) and to gauge evaporations and distillations. Rogers reported that the boiling-point thermometer "promises to afford us great facilities . . . and enable us to continue our tracings with all the accuracy and expedition that could be desired."[45] He estimated that "*altitude* [boiling-point] thermometers . . . from their portable form, and the ease of observing with them, were found to be particularly valuable especially in districts of a very rugged topography."[46] Furthermore, he claimed that the "employment of the thermometer" was able "to facilitate some of the most difficult explorations" the team was called upon to make.[47]

But it was not so easy. In 1839, James Rogers wrote to his brother and boss not only that his horse was sick and the weather was bad, but that his boiling-point thermometer was broken. In consecutive letters he pleaded with William for a shipment of new thermometers. In May 1840, the youngest survey assistant, Caleb Briggs, made special mention of the successful transport of his boiling-point thermometer to the northern part of the western zone, while another assistant, Israel Slade, a month later and farther south, was forced to call for a new one when his broke.

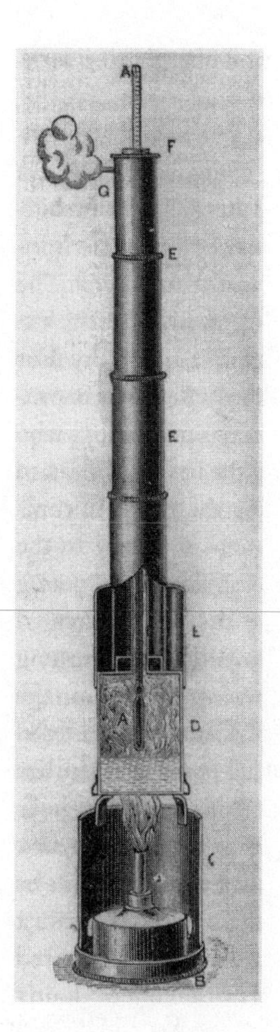

FIGURE 5.4. A hypsometer, or boiling-point apparatus, common to the era. An earlier version of this boiling-point thermometer was used by Rogers and his assistants during the survey. Freshfield and Wharton, eds., *Hints to Travelers,* 95.

Briggs's luck was limited though, since a year later he too broke his thermometer—twice. Thomas Ridgeway, deep in the mountains of western Virginia, outright bemoaned the state of his fragile "Boiling Point Thermometer," but perhaps his irritation was understandable since the instrument broke after a hornet had stung Ridgeway on the ear, causing him to knock it over.[48]

The thermometer, as a mundane instrument of analysis, provides just one example of problems with instruments in the field. Besides illustrating the everyday troubles of field workers in the backwoods of Virginia that belie Rogers's confident gloss on their ease of use, the thermometers provide a glimpse into one aspect of how the survey and its science were produced as the space between instruments and the land. Rogers's marl analyzing apparatus, which he claimed to use as frequently as the thermometer, provides another.

Rogers specifically designed the marl analyzing instrument for convenience, ease, and accuracy. Aimed at determining fertilizer properties, it was used toward different ends than the thermometer. It was also based at Rogers's lab, a location more amenable to glassware than the field. Improvers and chemists had been developing equipment to analyze fertilizers for some time. That brief lineage of analytical equipment shows analytical techniques tracking along with the rise of materialist theories of fertility into the 1840s and 1850s. In short, those actively looking for soil measurement strategies over the decades of the first half of the nineteenth century came to supplant earlier qualitative assessments with increasingly technical and quantitative methods.

The rural press carried articles and letters that elucidated some technique or other for identifying the quality of marl, often based on horticultural indicators. One farmer wrote to *The Farmer's Register* about using a plant of known genus, *Veronica,* but unknown species—the author recommended a new name, *Veronica Ruffinia*—whose growth was an indicator of nearby high marl quality.[49] The *Veronica,* he suggested, could be broadcast to readers as a valuable tool for identifying the rich marl. Humphry Davy had devised his own equipment for doing the same, which Ruffin used, although it presented numerous problems (figure 5.5). Ruffin noted that it was "complicated" and "expensive" in the editorial footnotes written to accompany the publication of Rogers's apparatus in 1834. By contrast, Rogers's apparatus was convenient, easy, and accurate, making it an attractive and superior substitute for Davy's method.[50] Its technical appearance also seemed to outshine the horticultural indicators.

Figure 5.6 shows the apparatus Rogers used for his marl analyses. It consisted of a bulb of light glass ("A") into which the sample was placed, a piston-like cork ("C") used to inject acid to drive off the gaseous components (carbonic acid) locked into the marl, and a third tube ("B") extending from the bulb to filter escaping gas. The entire apparatus was countered by a balance which would gauge the change in weight affected by the escaped gas. After performing the procedure, Rogers or "any farmer who uses calcareous manures" could calculate the amount of carbonate of lime. The operator would put a known quantity of powdered marl into the

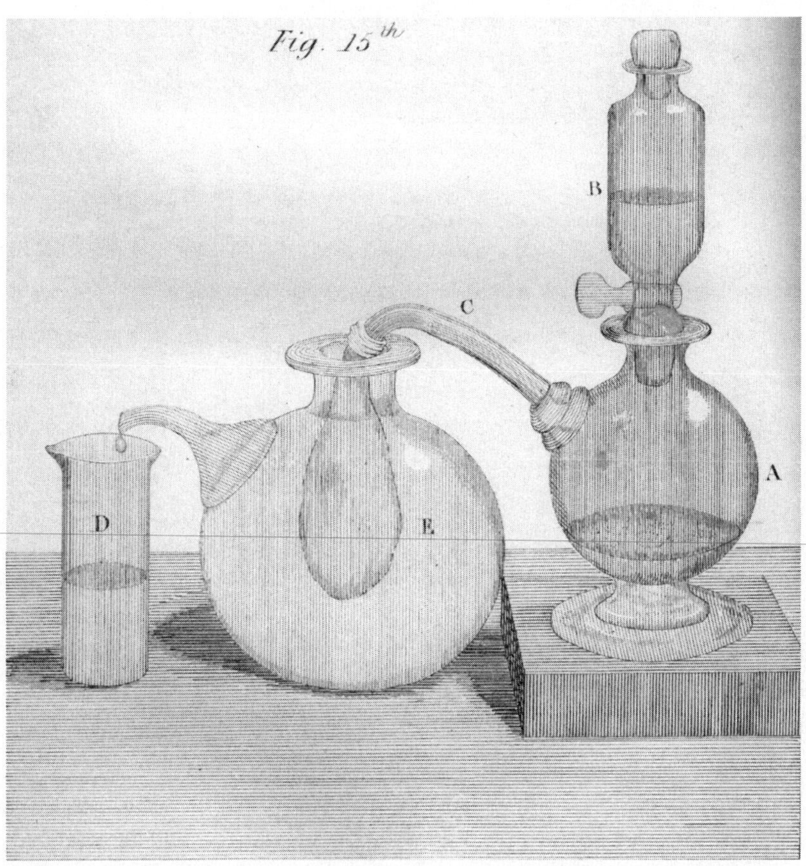

FIGURE 5.5. Humphry Davy's marl analysis apparatus. This is the equipment Ruffin would have been familiar with from his early years of analysis. Davy, *Agricultural Chemistry*, 311.

glass bulb with "a little water." Gas was then injected into the bulb drop by drop via the cork—Rogers used muriatic acid, or hydrochloric acid, as it is known today. The effervescing sample was left to rest for an hour or until all the gas had escaped into tube "B." By measuring that weight loss from the original marl sample, one could back calculate how much carbonate of lime had been in the original marl sample.[51]

The apparatus underlying Rogers's domain was laboratory equipment. But as an agent of the state assembly, the university, and the agriculturalists who had sent him samples, Rogers was not studying nature in a disembodied, abstract sense. Science, in its most general definition, is the

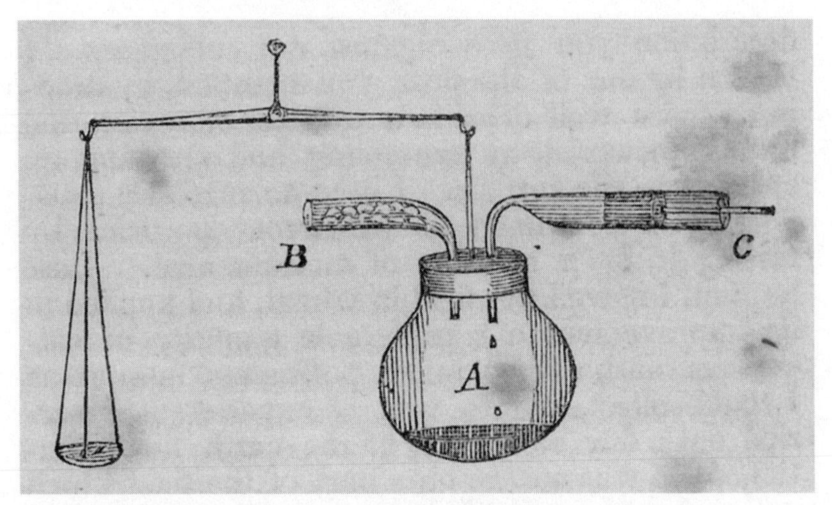

FIGURE 5.6. "Apparatus for Analyzing Marl and Carbonates in General." Rogers invented this apparatus for use in the 1830s. It was first presented in an 1834 article for *The Farmer's Register* and later reprinted in Rogers, *Geology of the Virginias*, 10.

study of nature, but describing it with such generality loses the direct, physical referent of "nature" and drifts into an idealized level. Like Jefferson in his *Notes on the State of Virginia,* however, Rogers was studying nature as physical matter, as samples in his lab that were the source of agricultural production, the very basis for Virginia's political economy. His technique and instrument were self-advertised as enabling "the operator [the farmer] to proceed with great accuracy and despatch [*sic*]," values that were part of the ethic of agrarian citizens and, in a georgic context, lent themselves to equal suitability in the barnyard or the lab.[52] Here Rogers tended toward blurring the two senses of nature, the abstract idea and the physical material, as he offered the apparatus to become part of the farmer's daily practice.

LEGACIES DIRECT AND CONCEPTUAL

The Geological Survey offers a multifaceted site of analysis, one that helps demonstrate how and why antebellum Americans were developing a science of place. As with the practice of fertilizer experiments on farms across the early Republic, the process of participating in survey science was becoming an element of the working experience of farmers and planters in Virginia. The survey's product, an account of the geological, agricultural, and mineralogical details of Virginia's land, fits into a narrative about increasing the circulation of scientific means for attaining and communicating environmental knowledge. Its operation by William Barton Rogers and his team of assistants illuminates the process by which scientific work produces new views of the land. In this case, the samples of water, soil, and marl sent by farmers for analysis came back as scientific specimens, quantified, cataloged, and subsumed as part of a greater mission of systematically defining the land.

With organizational, rhetorical, and other technological features, Virginia's Geological Survey exemplified a process by which scientific views of the state's land could be produced. Governor Floyd, the General Assembly, and Rogers intended their project to provide a grid of the territory for the purpose of controlling and improving that land. The era of internal improvements which helped shape this project was an effort in strengthening the nation morally, materially, and economically. Nature and its resources lay at the root of nearly every one of those improvement schemes and knowledge of that nature was essential for the success of the improvement plans. The scientific surveys aided this cause by producing new descriptions of different agrarian forums: land already cultivated but improvable; land not yet cultivated but capable of being so; and land bearing theretofore unknown wealth under its surface. Acting as a lens onto the landscape, survey science helped Virginians interact with their environment as a site of systematic and quantifiable analysis. Rather than passive bystanders, Virginians themselves, the everyday practitioners who helped Rogers and his team, were active agents in the production of scientific means for georgic improvement ends.

The legacy left by Virginia's survey cannot be characterized in a straightforward manner. It is instead best to evaluate it practically and conceptually. In practical terms, the first Geological Survey experienced a distinct dénouement. The state assembly voted in its 1841 session to repeal the survey's funding as of January 1, 1842. This abrupt end was harsh for Rogers and his patrons in the state assembly. The closing down of the near-seven-year project was not, however, unexpected. Rogers knew from constant conversation with his brother Henry that surveys all around the Union were in as much danger by the later 1830s as they were in favor earlier that decade. A broader 1830s economic setting speaks to the state's financial troubles, as an 1837 recession led to tight funding structures in all state budgets during a time when science was not quite the secure investment it might later become. Ohio had cut its survey short in 1838; Indiana and Massachusetts suspended theirs in 1839.[53]

Rogers wanted his most notable legacy to be his authorship of a final, comprehensive report. He considered it "the crowning work of the survey, from which alone a just estimate of its high economical and scientific value can be formed." But he never completed it. Historians of geology have assessed his tempered success with that metric of achievement, noting that the survey's "failure" was its lack of that final report.[54] Throughout the execution of the survey Rogers was also involved in a constant campaign of funding requests through the Board of Public Works in anticipation of having the resources to write that last report, always aware of the tenuous status of each year's allocation. His annual reports alluded to the hoped-for final one, though in his pleas was the implicit awareness that it might never arrive. That hope persisted into the early 1840s as his 1841 report (the last annual one) began with a tone of determination. He kept it short at just eight pages, compared to the 125 pages of 1840; four of those eight were devoted entirely to the case for extended funding. Emphasizing the "value," the "labor," and the "patriotic" purpose of his work, Rogers claimed that he was actually $5,000 under budget for the six-year duration and would need at least two years to properly finish the comprehensive report. He was granted an additional $2,500 for that year, but it was not enough. Increasing tension at the university in Charlottesville—Rogers had been witness to student riots there in the 1830s and early 1840s—and long-term planning with Henry that would eventually take him to Boston prevented the authorship of the report.

Another direct measure of the survey's success would be how much actual land improvement was afforded or caused by the survey itself. This would address an initial basic justification of the project by answering whether or not Virginia's productivity improved. Virginians often thought so, claiming at agricultural society meetings and in the rural press that the new knowledge of agricultural improvement of which survey science was a part had made old lands more productive.[55] Bolstering such qualitative outlooks, census figures from the antebellum period recorded the acreage of farmland along with acreage of "improved" farms, data that might be useful for assessing such scientific influence. State agents likely had such figures in mind—10,360,135 acres of "improved" farmland, as measured in the 1850 census. However, while those numbers were suitable for citizens and politicians of the time, they offer little help to a full analysis today: there was no pre-survey benchmark against which to compare, since the mid-century census was the first to record such statistics. At the same time, assessments of productivity were always imbued with political interest, not least of which were competing assessments from local Southerners and visiting Northerners interested in demonstrating the success or failure, respectively, of the slave-based agricultural system.

A preponderance of conflicting anecdotal and statistical evidence shows that using the survey to argue that Virginians immediately changed their farming practices or that, furthermore, the legacy of the survey was to impose a standard scientific protocol onto land management as measurable by acreage productivity addresses the wrong issues and asks the wrong questions.[56] As with the example of book farming, the survey more helpfully illustrates the availability of new scientific terms and practices, not their ultimate and unquestioned acceptance. The survey also speaks to a new mode of interaction *between* people and their land more than a revolutionary mode of production *on* that land. Its lasting legacy was more subtle than such direct metrics can demonstrate.

State and county surveys contributed to far-reaching conceptual and philosophical legacies influencing how Americans act with and define the land. The state survey left a record of cataloging "the native resources of our common country," in Rogers's words, of producing a summary of the state's natural wealth.[57] It provided this catalog through rhetorical representations in annual reports, with analytical results of hundreds of specimen evaluations, by the visual representation of rock strata and other geological features in diagrams, composite sketches, and other illustrations accompanying the report, and in the cabinet of specimens accrued over the course of the project.

The Rogers survey mapped the territory, providing an account of the state's native and natural resources. It incorporated the active participation of everyday farmers across the state whose interests were not only in obvious personal gain, but were enfolded within the greater mission of cultural improvement. It was part of a transition in which, as Carolyn Merchant has written, "the field had been reconstructed as a laboratory" and the environment could be defined with scientific practices. Rogers represents that reconstruction and the node through which nonspecialist farmers joined the process of quantified redefinition. This was, perhaps, the culmination of the antebellum georgic project.[58]

In more philosophical terms, the survey's execution and results also indicate the prominence of particularistic, materialistic concepts of nature. One major point of the survey work, Rogers said, was that the "researches" contained therein were directed at "the determination of the *composition* and consequent *value* of numerous specimens."[59] The survey established the "value" of soil and fertilizer by directly tying them to their chemically determined composition, a significant achievement and one that relied on a mechanistic reduction during translation from field to lab and from lab to report. Furthermore, by arbitrating the relationship between local citizen and local sample—no longer a jug of water, but now a "specimen" of chemical substances, no longer a piece of clayey marl but 21.5 percent carbonate of lime—Rogers further circulated the authority of scientific mediation for producing practical knowledge of the soil.

Rogers also maintained and elevated his own prominence in the state, a point that goes beyond simple biography and into the realm of scientific agricultural history. His increasing fame suggests a higher degree of visibility for scientific solutions to agroeconomic problems. A writer to *The Southern Planter* in 1841 thought the problems of agriculture then under debate could be resolved if they simply asked "Professor Rodgers [*sic*] to turn the light of science upon this [issue of fertilizer], or any other important subject of agriculture." The next year, another writer explained in an article titled "Agricultural Analysis," "Every farmer should understand the nature and composition of his soils, and may do so with little time, and at a mere trifle of expense." He was referring to analytical instruments Rogers used. The statewide Virginia Agricultural Society formed in 1845 to give further systematic attention to improvement polices, growing from the context of activity shaped by Rogers and county-based activities for several decades by that point. By the 1850s, states were employing agricultural chemists dedicated to analyzing fertilizer samples and other agricultural products.[60]

The state survey re-addressed and expanded issues brought out by the county surveys. These included the means by which citizens understood and worked their land, the tension between local experience-based notions of land knowledge and general, universal definitions of that nature, and the ways in which the surveys laid down conditions of possibility for the later scientific authority that provided practical and relevant knowledge of the environment.

"Knowing" the environment for antebellum citizens was based most directly on working the land, because agriculture was the primary occupation and mode of production. The labor required to plow fields, sow seeds, manure soil, cut and bail hay, harvest crops, and prepare the farm for the next season left little time for disengaged observation or sublime appreciation. Virginians knew their land by expending their energy in the dirt and through the slow pattern of the harvest. Even when they did transcend the rigors of labor—or as plantation masters escaped it outright by overseeing the labors of slaves and tenants and drawing on work-based values for their rhetorical significance alone—their notions of nature were either particularized within local regions, or defined by religious doctrine. The kinship networks that defined most farming communities helped promote a sense of locality and identity, so that extant practices of improvement, like manuring, crop rotation, and experimental planting systems, were common and popular within close-knit religious communities or extended-family towns. Almanacs fit squarely into this kind of agrarian context as guides to structuring daily practice. They held and promoted agricultural knowledge but fell prey to the rising plea for systematization, experimentation, and method the book farming and survey promoters were advocating. By virtue of new and self-described scientifically guided practices, American agrarians were changing the ways they interacted with their lands with means not defined by local knowledge alone.[61]

Codified, universalized definitions of the land remained in tension with the local views. The breadth and depth of the state survey's geological and agricultural analyses characterized a new and different way to see the land. Having state support, broad geographical aims, and a common method and purpose meant that the environment could be described and acted upon in a more totalizing manner. According to the historian Michele Aldrich, surveyors and their supporters made the view that they could provide a "unified, clear picture of Nature" a guiding assumption of the entire state survey era. With this assumption, improvement advocates introduced a scientific element conceived to interpret the soil, the land, and nature in a standardized way. When Robert Mills, for example, an engineer, architect, hopeful mineral prospector, and land-holder in northwestern Virginia, wrote to Rogers asking for a copy of the reports he had heard advance notice of, he was tapping into a new method for improving his lands. He assumed that whatever Rogers had collected in the report would have a direct effect on his property, the need for translation to local conditions being unnecessary. George Perkins Marsh observed much the same in his plea for "emancipation from . . . local causes." Rogers's, Mills's, and Marsh's goal was to make local places scientifically known and thus erase the restrictions of unique and uncodifiable land.[62]

In these efforts, the survey project, a collection of technical activities, became an instrument itself. It both enabled improvement and chronicled it. Survey science offered a complementary way to interact with the land—with analysis, with specimen acquisition, with fertilizing experiments—and so come to know it through different means. The process of codifying nature resulting from this new interaction appears straightforward in retrospect, suggesting a sort of inevitable accrual of measurements, observations, and analyses that leads to a scientifically describable environment. But the process itself reveals nature's resistance to this codification. The broken thermometers, undisciplined networks of communication, damaged roads, thick-brushed hillsides, realities of inconvenience and unreliability were the local natural factors that had to be overcome, ignored, or glossed over to report a nature that *becomes* uniform and thus uniformly improvable.

In retrospect, the local agricultural surveys conducted by county societies and planters might appear as small-scale precedents for the larger surveys. But they were not designed as test grounds for the state surveys any more than the state surveys were formulated as test grounds for the later federal surveys. Even so, they are connected by their conceptual aims and through their practical executions. In much the same way that historians have seen the early nineteenth century as preparing the conditions for a modern scientific worldview—the same one Dostoevsky bemoaned—the agrarian improvement schemes of nonspecialized citizens and their sometime partners the authorized scientific agents were important sites for deploying scientific means for agricultural and environmental ends. The survey was a version of what had been georgic science, a version more suitable to the federal ambitions of the second half of the century than to the end of the prior century, a version that could allow for a later scientifically industrialized agriculture.

CHAPTER 6

Agriculture, Ethics, and the Future of Georgic Science

No society sets itself tasks for whose accomplishment the necessary and sufficient conditions do not either already exist or are not at least beginning to emerge and develop.

—Antonio Gramsci, *Selections from the Prison Notebooks,* 1922

CONVERSATIONS ABOUT AGRICULTURE are also fundamentally about food, and conversations about food are simultaneously about culture. Today's farmers' markets, community-supported agriculture (CSA), food co-ops, organic markets, and movement toward localism attest to the cultural awareness of food choices. With the associated matters of affordability, access, security, and health, terms like "sustainable agriculture," "food miles," and "environmental justice" have likewise become part of a lively debate. Food in the twenty-first century, that is, has come to the center of environmental conversation, bringing to mind Wendell Berry's observation that "how we eat determines, to a considerable extent, how the world is used." At a broader political scale, when the United States Department of Agriculture (USDA) proposes its Farm Bill every five years or so, it speaks not just to the production and distribution of food or the isolated practice of agriculture but to a premier example of how humans will treat nature.[1]

This book is about food, but in an indirect sense. It more squarely addresses the soil in which food grows, how we know what it is, and why we do so. These are questions as much cultural as material. I have peered before and beneath the end product of an agricultural process—the food—to examine how Americans long ago developed new ways to know what they were doing *in* that process. Soil was the central locus of the early

American Republic, where politics, practice, and culture met; soil science was soon its actionable item, the thing one could do for the soil and for society. The means for actively applying principles of soil science across governed areas—say, Virginia, or New York—would become crucial issues in the decades before the Civil War and have since remained crucial issues throughout the world. Our modern agricultural system dramatically changed the way humans produce food and, with that, a crucial way we use the natural world. That system was built on concepts of soil identity developed across the decades of the early Republic.

By the mid-nineteenth century the sciences of agriculture had a place in rural American culture. They were a visible, viable, though still unstable lens or prism through which citizens could view their land. For centuries and perhaps millennia before, animal manure had been a well-understood source of nutrition for the soil. Along with that, gypsum, lime, and various other fertilizers, mineral and organic, had been part of continuing efforts toward achieving soil nutrition and fertility. Those practices of augmentation, manipulation, and fertilization were understood through the virtue-based experience and moral traditions of agrarian life, remaining mostly uncodified in text and thus centered in fields, not labs, of knowledge production. In the late 1840s, Liebig's theory of soil science became a prime reference point in the rural press and, soon, the basis for new agricultural policies, but Liebig sits at the end of this book, not the beginning. He represents the accomplishment of scientific credibility by the mid-nineteenth century, not its explanation. How then did such familial and experiential means for knowing the soil become "improved" by a new set of scientific practices? Understood within an agrarian cultural context that valued the virtue of experience, science and agriculture were brought together neither passively, inevitably, nor without resistance, but as shaped by an ideology of work, knowledge, and citizenship.

Just as interesting as where that combination of science and agriculture came from is what helped make it possible. This book's examination of agricultural science's place in the early Republic reveals the production of an important set of conditions that allowed later scientific developments across the land to have cultural meaning and environmental significance: forms of communication, precedents of organization, field-tested modes of analysis, a tradition of improvement and experimentation, the long-standing search for solutions to soil exhaustion, increasingly mechanistic philosophies of soil composition, a market force to drive all of these, and a unique American political and agricultural environment in which the above could take shape. The case offers insight into the shifting place of science in agricultural practices, but it furthermore suggests the process by which humans developed new ways to interact with nature in a broader sense. In that larger story, the rise of scientific practices on and for the soil shows how new ways of interacting with nature go hand in hand with new ways to understand and define the very dirt under our feet. Putting that circulation of agricultural science in the context of improvement-minded agents in the early Republic better locates agrarian American culture in a post-Enlightenment setting, better equips us to recognize how everyday citizens came to treat scientific practice as a legitimate means of interacting with their lands, and better develops a picture of the practice-based synthesis of morality, materiality, and theory.

The mid-century serves as my end point not because an inexorable logic of agricultural practice was then set in place, but because by the eve of the Civil War the cultural basis that allowed later developments to have meaning was gaining credibility. I would not claim from this that scientific practices of fertilization, soil analysis, or even book farming were dominant in all regions of the highly sectionalized nation in the same way and to the same degree. More pertinent, rather, is my argument that scientific practices were not only circulating, but were doing so as a subset of cultural practices lived and debated by rural and agrarian communities. Though the full rise of industrial agriculture is more properly a story of the early twentieth century, when, as the historian Deborah Fitzgerald has shown, the farm became a factory, it should indeed make reference to those conditions from the mid-nineteenth century. And though the place of chemistry in agricultural history has more frequently been set as part of a twentieth-century saga, this book helps show the early nineteenth-century location of chemical practices and insights that would later become more complex. The government formed the USDA in 1862, in part a political product of Lincoln's federal success. But it was also a conceptual product of unifying principles of soil identity and management from the prior decades. The consequences of introducing the sciences of agriculture to improvement-era America thus go beyond the mere temporal scope of those early nineteenth-century decades, as they stand for some of the necessary conditions that Gramsci observed always exist before the emergence of a new task, in this case not just the redefinition of soil but the scientization of nature.[2]

Notes from the Ground has in large part been concerned with highlighting the circulation of experimental practices outside (and in conjunction with) the sphere of professional science in places as disparate as Maine, Pennsylvania, and Virginia. My focus on Virginia, especially in the latter half of the work, should be read as looking to the hard case, one that works beyond historical stereotypes that paint the South as backward and dogmatically resistant, and reveals instead that, even in a slave-based economy trying to further a failed social system and overcome the soil-exhausting legacy of a fading tobacco regime, we find active practices of systematic and scientific investigation. Advocates and opponents argued then and still argue now over the value of science for farming; the basis of authority for defining what the soil is, however, shifted thereafter from the dominant culture of nonscientists to the rising epistemic culture of a professional scientific and government-backed class. In the early American Republic, georgic science helps show, the debate was not simply about whether or not to accept and promote science. It was more truly about which science would fit the improvement trope, as developed how, and by whose direction. Even today, we cannot approach agricultural issues with a binary of "science" or "no science." Instead, we see arguments made that promote a more ecologically informed means for working the land or a more reductively industrial logic. The entirety of our modern scientific worldview cannot be traced to the activities of the early American Republic's agrarian class, but this historically situated example of rural science and agricultural improvement helps illustrate what it takes to make a scientific worldview.

At the same time, this study was born of and speaks to a twenty-first-century setting at which my final comments are directed. I mean this in two directions, one being toward the way we study relationships now between science and the environment in the past, the other toward how we might promote a different emphasis in future environmental action itself. On the former, this book offers a corrective for historical scholars by suggesting that the rise of a world defined scientifically has more to do with the everyday workings and goals of nonscientists than has been generally acknowledged. Resistance to those developments was not necessarily or always irrational or ignorant; experiments and debates about soil treatment did not always simply follow dictates of the latest scientific theoretician. Additionally, *Notes from the Ground* shifts attention from idealized pastoral to hard-worked agrarian settings of the early Republic as a locus for origins in environmental thought. After recovering the georgic ethic from Classicist circles for its descriptive and explanatory utility in agrarian studies, I have also sought to enunciate an environmental history of the early Republic that looks to sites of work and development, not sublime appreciation or disconnected commentary. By calling attention to the value that agricultural work held during the early national period, I maintain that the georgic ethic best expresses the relationship Americans had to their environments during that period when agriculture was central to political economy, cultural identity, and national exceptionalism. With labor as the center point, the georgic ethic usefully forces the connection between work in the environment and the moral dictates to strengthen social fabric by reference to lived experience. One final implication of this book's narrative is that it resituates how we assess the historical combination of science and the environment by attention to working practices while further pushing the basis from which we might deal with that combination in the future.

Perhaps more significant in this story's meaning for the present, in my view, is the potential for a renewed kind of environmental georgic ethic that aims at a reconfiguration of that combination. In a recent translation of the *Georgics*, Janet Lembke argues that Virgil's composition is "a poem for our time" because it advocates "caring without cease for the land and for the crops and animals it sustain[s]." I read the *Georgics* and the ethic it expresses as relevant to our age for reasons of care as well, but for reasons that do not rely upon a return-to-the-land philosophy. Rather, the georgic ethic is an ethic for our day because it places the connection of humans to the land back at the center of an agricultural ethic while emphasizing that environmental knowledge is constituted in practice. As the agricultural writer Laura Sayre has put it, "Georgic . . . *is what we need now.*"[3]

My aim here has been to tie this georgic sensibility to the history of scientific practices on the land. I believe that we can also promote the same terms of a georgic science today to undergird building new scientific and technological practices that reveal rather than conceal our connections to the land. Across this book, I have used georgic science to help emphasize that new soil manipulation practices gained credibility by reference to prior experience, but also to reintroduce a view of the relationship between humans and the land that relies upon sustained and intentional interaction, not disconnection. My hope is that new formulations of a georgic science can be fostered, offering an integration of moral and material improvement in the coming years of debate about sustainable agriculture and ecological health. I see this happening already with a chorus of environmental humanities scholars seeking more place-based narratives that emphasize the value of knowing a place through lived cultural experience; environmental and agricultural advocates seeking more attention to local and regional foodsheds and policies that allow community (and place-based) values to factor into the conversation on improving environmental and human health; and politicians along with them who need more help crafting those policies so that embedded in them are the values drawn from the epistemic practices of everyday life.

In particular, for those engaged in public debate about land and farm use the georgic's cultural and epistemological possibilities offer an alternative to the pastoral ethic often distinguished by analysts of nineteenth-century America, where the relevant contrast is a difference between understanding life as labor and life as leisure. With a georgic approach, we can begin conversations about environmental policy and action by assuming humans are active members of that environment, not passive observers. Drawing out the georgic helps show that values based on interaction dominate the way people view their surroundings and dictate what they do with and in those surroundings. Connection, interaction, and labor stand as the basis for georgic environmental sensibilities in our time just as they did with my understanding of science in the American agrarian world of the past. The georgic gives us the terms to highlight the observation that knowledge is constituted in practice; if our practices are to be scientific, then we should consider how better to use them as means for engagement, not concealment.

The agricultural researcher Paul Richards, studying indigenous agricultural revolutions in West Africa, found that farmers' experience of working the land provided the basis for new agricultural knowledge. "The changes in agriculture," he wrote, "were brought about by farmers, not scientists." The experience preceded and led to scientific knowledge; it was constituted through lived culture first, codified and put into disembodied principles second. In the same way, the rural sociologist Jack Kloppenburg has written that farmer-based knowledge should play a larger role in the construction of new agricultural policies, where "such knowledge is local in the sense that it is derived from the direct experience of a labor process which is itself shaped and delimited by the distinctive characteristics of a particular place with a unique social and physical environment."[4]

I highlight and want to help open up the same space Richards, Kloppenburg, and a range of others have observed to consider the epistemic importance of new agricultural practices in America and also, and perhaps more crucially, across the world.[5] This requires more stark attention to knowledge-making practices as they follow from forms of human-land interaction. Sustainability initiatives of the new century proceed at the peril of short-sightedness if they do not consider the cultural conditions that make such new practices and interactions possible. Viewing sustainable agriculture as a new form of agricultural improvement can help frame the point that decisions about soil treatment can be thought about as practices both moral and material. Toward that end, this book suggests that the possibilities for sustainable agriculture will be better produced if defined through a georgic framework that assumes human work on the land and seeks to keep the experience of that work locally relevant and meaningful.

In the end, the lesson of this book is not to go back to a prior era where most of us would have been farmers, but to suggest that when working to promote new, more sustainable agricultural systems we should pay closer attention to how those systems can connect, rather than disconnect, us from that land. We might also pay more attention to the cultural basis from which such debates about new agricultural systems arise, finding that preexisting cultural commitments explain better why some people resist new practices on the land while others accept them. The situation today and in the future is as much about cultural authority as it is about technical acceptability, just as it was in the nineteenth century. Future arguments about the conceptual and material bases for seeing nature in a new, perhaps more ecologically healthy way can thus gain strength and plausibility by reference to the workings of antebellum agents who first sought to use science to define their own soil and then the modern agricultural world.

NOTES

ABBREVIATIONS

AF	*American Farmer*
FR	*Farmer's Register*
GF	*Genesee Farmer*
NEF	*New England Farmer*
PB	*Plough Boy*
SP	*Southern Planter*
VHS	Virginia Historical Society

INTRODUCTION

1. Jefferson, *Notes on the State of Virginia,* 48.
2. Pollan, *The Omnivore's Dilemma,* 146.
3. McNeill and Winiwarter, eds., *Soils and Societies;* Gregory Cushman, Personal Correspondence, November 6, 2007.
4. Eppes, in his diary, as quoted in Kirby, *Poquosin,* 37; Franklin Minor, "An Address Delivered before the Virginia State Agricultural Society, at its Annual Fair," *SP* (December 1855): 368–380, on 376; Ewbank, *The World a Factory,* v; Campbell, before Congress, as quoted in Philips, "Antebellum Agricultural Reform, Republican Ideology, and Sectional Tension," 817–818; Nichols, *Chemistry of the Farm and the Sea,* 45.
5. Dowling, *Poetry and Ideology in Revolutionary Connecticut,* 36.
6. Hamilton, "Report on Manufacturers [1791]," as reprinted in Smith and Clancey, eds., *Major Problems in the History of American Technology,* 111; Adams, as quoted in Loehr, "Arthur Young and American Agriculture," 46.
7. Several authors in recent years have helped broaden the agroenvironmental

character of the early Republic and eighteenth century before that. Stoll, *Larding the Lean Earth,* Valencius, *The Health of the Country,* and Donahue, *The Great Meadow,* provide three excellent examples.

8. White, "Are You an Environmentalist or Do You Work for a Living?," 171. For a broader development of the point see White, *The Organic Machine.* This notion of work-as-producing-knowledge is consistent with the pragmatist theories of John Dewey and Charles Sanders Peirce showing that doing-is-knowing. As related to that, my interest in the relationships between human and nonhuman nature is not just a way to study environmental history, but simultaneously part of a larger philosophical approach to science and society. To bridge the gap between the connection of practice and ideas in science studies and the payout for environmental historiography, I focus on the relational meaning between nonhuman nature and the human practices in and on it. In this regard, James Scott's (1998) treatment of "mētis," a form of "practical knowledge" meant in contrast to "more formal, deductive, epistemic knowledge" (6), helps give substance to the experience-knowledge relationship. See Scott, *Seeing Like a State,* especially 11–52 and 309–357. Also see Thompson and Hilde, eds., *The Agrarian Roots of Pragmatism,* for more on Peirce, Dewey, and the relationship between pragmatic thought and agrarian principles. More recently, Sachs, *The Humboldt Current,* argues that the very origins of modern environmentalism should be understood in relation to practices of exploration and practical interventions in the landscape.
9. Sweet, *American Georgics;* McCoy, *The Elusive Republic.* For an analysis of the similarly multifaceted improvement ethic already under debate in Britain earlier in the eighteenth century, see Chaplin, *An Anxious Pursuit,* 23–37. In the twentieth century, Wendell Berry's work forwards a neo-georgic ethic in his call for work-based agrarian virtue. Among a lifetime of notable works, see Berry, *The Unsettling of America.*
10. On this lack of attention to Virginia specifically, see Kirby, "Virginia's Environmental History."
11. A roster of environmental and scientific historians from the late twentieth century offer conclusions similar to their predecessors from the nineteenth. See in particular Merchant, *Ecological Revolutions,* 206 and 220, and, for more on the nineteenth-century context of American science, Rosenberg, *No Other Gods;* Oleson and Brown, eds., *The Pursuit of Knowledge in the Early American Republic;* Marcus, *Agricultural Science and Its Quest for Legitimacy;* Rossiter, *The Emergence of Agricultural Science;* Kohlstedt, *The Formation of the American Scientific Community;* Greene, *American Science in the Age of Jefferson;* and Bruce, *The Launching of Modern American Science.* Stilgoe, *Common Landscape of America,* also discusses the increasingly dominant ideals of productivity and efficiency in the practices of agriculture.
12. Stoll, *Larding the Lean Earth,* 168.

CHAPTER 1. DISTINGUISHING THE GEORGIC

1. Spurrier, *The Practical Farmer,* 28 and 56.
2. Home, *The Gentleman Farmer, iii;* Washington, "Address to U.S. Congress, 7 Dec 1796," in *Diaries of George Washington,* 15. On Virgil, see Lembke's notes in her translation of *The Georgics.*
3. Spurrier, *The Practical Farmer,* 14–16.
4. Nash, *Wilderness and the American Mind;* Marx, *The Machine in the Garden;* Buell, *Environmental Imagination,* 439; and Worster, *Nature's Economy,* 3–55.
5. Snell, as quoted in Martindale, "Green Politics," 110.
6. The Thoreau scholar James Tillman, writing about *Walden* and Thoreau's mixture of pastoral and georgic ideals, casts the pastoral as "essentially characterized by *otium,* pleasure, and the enjoyment of poetry and contemplation" while the georgic "is characterized by labor, painstaking forethought, and respect for science and common sense." In Tillman, "The Transcendental Georgic in *Walden,*" 137.
7. See Sweet, *American Georgics.* Though limited in number, scholars from a range of disciplines have found the approach useful, emphasizing the georgic role of work in environmental studies. In addition to Sweet, see, for example, Buell, *Environmental Imagination;* Dowling, *Poetry and Ideology in Revolutionary Connecticut;* Grammer, *Pastoral and Politics in the Old South;* Marx, "Pastoralism in America"; Sayre, "Farming by the Book"; and Schama, *Landscape and Memory.*
8. For entry points to Berry's work, see Berry, *Recollected Essays, 1965–1980;* idem, *The Unsettling of America;* and Smith, *Wendell Berry and the Agrarian Tradition.* Also see Jackson, *New Roots for Agriculture,* and, on pragmatist-oriented environmental ethics, Light and de Shalit, eds., *Moral and Political Reasoning in Environmental Practice;* Minteer, *The Landscape of Reform;* Thompson, *The Spirit of the Soil;* and Thompson and Hilde, eds., *The Agrarian Roots of Pragmatism.*
9. This labor-based mode of orientation also meshes well with recent and innovative environmental historiography. In particular, Richard White's environmental narratives have encouraged attention to the modes of interaction with our surroundings—like labor, or work—as a way to explore more realistically how human concepts and attitudes about nature develop and become entrenched. All people live in nature, but our environmental ethics do not always acknowledge the connections we have with nonhuman nature as active agents. White's work, with *The Organic Machine* as a prime example, draws our attention to twentieth-century developments and the issue of industrialized nature. See White, *The Organic Machine,* and idem, "Are You an Environmentalist or Do You Work for a Living?" Also see Scott, *Seeing Like a State.*
10. See Virgil, *Georgics,* ed. Mynors, for further modern commentary on the poem. Also see Martindale, ed., *The Cambridge Companion to Virgil,* for the breadth of Virgilian interpretation, and Batstone, "Virgilian Didaxis," for par-

ticular attention to the Georgics and a concise commentary on modern exegeses of the poem.

11. Sayre, "Farming by the Book," 10.
12. Virgil, *Georgics,* trans. Lembke, 22. See lines 226–258 for hands-on "guidelines on how to recognize each type" of soil.
13. For the recent reconsideration of Young's influence, see Brunt, "Rehabilitating Arthur Young."
14. Home, *The Gentleman Farmer,* iii; Cochrane, *A Treatise Shewing the Intimate Connection that Subsists between Agriculture and Chemistry,* 3–4; Hunter, *Georgical Essays.* Hunter was also a Fellow of the Royal Society who wrote numerous pieces on medicine and agriculture. For further details on Hunter (1729–1809), see the *Dictionary of National Biography,* vol. 10, 283–284.
15. Spurrier, *The Practical Farmer,* 18.
16. Chaplin, *An Anxious Pursuit,* provides further detail about the eighteenth-century improvement ethos. For a view of the shifting relationship between technology and improvement beginning in the same period, see Marx, "Does Improved Technology Mean Progress?" Marx's central move in this brief article is to add the question, "progress toward what?" and, relatedly, "progress for whom?" to any discussion of technology and improvement. The discussion in this book takes a similar tack with respect to agricultural science. Also see Marx and Mazlish, eds., *Progress.*
17. Marcus, "The Wisdom of the Body Politic," 7. See also Merchant, "Farms and Subsistence."
18. The French thinkers Anne-Robert-Jacques Turgot and François Quesnay were representative Physiocrats. Jefferson's thoughts on Physiocracy come across most clearly in his correspondence with Pierre Samuel Du Pont de Nemours, in *The Correspondence of Jefferson and Du Pont de Nemours.* For more on the Physiocrats, see Elizabeth Fox-Genovese, *The Origins Of Physiocracy,* and Baker, *Condorcet.*
19. Hamilton, "Report on Manufacturers [1791]," as reprinted in Smith and Clancey, eds., *Major Problems in the History of American Technology,* 111.
20. Federalists proposed a different emphasis on the political economic triumvirate of agriculture, commerce, and manufacturing when they suggested the view, usually understood as the expression of Alexander Hamilton, Tench Coxe, and John Adams, that the manufacturing sector was of greatest importance. But even then, manufacturing was the end point of a conversion process from agricultural and natural resources to fabricated ends and not a mode of production that denied the importance of the farm.
21. Jefferson's apparent retraction of this view in the early 1800s, writing in 1816 that "we must now place the manufacturer by the side of the agriculturalist," was not truly a wholesale reversal. It was, given the new post–War of 1812 era, a call to shift emphases within the agriculture-commerce-manufacturing triumvirate of political economy. As quoted in Smith and Clancey, eds., *Major Problems in the History of American Technology,* 119.

22. By tying Jefferson to the georgic ethic, my analysis stands apart from traditional evaluations of Jefferson's pastoral identity. Leo Marx specifically distinguishes between an agrarian and pastoral identity when he discusses Jefferson's moral and political concerns. Jefferson, Marx (1999) writes, was concerned primarily with "rural virtue" (a sign of the pastoral) and not agrarianism (a mere economic descriptor) (125–128). In the cases I address, a georgic conception better explains the non-distinction at the time between virtue and economy, between moral improvement and material progress. Marx also ends *The Machine in the Garden* with a final chapter, "Two Kingdoms of Force," in which he poses a contrast in American literature between the natural, organic, life-affirming pastoral and abstract, dehumanizing technology. The georgic is also a useful bridge between those two supposed poles, allowing for the acceptance of technological means (as work) and the life-affirming substance of agricultural virtue (as work). Meikle, "Leo Marx's *The Machine in the Garden*," has provided recent commentary on Marx's influential writing that touches on a similar point.
23. Even Thoreau's *Walden* has been characterized in terms of its georgic connotations—though Thoreau quite famously eschewed the rigor of a schedule, favored the leisurely saunter, and promoted "useless knowledge" as a counterpoint to the frivolities of spiritually empty labor, he did not avoid or bemoan the merits of work itself. As one scholar has written, Thoreau seemed "particularly responsive to the georgic ideal whenever he makes his own labors and piety an exemplum of true husbandry" (Tillman, p. 137).
24. Deane, *The New England Farmer; Or Georgical Dictionary*, 265; Hunter, *Georgical Essays*. See also Samuel Austin Allibone, *Critical Dictionary of English Literature*, 922. The Reverend Dr. John Walker's "Georgical Lectures" of 1790 are discussed in Withers, "A Neglected Scottish Agriculturist," and idem, "Improvement and Enlightenment." Notably, Baltimore's *American Farmer* had offered a recommendation of Hunter's essays to its readers, there following notice given by the Philadelphia Agricultural Society. See "Notices for a Young Farmer," *AF* (July 2, 1819): 105.
25. See "Samuel Deane," *Dictionary of American Biography*, vol. 4, 172–173.
26. Adams, *The Agricultural Reader;* quote from Deane, *Georgical Dictionary*, 213. In colonial America, a new husbandry was already being developed with increased attention to mechanical implements. See Eliot, *Essays Upon Field Husbandry in New England*. In England, see Tull, *The Horse-hoeing Husbandry*.
27. Agricultural improvement in general was popular among men of leisure, wealth, and education. The fashion of agriculture among the gentry—in England and colonial and early Republic America—and the political economy of agrarianism both made common reference to the virtues of agriculture. See Sweet, *American Georgics*, especially "Ideologies of Farming," 97–121, and John Fea, "The Way of Improvement Leads Home," for an analysis within the Republic of Letters.
28. Spurrier, *The Practical Farmer*, 28 and 56.

29. Deane, *Georgical Dictionary,* quotes from 211 and 260, but also see 1–4, 9, 210–213, 243–244, 250–262, 264–66, 400–403, and 483–485 for other entries confirming the front-page advertisements' claim that the compendium suited the "practical as well as the scientific agriculturalist."
30. See Cochrane, *A Treatise Shewing the Intimate Connection that Subsists between Agriculture and Chemistry,* and Samuel Parkes, "Extract from *Parke's* [sic] *Chemical Essays,*" *NEF* (May 29, 1824): 349. For more on Parkes see Kurzer, "Samuel Parkes." Also note the chemically informed work of Home, *The Principles of Agriculture and Vegetation;* Home, *The Gentleman Farmer;* and Fothergill, "On the application of chemistry to agriculture." The historian Jan Golinski characterized Home's view as suggesting that "improving farmers were natural patriots, because in seeking to raise their own productivity they were simultaneously serving national and civic needs." Golinski, *Science as Public Culture,* 35. William Cullen, one of the foremost chemists of the Scottish Enlightenment, prepared an unpublished manuscript called "Lectures on the chemical history of vegetables" as early as 1755. See Withers, "William Cullen's Agricultural Lectures," for further citations to Cullen's work. Golinski, *Science as Public Culture,* 32–35, also discusses the correspondence between Home and Cullen.
31. Jefferson to Thomas Jefferson Randolph, 3 January 1809, as quoted in Miller, *Jefferson and Nature,* 45.
32. Ewell, *Plain Discourses on the Laws or Properties of Matter,* 22.
33. Eli Davis, "For the American Farmer," *AF* (November 12, 1819): 265.
34. See Cooter, "Separate Spheres and Public Places"; Secord, "Science in the Pub"; Shapin, " 'Nibbling at the Teats of Science.' " Burnett, *Trying Leviathan,* offers a notable example of working around divisions of public and philosophical (scientific) in a case also set within the early Republic. He considers the views of natural philosophers, whalers, artisans and manufacturers, and the general public during a court case about the taxonomic status of whales.
35. "Introduction," *The Diaries of George Washington,* xxxi.
36. George Washington, *Letters from His Excellency George Washington, to Arthur Young. . . .* The quote is from Fussell, "Science and Practice in Eighteenth-Century British Agriculture," 15. Also see Loehr, "Arthur Young and American Agriculture." Young had come to prominence by writing about agricultural practices in his popular Tours. In the 1780s, Young founded the periodical *Annals of Agriculture,* which he edited through 1808. In the early 1790s, with John Sinclair (president of the Board of Agriculture), he was one of the best known agricultural writers in England. Individuals as wide-ranging as King George III and Jeremy Bentham published in the *Annals of Agriculture.* See Young, *A Six Months Tour through the North of England;* idem, *Travels During the Years 1787, 1788, & 1789;* and Brunt, "Rehabilitating Arthur Young."
37. See Washington's diary entries of 21 April 1787, 25 October 1787, 15 March 1788, 18 March 1788, and 4 December 1788, for his commentary on fertiliz-

ing and correspondence with Arthur Young; also, letters from John Sinclair to George Washington, 29 March 1797, 15 July 1797, and 6 June 1798, and from Washington to Sinclair on 6 November 1797 and 28 November 1797. All correspondence in *The Diaries of George Washington.* For Jefferson, see Betts, ed., *Thomas Jefferson's Garden Book, 1766–1824* and idem, *Thomas Jefferson's Farm Book.* For more on the relation between agricultural improvement and fear of westward migration, see Stoll, *Larding the Lean Earth.*

38. Golinski, *Science as Public Culture,* 11–49, elaborates the context of those studies. For more on the association of georgic values with agricultural chemistry, see Withers's articles on the Scottish Enlightenment: Withers, "A Neglected Scottish Agriculturist"; idem, "Improvement and Enlightenment"; and idem, "On Georgics and Geology." Fussell, "Science and Practice in Eighteenth-Century British Agriculture," draws out the British setting of science and improvement.

39. On his consideration of Virgil's *Georgics,* see Richard Peters, "Usefulness, to Husbandmen, of attending to natural Phenomena, by Richard Peters." *Memoirs of the PSPA* 3 (1814): 337–352.

40. Thomas Jefferson, March 16, 1816, in *Thomas Jefferson's Garden Book,* 199. As a commentary on a lack of political sectional specificity in the late eighteenth century, I should note that these experiments were not relegated only to the South and mid-Atlantic either—as one example, Robert Livingston presented the results of his experiments in the 1790s with "calcareous and gypseous earths" and then "gypsum and lime" to the Society of the Promotion of Agriculture, Arts, and Manufactures of New York, of which he was president, and in their Proceedings. See *Transactions of the Society for the Promotion of Agriculture, Arts, and Manufactures of New York* (1801): 34–56 and 330–334. Like Washington, Livingston also corresponded with Arthur Young. See the reprint of his letter in the *Transactions of the Society for the Promotion of Agriculture, Arts, and Manufactures of New York* (1801): 163–167.

41. Washington to John Sinclair, 6 March 1797, and Richard Peters to Washington, 12 May 1796, in *The Papers of George Washington, Retirement Series.* Also, see Wines, *Fertilizer in America,* 15. The American-British nexus ran straight through Peters and the PSPA, as when Washington personally delivered a gift from Arthur Young of Young's six-volume *Annals of Agriculture* to the Society. See Baatz, *"Venerate the Plough,"* 119.

42. Jefferson, in fact, was so enamored of Binns's work that he sent John Sinclair copies in England. Sinclair wrote back to say the plaster did not work there; he conjectured from that failure that the plaster worked by "attracting moisture from the atmosphere," which was already plentiful in the soil of England. Letter to Jefferson from Sinclair, 1 January 1804, in *Thomas Jefferson's Farm Book,* 197. Under his influence, and with Binns's local experience recommending it, Jefferson literally bought tons of plaster over the years from dealers in Baltimore, continuing to encourage his friends in the practice.

43. See Spurrier, *The Practical Farmer,* v and x.
44. Ibid., iii–iv; Ewell, *Plain Discourses on the Laws or Properties of Matter,* was likewise dedicated to Jefferson. Like his contemporaries, Ewell saw his work as a discourse on modern chemistry "connected with domestic affairs," believing that "agriculture is the most intimately connected with chemistry" (22).
45. For background, see Goodman, *The Republic of Letters;* Ferguson, *The American Enlightenment;* Ostrander, *Republic of Letters,* chapter 1 in particular; and Daston, "The Ideal and Reality of the Republic of Letters in the Enlightenment."
46. Sayre, "Farming by the Book," 27.
47. Shalhope, *John Taylor of Caroline.* Shalhope is representative of Taylor scholarship in that he presents Taylor as a pastoral figure.
48. Taylor, *Arator,* 313.
49. The classic work on the subject of soil exhaustion is Craven, *Soil Exhaustion as a Factor in the Agricultural History of Virginia and Maryland.* Kirby, ed., *Nature's Management,* xiii–xxv, provides concise commentary on the literature on soil exhaustion, as does Worster, ed., *The Ends of the Earth.*
50. As Kirby, *Poquosin,* has noted, after the agrarian crisis of the 1790s, Taylor "linked his agrarian politics with agricultural reform" (67). Stoll, *Larding the Lean Earth,* writes that Taylor "wrote a hard pastoral, insisting on agriculture as moral instruction, not Arcadian escape" (71). Also see Grammer, *Pastoral and Politics in the Old South,* 33–35, and Bradford, "A Virginia Cato," 36–37. Though initially he published anonymously, Taylor signed the updated version of 1814. Four more editions were printed, concluding with Edmund Ruffin's 1840 reprint in *FR,* before it was again published (as the seventh edition) as a historical text in 1977. For the agrarian crisis of the 1790s, see Sellers, *The Market Revolution,* 17–18 and the notes therein.
51. The relation between agricultural literature and civic organizations was also tied through the later rural press. For example, Baltimore's *American Farmer* had the Agricultural Society of Virginia's members subscribe. Albany's *Cultivator,* founded in 1834, was the mouthpiece of the New York Agricultural Society.
52. Daniel Adams, "An Address Delivered before the Hillsborough Agricultural Society at their Cattle Show and Fair, at new Boston, Sept. 22, 1825," *NEF* (December 2, 1825): 148–150. Italics in original.
53. See Baatz, *"Venerate the Plough,"* for a treatment specifically of the Philadelphia Society for the Promotion of Agriculture and its publishing ties; for general background, see Gray, *History of Agriculture in the Southern United States to 1860,* vol. 2, 779–810; Demaree, *The American Agricultural Press;* and Gates, *The Farmer's Age.*
54. Grammer, *Pastoral and Politics in the Old South,* 42.
55. Taylor, *Arator,* 53.
56. Bradford, "Introduction to *Arator,*" 37.

57. This is not to say Taylor and Davy were writing the same thing, in the same style, and with the same skill, but both works were the end result of years of reflection on agricultural attention. The fact of their quite different cultural settings and biographies is instructive for recognizing the public popularity of agrochemical pursuits.
58. With *Arator* behind him, he then began to turn his attention to the future, participating in the nascent agricultural societies of his home state by addressing the ever-increasing demands for local improvement. He had disagreed with the land-grabbing justification for the War of 1812, thereafter becoming further attentive to the needs for local improvement as a way to conserve social stability in the East, rather than promote emigration to the fertile West.
59. John Taylor, "President's Address to Agricultural Society of Virginia," *Proceedings of the Agricultural Society of Virginia* 1 (June 8, 1818): 28–33, on 28. He further noted, "The chief obstacle to the success of [agriculture] in Virginia, in my view, is the morbid aversion to writing on that subject for publication." Taylor's approach was well established. As Sweet, *American Georgics,* notes, as early as 1758 a writer using the pseudonym "Agricola" recommended "a course of improvement that agrarian writers would soon develop in greater detail. . . . He proposed that farmers form agricultural societies 'to converse about every method of improving the lands' and 'to make new experiments,' which might be reported in periodicals such as the *American Magazine*" (100). This plea was repeated nearly every time a new civic organization formed to improve the common lot of its members. The greater social prominence and local authority of Taylor and the early national context of his address, in particular, give his similarly styled ideas a stronger and, as such, different legacy.
60. John Taylor, "The Necessities, Competency and Profit of Agriculture," *Proceedings of the Agricultural Society of Virginia* (October 19, 1818): 79–87. Wilson Carey Nicholas, "Address to the Agricultural Society of Virginia," *Proceedings of the Agricultural Society of Virginia* (March 10, 1818): 25–28, announced in his address that the agricultural society was valuable as a source for local knowledge, while providing the credibility (through organization) that locality did not have.
61. Spurrier, *The Practical Farmer,* 15.

CHAPTER 2. "THE SCIENCE OF AGRICULTURE AND BOOK FARMING"

1. The secondary literature on book farming is vast. Yet almost unanimously, none of those scholarly works address the social dimension of science, instead taking it as given that science was a force on the horizon, inevitably to be accepted and utilized. To begin with, see Demaree, *The American Agricultural Press;* idem, "The Farm Journals, Their Editors, and Their Public, 1830–1860"; Gray, *History of Agriculture in the Southern United States to 1860;*

Gates, *The Farmer's Age;* Abbott, "The Agricultural Press Views the Yeoman"; Danhof, *Change in Agriculture;* Rubin, "The Limits of Agricultural Progress in the Nineteenth-Century South"; Marti, "Agricultural Journalism and the Diffusion of Knowledge"; Bullion, "The Agricultural Press"; and McMurry, "Who Read the Agricultural Journals?"

2. "Agriculture," *AF* (1819): 210, and "Agricultural Improvement," *AF* (1831): 359–360.
3. Anti-Philosopher, "Philosophy," *SP* (1842): 177.
4. Anonymous contributor to *The Northwestern Farmer,* as quoted by Throne, " 'Book Farming' in Iowa," 117.
5. Grammer, *Pastoral and Politics in the Old South,* 41; Spurrier, *The Practical Farmer,* 18.
6. Chapter 3 addresses the different philosophies of nature associated with these forms of interaction more directly, philosophies that follow or are led by the use of scientific analysis to interpret the land.
7. "Agriculture," *AF* (1819): 209–210.
8. By comparison, *The Richmond Enquirer,* an influential urban paper, had 5,000 subscribers in 1829. Amber, *Thomas Ritchie,* 121. Rural press circulation statistics are from Demaree, *The American Agricultural Press,* 36, and Danhof, *Change in Agriculture,* 56. Demaree also notes that *The Cultivator,* founded in 1834 and published in Albany, began with 15,000 subscribers. Lewis Gray, *History of Agriculture in the Southern United States to 1860,* 788, notes that the "first purely agricultural paper in the United States [was] the *Agricultural Museum,*" established in 1810 in Georgetown. It lasted but two years. The classic work on the rural press is Demaree, *The American Agricultural Press.* Idem, "The Farm Journals, Their Editors, and Their Public, 1830–1860," provides a succinct review of the salient themes in his book. McMurry, "Who Read the Agricultural Journals?" offers the most recent useful account of the meaning of the press, in the process providing references to the main secondary sources dealing with agricultural journalism before the Civil War.
9. John Skinner of *The American Farmer* was grateful to have members of the Agricultural Society of Albemarle (ASA) as subscribers—good for business and good for the diffusion of farming knowledge—so he sent complimentary copies of his paper to the attention of James Madison, then president of the ASA. At an 1819 meeting, the ASA minutes noted, "In consequence of the polite offer of Mr. Skinner . . . it was further resolved, that the Society would make use of the *American Farmer,* as a medium of communicating its proceedings to the public." *Minute Book of the Agricultural Society of Albemarle County,* Sheet 4, Mss4.
10. After his editorship ended at *American Farmer* in 1829, Skinner published and edited *The American Turf Register and Sporting Magazine* from 1829 to 1839, edited a series published by Horace Greeley and Charles McElrath called *The Farmer's Library and Monthly Journal of Agriculture,* and then re-

turned to the agricultural press to edit *The Plough, the Loom, and the Anvil* from 1848 until his death in 1851. He also had a prominent interest in Latin American affairs. Connections in South America first introduced him to guano—bird dung, a highly effective fertilizer—in 1824. See Wines, *Fertilizer in America*. For further biographical details, consult *Appleton's Cyclopedia of American Biography*, vol. 6, and Bierck, "Spoils, Soils, and Skinner."

11. A New York Democrat before Democrats were well organized, Southwick later promoted the anti-Masonic cause during the decades he spent in publishing. *Appleton's Cyclopedia*, vol. 6, *Dictionary of American Biography*, vol. 9, 413–414, and the *American National Biography*, vol. 20, 396–397, offer further biographical detail.
12. See Christopher Caustick [Thomas Greene Fessenden], *The Modern Philosopher or Terrible Tractoration!* An old New England Federalist, staunchly anti-Jeffersonian, Fessenden had turned from satire and literature to newspaper and then agricultural journal editing. In 1804, under the pseudonym Christopher Caustick, he polemicized against the pretensions of Jeffersonian-style science within the framework of the American Philosophical Society. There, as Elizabeth Hayes, "Science, Politics, and Satire," noted, he exposed "the absurdity of the rhetoric of 'useful knowledge,' suggesting that truly 'useful' knowledge is . . . the product of benevolent and pious and thoughtful men, 'men of real science,' who are interested primarily in social stability, morality, and truth." Fessenden thought, as Hayes described it, that APS facts rested "on dubious authority, but which, by the mere fact of their printing, are lent a kind of legitimacy that can only be undermined through careful criticisms by people who have the knowledge and the inclination to set the record straight" (10). For biographical details of Fessenden, see McCorrison, "Thomas Green Fessenden, 1771–1837," and Demaree, *The American Agricultural Press*, 321–326.
13. For a sampling, refer to "An Address to the Members of the Maryland Agricultural Society," *AF* (April 23, 1819); "Notices for a Young Farmer [in Philadelphia]," *AF* (May 21, 1819); "Paper laid before the Agricultural Society of Virginia," *AF* (June 11, 1819); "Mr. Madison's Address: Delivered before the Agriculture Society of Albemarle on Tuesday, May 12, 1819," *AF* (August 20, 1819) (Madison's address was reprinted even before that in *PB*, June 12, 1819); "Treatise on Agriculture," *AF* (October 1, 1819); "Address to the Cumberland [Tennessee] Agricultural Society," *AF* (October 8, 1819), and the Constitution and first presidential Address of the "South Carolina Agricultural Society," *AF* (October 8, 1819). The minutes of the PSPA, to provide another example, indicate a similar pattern of wide geographical correspondence, with letters to South Carolina, Connecticut, New York, and Canada. See Peskin, *Manufacturing Revolution*, 125 and 258, for references to the PSPA's minutes.
14. Thomas Fessenden, "The Science of Agriculture and Book Farming," *NEF* (August 3, 1822): 14–15. This logic had already been laid out between Jeffer-

son and Washington when Jefferson wrote in 1793 that "manure does not enter [into a good farm] because we can buy an acre of new land cheaper than we can manure an old acre." Letter to George Washington, July 28, 1793, excerpted in Jefferson's *Farm Book,* p. 194. For further discussion of the early dimensions of sectionalism see Grammer, *Pastoral and Politics in the Old South,* and Stoll, *Larding the Lean Earth,* 37–48. Grammer looks to John Taylor as one of the earliest authors of a sectional doctrine.

15. See note 9 above.
16. As quoted by Thomas Griffin, "Paper Laid Before the Agricultural Society of Virginia," reprinted in *AF* (June 11, 1819): 83.
17. For a sampling of the extensive references to the issue of lime and plaster for improved fertility, see "Notices for a Young Farmer," *AF* (May 21, 1819): 57, an article from the Memoirs of the Philadelphia Agricultural Society advising that farmers always add "appropriate manures—Lime, Plaster, Marle, etc." as needed; "To the editor," *AF* (June 11, 1819): 84, a reader's response to Skinner's query about the effects of plaster on fields, enthusiastically noting the benefits of both lime and gypsum; and "Agricultural Chemistry, No. 3," *AF* (February 11, 1820): 366, the third in a series of contributions from "A.B.M." delineating the components of marl (carbonic acid and lime) and plaster (sulphuric acid and lime) and advising on their application.
18. Wines, *Fertilizer in America.*
19. Adams, *The Agricultural Reader,* 33.
20. Both Taylor's and Peters's works, specifically, offer clear bridges between the earlier prominence of agricultural treatises and later years of society minutes and addresses. They stand across the shift from georgic tours to book farming, using principles of agricultural science as their element of consistency. The press was a continuation of the earlier agricultural treatises, only more succinct and periodical.
21. Gideon Ramsdell, "My Motto is—Wheat Will Turn to Chess," *GF* (October 13, 1832): 326.
22. Anonymous, "Letter to the Editor," *FR* (March 1837): 659. This expression of values was often just that, an expression. Many Americans who participated in book farming conversations, for and against, were not the quotidian farmers that the georgic ideal defined. When fighting for ideals of improvement, wealthy planters knew enough to aim for the georgic ideal in their argumentative strategies. My interest is in the perceptions and claims of the improvement advocates more than the veracity of the basis for those claims.
23. See Feller, *The Jacksonian Promise.*
24. Cunningham and Williams, "De-Centring the 'Big Picture' "; James, ed., *The Development of the Laboratory*; Yeo, *Defining Science*; Hankins, *The Second Great Awakening and the Transcendentalists.*
25. Thomas Fessenden, "The Science of Agriculture and Book Farming," *NEF* (August 10, 1822): 14–15.
26. "Whim-whams" were everywhere in the early nineteenth century. See, for ex-

ample, Washington Irving and James Kirke Paulding's New York series, *Salmagundi; Or, The Whim-Whams and Opinions of Launcelot Langstaff*.

27. *The Plough Boy* exemplified the common style of the rural journals by reprinting agricultural addresses from across the country (Mr. Madison's Address to the Albemarle Society; minutes from a meeting of the Agricultural Society of Virginia) and extracts from scientific works (Davy), by pronouncing on the "Utility of Agricultural Societies and Agricultural papers," and by promoting the general view that communication was crucial for the improvement of the agricultural lifestyle.
28. Arator, "Utility of Agricultural Societies and Agricultural Papers," *PB* (August 7, 1819): 75.
29. Henry Homespun, "The Moral Plough Boy," *PB* (June 12, 1819): 9.
30. "Dandies," *PB* (June 19, 1819): 23.
31. Zakim, "Sartorial Ideologies"; Dwight, *Greenfield Hills,* part 1, lines 42–65. Peskin, *Manufacturing Revolution,* 123–132, comments on the clearly gendered dimensions of the masculine homespun ethic. Peskin's interpretive view is consistent with Southwick's editorial perspective and adds to the dandy-homespun contrast a sense of a feminine-masculine split.
32. "The Plough Boy and the Dandy: A Fable," *PB* (October 30, 1819): 169.
33. With anti-Masonic and pro-temperance inclinations that undermined his claim that the paper would avoid factionalism or party orientation and ideological stances that promoted certain values and denigrated others, Southwick in fact was much like Skinner and Fessenden with his paper carrying subtle political intonations. They forwarded stark moral commitments about the promotion of certain codes of proper living. John Skinner, "Once and For All," *AF* (May 14, 1819): 55, claimed, "Once and for all, then . . . not a word of party politics will ever be allowed to enter its columns." Fessenden, "Prospectus of the New England Farmer," *NEF* (August 3, 1822): 1, claimed that "party politics and polemical divinity shall be . . . excluded" from its columns. Southwick's *Plough Boy* was already defunct by the time an actual anti-Mason party was formed, later in the 1820s, but Southwick's political character was still apparent in the paper.
34. This scientific hagiography follows from a historiographical tradition set out by Craven, *Soil Exhaustion;* Gray, *History of Agriculture in the Southern United States to 1860;* Demaree, *The American Agricultural Press;* and Gates, *The Farmer's Age*. Craven, for example, writes that "such men as Washington, Jefferson, Madison, John Taylor, J. M. Garnett, etc., formed what might be called a school of gentleman farmers who had run counter to the general backward drift" (128). Danhof, *Changes in Agriculture,* provides a notable exception to the assumption that resistance equaled ignorance by showing the rationality of resisting change. McMurry, "Who Read the Agricultural Journals?" also usefully questions the assumptions of prior analyses of book farming. She points out that "as sources of authority, the 'book' and 'tradition' have often been treated as polar opposites, by the farm journals them-

selves and by historians. . . . But in the nineteenth century, when the modern research system was not yet established, the evidence indicates that there was not such a tidy contrast" (15). Rossiter, *The Emergence of Agricultural Science,* equated book farming with "the science of agriculture" (7); McMurry suggests that historical scholarship about book farming makes this equation neat, though she does not commit herself to it; and some rural press editors, such as Thomas Fessenden in his "Book Farming and the Science of Agriculture," did not assume the two were the same.

35. The unease between speculative and fact-based advice and between repeating tired methods and accepting new ones was prominent not just in the view of the editors but also in that of small, practical farmers. Writing from Hagerstown in western Maryland, an anonymous "practical" farmer was concerned that "our whole system of agriculture depends on the phases of the moon, and the signs of the zodiac, as marked in the large Dutch Almanac." This opinion looked from tradition—"though we make excellent crops, every son treads in his father's precise footsteps"—toward progress, but without direct or specific appeal to science or chemistry or book farming. It was merely encouragement to think beyond the pattern of agricultural practice by inheritance. "To the editor," *AF* (May 28, 1819): 68. Others wrote to report successes in using plaster and lime. Still others commented more generally on the possible benefits of chemistry, but also wondered what it would ultimately achieve for them. Eli Davis, a farmer of limited means in South Carolina, wrote to Skinner about "the importance of chemistry as connected with Agriculture." "For the American Farmer," *AF* (October 8, 1819): 265. "A SUBSCRIBER," *AF* (January 14, 1820): 335, claimed to have "a little smattering of chemistry" with which he "had analised cobs of corn," hoping it would advance his cultivation.

36. "An Apology for Book Farmers," *FR* (June 1834): 17–19. Also see Ruffin, "Anonymous Contributions to *The Farmer's Register.*" In the 1950s, J. G. de Roulhac Hamilton found a document from Ruffin to Nathan Cabell, probably from the 1850s, where Ruffin listed the authors of all anonymous contributions to his journal, a plurality of which were Ruffin's own writings. More generally, Ruffin studies have become a kind of cottage industry in Virginia history. The most recent of that literature includes Mathew, *Edmund Ruffin and the Crisis of Slavery in the Old South,* and Allmendinger, *Ruffin.* Jack Temple Kirby provides a concise review of the interpretations given by Ruffin's principal biographers in his introduction to Ruffin, *Nature's Management,* xvii–xix. While recent work has been less hagiographical than biographies and studies earlier in the century, there is still a consensus that *The Farmer's Register,* which Ruffin edited for about a decade, his work on promoting marl as a manure, specifically in his *Essay on Calcareous Manures,* and his legend as the first to fire a shot in the Civil War mark him as a crucial antebellum figure.

37. Those records are also located within distinct regional settings. As with the

practices of agricultural improvement advocates earlier in the century, though, the records reveal a significant degree of overlap between North and South. On a related point, some scholars have suggested that print circulation statistics from the rural press are misleading and that the intellectual circulation of science, theory, and mechanical implements was less significant than nineteenth-century advocates claimed (see references in note 34, above). However, the social dynamics of the early American rural citizenry, based on a strong collective nature of agricultural communities, made the conveyance of ideas and practices through informal avenues more widespread than can be measured by simple census and circulation figures. Kinship networks and local, familial ties in the antebellum period, made even more prevalent in rural culture, indicate that subscription rates to the press cannot adequately gauge readership or influence. Book knowledge and tradition, as McMurry, "Who Read the Agricultural Journals?," 4 and 15, has pointed out, "intermingled among the same people," people who "traded, went to church, and exchanged work." For more commentary, see Merchant, *Ecological Revolutions;* Bushman, *In Old Virginia;* and Majewski, *A House Dividing.* Data for census and literary statistics are available at http://fisher.lib.virginia.edu/collection/stats/histcensus (accessed 26 February 2008).

38. Correspondences and diaries are in the Wickham Family Papers, Mss1 W6326 c FA2, Series 4, Boxes 5 and 6. Also see the Virginia Heritage Database for archival details about the Wickham collection. Cashin, "Landscape and Memory in Antebellum Virginia," 493–496, and Michel, "From Slavery to Freedom," also discuss William Fanning Wickham. On Wickham's anonymous contributions, see Ruffin, "Writers of Anonymous Articles in *The Farmer's Register,*" which references the contributions by Wickham.
39. Wickham's father, John (1763–1839), had also been a prominent planter, a lawyer, and a participant in the rural press culture, contributing to Ruffin's *Farmer's Register* in the 1830s. Famous for providing the defense in Aaron Burr's treason trial, John Wickham was well known throughout the Republic. He was also a former Loyalist, a southern Federalist, and a critic of Jefferson's political and agrarian skills, commenting in passing at one point that "Mr. Jefferson's reputation does not rest on his knowledge of agriculture." "Tobacco and Wheat Culture Compared," *FR* (January 1836): 513–514. For further details about John Wickham, see the Wickham Papers, Mss1 W6326 c FA2, Series 3, Box 1.
40. Diary entries for 27 April 1828; 25 December 1828; 3 March 1829; 24 January 1830; and 25 March 1830, provide early comments on quantities of marl. Wickham Papers, Mss1 W6326 c FA2, Series 4, Box 5, Diary Volume I; correspondences in Box 6.
41. 18 April 1830; 17 June 1830; August 1832 (passim) in Wickham Papers, Box 5, Volumes I and II.
42. This view is seconded in Michel, "From Slavery to Freedom," an exploration of the Wickham family's experience with slavery before and after the Civil

War. It counters the still-prevalent claim by historians that slave culture *necessarily* discouraged science or experimentation. See Montgomery, *Dirt,* chapter 6, for one example, which follows a longer tradition of historical work on science and the South in this regard.

43. Cashin, "Landscape and Memory in Antebellum Virginia," 487.
44. "Fall Ploughing," *SP* (October 1841): 189; A Hanoverian, "To the editor of the Southern Planter," *SP* (October 1841): 190–191; Za. Drummond, "Manuring the year before tillage," *SP* (December 1841): 243–244.
45. Charles Botts, "[Reply to A Hanoverian]," *SP* (October 1841): 191. Also see Botts, "Untitled," *SP* (September 1841): 158. In a follow-up issue, yet another respondent attempted to arbitrate between "W.W" and "A Hanoverian," noting that he was persuaded to follow Wickham's method "by personal observation" and that Wickham's methods seemed to avoid losses of fertility. Wickham's technique was "infinitely preferable" for reasons of convenience, so long as there was "no greater loss or expenditure of the fertilizing principles of manure." J.R.G., "Top Dressing Pasture Lands," *SP* (December 1841): 228–230.
46. See Wickham Papers, Mss1 W6326 c FA2, Box 5, Series 4, Volumes I–III of Wickham Diary.
47. The traditional distinction between a planter and a farmer is the number of slaves owned, with those having over 20 deemed planters. Although by simple numerical count Walker would thus be a planter, by virtue of his familial context he is more likely thought of as a farmer.
48. Bushman, *In Old Virginia.*
49. Ibid., 52–55.
50. Ibid., 52.
51. Ibid., 48.
52. The Fredericksburg Agricultural Society met in the county east of Lewis's Spotsylvania home. Garnett was also a longtime contributor to *The American Farmer* and in the 1840s the president of the Virginia Agricultural Society.
53. John Lewis Commonplace Book, Mss3 L7702 a 5.
54. Some historians have noted that the definitive differences in book farming were cast solely along socioeconomic lines, that, for example, wealthy planters had less at stake in failed experiments and thus took a different and perhaps more cavalier approach. See, for example, Fussell, "Science and Practice in Eighteenth-Century British Agriculture." The sources I have reviewed challenge such an analysis.
55. Peters, "Notices for a Young Farmer," *Memoirs of the PSPA* 4 (1818): vi–lii, on xliii; Skinner, "Ruta Baga," *AF* (May 21, 1819): 62; Nicholas, "Address to the Agricultural Society of Virginia," *Proceedings of the Agricultural Society of Virginia* (March 10, 1818): 25–28.
56. Undermining a simple faith in progress and novelty, Fessenden had said a farmer "should exercise his own good sense on every proposed improvement, and neither consider that it must be useful because it is new and has

the sanction of some great names, nor let its novelty be an insuperable objection to its adoption." "Many . . . theories relating to agriculture . . . considered as very useful and meritorious, are now deservedly sunk in public estimation." Fessenden, "The Science of Agriculture and Book Farming," 15.

57. Southwick, "Remarks," *PB* (June 19, 1819): 19, argued that "sound and extensive science [is] acquired only by . . . laborious study and research." Skinner, "Ruta Baga," *AF* (May 21, 1819): 62, asked his readers to "lay aside prejudice and listen to the suggestions of those experienced *in the thing,* and *then judge,*" as a general policy on interpreting results. Elsewhere he emphasized the direct connection between systematic attention and improvement, writing that "the *systems* pursued, if described in detail, could not fail to promote . . . improvement." "Note," *AF* (May 28, 1819): 68. Ruffin's *Farmer's Register* reprinted a report noting that the "most plausible reasoning upon the operations of nature, without accompanying proof deduced from facts, may lead to a wrong conclusion, and it is often difficult to separate that which is really useful from that which is merely visionary." This was in a subsection titled "Book Farming" of "Extracts from British Husbandry Published by the Society for the Diffusion of Useful Knowledge," *FR* (May 1834): 741–744, on 743.

58. Ihde, *The Development of Modern Chemistry,* 420–421. Brock, *The Norton History of Chemistry,* 185–207, further explores the new emphasis on method, apparatus, and organic chemistry.

59. Samuel Smartwood, "On Reclaiming Salt Marshes," *AF* (November 26, 1819): 277–278, on 278.

60. "Theories," *SP* (June 1842): 126; Charles Botts, "[editorial response]," *SP* (November 1842): 257–258, on 258; "Agricultural papers and agricultural writers," *SP* (January 1845): 23–24, on 23.

61. As an example of how the tension between a science of agriculture and actual chemists (scientists) was portrayed in the agricultural press, consider "Correspondents," *The Southern Cultivator* (May 1845): 72. On one page, the editor explained: "It would not be a very difficult task to make the *Cultivator* a scientific paper, filling it with speculations of philosophy, and extracts from Chaptal, Davy, Liebig, Johnston, Boussingault and others. But this is not what the country wants just now. . . . What we want now, is a plain account of the experiments of men of plain common sense." The claim for plain advice from plain men was meant to stand in opposition to philosophical or speculative advice. Yet, four pages earlier a column reported, "Chemistry, the patron-genius of agriculture, is now lending its aid . . . to dispel the darkness which has too long enveloped the farmer in his pursuit." Chemistry could be shunned as speculative, while lauded as a patron genius. "Necessity of supplying the Soil with the Constituents of the Crops grown on it," *The Southern Cultivator* (May 1845): 68.

62. "Theories," *SP* (June 1842): 126; H. Honor, Wm. W. Minor, and Frank Carr "Experiments," *SP* (April 1844): 81–83. In another editorial applauding a

speech of James Garnett's, Charles Botts, "Untitled," *SP* (November 1842): 257–258, on 257, defined the common tenets of natural philosophy and chemical science by saying "Philosophy is the generalization of well ascertained facts, and without the facts, there can of course be no philosophy." Another farmer writing from Kentucky believed only in "well ascertained and established principles deduced from facts." John Lewis, "Letter for The Southern Planter," *SP* (April 1844): 77.

63. As reprinted in *The Southern Planter,* "Liebig," (February 1844): 27–28, a French professor, Dr. Mohl, wrote that Liebig had "not formed his conclusions on the detailed facts of vegetable phenomena, but on random observations, or vague operations on a large scale." In so doing, Mohl provided the critique a practical farmer would want to hear. Also see Rossiter, *The Emergence of Agricultural Science;* Brock, *Justus von Liebig;* and Hall, *A History of the Yorkshire Agricultural Society, 1837–1987.*
64. For more on the development of laboratory chemistry, see Brock, *Justus von Liebig;* James, ed., *The Development of the Laboratory;* Holmes and Levere, eds., *Instruments and Experimentation in the History of Chemistry;* and Brock, *Norton History of Chemistry,* 194–207 and 427–435.
65. Eli Davis, "For the American Farmer," *AF* (October 8, 1819): 265, wrote that an Agricultural Society should be set up because "By this means the great laboratory of nature would be gradually opened to the view of many who, perhaps, know but little of the very intimate connexion that subsists between chemistry and agriculture"; Robert Smith, "An Address to the Maryland Agricultural Society," *AF* (October 13, 1820): 228–229; Eppes, as quoted in Kirby, *Poquosin,* 37. In *The Southern Planter* of the 1840s, Charles Botts, "Agricultural Chemistry," *SP* (August 1842): 186–187, on 187, argued against book farming because the use of "laboratory" had lost its attachment to nature—"The philosopher must exchange his laboratory for the open field." A local leader, Dr. Ezekiel Holmes, told the Lewiston Cattle Show and Fair that "the earth is the great laboratory and the farmer is the chemist," as quoted in Merchant, *Ecological Revolutions,* 210. This topic unfolds into discussion of laboratory versus field science, the full force of which I can only allude to here. For discussions of the lab-field border see Kohler, *Labscapes and Landscapes,* and Kuklick and Kohler, eds., "Science in the Field."
66. Campbell, as quoted in Philips, "Antebellum Agricultural Reform," 817–818. Also see the *DAB,* vol. 3, 455.

CHAPTER 3. KNOWING NATURE, DABBLING WITH DAVY

1. Liebig, quoted in Brock, *Justus von Liebig,* 145.
2. Science in the Romantic era, if treated as a whole, appears as a confusing assemblage of competing metaphysical and conceptual commitments. See Cunningham and Jardine, eds., *Romanticism and the Sciences;* Melhado and Frangsmyr, eds., *Enlightenment Science in the Romantic Era;* Richards, *The Romantic Conception of Life;* Ramberg, "The Death of Vitalism and the Birth of

Organic Chemistry"; and Klein, *Experiments, Models, Paper Tools.* E. Benton, "Vitalism in 19th-Century Scientific Thought," long ago made the sensible presentation of vitalism as a label for a variety of positions, and I do not mean to reify the term to treat it as one doctrine. Historians of chemistry have contributed widely to questions about popular chemistry in Europe and America and chemical education, but mostly without attention to agricultural chemistry or the farming classes and, in the best work, with use of Jane Marcet's popular texts as the historical source. For suggestive accounts of these topics, see Knight, "Accomplishment or Dogma"; Lindee, "The American Career of Jane Marcet's Conversations on Chemistry"; Myers, "Fictionality, Demonstration, and a Forum for Popular Science"; and Lewenstein, " 'To Improve Our Knowledge in Nature and Arts.' "

3. As quoted in Levere, "Coleridge, Chemistry, and the Philosophy of Nature," 358.
4. Merchant, *Ecological Revolutions,* 147–231. Also see idem, *The Death of Nature.*
5. Davy showed that animal and vegetable matter provided the necessary food for plant growth, ideas that would lead farmers to increase animal manure usage and vegetable composting while pursuing a general farming strategy of mixing the two. Farmers were already doing that, for the most part, so Davy's theories were often easily received in the rural community. Put simply, as the historian of chemistry William Brock has said, "Davy's advice was uncontroversial." Brock, *Justus von Liebig,* 146. For similar assessments of Davy's work, see Knight, "Agriculture and Chemistry in Britain around 1800"; Orwin and Whetham, *History of British Agriculture, 1846–1914;* and Hall, *A History of the Yorkshire Agricultural Society, 1837–1987,* 60.
6. Liebig, in fact, was well known for his later opposition to the vitalist *naturphilosophie* setting in which he was educated. On his philosophy of nature, reaction to *naturphilosophie*'s Romantic excesses, and place within nineteenth-century science, see Brock, *Justus von Liebig;* Kirschke, "Liebig"; Schling-Brodersen, "Liebig's Role in the Establishment of Agricultural Chemistry"; Munday, "Liebig's Metamorphosis"; and Findlay, "The Rehabilitation of an Agricultural Chemist."
7. Deborah Fitzgerald makes a similar comment with reference to a later, early-twentieth-century era of farming labor. See Fitzgerald, "Farmers Deskilled," 327–328. Fitzgerald's discussion is directed at how the introduction of hybrid seeds deskilled farmers by removing the need for the thoughtful analysis of corn planting techniques. See also idem, *Every Farm a Factory.*
8. Among some of the other prominent rural works and agricultural treatises that broached the subjects of chemistry, Davy, and improvement were Butler, *The Farmer's Manual* and Drown, *Compendium of agriculture.* Another popular account in the spirit of the georgic tour, written from travels in Nova Scotia and cited in, among others, Daniel Adams's book, is Young, *The Letters of Agricola on the Principles of Vegetation and Tillage.* Jesse Buell, perhaps

Ruffin's New York counterpart, published his (1839) *Farmer's Companion* as a compendium of his own experiences in the previous decades. Charles Squarey, a chemist, likewise sought to write a book "adapted for general perusal" in his *Popular Treatise on Agricultural Chemistry.* Also see Dana, *A Muck Manual for Farmers.* Fessenden, *The Complete Farmer and Rural Economist,* compiled decades of his own observations and writings from *The New England Farmer* and beyond, including his many satirical accounts of science and agriculture from earlier in the century.

9. The distinction between these three and the dominant scientific authors of the time is the difference between those who were "men of science" and those who were men of agriculture, not one of upper versus lower class. The three examples associate those two realms of thought and practice in the direction *from* the social sphere of agriculture *to* science, not from science *to* agriculture.
10. Jack Temple Kirby provides a concise review of the interpretations given by Ruffin's principal biographers in his introduction to Ruffin, *Nature's Management,* xvii–xix. In addition to the general biographical overviews, the historian of chemistry Aaron Ihde has written specifically of Ruffin's chemical acumen. See Ihde, "Edmund Ruffin, Soil Chemist of the Old South."
11. Rossiter, "The Organization of Agricultural Improvement in the United States, 1785–1865," 289.
12. See Mathew, *Edmund Ruffin and the Crisis of Slavery in the Old South;* Allmendinger, *Ruffin;* Kirby, *Poquosin,* 61–91; and Stoll, *Larding the Lean Earth,* 151–159.
13. J. Carlyle Stitterson, in a 1960 reprint of the *Essay,* estimated that 5,000–6,000 copies were ultimately printed. See "Editor's Introduction," in Ruffin, *An Essay on Calcareous Manures.*
14. Parenthetical references in this discussion of Ruffin refer to the page numbers in the 1961 edition of *Essay on Calcareous Manures,* which is a reprint of the 1832 edition.
15. While later in his life Ruffin devoted many addresses to the tenets of slavery, in his *Essay* he was less concerned with the economy of slaveholding, for the sake of the slaves, than with the cause of increased food production to sustain increasing slave populations (see 162–164). Just as he sought to improve soil to keep free white Virginians comfortably at home, he also thought nothing could address "the forced emigration of blacks, and the voluntary emigration of whites, except increased production of food, obtained by enriching our lands" (164).
16. In the following I do not emphasize the health metaphors Ruffin subscribed to in his own life and applied to his land, although such investigations have recently proven insightful in environmental and science studies. In particular, for more on the depth of the health metaphor in antebellum concepts of nature, see Valencius, *The Health of the Country.*
17. For example, the land Ruffin sought to improve near the falls of the James

River was already being planted in wheat, oats, and corn as Virginia's tobacco shifted to the south and west by the early 1800s. Nonetheless, the legacy of tobacco's soil-exhausting tendencies remained the primary culprit for the sense of decline in upper South agriculture for decades to come—and the story of tobacco cultivation is also an important backdrop for Ruffin's work.

18. By late in his career, Allmendinger notes, Ruffin was comfortable with the georgic genre of the agricultural tour, but this familiarity and interest in the Arthur Young–like style was evident already in select passages from his *Essay.* The similarities were not random—Ruffin had read Young's agricultural Tours and was familiar with the more recent and similarly styled writings of William Cobbett. Allmendinger, *Ruffin,* 38–44.
19. Ibid., 23.
20. In his editing of Ruffin's writings on landscape, Jack Temple Kirby seeks to place Ruffin as a proto-conservationist, identifying nature's course so as to better emulate it. But this is only part of the story. As the historian David Spanagel has written in review of Kirby's edited work, Ruffin's "sensitivity to the 'wisdom of nature' extended only to questions of efficacy, not to intent," and as such Ruffin's relationship to the land is more complicated and harder to gauge than the imposition of a later environmentalist framework could make clear. See Ruffin, *Nature's Management,* and Spanagel, "Review of Edmund Ruffin, *Nature's Management.*"
21. Even this fact is not without dispute—the *Encyclopedia Britannica* claims he was born in England, though records from the Historical Society of Pennsylvania indicate he was born in Maryland. Stoll, *Larding the Lean Earth,* notes that he was born on the Eastern Shore of Maryland of recent English immigrants (97). For more biographical commentary see Sweet, *American Georgics,* 110–119, and Jehlen, *American Incarnations,* 66. The *Encyclopedia Britannica* summarizes Lorain's life work with the mere mention of corn hybrid experiments he reported in *Nature and Reason,* calling him a pioneer in cross-breeding seeds. Many of the details here about Lorain follow Stoll, *Larding the Lean Earth,* 96–108.
22. See Lorain's contributions to the *Memoirs of the Philadelphia Society for Promoting Agriculture,* which all appeared in the second and third volumes: "On soiling cattle: mixed cultivation of Corn and Potatoes," 2 (1811): 200–205; "Profit of soiling cattle," 2 (1811): 319–324; "Farther remarks on mixed crops of Corn and Potatoes," 2 (1811): 330–349; "On the Agriculture of England, on manures, convertible husbandry, and soiling," 3 (1814): 32–36; "Observations upon the agriculture and roads of new settlements in Pennsylvania, with hints for improvement," 3 (1814) 84–97; "Observations on the comparative value of soils," 3 (1814): 98–102; "Account of the modes of clearing land pursued in Pennsylvania, and on the fences in new settlements," 3 (1814): 112–119; "Observations on Indian corn and potatoes," 3 (1814): 303–325; and "On grass lays, manures, etc.," 3 (1814) 326–336.
23. Skinner's reprint of Lorain's work appears in "Agriculture," *AF* 7 (May 6,

1825): 49–50. Stoll, *Larding the Lean Earth,* 102. Also, see Sweet, *American Georgics,* and Jehlen, *American Incarnations.*

24. Parenthetical citations in this section of the chapter refer to page numbers in Lorain's *Nature and Reason.*
25. This is not to say Lorain comments only on vitalism, or that his work is reducible to just that point. Rather, given the immense scope and length of the book, Lorain's discussions of vital principles serve as a useful focal point with which to touch on the many ways he describes and conceptualizes nature and, specifically, the soil. They also provide a reference point against which Ruffin's views on nature and soil can be compared and contrasted.
26. Other descriptions of nature's economy abound in Lorain's work. He elsewhere describes how seeds are dispersed, by wind and water, and scattered by "birds and quadrupeds" (72). He also tells a story of his season of sowing wheat on newly cleared land. That endeavor required him to divert water through furrows and to find that ditching was unnecessary and costly (193). He was glad to have decided against cutting "off the springs" to form a ditch for irrigation, because his later success proved that was unnecessary.
27. Dickens, *American Notes,* 152–153.
28. See *Appleton's Cyclopedia,* 1 (1888): 13, and *DAB, vol.* 1: 54–55. In addition, Adams had already put his *Scholar's Arithmetic* through a third edition by 1805.
29. In his presidential address to the Hillsborough Agricultural Society he excerpts Henry Home as follows (and as quoted in chapter 1 of this book): "In a political view, [agriculture] is perhaps the only firm and stable foundation of national greatness. As a profession, it strengthens the mind without enervating the body. In morals, it has been well observed, it leads to increase of virtue, without introducing vice. In religion, it naturally inspires devotion and dependence on Providence." Adams, "An Address Delivered before the Hillsborough Agricultural Society," 148.
30. Parenthetical citations in this section of the chapter refer to page numbers in Adams, *The Agricultural Reader.*
31. Other contemporary chemical textbooks for schoolchildren, ones specifically scientific though not agricultural—like Jane Marcet's *Conversations on Chemistry, Intended more Especially for the Female Sex*—followed that question-and-answer format. By the 1830s this style was losing favor, though, as exemplified by Comstock, *Elements of Chemistry.* For a useful elaboration on textbooks and shifts in science education in the era, see Tolley, *The Science Education of American Girls.*
32. On the disciplining tendencies of textbooks, see Lundgrens and Bensaude-Vincent, eds., *Communicating Chemistry.*
33. For concise and accessible reviews of natural theology, see Bowler, *Evolution,* and idem, *The Norton History of the Environmental Sciences.*
34. Adams's belief in a divine intelligence and his perspective on the economy of nature were not unrelated, instead fitting together in a complex worldview. Adams quotes the Letters of Agricola to say, "I know of no indication of

greater skill in the Divine Intelligence, nor a more indubitable mark of his care and goodness, than the contrivance of resolving all dead animal and vegetable matter into elementary principle; that, in the first place, he might relieve the earth of such loathsome incumbrances [*sic*], and in the next place, be supplied with fresh materials out of which to form and sustain the new and successive families of plants" (50).

35. As quoted in Loehr, "Arthur Young and American Agriculture," 46.
36. This was quite similar to the "law of minima" expressed later by Liebig, meaning that there is a minimum quantity of specific and detectable elements necessary for plant growth and that when those specific components were provided, the requirements of plant nutrition could be met. See Brock, *Justus von Liebig;* and Black, *Soil Fertility Evaluation and Control.*
37. To make a common science studies reference, the agriculturalist in the field becomes the obligatory passage point for knowledge of nature. For chemistry to be that passage point it must be aligned with that agrarian citizen. See Callon, "Some Elements of a Sociology of Translation," and actor-network theory in general, as developed by Bruno Latour, John Law, and Callon.
38. I here return to citing Ruffin's *Essay on Calcareous Manures* with parenthetical page references.
39. The reference to "true philosophical mode" comes from William Boulware's biographical sketch of Ruffin: "Edwin Ruffin, of Virginia, Agriculturist, Embracing a View of Agricultural Progress in Virginia for the Last Thirty Years, With a Portrait," *DeBow's Review* 11 (1851): 431–436, reprinted in Ruffin, *Incidents of My Life,* 171. Boulware is responsible for misstating Ruffin's first name as "Edwin" instead of "Edmund."
40. Edmund Ruffin, "First Views Which Led to Marling in Prince George County," *Farmer's Register* 7 (1839): 659–667, on 665.
41. Ruffin would have read the 1815 American edition of Davy's 1813 *Elements of Agricultural Chemistry.*
42. With the belief that "nothing is more wanting in the science of agriculture than a correct nomenclature of soils," Ruffin listed six extant definitions of "loam," his own and Davy's included, to demonstrate its inconsistent usage. That is, Ruffin understood that soil taxonomy was a problem greater than he alone could solve. *Essay on Calcareous Manures,* Appendix B, 160–162.
43. On boundary objects, see Star and Greisemer, "Institutional Ecology, 'Translations' and Boundary Objects."
44. I here return to citing Lorain's *Nature and Reason* with parenthetical page references.
45. Whether or not Lorain was warranted to speak as a practicing farmer—since he had been a merchant and was new to farming when he moved out of Philadelphia—is interesting in hindsight. It would have been irrelevant, however, to a consideration of his rhetorical purpose of contrasting Davy's lab-based theory with field-based truth.
46. Lorain further defines that clear distinction between chemistry as a body of

theories (which must be questioned before being relied upon) and chemistry as an aid to nature, or a practice of analysis, when he uses analytical results from Davy's *Elements of Agricultural Chemistry*, such as ash analyses, without qualms or skepticism. To be sure, Lorain's understanding of chemistry for studying and then improving nature was complex, a fact I do not want to conceal with the above synopsis. For most of his work, he treats it not as a thing to be used, a material and diagnostic tool, but as a form of reason to be invoked. Yet at times he did rely upon it as a theoretical approach to understanding the economy of nature. He overplayed his hand when debunking others' theories, claiming that Davy was wrong because he denied the vital principle, when in fact Davy did no such thing. He was also unclear and contradictory about the relation between science and reason, at times equating them (see 8, 10, 31, 33, 196, 362), at other times claiming the two were distinct (see 62, 70, 170, 352).

47. I here return to citing Adams's *Agricultural Reader* with parenthetical page references.
48. His sense of chemical history is noteworthy, in some parts, since Adams clearly writes in what was called the post-pneumatic phase of chemical research, giving reference in that term to Joseph Priestley's 1780s work on the chemistry of the airs. (This provided an important signpost for the science of agriculture since it indicated attention to the atmosphere for plant nutrients.) He also introduces concepts like "chemical affinity" and "caloric" to his students, though in those definitions he falls decades behind the accepted theory of chemists themselves (15–21). Furthermore, when introducing his theme on lime and the action of quick lime as a fertilizer, Adams lists four separate views on the matter—those of Davy, Peters, Robert Smith of Maryland (then president of the Maryland Agricultural Society), and Fessenden—before avoiding a final conclusion on the issue. Instead of that conclusion, he broadens his theme to discuss Arthur Young's belief in the value of manure in general, thus escaping a commitment to any one theory (59–61). In a publication by a textbook author aiming for wide and uncontroversial appeal, the superficial appropriation of standard works is not extraordinary. In fact, the opposite, if Adams had introduced agricultural works for the point of refuting them, would be more unlikely. His book is a primer on agriculture, not specifically or only a text on agricultural chemistry. At once, then, Adams's views and uses of chemistry can be read as somewhat typical—considering the uncontroversial to be typical, in this instance—while his representations of chemistry and soil emerge as overly general and perhaps uninteresting.
49. Daniel Adams, "An Address Delivered before the Hillsborough Agricultural Society at their Cattle Show and Fair, at New Boston, Sept. 22, 1825," *NEF* (December 2, 1825): 148–150, on 149. The address was a pastiche of the extant agricultural works of the day, again borrowing heavily from Fessenden, Henry Home, and the letters of George Washington.
50. This is not to say that Adams was copying Lorain (whose text preceded his).

In fact, most of Adams's views of the soil were cribbed from his peer Thomas Fessenden, editor of *The New England Farmer.* Fessenden was not critical of Davy and so Adams's use of Fessenden to present Davy was twice as glossy, two steps removed from an individual expression of the ideas about soil.

51. Adams, "An Address Delivered before the Hillsborough Agricultural Society," 148–149.
52. But while at one point he claims that analysis is "the business of chemists," suggesting deference to an established group of professional philosophers, he indicates more broadly, as I noted above, that agricultural chemistry is a tool that can be used by any farmer interested in improvement.
53. Adams, "An Address Delivered before the Hillsborough Agricultural Society," 149.
54. Lorain, *Nature and Reason,* 99 and 403.
55. That natural state is not necessarily, or even often, completely uncultivated virgin soil. "Natural fertility" is measurable against the productive abilities of the soil after it has been cultivated for a few seasons. Before then, "several temporary causes . . . operate either to keep down, or to augment the product," such as the leftover roots and consequent bad tillage that follow the first clearing of woodlands. Ruffin, *Essay on Calcareous Manures,* 22.

CHAPTER 4. THE AGRICULTURAL SOCIETY, THE PLANTER, AND THE SLAVE

1. Madison, "Address to the Albemarle Agricultural Society," available at http://www.jmu.edu/madison/center/main—pages/madison—archives/life/retirement/ag.htm (accessed 25 April 2008).
2. James Madison was chosen first president of the Agricultural Society of Albemarle on October 7, 1817. See "Minute Book of the Agricultural Society of Albemarle," Mss4 Ag832 a 1, at the VHS, Richmond, VA. The society was sometimes referred to as the "Albemarle Agricultural Society" and other times as the "Agricultural Society of Albemarle." The original Minute Book of the Society gives its name as the "Agricultural Society of Albemarle," and I retain this reference as "ASA." Rodney True reprinted a more accessible version of the Minutes and labeled it, confusingly, "Minute Book of the Albemarle Agricultural Society." See True, "Early Days of the Albemarle Agricultural Society." While I refer to the Society as the ASA in my text, I mainly refer to True's reprinted version of the Minutes in the notes. The citation above to Madison's election appears as pages 263–271 in True's reprint.
3. Thomas Jefferson to John Quincy Adams, November 3, 1807; Schlotterbeck, "Plantation and Farm," 260.
4. Madison, "Address to the Albemarle Agricultural Society" [no pagination]; "Minute Book of the Albemarle Agricultural Society," 289.
5. Despite its rich historiography, Virginia has nonetheless suffered from relatively little attention to its environmental history. On this lack of attention, see Kirby, "Virginia's Environmental History." On the dearth of southern en-

vironmental history in general, see Graham, "Again the Backward Region?" and Stewart, "Southern Environmental History." Notable exceptions to this claim are Cowdrey, *This Land, This South;* Silver, *A New Face on the Countryside;* Kirby, *Poquosin;* Stewart, *"What Nature Suffers to Groe";* and Curtin, ed., *Discovering the Chesapeake.* For a view of seventeenth-century Virginia, see Adams, *The Best and Worst Country in the World.* For a view of eighteenth-century Virginia, see Isaac, *The Transformation of Virginia, 1740–1790.* For other work that either aims directly at or touches on the environmental history of Virginia, see Cashin, "Landscape and Memory in Antebellum Virginia"; Davis, *Where There Are Mountains;* Lewis, *Transforming the Appalachian Countryside;* and Hofstra, *The Planting of New Virginia.*

6. James Mercer Garnett to John Randolph, 16 October 1827, quoted in Fisher and Kelly, *Bound Away,* 204. See that volume and Shade, *Democratizing the Old Dominion,* for more on the spirit of Garnett's critique. Virginia Board of Public Works referenced in Larson, *Internal Improvement,* 94–95; Benjamin Watkins Leigh as quoted in Rasmussen and Tilton, *Old Virginia,* 68.
7. For Pennsylvania, see Baatz, *"Venerate the Plough";* for New York, see Aldrich, *New York State Natural History Survey.* The cultural "Virginia"—often referred to as "Old Virginia"—was that which was dominated by planter elites, old tobacco grounds, and slave-labor dynamics. The geographical "Virginia" extended far to the west and received many of the émigrés from the east. The Scots-Irish, Germans, and Quakers of the Valley and west—many of whom had moved south from Pennsylvania to fill the Shenandoah Valley—lived different lives and told different foundation stories. Their impetus for improvement may have been tied to material agricultural goals more than deep cultural ones.
8. The details in this paragraph were drawn from Shade, *Democratizing the Old Dominion,* 31–38. See Koons and Hofstra, eds., *After the Backcountry,* for further analysis of the many nineteenth-century contexts of the Shenandoah Valley, including several discussions of the place of wheat in the social dynamics of that region. Also see Hofstra, *The Planting of New Virginia.*
9. Faust, *A Sacred Circle.* For entry points into further discussion of the matter, see Wallenstein, *Cradle of America: Four Centuries of Virginia History,* and, with specific attention to Ruffin, Mathew, *Edmund Ruffin and the Crisis of Slavery in the Old South.* On the particulars of slavery, Shade, *Democratizing the Old Dominion,* 43–71, offers the following summative and statistical view: In 1830, 39 percent of Virginia's 1.2 million people were enslaved; in 1850, the figure was 33 percent of 1.4 million. This slave population was disproportionately skewed to eastern and southern counties. More than 60 percent of the land holdings in counties of the south-central Piedmont had large slaveholdings; only 40–60 percent of the farms in counties of the central Piedmont, mixed with plantations and yeoman farmers, had slaves; and, but for four counties, landowners to the west of the Blue Ridge mountains were 1–19 percent slaveholding. Mapping compiled by Shade shows that the per-

centage of slaves closely matched the political valence of the state's constituents, with conservative counties holding the most, reformers the least. Furthermore, the geological map of the state shows that slaveholding lands were directly related to elevation—the further up the Piedmont, the fewer the slaves.

10. Shade, *Democratizing the Old Dominion*, 193.
11. Cocke, as quoted in Fisher and Kelly, *Bound Away*, 204.
12. Olmsted, *Journey in the Seaboard Slave States*, 278–279. A further reading of this work provides more commentary on the state of agriculture in antebellum Virginia, usually through a strictly economic calculus.
13. *The Richmond Enquirer*, 2 February 1837, Richmond, VA; Shade, *Democratizing the Old Dominion*, 3.
14. Tocqueville, *Democracy in America*, vol. II, 181.
15. The membership list for the ASV is in the *Memoirs of the Society of Virginia for Promoting Agriculture* (Richmond: Shepherd and Pollard, 1818), xii–xiii. The membership list for the ASA is in "Minute Book of the Albemarle Agricultural Society," 263 and 269.
16. Merchant, *Ecological Revolutions*, 163–166. Merchant's New England observations match the mid-Atlantic as well.
17. Madison, "Address to the Albemarle Agricultural Society"; Chevalier, *Society, Manners, and Politics in the United States*, 316.
18. *Memoirs of the Society of Virginia for Promoting Agriculture*, iii–iv.
19. *Memoirs of the ASV*, iv; *Memoirs of the ASV*, as quoted in Schlotterbeck, "Plantation and Farm," 271.
20. "Minute Book of the Albemarle Agricultural Society," 249. Also see chapter 1.
21. "Minute Book of the Albemarle Agricultural Society," 265.
22. See "A Report of the Committee of the Agricultural Society of Albemarle, on Farms, made to the Society at their Show and Fair, the 21st October, 1828," as reprinted in the *AF* 12 (August 20 and 27, 1830): 177–178 and 185–187.
23. Dabney Minor, "On the Importance and advantages of keeping a regular diary of operations of Husbandry," *AF* 2 (December 1, 1820): 285.
24. See Joseph Cabell's agricultural diary in the Cabell Family Papers, MSS 38–111, Box 3, Alderman Library Special Collections, University of Virginia, Charlottesville. The diary is also available in electronic form, at http://www.lib.virginia.edu/speccol/collections/cabell/digitalarchive/index.html (accessed 26 February 2008).
25. Seed trading was not new. Exchanging plant seeds had been a mainstay of transatlantic correspondence from the later eighteenth century. But, as with almost all matters, the forum of the Society allowed greater coordination and promotion of the activity. The "Minute Book of the ASA" offers representative examples, including wheat seeds from the Black Sea, meeting of 10 October 1820 (290); Paris, meeting of 4 February 1822 (296); the Pacific Ocean, meeting of 7 October 1822 (298); China, meeting of 11 October 1824 (304).
26. Fluvanna County was formed out of the southeastern part of Albemarle

County in 1777. See letters from John Moody to JHC, 15 June 1824, William Shepard to JHC, 28 June 1824, General J. Swift to JHC, 10 July 1824, H. E Watkins to JHC, 12 July 1824, John Tandy to JHC, 3 August 1824, and David Patterson to JHC, 13 September 1824, all in Cocke Family Papers, Accession Number 640, Box 41, Alderman Library Special Collections, University of Virginia, Charlottesville [hereafter JHC Papers]; also see Edmund Wickham to JHC, 29 June 1825, in JHC Papers, Box 43 and George Love to JHC, 31 July 1834, in JHC Papers, Box 41.

27. See the meetings of 3 and 4 November 1817 and 2 March 1818 in "Minute Book of the Albemarle Agricultural Society," 248, 271, 275, and 277–78, and Pauly, "Fighting the Hessian Fly." Also see J. H. Cocke "On Hessian Fly—No. 1," *AF* 1 (1819): 296–297, for the first in the series of reports he published in the *Richmond Enquirer.*
28. See James Garnett to JHC, Letter of 11 December 1819, in JHC Papers, Box 30.
29. On the Rivanna River projects, see "Minute Book of the Albemarle Agricultural Society," 312, 324, 325, and 333–334. Also see McGehee, "The Rivanna Navigation Company," and Majewski, *A House Dividing,* 28–32. Yet another example of partisan political agitation was Edmund Ruffin, "An Address to the Public from the Delegation of the United Agricultural Societies of Virginia," read at the May 8, 1820, meeting and published in *AF* 2 (May 19, 1820): 57–59.
30. "Minute Book of the Albemarle Agricultural Society," 253. Meeting of 16 December 1826. See also Majewski, *A House Dividing,* 31.
31. The institutionalization of county-wide agricultural societies took hold far more readily after the end of the War of 1812, under a more stable political environment. See Rossiter, "The Organization of Agricultural Improvement in the United States, 1785–1865," and Baatz, *"Venerate the Plough."*
32. Bordley quoted in Peskin, *Manufacturing Revolution,* 123.
33. There are no substantial biographical works on Cocke, though there are treatments of his estate at Bremo and his involvement in founding freedmen slave colonies in Africa. See Miller, ed., *"Dear Master";* Rogers, "John Hartwell Cocke (1780–1866)"; and Helliar-Symons, "John Hartwell Cocke and Bremo." Cocke's canal ambitions were spurred on by correspondence with another wealthy antebellum political figure, DeWitt Clinton of New York, who was contemporaneously building the Erie Canal. See Clinton to JHC, 20 July 1824, in JHC Papers, Box 41.
34. The role of quantification in the establishment of a "modern" scientific world is one of great complexity, and I cannot do the subject proper justice with a single footnote. For two works that offer more insight into quantification and science, see Porter, *Trust in Numbers,* and Poovey, *A History of the Modern Fact.*
35. Cocke's report was delivered to the ASA in May 1818, per the Minute Book, without details. Actual details can be found in "Albemarle Agricultural Society Report," JHC Papers, Box 182.

36. JHC to JHC Jr., 2 January 1836, JHC Papers, Box 84.
37. Ibid.
38. Majewksi, *A House Dividing,* 32.
39. "Extract from *The Farmer's Pocket Guide,*" 21 August 1804, JHC Papers, Box 4.
40. "Recipe for detecting adulterated wines," 21 November 1822, JHC Papers, Box 37.
41. To be fair, the same assessment of unimpressive practical chemical facility had been leveled against Virginia's most famous son, Thomas Jefferson. M. M. Robinson wrote to Cocke that Jefferson "is an ornament to his country, as a gentleman and a man of letters, but in whom, as a practical man, either in state affairs or in common concerns of life, I have no confidence." Robinson to JHC, 22 February 1822, JHC Papers, Box 33.
42. See Nathan Cabell to JHC, "Cabell's opinion of John Hartwell Cocke's Essay on Agriculture," in JHC Papers, [no date], Box 182. Given the other documents in Box 182 and the tenor and goals of Cocke's essay, it is likely from the 1820s. Cocke was in fact proposing nearly the same theory as the chemist Van Helmont two centuries earlier, where Van Helmont "proved" the source of plant nutrition with a famous tree experiment, showing that his tree gained 164 pounds in five years with the addition of nothing but water. Brock, *The Norton History of Chemistry,* xxi.
43. "Draft of a letter to the Farmer's Register, April 15, 1835," JHC Papers, Box 1.
44. Robert Rogers to JHC, "Soil Analysis of Bremo Soils," June 1846,Box 117, and "A report on Guano, by the Maryland State Ag Society Chemist, David Stewart," 9 February 1859, JHC Papers, Box 157.
45. For brief commentary on John Wickham, see Turner, *Virginia's Green Revolution.*
46. The Carter family, beginning with Robert "King" Carter, is well known in Virginia history for their large landholdings, prominent slave ownership, and, giving us the means to study it all, diary-writing tradition. See Isaac, *Landon Carter's Uneasy Kingdom,* for a detailed exploration of Landon Carter's diary and times. Also consult the Carter Family Papers, listed as the Sabine Hall Papers, Accession Number 1959 and 2658, Alderman Library Special Collections, University of Virginia Library, Charlottesville.
47. See Wickham Diary entries for 9 June 1837, 10 May 1838, 17 June 1840, 15 March 1841, in Volume III, Box 5, Series 4 of the Wickham Family Papers, Mss1 W6326 c FA2, Virginia Historical Society, Richmond [hereafter Wickham Papers]. Also see chapter 2, note 38, for further references to works on the Wickham family.
48. There are no records of Wickham having completed a questionnaire from one of the regional societies or other local reconnaissance efforts.
49. M. A. Miller, "Report on Survey of 'Hickory Hill' plantation, Hanover Co., Virginia, 1878," Wickham Papers, Box 5.
50. Wickham was of course not alone in these endeavors. He was, rather, representative of a more general shift in ways of seeing—representing—the land.

For other examples, see the papers of Robert Baylor, in particular his commonplace book detailing experiments on "Lime, Magnesia, Plaster" and his "Address circa 1840 to the Agricultural Society of Essex County," Baylor Family Papers, Mss1 B3445 e FA2, Series VIII, Boxes 5 and 6, at the VHS. Also see the records of Richard Eppes detailing two decades of marling experiments: Richard Eppes Diary, 1840, pp. 160–161, in the Eppes Family Papers, Mss1 Ep734 d353, at the VHS; and the Papers of Robert Wormeley Carter, another member of the famous Carter family descending from Robert "King" Carter, into which Wickham married. Carter tabulated the cartloads of marl brought to his fields, contributed (though anonymously) to *The Farmer's Register,* and had his crops chemically analyzed to identify the "constituents of different grains." Robert Wormeley Carter Papers, Mss 1959-a, Boxes 2–3, at the VHS.

51. *Memoirs of the Society of Virginia for Promoting Agriculture,* iv. For a study in general of the actual labor performed by slaves, see Berlin and Morgan, eds., *Cultivation and Culture;* for slave labor in the environmental history of the Georgia coast, along with an extensive bibliography on race and labor in sociohistorical context, see Stewart, *"What Nature Suffers to Groe";* for an entry in the vast field of indigenous knowledge studies, especially as related to southern agricultural knowledge, see Carney, *Black Rice;* for slave labor in southern industry see Dew, *Bond of Iron;* for the slave labor of an area cutting across agricultural and industrial regimes, the naval store, see Outland, "Slavery, Work, and the Geography of North Carolina Naval Stores Industry, 1835–1860."
52. "Minute Book of the ASA," 274.
53. *AF* 12 (1831): 177
54. *AF* 12 (1831): 186. For more on Meriwether, see Majewski, *A House Dividing,* 31.
55. This was the debate reviewed in chapter 2.
56. Ruffin, *An Essay on Calcareous Manures,* 150–151. This line is also quoted in Allmendinger, *Ruffin,* 33.
57. Wickham Diary, 13 January 1841, in Wickham Papers, Volume III, Box 5, Series 4.
58. Ruffin, "On the Composition of Soils," 319, as quoted in Allmendinger, *Ruffin,* 33.
59. This paragraph follows from Allmendinger, *Ruffin,* 32–34.
60. G. P. Marsh, "The Study of Nature," *The Christian Examiner* (January 1860), as reprinted in Worster, ed., *American Environmentalism,* 16.
61. John Taylor, "Presidential Address to the Agricultural Society of Virginia, June 8th, 1818," *Memoirs of the ASV* (1818): 28–29.
62. James Garnett, "Agricultural Notices, September 8, 1818," *Memoirs of the ASV* (1818): 116–119.
63. The longtime ally of Virginia reform, John Skinner, later expanded his publishing ambition to promote a series of monographs, the *Farm Library.* The

Book of the Farm (1845) was the first of this series, with *Lectures to Farmers on Agricultural Chemistry*, by Alexander Petzholdt, following the next year. Daniel Webster and Horace Greeley had their hands in those efforts. Webster offered well-worn advice to Skinner that he should republish European sources but note their differences in practice, since "in addition to difference of soil and climate, the higher rate of wages, which, fortunately for the general good, exists with us, must itself make a material change in all agricultural calculations." These exchanges reinforced a sense that tensions between local and global were only heightened through the confluence of science and agriculture. See Letter of 4 November 1841, from Webster to Skinner, in the Papers of John Stuart Skinner, Mss2 Sk352b, at the VHS.

64. As quoted in Schlotterbeck, "Plantation and Farm," 282.
65. W. C. Nicholas, "Vice-Presidential Address to the Agricultural Society of Virginia, March 18th, 1818," *Memoirs of the ASV* (1818): 26.
66. Organizations alone did not *cause* the changes. Rather, the introduction of scientific improvement through their socially credible forum helped shape the context within which farmers developed new views of the soil under their feet.
67. *Memoirs of the ASV*, iii–iv.
68. John Majewski offers an exception to this by noting the "emphasis on rational land use, increasing productivity, and detailed accounting procedures" prevalent by the 1840s: Majewski, *A House Dividing*, 22.

CHAPTER 5. THE GEOLOGICAL SURVEY, THE PROFESSOR, AND HIS ASSISTANTS

1. Hezekiah Daggs to W. B. Rogers, 23 March 1837, Geological Survey Papers, Accession Number 24815, Box 1, Board of Public Works Records, Record Group 51, Library of Virginia, Richmond [hereafter cited as Geological Survey Papers, LVA]. Daggs operated the Virginia Hot Springs where a mineral analysis could be useful for business. See Editors of *The William and Mary Quarterly*, "Notes on Hollins College."
2. William Barton Rogers, "Analyses of Marl, Sand, and Soils," 1835, in Box 4, Folder 2, Geological Survey Papers, LVA.
3. Stoll, *Larding the Lean Earth*, 9.
4. This is true despite a general approach by historians to study the state scientific surveys mainly for their contribution to the history of geology. The surveys, as embedded in the era of agricultural improvement, can also be examined for their part in producing new ideas about agrarian land. On the broader state survey movement see Aldrich, "American State Geological Surveys, 1820–1845"; Millbrooke, "State Geological Surveys of the Nineteenth Century"; Socolow, ed., *The State Geological Surveys;* Corgan, ed., *The Geological Sciences in the Antebellum South;* Adams, *Old Dominion, Industrial Commonwealth;* and Aldrich, *The New York Natural History Survey, 1836–1842*. Two other articles offer additional context with their focus on, respectively, the

Pennsylvania (organized by Henry Darwin Rogers) and North Carolina surveys: Boscoe, " 'The Insanities of an Exalted Imagination' " and Smith, "The Conflict between 'Practical Utility' and Geology." Earlier compendiums on geological surveys include Merrill, *The First One Hundred Years of American* Geology, and idem, ed., *Contributions to a History of American State Geological and Natural History Surveys.* Slotten, *Patronage, Practice, and the Culture of American Science,* provides a good example of surveys as aiding the growing cultural authority of science in American society by offering places of employment, opportunities to refine technique, and venues in which to develop and test theories. The U.S. Coastal Survey serves as Slotten's example in this regard. Turner, "The Survey in Nineteenth Century American Geology," examines the survey as the first legitimate source of patronage for American scientists. In environmental history, the surveys have been studied more broadly in their federal form, and as part of the story of the development of the West. See, for example, Worster, *A River Running West,* and Goetzmann, *Exploration and Empire.*

5. *The Richmond Whig,* 11 March 1854, as quoted in Adams, *Old Dominion, Industrial Commonwealth,* 148. Rogers, *Geology of the Virginias,* 754–762, and 172. See also Aldrich and Leviton, "William Barton Rogers and the Virginia Geological Survey, 1835–1842," 83–89.
6. Improvement and the survey were more intimately related than most historical studies reflect. For example, the 1843 report *Improvements in Agriculture and Arts* authored by the commissioner of patents "only once mentioned machines or implements in connection with improvement; instead, [it] named geological surveys, agricultural societies, rural periodicals, and sobriety as pillars of improvement," as noted by Stoll, *Larding the Lean Earth,* 192.
7. The role of slave labor in the survey is difficult to track, since the official records of the project, at best, relate details of conversation between farmers and the survey staff. We can presume, though, given what we know about the labor used to actually extract samples and prepare fertilizers from chapter 4, that behind the many local contributors who offered rock, soil, and water samples to Rogers were slaves actually doing the uncredited work of sample preparation. The difference between the county and state surveys in this regard appears more as a contrast in the scholar's access to information than in the slave's participation in the scientific activity.
8. Almost *all* improvement projects were state efforts in the 1820s and 1830s, a time when support for federal projects was limited in the strong states rights milieu of the early Republic. Larson, *Internal Improvement,* 69, provides a review of state and federal political dynamics of improvement projects. He notes, in particular, that "Madison's veto of the [1817] Bonus Bill [proposed by John Calhoun to use money from the new National Bank to support Federal Internal Improvements] effectively spread the burden of internal improvements, at least for the moment, on the backs of the states or private enterprise."

9. Connections between New York and Virginia were more than incidental. Leaders from the two states were in correspondence about the opportunities and plans for improvement schemes like canals and surveys. John Hartwell Cocke was a frequent correspondent of DeWitt Clinton, one of the champions of such improvement projects in New York. Their letters show the Virginia planter eager to replicate the Erie Canal's success with his state's nascent James River and Kanawha Canal project while introducing all manner of improvement schemes throughout the state. Clinton to JHC, 20 July 1824, in JHC Papers, Box 41. Browne, quoted in Boscoe, " 'The Insanities of an Exalted Imagination,' " 291.
10. Historians have given the era several names. Charles Sellers called it The Market Revolution; before that, George Rogers Taylor called it The Transportation Revolution. Millbrooke and Aldrich among others note that it was the era of scientific surveys. What is more, all of these fall under the broad political category of Jacksonian America. Within the history of science, scholars have also looked to projects like state surveys as markers of professionalization and organization, of a nascent scientific community in America. The American Association for the Advancement of Science (AAAS), for example, was founded in 1848 as an outgrowth of the American Association of Geologists (AAG). Their members had been primary figures in the state survey movement. Rogers and his brother, the geologist Henry Darwin Rogers, helped found the AAG, and then acted as partners to the AAAS's inception. See Slotten, *Patronage, Practice, and the Culture of American Science,* and Kohlstedt, *The Formation of the American Scientific Community.*
11. This political impetus for the survey is consistent with the justifications of other eastern states, as shown in Millbrooke, "State Geological Surveys of the Nineteenth Century," 32–133, and Stoll, *Larding the Lean Earth,* 69–169. Also, see Aldrich and Leviton, "William Barton Rogers and the Virginia Geological Survey, 1835–1842," 88, for a map of county voting patterns on funding the survey. See Stoll, *Larding the Lean Earth,* 13–66, Shade, *Democratizing the Old Dominion,* 30–49, and Kirby, *Poquosin,* passim, for more development on the general political and economic context. Each author refers to and moves beyond Craven, *Soil Exhaustion as a Factor in the Agricultural History of Virginia and Maryland, 1606–1860,* as an early source on the relation of soil exhaustion to demographic shift. Also see Kennedy, *Mr. Jefferson's Lost Cause,* for more detail about the tensions in economic planning on and for the land of Virginia after Jefferson's presidency.
12. Jefferson, *Notes on the State of Virginia,* 36; *The Virginia Argus,* 1 February 1816, Petersburg, VA.
13. For a fuller exegesis of the survey see the annual reports published by his widow Emma Savage Rogers later in the century (1884), cited above as Rogers, *Geology of the Virginias.* Archival material for the survey is available at the Library of Virginia and the MIT Institute Archives and Special Collections, MC-1 (hereafter cited as *Rogers Family Papers,* MIT). Emma Rogers edited and pub-

lished William's letters in *Life and Letters of William Barton Rogers.* The Virginia survey, with cursory biographical details of Rogers, has also been discussed by Ernst, "William Barton Rogers"; Aldrich and Leviton, "William Barton Rogers and the Virginia Geological Survey, 1835–1842"; Milici and Hobbs, "William Barton Rogers and the First Geological Survey of Virginia, 1835–1841"; and Adams, "Partners in Geology, Brothers in Frustration."

14. Rogers, *Geology of the Virginias,* 754–762.
15. See William Barton Rogers, "On the Discovery of Green Sand in the Calcareous Deposit of Eastern Virginia, and on the Probable Existence of this Substance in Extensive Beds near the Western Limits of our Ordinary Marl," *FR* 2 (1834): 134–137; idem, "Apparatus for Analyzing Marl and Carbonates in General," *FR* 2 (1834): 364–365; and idem, "Further Observations on the Green Sand and Calcareous Marl of Virginia," *FR* 3 (1835): 68–71.
16. Littleton Tazewell, "To the House of Delegates" (11 January 1836), as reprinted in Rogers, *Geology of the Virginias,* 22. On Henry Darwin Rogers, see Gerstner, *Henry Darwin Rogers, 1808–1866.*
17. Historians Michele Aldrich and Alan Leviton have studied the political patronage of the legislative bill by examining county-wide voting patterns for the survey. They could not find any guiding pattern of aye/nay votes between eastern and western regions or Whigs and Democrats. Even so, the county delegate votes do show variations in economic bases across the state. The counties of heavy agricultural society involvement—including Albemarle and its neighboring counties—voted for the survey, although many Tidewater and peninsular counties did not; several large counties in the western coal region voted in favor, although many did not cast a vote at all. See Aldrich and Leviton, "William Barton Rogers and the Virginia Geological Survey, 1835–1842," 85.
18. One easy example of the co-produced social and agroeconomic relationship comes during the Civil War, when many western counties in the state voted to remain in the Union while present-day Virginia counties voted to secede. Those non-seceding counties were admitted to the Union in 1863 as the free state of West Virginia.
19. The geological demarcations of the map also closely trace the political regions of the 1830 state constitution. See Shade, *Democratizing the Old Dominion,* 20.
20. Rogers, *Geology of the Virginias,* 156 and 412.
21. See Aldrich and Leviton, "William Barton Rogers and the Virginia Geological Survey, 1835–1842," 92–93, for a synopsis of the assistants to the survey.
22. See Rogers, *Life and Letters,* 179, and Merrill, *Contributions to a History of American State Geological and Natural History Surveys,* 511.
23. As quoted in Merrill, *Contributions to a History of American State Geological and Natural History Surveys,* 509.
24. One thousand copies of the 1837 report were printed in 1838. See James Brown to W. B. Rogers, 18 April 1838, Geological Survey Papers, LVA, Box 1. The next year, Brown told Rogers that the 1838 reports had been printed and

were in hand, though he did not specify how many. James Brown to W. B. Rogers, 15 April 1839, Geological Survey Papers, LVA, Box 1. (Details are unavailable for the other years.) While the annual reports were printed, the final report, which Rogers had promised would be a more substantial and comprehensive document, was never funded. Rogers recognized the value of the summary statements in the reports, so, anticipating problems with final funding, he would include the past year's data in the new year's report. See, for instance, Rogers, *Geology of the Virginias,* 384–385. As another end-around to publication problems, Rogers had the predominantly agricultural focus of the first year's report reprinted in its entirety in *FR* 4 (April 1, 1837): 713–721.

25. Problems of travel, coordination, and geography have been addressed by geographers and historical geographers in great detail. For an entry point to those discussions, see Livingstone, *Putting Science in Its Place.* Issues of enrollment, standardization, and network creation and deployment have been prime territory for science studies scholars, especially within the purview of laboratory studies and the notion of metrology. See Latour, *Science in Action.* Also see idem, *Pandora's Hope,* chapter 2, where he explicates more thoroughly the processes of transforming physical soil into words and concepts and then back (or, "circulating references"). Also see O'Connell, "Metrology." Each of those studies, however, generally starts with the premise that the scientific success of redefining nature needs to be explained. My concern in this chapter starts with the premise that science has already been represented as succeeding, leaving us to wonder what this means for what antebellum agents thought of nature.
26. Taylor, *The Transportation Revolution, 1815–1860.*
27. See letters from C. Hayden to W. B. Rogers, for example, 23 May 1836, 9 May 1837, 23 May 1838, and 25 August 1838, in Geological Survey Papers, LVA, Box 1. The letters of Briggs, Slade, Boyd, Aikin, and James Rogers are spread throughout the Survey Papers.
28. See Rogers, *Geology of the Virginias,* 157.
29. Charles Hayden to W. B. Rogers, 25 August 1838; Israel Slade to W. B. Rogers, 20 June 1839; W. E. A. Aikin to W. B. Rogers, 11 June 1838; and James Rogers to W. B. Rogers, 4 and 5 June 1839, all in Geological Survey Papers, LVA, Box 1. Adams, *Old Dominion, Industrial Commonwealth,*" 141–143, provides a concise overview of these field problems. Also, I do not want to downplay here William Rogers's role as a field worker too—he also traveled the state, though his final responsibility was centered around his home base. His role in redefining the land was more complex, as both the organizer of and a contributor to the survey's data bank.
30. W. B. Rogers to H. D. Rogers, 30 November 1834, as reprinted in Rogers, *Life and Letters,* 113; Rogers, *Geology of the Virginias,* 186.
31. See W. B. Rogers, "Report of the Progress of the Geological Survey of the State of Virginia, for the Year 1836," *FR* 4 (April 1, 1837): 713–721. For W. B.

Rogers to Henry Darwin Rogers, see 27 February 1835, in *Life and Letters,* 118. The *Richmond Compiler and Semi-Weekly Compiler,* 16 January 1836, *Richmond Enquirer,* 2 February 1837, and *Richmond Whig and Public Advertiser,* 3 February 1837 devoted space to reports about and comments on the survey, as noted by Adams, *Old Dominion, Industrial Commonwealth,* 130 and 135. See also Ernst, "William Barton Rogers," 15, for Rogers's friendship with Cabell.

32. "An act to provide for a geological survey of the State, and for other purposes, passed February 29, 1836," as reprinted in Merrill, ed., *Contributions to a History of American State Geological and Natural History Surveys,* 509.
33. Rogers, *Geology of the Virginias,* 193.
34. See, for example, the field notebooks of Thomas Ridgeway, Geological Survey Papers, LVA, Box 3.
35. For the assistants' assignments, see Rogers, *Geology of the Virginias,* 247–248, 413, and 539–541.
36. For a compilation of Rogers's early analyses, see William Barton Rogers's one-volume, 43-page notebook, "Analyses of Marl, Sand, and Soils," (1835) in Geological Survey Papers, LVA, Box 4.
37. Edmund Ruffin, "On the Composition of Soils, and Their Improvement by Calcareous Manures," *AF* 3 (1822): 313–320; Ruffin, *An Essay on Calcareous Manures.*
38. Rogers, *Geology of the Virginias,* 172 and 385–389.
39. Ibid., 150 and 169.
40. Ibid., 153. John Pollard's Account Book for 1830–1843, with details of farming activities, is available at the VHS, Accession Number Mss5:3 P7624:2. Also see the Bagby Family Papers, VHS, Accession Number Mss1 B1463.
41. Rogers, *Geology of the Virginias,* 186.
42. For an examination of the social milieu of the Virginia springs in the antebellum period see Lewis, *Ladies and Gentlemen on Display.*
43. Hezekiah Daggs to W. B. Rogers, 23 March 1837, Geological Survey Papers, LVA, Box 1. The letter from Black Sulphur Springs is to W. B. Rogers, 4 February 1836, *Rogers Family Papers,* MIT, MC-1, Box 13. Although they do not reference any possible legal statutes, see also the letters from John Sites to W. B. Rogers asking for a Spring Water analysis, 26 May 1838, and William Seymour to W. B. Rogers asking for the same, 30 May 1838, and a letter from Samuel McCamant to W. B. Rogers asking for chemical analysis results for "Grayson Sulphur Springs Company," 3 June 1838, Geological Survey Papers, LVA, Box 1.
44. See notebooks of William Rogers and Robert Rogers, from the years 1836–1839, in Geological Survey Papers, LVA, Box 4.
45. Rogers, *Geology of the Virginias,* 346–347.
46. Ibid., 193.
47. Ibid., 416.
48. James Rogers to W. B. Rogers, 4 June and 5 June 1839; Caleb Briggs to W. B. Rogers, 28 May 1840; Israel Slade to W. B. Rogers, 29 July 1840; Briggs to W. B. Rogers, 1 June 1841; Thomas Ridgeway to W. B. Rogers, 27 June 1841, in

Geological Survey Papers, LVA, Box 1. See also a letter, undated, where Briggs and Ridgeway, who were working together at the time, are concerned about their broken thermometer: Letter 156, in Geological Survey Papers, LVA, Box 1.

49. As reprinted in Dr. William Westmore, "The Marl Indicator," *SP* 4 (1844): 221–223.
50. W. B. Rogers, "Apparatus for Analyzing Marl and Carbonates in General," *FR* 2 (1834): 364–365. The report was republished nine years later in *SP* 3 (1843): 203–205, and again decades later in *Geology of the Virginias*, 9–11.
51. Rogers, "Apparatus for Analyzing Marl and Carbonates in General," 364–365. One would do the calculation by assuming, first, that carbonate of lime is comprised of a constant ratio of lime and carbonic acid, at 56-to-44. Thus carbonate of lime contains a fixed percentage of 44 percent carbonic acid; if 2.91 grains of carbonic acid were driven out of the sample, one would multiply 2.91 by 100/44, resulting in, for this example, 6.61 grains of carbonate of lime (about 66 percent of a 10-grain sample).
52. Ibid., 364.
53. See Socolow, ed., *The State Geological Surveys*, and Corgan, ed., *The Geological Sciences in the Antebellum South.*
54. See sources in note 4 above.
55. These speeches were frequent, growing in the years after the 1830s. Lewis Gray and Avery Craven provide useful entry points to references for the 1830s through the 1850s. See Gray, *History of Agriculture in the Southern United States to 1860*, 800–810 and Craven, *Soil Exhaustion as a Factor in the Agricultural History of Virginia and Maryland, 1606–1860*, 122–162.
56. Tying the improvement of an acre of land to the work of the survey is a difficult task; no amount of statistical exegesis could persuasively link the two factors in a reductive cause-effect binary. Property value figures might also be considered useful here—$207 million in 1817, $90 million in 1829, and then $216 million in 1850—but currency fluctuations and variable socioeconomic considerations make conclusions from cross-decades comparisons difficult. These figures are quoted in Fischer and Kelly, *Bound Away*, 202, and from antebellum census statistics available on-line at University of Virginia Library, "Historical Census Browser." Within the scope of Virginia history, it seems that scholars more than a century and a half after the survey still do not present a unanimous view about the broader question of whether Virginia recovered from its declining agricultural status in the antebellum period—not to mention whether specifically the survey, only one of many projects in that period, reversed downward trends. The historian William Shade states his view clearly: "The three decades after Jefferson's death [that is, from 1826 to the 1850s] were characterized by agricultural revival, rather than decline." See Shade, *Democratizing the Old Dominion*, 43. For different interpretations, see references in note 11, above, as well as Mathew, *Edmund Ruffin and the Crisis of Slavery in the Old South* and Allmendinger, *Ruffin.*

57. Rogers, *Geology of the Virginias,* 546.
58. Merchant, *Ecological Revolutions,* 210.
59. Rogers, *Geology of the Virginias,* 192.
60. "B.," "Manure," *SP* 1 (1841): 11; and the Rev. Henry Coleman, "Agricultural Analysis," *SP* 2 (1842): 42–43, reprinted from *The Genesee Farmer,* as noted by the editor, Charles Botts.
61. For statistics on Virginia's workforce, see Shade, *Democratizing the Old Dominion,* 30–46. For an assessment of the sources of knowledge about nature see Merchant, *Ecological Revolutions,* 133. Bushman, *In Old Virginia,* an account of a Methodist farmer of eastern Virginia, provides an excellent example of the local context of agricultural knowledge and farming practice.
62. Aldrich, *New York State Natural History Survey,* 54; Robert Mills to W. B. Rogers, 25 April 1840, Geological Survey Papers, LVA, Box 1; G. P. Marsh as reprinted in Worster, ed., *American Environmentalism,* 16.

CHAPTER 6. AGRICULTURE, ENVIRONMENT, AND THE FUTURE OF GEORGIC SCIENCE

1. Wendell Berry, *What Are People For?,* 149.
2. Fitzgerald, *Every Farm a Factory.* See Danbom, *Born in the Country,* and Hurt, *American Agriculture,* for entry points into general American agricultural history.
3. Lembke, Virgil's *Georgics,* xiii; Sayre, "Cultivating Georgic," 189.
4. Richards, *Indigenous Agricultural Revolution,* 117; Kloppenburg, "Social Theory and De/Reconstruction of Agricultural Science," 528.
5. Chambers, Pacey, and Thrupp, eds., *Farmer First*; Fischer, *Citizens, Experts, and the Environment*; Thompson and Hilde, eds., *The Agrarian Roots of Pragmatism*; Leach, Scoones, and Wynne, eds., *Science and Citizens.*

BIBLIOGRAPHY

PRIMARY SOURCES

Archival Material

Bagby Family Papers. VHS, Richmond, VA.

Baylor Family Papers. VHS, Richmond, VA.

Joseph Cabell, Agricultural Diary. University of Virginia Special Collections Library, Charlottesville, VA.

Robert Wormeley Carter Papers. VHS, Richmond, VA.

John Hartwell Cocke Papers. University of Virginia Special Collections Library, Charlottesville, VA.

Eppes Family Papers. VHS, Richmond, VA.

John Lewis Commonplace Book, 1806–1823. VHS, Richmond, VA.

"Minute Book of the Agricultural Society of Albemarle." VHS, Richmond, VA.

John Pollard Account Book, 1830–1843. VHS, Richmond, VA.

William Barton Rogers Papers. MIT Institute Archives and Special Collections, Cambridge, MA.

John Stuart Skinner Papers. VHS, Richmond, VA.

Virginia Geological Survey Records, 1834–1903. The Library of Virginia, Richmond, VA.

William Fanning Wickham Papers. VHS, Richmond, VA

Periodicals and Journals

American Journal of Science.

The Literary Magazine and American Register.

Memoirs of the Philadelphia Society for Promotion of Agriculture.

Memoirs of the Society of Virginia for Promoting Agriculture.
Proceedings of the Agricultural Society of Virginia.
Transactions of the American Philosophical Society.
Transactions of the Society for the Promotion of Agriculture, Arts, and Manufactures of New York.

Primary Works

Adams, Daniel. *The Agricultural Reader, Designed for the Use of Schools.* Boston: Richardson and Lord, 1824.

Adams, Daniel. *The Scholar's Arithmetic.* Leominster, MA: Printed by Salmon Wilder, 1805. 3rd edition.

Adams, Daniel. *School Atlas to Adams' Geography.* Boston: Lincoln and Edmands, 1823.

Allibone, Samuel Austin. *Critical Dictionary of English Literature, and British and American Authors, Living and Deceased, from the Earliest Accounts to the Middle of the Nineteenth Century.* Philadelphia: J. B. Lippincott, 1859–1871.

Binns, John. *A Treatise on Practical Farming; Embracing Particularly the Following Subjects, Viz. the Use of Plaister of Paris, with Directions for Using It: And General Observations on the Use of Other Manures.* Fredericktown, MD: Printed by John B. Colvin, 1803.

Boussingault, Jean Baptiste. *Rural Economy, in its Relation with Chemistry, Physics, and Meteorology, or, Chemistry Applied to Agriculture.* New York: D. Appleton; Philadelphia: G. S. Appleton, 1850.

Buel, Jesse. *The Farmer's Companion; or, Essays on the Principles and Practice of American Husbandry.* Boston: Marsh, Capen, Lyon, and Webb, 1840.

Butler, Frederick. *The Farmer's Manual: Being a Plain Practical treatise on the Art of Husbandry Designed to promote an Acquaintance with the Modern Improvements in Agriculture.* Hartford, CT: Samuel Goodrich, 1819.

Cato. *De Agricultura.* Totnes, UK: Prospect, 1998.

Chevalier, Michel. *Society, Manners, and Politics in the United States: Letters on North America.* Gloucester, MA: P. Smith, [1839] 1967.

Cobbett, William. *A Year's Residence in the United States of America: Treating of the Face of the Country, the Climate, the Soil, the Products, the Mode of Cultivating the Land, the Prices of Land, of Labour, of Food, of Raiment.* New York: A. M. Kelley, [1819] 1969.

Cochrane, Archibald, Earl of Dundonald. *A Treatise Shewing the Intimate Connection that Subsists between Agriculture and Chemistry.* London: For the Author, 1795.

Comstock, John Lee. *Elements of Chemistry.* New York: Robinson, Pratt, 1839.

Dana, Samuel. *A Muck Manual for Farmers.* Lowell, MA: James P. Walker, [1842] 1851. 3rd edition.

Darwin, Erasmus. *Phytologia; or The philosophy of agriculture and gardening, with the theory of draining morasses and with an improved construction of the drill plough.* London: J. Johnson, 1800.

Davy, Humphry. *Elements of Agricultural Chemistry*. In John Davy, ed., *The Collected Works of Sir Humphry Davy*. London: Smith, Elder, [1813] 1840. Volume VII.

Deane, Samuel. *The New England Farmer; or Georgical Dictionary*. New York: Arno, 1972. Reprint of 3rd edition, Boston: Wells and Lilly, [1790] 1822.

Dickens, Charles. *American Notes*. New York: St. Martin's Press, [1842] 1957.

Dostoevsky, Fyodor. *Notes from Underground*. In *Existentialism from Dostoevsky to Sartre*, ed. Walter Kaufmann, 52–82. New York: Penguin Books, 1975.

Drown, William. *Compendium of Agriculture: or, The Farmer's Guide, in the Most Essential Parts of Husbandry and Gardening, Compiled from the Best American and European Publications, and the Unwritten Opinions of Experienced Cultivators*. Providence: Field and Maxcy, 1824.

Dwight, Timothy. *Greenfield Hills: A Poem*. New York: Childs and Swaine, 1794.

Eliot, Jared. *Essays Upon Field Husbandry in New England, and Other Papers, 1748–1762*. Ed. Harry Carman and Rexford Tugwell. New York: Columbia University Press, [1762] 1934.

Ewell, Thomas. *Plain Discourses on the Laws or Properties of Matter*. New York: Brisban and Brannan, 1806.

Fessenden, Thomas. *The Complete Farmer and Rural Economist; Containing a Compendious Epitome of the Most Important Branches of Agriculture and Rural Economy*. Boston: Lilly, Wait, and Company, and G. C. Barrett, 1834.

[Fessenden, Thomas] Christopher Caustick. *The Modern Philosopher or Terrible Tractoration!* Philadelphia: E. Bronson, 1806.

Forbes, Stephen. "The Lake as a Microcosm." *Bulletin of the Peoria Scientific Association*. 87 (1887): 77–87.

Fothergill, A. "On the application of chemistry to agriculture." *Letters and Papers of the Bath Society* 3 (1791): 54–62.

Freshfield, Douglas, and W. J. L. Wharton, eds. *Hints to Travelers Scientific and General*. London: The Royal Geographical Society, 1893.

Home, Francis. *The Principles of Agriculture and Vegetation*. Edinburgh: Sands, Donaldson, Murray, and Cochran, 1756.

Home, Henry (Lord Kames). *The Gentleman Farmer: Being an Attempt to Improve Agriculture by Subjecting it to the Test of Rational Principles*. Edinburgh: W. Creech, 1776.

Hunter, Alexander. *Georgical Essays*. London: A. Ward, 1770–1774. 8 volumes.

Irving, Washington, and James Kirke Paulding. *Salmagundi; Or, The Whim-Whams and Opinions of Launcelot Langstaff, Esq and Others*. New York: David Longworth, 1807–1808.

Jefferson, Thomas. *The Correspondence of Jefferson and Du Pont de Nemours*. Baltimore: The Johns Hopkins Press, 1931.

Jefferson, Thomas. *Notes on the State of Virginia*. New York: Penguin, [1787] 1999.

Jefferson, Thomas. *Thomas Jefferson's Farm Book*. Ed. Edwin Morris Betts. Charlottesville, VA: University Press of Virginia, 1976.

Jefferson, Thomas. *Thomas Jefferson's Garden Book, 1766–1824*. Ed. Edwin Morris Betts. Philadelphia: The American Philosophical Society, 1944.

Johnston, J. F. W. *Notes on North America, Agricultural, Economical, and Social.* Boston: C. C. Little and J. Brown, 1851.

Johnston, J. F. W. "The Present State of Agriculture in its Relation to Chemistry and Geology." *Journal of the Royal Agricultural Society of England* 9 (1848): 200–236.

Liebig, Justus. *Chemistry in its Application to Agriculture and Physiology.* Ed. Lyon Playfair. London: Printed for Taylor & Walton, [1840] 1842. 2nd edition.

Lorain, John. *Nature and Reason Harmonized in the Practice of Husbandry.* Philadelphia: H. C. Carey and I. Lea, 1825.

Madison, James. "Address to the Albemarle Agricultural Society," May 1818. Electronic edition. Accessed 25 February 2008 at http://www.jmu.edu/madison/center/main—pages/madison—archives/life/retirement/ag.htm.

Marcet, Jane. *Conversations on Chemistry, Intended More Especially for the Female Sex.* Greenfield, MA: Denio and Phelps, [1807] 1818. 4th edition.

Marsh, George. *Man and Nature.* Ed. David Lowenthal. Seattle: University of Washington Press, [1864] 2003.

Marsh, George. "The Study of Nature," *The Christian Examiner.* 1860. [n.p.]

Nichols, James R. *Chemistry of the Farm and Sea, with Other Familiar Chemical Essays* New York: Arno, [1867] 1970.

Olmsted, Frederick Law. *A Journey in the Seaboard Slave States; With Remarks on Their Economy.* New York: Dix and Edwards, 1856. Electronic Edition. Accessed 20 October 2007 at http://docsouth.unc.edu/nc/olmsted/olmsted.html.

Peters, Richard. *Agricultural Enquiries on Plaister of Paris.* Philadelphia: PSPA, 1797.

Petzholdt, Alexander. *Lectures to Farmers on Agricultural Chemistry.* New York: Greeley and McElrath, 1846.

Pritts, Joseph. *The Farmer's Book and Family Instructor.* Chambersburg, PA: Printed for Purchasers, 1845.

Rogers, William Barton. *Life and Letters of William Barton Rogers.* Ed. Emma Rogers. New York: Houghton Mifflin, 1896. 2 volumes.

Rogers, William Barton. *A Reprint of Annual Reports and Other Papers, on the Geology of the Virginias.* New York: D. Appleton, 1884.

Ruffin, Edmund. *Agriculture, Geology, and Society in Antebellum South Carolina: The Private Diary of Edmund Ruffin, 1843.* Ed. William Mathew. Athens: University of Georgia Press, [1843] 1992.

Ruffin, Edmund. *An Essay on Calcareous Manures.* Ed. J. Carlyle Stitterson. Cambridge, MA: Harvard University Press, [1832] 1961. 4th edition.

Ruffin, Edmund. *Nature's Management: Writings on Landscape and Reform, 1822–1859.* Ed. Jack Temple Kirby. Athens: University of Georgia Press, 2000.

Ruffin, Edmund. "Writers of Anonymous Articles in *The Farmer's Register.*" *Journal of Southern History* 23 (1957): 9–102.

Smith, Joseph. *Productive Farming: Or, a Familiar Digest of the Recent Discoveries of Liebig, Johnston, Davy, and other Celebrated Writers on Vegetable Chemistry.* New York: Wiley and Putnam, 1843.

Spurrier, John. *The Practical Farmer, Being a New and Compendious System of Husbandry, Adapted to the Different Soils and Climates of America. Containing the Mechanical, Chemical, and Philosophical Elements of Agriculture.* Wilmington, DE: Brynberg and Andrews Printers, 1793.

Squarey, Charles. *A Popular Treatise on Agricultural Chemistry: Intended for the Use of the Practical Farmer.* Philadelphia: Lea & Blanchard, 1842.

Stephens, Henry. *Book of the Farm.* New York: Greeley and McElrath, 1847.

Strickland, William. *Observations on the Agriculture of the United States of America.* London: W. Bulmer, 1801.

Taylor, John. *Arator: Being a Series of Agricultural Essays, Practical and Political, in Sixty-Four Numbers.* Ed. M. E. Bradford. Indianapolis: Liberty Fund, [1813] 1977.

Tocqueville, Alexis de. *Democracy in America.* Ed. Andrew Hacker. New York: Washington Square Press, [1835–1840] 1964. Volume II.

Tull, Jethro. *The Horse-hoeing Husbandry; or, An Essay on the Principles of Tillage and Vegetation.* London: Printed for the Author, 1733.

Virgil. *Georgics.* Ed. Roger Mynors. New York: Oxford University Press, 1990.

Virgil. *Georgics.* Trans. Janet Lembke. New Haven, CT: Yale University Press, 2005.

Washington, George. *The Diaries of George Washington.* Ed. Donald Jackson and Dorothy Twohig. Charlottesville, VA: University Press of Virginia, 1976.

Washington, George. *Letters from His Excellency George Washington, to Arthur Young . . . and Sir John Sinclair . . . Containing an Account of His Husbandry, with His Opinions on Various Questions in Agriculture; and Many Particulars of the Rural Economy of the United States.* Alexandria: Cottom and Stewart, 1803.

Washington, George. *The Papers of George Washington, Retirement Series.* Ed. Dorothy Twohig. Charlottesville, VA: University Press of Virginia, 1998.

Young, Arthur. *The Farmer's Tour through the East of England.* London: Printed for W. Strahan, 1771. 4 volumes.

Young, Arthur. *A Six Months Tour through the North of England, Containing, an Account of the Present State of Agriculture, Manufactures and Population, in Several Counties of this Kingdom.* 4 volumes. London: Printed for W. Strahan, 1771.

Young, Arthur. *Travels During the Years 1787, 1788, & 1789.* London: W. Richardson, 1794.

Young, John. *The Letters of Agricola on the Principles of Vegetation and Tillage.* Halifax, NS: Printed by Holland and Company, 1822.

REFERENCES

American National Biography. Ed. John Garraty and Mark Carnes. New York : Oxford University Press, 1999. 24 volumes.

Appleton's Cyclopedia of American Biography. Ed. James Grant Wilson and John Fiske. New York: D. Appleton, 1887–1889. 6 volumes.

Dictionary of American Biography. Ed. Allen Johnson, Dumas Malone, Marris E. Starr, and Robert Livingston Schuyler. New York: Scribner, 1943–1958. 22 volumes.

Oxford Dictionary of National Biography [British]. Ed. H. C. G. Matthew and Brian Harrison. New York : Oxford University Press, 2004.

SECONDARY SOURCES

Abbot, Richard. "The Agricultural Press Views the Yeoman: 1819–1859." *Agricultural History* 42 (1968): 35–44.

Adams, Sean Patrick. *Old Dominion, Industrial Commonwealth: Coal, Politics, and Economy in Antebellum America*. Baltimore: Johns Hopkins University Press, 2004.

Adams, Sean Patrick. "Partners in Geology, Brothers in Frustration: The Antebellum Geological Surveys of Virginia and Pennsylvania." *The Virginia Magazine of History and Biography* 106 (1998): 5–33.

Adams, Stephen. *The Best and Worst Country in the World: Perspectives on the Early Virginia Landscape*. Charlottesville, VA: University Press of Virginia, 2001.

Aldrich, Michele. "American State Geological Surveys, 1820–1845." In *Two Hundred Years of Geology in America*, ed. Cecil Schneer, 133–144. Hanover, NH: University Press of New England, 1979.

Aldrich, Michele. *The New York State Natural History Survey, 1836–1842*. Ithaca, NY: Paleontological Research Institute, 2000.

Aldrich, Michele, and Alan Leviton. "William Barton Rogers and the Virginia Geological Survey, 1835–1842." In *The Geological Sciences in the Antebellum South*, ed. James Corgan, 83–89. Tuscaloosa: University of Alabama Press, 1982.

Allmendinger, David. *Ruffin: Family and Reform in the Old South*. New York: Oxford University Press, 1990.

Alpers, Paul. *What Is Pastoral?* Chicago: University of Chicago Press, 1996.

Amber, Charles Henry. *Thomas Ritchie: A Study in Virginia Politics*. Richmond, VA: Bell Book and Stationery, 1913.

Aulie, Richard. "Boussingault and the Nitrogen Cycle." *Proceedings of the American Philosophical Society* 114 (1970): 433–479.

Baatz, Simon. *"Venerate the Plough": A History of the Philadelphia Society for the Promotion of Agriculture, 1785–1985*. Philadelphia: Philadelphia Society for the Promotion of Agriculture, 1985.

Baker, Keith. *Condorcet: From Natural Philosophy to Social Mathematics*. Chicago: University of Chicago Press, 1975.

Batstone, William. "Virgilian Didaxis: Value and Meaning in the *Georgics*." In *The Cambridge Companion to Virgil*, ed. Charles Martindale, 125–144. Cambridge, UK: Cambridge University Press, 1997.

Bedell, Rebecca. *The Anatomy of Nature: Geology and American Landscape Painting, 1825–1875*. Princeton, NJ: Princeton University Press, 2001.

Bedini, Silvio. *Thomas Jefferson: Statesman of Science*. New York: Macmillan, 1990.

Bensaude-Vincent, Bernadette, and Isabelle Stengers. *The History of Chemistry*. Cambridge, MA: Harvard University Press, 1995.

Benton, E. "Vitalism in 19th-Century Scientific Thought: A Typology and Reassessment." *Studies in History and Philosophy of Science* 5 (1974): 17–48.

Bercovitch, Sacvan, and Myra Jehlen, eds. *Ideology and Classic American Literature*. New York: Cambridge University Press, 1986.

Berlin, Ira, and Philip Morgan, eds. *Cultivation and Culture: Labor and the Shaping of Slave Life in the Americas*. Charlottesville, VA: University Press of Virginia, 1993.

Berry, Wendell. *Recollected Essays, 1965–1980*. San Francisco: North Point, 1980.

Berry, Wendell. *The Unsettling of America: Culture and Agriculture*. San Francisco: Sierra Club Books, [1977] 1996. 3rd edition.

Berry, Wendell. *What Are People For?* New York: FSG, 1990.

Bierck, Harold A. "Spoils, Soils, and Skinner." *Maryland Historical Magazine* 49 (1954): 21–40, 143–155.

Black, Charles. *Soil Fertility Evaluation and Control*. Boca Raton, FL: Lewis, 1993.

Boscoe, Francis. " 'The Insanities of an Exalted Imagination': The Troubled First Geological Survey of Pennsylvania." *The Pennsylvania Magazine of History and Biography* 127 (2003): 291–308.

Bowler, Peter. *Evolution: The History of an Idea*. Berkeley, CA: University of California Press, [1984] 2003. 3rd edition.

Bowler, Peter. *The Norton History of the Environmental Sciences*. New York: Norton, 1992.

Bradford, M. E. "A Virginia Cato: John Taylor of Caroline and the Agrarian Republic." In John Taylor's *Arator*. Indianapolis: Liberty Fund, 1977.

Bressler, R. G. "The Impact of Science on Agriculture." *Journal of Farm Economics* 40 (1958): 1005–1015.

Brock, William. *Justus von Liebig: The Chemical Gatekeeper*. New York: Cambridge University Press, 1997.

Brock, William. *The Norton History of Chemistry*. New York: Norton, 1992.

Brown, Chandos Michael. *Benjamin Silliman: A Life in the Young Republic*. Princeton, NJ: Princeton University Press, 1989.

Brown, Ralph. "Agricultural Science and Education in Virginia before 1860." *William and Mary Quarterly* 19 (1939): 197–213.

Bruce, Robert. *The Launching of Modern American Science, 1846–1876*. New York: Knopf, 1987.

Brunt, Liam. "Rehabilitating Arthur Young." *Economic History Review* 56 (2003): 265–299.

Bryson, Michael. *Visions of the Land: Science, Literature, and the American Environment from the Era of Exploration to the Age of Ecology*. Charlottesville, VA: University Press of Virginia, 2002.

Buell, Lawrence. *The Environmental Imagination: Thoreau, Nature Writing, and the Formation of American Culture*. Cambridge, MA: Harvard University Press, 1995.

Bullion, Brenda. "The Agricultural Press: To Improve the Soil and the Mind." In *The Farm*, ed. Peter Benes, 74–94. Dublin Seminar for New England Folklife, Annual Proceedings 1986: Boston University Scholarly Publications, 1988.

Burnett, D. Graham. *Trying Leviathan: The Nineteenth-Century New York Court Case That Put the Whale on Trial and Challenged the Order of Nature*. Princeton, NJ: Princeton University Press, 2007.

Bushman, Claudia. *In Old Virginia: Slavery, Farming, and Society in the Journal of John Walker.* Baltimore: Johns Hopkins University Press, 2002.

Callon, Michel. "Some Elements of a Sociology of Translation: Domestication of the Scallops and the Fishermen of St Brieuc Bay." In *Power, Action and Belief. A New Sociology of Knowledge?*, ed. John Law, 196–233. London: Routledge and Kegan Paul, 1986.

Carney, Judith. *Black Rice: The African Origins of Rice Cultivation in the Americas.* Cambridge, MA: Harvard University Press, 2001.

Carter, Edward, ed. *Surveying the Record: North American Scientific Exploration to 1930.* Philadelphia: American Philosophical Society, 1999.

Cashin, Joan. "Landscape and Memory in Antebellum Virginia." *The Virginia Magazine of History and Biography* 102 (1994): 477–500.

Chambers, Robert, Arnold Pacey, and Lori Ann Thrupp, eds. *Farmer First: Farmer Innovation and Agricultural Research.* London: Intermediate Technology, 1989.

Chaplin, Joyce. *An Anxious Pursuit: Agricultural Innovation and Modernity in the Lower South, 1730–1815.* Chapel Hill, NC: University of North Carolina Press, 1993.

Cohen, I. B. *Science and the Founding Fathers: Science in the Political Thought of Jefferson, Franklin, Adams, and Madison.* New York: Norton, 1995.

Cohen, I. B., ed. *Jefferson and the Sciences.* New York: Arno, 1980.

Collins, Harry. *Changing Order: Replication and Induction in Scientific Practice.* Chicago: University of Chicago Press, 1985.

Collins, Harry, and Trevor Pinch. *The Golem: What You Should Know about Science.* Cambridge, UK : Cambridge University Press, 1998.

Conlogue, William. *Working the Garden: American Writers and the Industrialization of Agriculture.* Chapel Hill, NC: University of North Carolina Press, 2001.

Cooter, Roger. "Separate Spheres and Public Places: Reflections of the History of Science Popularization and Science in Popular Culture." *History of Science* 32 (1994): 237–267.

Corgan, James X., ed. *The Geological Sciences in the Antebellum South.* Tuscaloosa, AL: University of Alabama Press, 1982.

Cowdrey, Albert. *This Land, This South: An Environmental History.* Lexington, KY: University Press of Kentucky, 1996. 2nd edition.

Craven, Avery. *Soil Exhaustion as a Factor in the Agricultural History of Virginia and Maryland, 1606–1860.* Urbana, IL: University of Illinois Press, 1926.

Cronon, William. "Cutting Loose or Running Aground?" *Journal of Historical Geography* 20 (1994): 38–43.

Cronon, William, ed. *Uncommon Ground: Rethinking the Human Place in Nature.* New York: Norton, 1995.

Crosby, Alfred. "An Enthusiastic Second." *Journal of American History* 76 (1990): 1107–1110.

Cunningham, Andrew, and Nicholas Jardine, eds. *Romanticism and the Sciences.* Cambridge, UK: Cambridge University Press, 1990.

Cunningham, Andrew, and Perry Williams. "De-Centring the 'Big Picture': *The*

Origins of Science and the Modern Origins of Science." *British Journal for the History of Science* 26 (1993): 407–432.

Curtin, Philip, ed. *Discovering the Chesapeake: The History of an Ecosystem*. Baltimore: Johns Hopkins University Press, 2001.

Danbom, David. *Born in the Country: A History of Rural America*. Baltimore: The Johns Hopkins University Press, 1995.

Danhof, Clarence. *Change in Agriculture: The Northern United States, 1820–1870*. Cambridge, MA: Harvard University Press, 1969.

Daniels, George. *American Science in the Age of Jackson*. New York: Columbia University Press, 1968.

Daniels, George, ed. *Nineteenth-Century American Science: A Reappraisal*. Evanston, IL: Northwestern University Press, 1972.

Daston, Lorraine. "The Ideal and Reality of the Republic of Letters in the Enlightenment." *Science in Context* 4 (1991): 367–386.

Davidson, Arnold. *The Emergence of Sexuality: Historical Epistemology and the Formation of Concepts*. Cambridge, MA: Harvard University Press, 2001.

Davis, Donald. *Where There Are Mountains: An Environmental History of the Southern Appalachians*. Athens, GA: University of Georgia Press, 2000.

De Certeau, Michel. *The Practice of Everyday Life*. Berkeley, CA: University of California Press, [1974] 1984.

Demaree, Albert. *The American Agricultural Press, 1819–1860*. Philadelphia: Porcupine, [1941] 1974.

Demaree, Albert. "The Farm Journals, Their Editors, and Their Public, 1830–1860." *Agricultural History* 15 (1941): 182–188.

Demeritt, David. "Ecology, Objectivity, and Critique in Writings on Nature and Human Societies." *Journal of Historical Geography* 20 (1994): 22–37.

Dew, Charles. *Bond of Iron: Master and Slave at Buffalo Forge*. New York: Norton, 1994.

Dowling, William. *Poetry and Ideology in Revolutionary Connecticut*. Athens, GA: University of Georgia Press, 1990.

Dupree, A. Hunter. *Science in the Federal Government: A History of Politics and Activities*. Cambridge, MA: Harvard University Press, 1957.

Editors of *The William and Mary Quarterly*, "Notes on Hollins College." *The William and Mary Quarterly* 9 (October 1929): 330–331.

Epstein, Steven. *Impure Science: AIDS, Activism, and the Politics of Knowledge*. Berkeley, CA: University of California Press, 1995.

Ernst, William. "William Barton Rogers: Antebellum Virginia Geologist." *Virginia Cavalcade* 24 (1974): 13–21.

Faust, Drew Gilpin. *A Sacred Circle: The Dilemma of the Intellectual in the Old South, 1840–1860*. Baltimore: Johns Hopkins University Press, 1977.

Fea, John. "The Way of Improvement Leads Home: Philip Vickers Fithian's Rural Enlightenment." *Journal of American History* 90 (2003): 462–90.

Feller, Daniel. *The Jacksonian Promise: America, 1815–1840*. Baltimore: Johns Hopkins University Press, 1995.

Ferguson, Robert. *The American Enlightenment, 1750–1820*. Cambridge, MA: Harvard University Press, 1997.

Findlay, Mark. "The Rehabilitation of an Agricultural Chemist: Justus von Liebig and the Seventh Edition." *Ambix* 38 (1991): 155–167.

Fischer, David Hackett, and James C. Kelly. *Bound Away: Virginia and the Westward Movement.* Charlottesville, VA: University Press of Virginia, 2000.

Fischer, Frank. *Citizens, Experts, and the Environment: The Politics of Local Knowledge.* Durham, NC: Duke University Press, 2000.

Fitzgerald, Deborah. *Every Farm a Factory: The Industrial Ideal in American Agriculture.* New Haven, CT: Yale University Press, 2003.

Fitzgerald, Deborah. "Farmers Deskilled: Hybrid Corn and Farmers' Work." *Technology and Culture* 33 (1993): 324–343.

Forgan, Sophie, ed. *Science and the Sons of Genius: Studies on Humphry Davy.* London: Science Reviews, 1980.

Foster, George. *Traditional Societies and Technological Change.* New York: Harper and Row, 1973.

Fox-Genovese, Elizabeth. *The Origins of Physiocracy: Economic Revolution and Social Order in Eighteenth-Century France.* Ithaca, NY: Cornell University Press, 1976.

Fussell, G. E. "Science and Practice in Eighteenth-Century British Agriculture." *Agricultural History* 43 (1969): 7–18.

Gates, Paul W. *The Farmer's Age: Agriculture, 1815–1860.* New York: Harper Torchbook, 1960.

Gee, Brian. "Amusement Chests and Portable Laboratories: Practical Alternatives to the Regular Laboratory." In *The Development of the Laboratory,* ed. Frank James, 37–59. Basingstoke, UK: Macmillan, 1989.

Gerstner, Patsy. *Henry Darwin Rogers, 1808–1866: American Geologist.* Tuscaloosa, AL: University of Alabama Press, 1994.

Goetzmann, William. *Exploration and Empire: The Explorer and the Scientist in the Winning of the American West.* New York: Knopf, [1966] 1993.

Golinski, Jan. *Science as Public Culture: Chemistry and Enlightenment in Britain, 1760–1820.* Cambridge, UK: Cambridge University Press, 1992.

Goodman, Dena. *The Republic of Letters: A Cultural History of the French Enlightenment.* Ithaca, NY: Cornell University Press, 1994.

Graham, Otis. "Again the Backward Region? Environmental History in and of the American South," *Southern Cultures* 6 (2000): 50–72.

Grammer, John. *Pastoral and Politics in the Old South.* Baton Rouge, LA: Louisiana State University Press, 1996.

Gramsci, Antonio. *An Antonio Gramsci Reader.* Ed. David Forgacs. New York: Schocken Books, 1988.

Gray, Lewis. *History of Agriculture in the Southern United States to 1860.* New York: Peter Smith, [1933] 1941. 2 volumes.

Greene, John. *American Science in the Age of Jefferson.* Ames, IA: Iowa State University Press, 1984.

Hall, Vance. *A History of the Yorkshire Agricultural Society, 1837–1987: In Celebration of the 150th Anniversary of the Society.* London: Chrysalis Books, 1987.

Hankins, Barry. *The Second Great Awakening and the Transcendentalists.* Westport, CT: Greenwood Press, 2004.

Hayes, Elizabeth. "Science, Politics, and Satire: Reconsidering the American Philosophical Society in Its Political Contexts, 1771–1806." Paper delivered at the History of Science Society Annual Meeting, Cambridge, MA, November 20–23, 2003.

Headlee, Sue. *The Political Economy of the Family Farm: Agrarian Roots of American Capitalism*. New York: Praeger, 1991.

Helliar-Symons, Catherine. "John Hartwell Cocke and Bremo: A Study of Plantation Life." M.S., University of Wales, 2003.

Hofstra, Warren. *The Planting of New Virginia: Settlement and Landscape in the Shenandoah Valley*. Baltimore: Johns Hopkins University Press, 2004.

Holmes, Frederic, and Trevor H. Levere, eds. *Instruments and Experimentation in the History of Chemistry*. Cambridge, MA: MIT Press, 2000.

Holton, Gerald. "On the Jeffersonian Research Program." *Archives internationales d'histoire des sciences* 36 (1986): 325–336.

Hurt, R. Douglas. *American Agriculture: A Brief History*. West Lafayette, IN: Purdue University Press, 2002.

Ihde, Aaron. *The Development of Modern Chemistry*. New York: Harper and Row, 1964.

Ihde, Aaron. "Edmund Ruffin, Soil Chemist of the Old South." *Journal of Chemical Education* 29 (1952): 407–414.

Isaac, Rhys. *Landon Carter's Uneasy Kingdom: Revolution & Rebellion on a Virginia Plantation*. New York: Oxford University Press, 2004.

Isaac, Rhys. *The Transformation of Virginia, 1740–1790*. Chapel Hill, NC: University of North Carolina Press, 1982.

Jackson, Wes. *New Roots for Agriculture*. Lincoln, NE: University of Nebraska Press, 1985.

Jacobson, David. *Place and Belonging in America*. Baltimore: Johns Hopkins University Press, 2002.

James, Frank, ed. *The Development of the Laboratory*. Basingstoke, UK: Macmillan, 1989.

Jehlen, Myra. *American Incarnations: The Individual, the Nation, and the Continent*. Cambridge, MA: Harvard University Press, 1986.

Johnson, Hildegard. *Order upon the Land: The U.S. Rectangular Survey and the Upper Mississippi Country*. New York: Oxford University Press, 1986.

Judd, Richard. *Common Lands, Common People: The Origins of Conservation in New England*. Cambridge, MA: Harvard University Press, 1997.

Keeney, Elizabeth. *The Botanizers: Amateur Scientists in Nineteenth-Century America*. Chapel Hill, NC: University of North Carolina Press, 1992.

Kennedy, Roger. *Mr. Jefferson's Lost Cause: Land, Farmers, Slavery, and the Louisiana Purchase*. New York: Oxford University Press, 2003.

Kirby, Jack Temple. *Poquosin: A Study of Rural Landscape*. Chapel Hill, NC: University of North Carolina Press, 1995.

Kirby, Jack Temple. "Virginia's Environmental History: A Prospectus." *Virginia Magazine of History and Biography* 99 (1990): 449–488.

Kirsch, Scott. "John Wesley Powell and the Mapping of the Colorado Plateau, 1869–1879: Survey Science, Geographical Solutions, and the Economy of Environmental Values." *Annals of the Association of American Geographers* 92 (2002): 548–572.

Kirschke, Martin. "Liebig, His University Professor Karl Wilhelm Gottlob Kastner (1783–1857), and His Problematic Relation with Romantic Natural Philosophy." *Ambix* 50 (2003): 3–24.

Klein, Ursula. *Experiments, Models, Paper Tools: Cultures of Organic Chemistry in the Nineteenth Century*. Palo Alto, CA: Stanford University Press, 2003.

Kline, Ronald. *Consumers in the Country: Technology and Social Change in Rural America*. Baltimore: Johns Hopkins University Press, 2002.

Klonk, Charlotte. *Science and the Perception of Nature: British Landscape Art in the Late Eighteenth and Early Nineteenth Centuries*. New Haven, CT: Yale University Press, 1996.

Kloppenberg, James. "The Virtues of Liberalism: Christianity, Republicanism, and Ethics in Early American Political Discourse." *Journal of American History* 74 (1987): 9–33.

Kloppenburg, Jack. "Social Theory and De/Reconstruction of Agricultural Science: Local Knowledge for an Alternative Agriculture." *Rural Sociology* 56 (1991): 519–548.

Knight, David. "Accomplishment or Dogma: Chemistry in the Introductory Works of Jane Marcet and Samuel Parkes." *Ambix* 33 (1986): 88–98.

Knight, David. "Agriculture and Chemistry in Britain around 1800." *Annals of Science* 33 (1976): 187–196.

Knight, David. *Humphry Davy: Science and Power*. Cambridge, UK: Cambridge University Press, 1992.

Kohler, Robert. *Labscapes and Landscapes: Exploring the Lab-Field Border in Biology*. Chicago: University of Chicago Press, 2002.

Kohlstedt, Sally. "Curiosities and Cabinets: Natural History Museums and Education on the Antebellum Campuses." *Isis* 79 (1988): 405–426.

Kohlstedt, Sally. *The Formation of the American Scientific Community: The American Association for the Advancement of Science, 1846–1860*. Urbana: University of Illinois Press, 1976.

Kohlstedt, Sally. "The Nineteenth Century Amateur Tradition: The Case of the Boston Society of Natural History." In *Science and Its Publics: The Changing Relationship; Boston Studies in the Philosophy of Science*, ed. G. Holton and W. Blanpied, 173–190. Boston: D. Reidel, 1976. Volume 33.

Kohlstedt, Sally. "Parlors, Primers, and Public Schooling: Education for Science in Nineteenth Century America." *Isis* 81 (1990): 424–445.

Kohlstedt, Sally, and Margaret Rossiter, eds. "Historical Writing on American Science." *Osiris* 1 (1985): 1–321.

Koons, Kenneth, and Warren Hofstra, eds. *After the Backcountry: Rural Life in the Great Valley of Virginia, 1800–1900*. Knoxville, TN: University of Tennessee Press, 2000.

Kuklick, Henrika, and Robert Kohler, eds. "Science in the Field." *Osiris* 11 (1996): 1–265.

Kulikoff, Allan. *The Agrarian Origins of American Capitalism*. Charlottesville, VA: University Press of Virginia, 1992.

Kurzer, Frederick. "Samuel Parkes—Chemist, Author, Reformer; A Biography." *Annals of Science* 54 (1997): 431–462.

Larson, John Lauritz. *Internal Improvement: National Public Works and the Promise of Popular Government in the Early United States*. Chapel Hill, NC: University of North Carolina Press, 2001.

Latour, Bruno. *Pandora's Hope*. Cambridge, MA: Harvard University Press, 1999.

Latour, Bruno. *Science in Action*. Cambridge, MA: Harvard University Press, 1987.

Latour, Bruno, and Steve Woolgar. *Laboratory Life: The Construction of Scientific Facts*. Princeton, NJ: Princeton University Press, [1979] 1986.

Laudan, Rachel. *From Mineralogy to Geology: The Foundations of a Science, 1650–1830*. Chicago: University of Chicago Press, 1987.

Lawrence, Christopher. "The Power and the Glory: Humphry Davy and Romanticism." In *Romanticism and the Sciences*, ed. Andrew Cunningham and Nicholas Jardine, 213–227. Cambridge, UK: Cambridge University Press, 1990.

Leach, Melissa, Ian Scoones, and Brian Wynne, eds. *Science and Citizens: Globalization and the Challenge of Engagement*. London: Zed Books, 2006.

Levere, Trevor. "Coleridge, Chemistry, and the Philosophy of Nature." *Studies in Romanticism* 16 (1977): 349–379.

Levere, Trevor. *Transforming Matter: A History of Chemistry from Alchemy to the Buckyball*. Baltimore: Johns Hopkins University Press, 2001.

Lewenstein, Bruce. " 'To Improve Our Knowledge in Nature and Arts': A History of Chemical Education in the United States." *Journal of Chemical Education* 66 (1989): 37–44.

Lewis, Charlene M. Boyer. *Ladies and Gentlemen on Display: Planter Society at the Virginia Springs, 1790–1860*. Charlottesville, VA: University Press of Virginia, 2001.

Lewis, Ronald. *Transforming the Appalachian Countryside: Railroads, Deforestation, and Social Change in West Virginia, 1880–1920*. Chapel Hill, NC: University of North Carolina Press, 1998.

Light, Andrew, and A. de Shalit, eds. *Moral and Political Reasoning in Environmental Practice*. Cambridge, MA: MIT Press, 2003.

Lindee, Susan. "The American Career of Jane Marcet's *Conversations on Chemistry*." *Isis* 82 (1991): 8–23.

Livingstone, David. *Putting Science in Its Place: Geographies of Scientific Knowledge*. Chicago: University of Chicago Press, 2003.

Livingstone, David, and Charles Withers, eds. *Geography and Enlightenment*. Chicago: University of Chicago Press, 1999.

Loehr, Rodney. "Arthur Young and American Agriculture." *Agricultural History* 43 (1969): 43–67.

Lowery, Charles. "James Barbour, a Progressive Farmer of Antebellum Virginia."

In *America, The Middle Period: Essays in Honor of Bernard Mayo*, ed. John Boles, 168–187. Charlottesville, VA: University Press of Virginia, 1973.

Lucier, Paul. "Commercial Interests and Scientific Disinterestedness: Consulting Geologists in Antebellum America." *Isis* 86 (1995): 245–267.

Lundgrens, Anders, and Bernadette Bensaude-Vincent, eds. *Communicating Chemistry: Textbooks and Their Audiences, 1789–1939*. Canton, MA: Science History Publications, 2000.

Majewski, John. *A House Dividing: Economic Development in Pennsylvania and Virginia before the Civil War*. Cambridge, UK: Cambridge University Press, 2002.

Mårand, Erland. "Everything Circulates: Agricultural Chemistry and Recycling Theories in the Second Half of the Nineteenth Century." *Environment and History* 8 (2002): 65–84.

Marcus, Alan. *Agricultural Science and Its Quest for Legitimacy: Farmers, Agricultural Colleges, and Experiment Stations, 1870–1890*. Ames, IA: Iowa State University Press, 1985.

Marcus, Alan. "The Wisdom of the Body Politic: The Changing Nature of Publicly Sponsored American Agricultural Research since the 1830s." *Agricultural History* 62 (1988): 4–26.

Marti, Donald. "Agricultural Journalism and the Diffusion of Knowledge: The First Half Century in America." *Agricultural History* 54 (1980): 37.

Martindale, Charles. "Green Politics: The *Eclogues*." In *The Cambridge Companion to Virgil*, ed. Charles Martindale, 107–124. Cambridge, UK: Cambridge University Press, 1997.

Marx, Leo. "Does Improved Technology Mean Progress?" *Technology Review* 90 (1987): 33–41.

Marx, Leo. *The Machine in the Garden: Technology and the Pastoral Idea in America*. Oxford: Oxford University Press, [1964] 1999. 2nd edition.

Marx, Leo. "Pastoralism in America." In *Ideology and Classic American Literature*, ed. Sacvan Bercovitch and Myra Jehlen, 36–69. Cambridge, UK: Cambridge University Press, 1986.

Marx, Leo, and Bruce Mazlish, eds. *Progress: Fact or Illusion?* Ann Arbor, MI: University of Michigan Press, 1996.

Mathew, William. *Edmund Ruffin and the Crisis of Slavery in the Old South: The Failure of Agricultural Reform*. Athens, GA: University of Georgia Press, 1988.

McCall, A. "The Development of Soil Science." *Agricultural History* 5 (1931): 43–56.

McClelland, Peter. *Sowing Modernity: America's First Agricultural Revolution*. Ithaca, NY: Cornell University Press, 1997.

McCook, Stuart. *States of Nature: Science, Agriculture, and Environment in the Spanish Caribbean, 1760–1940*. Austin, TX: University of Texas Press, 2002.

McCorrison, Marcus. "Thomas Green Fessenden, 1771–1837: Not in *BAL*." *Proceedings of the Bibliographic Society of America* 89 (1995): 5–59.

McCoy, Drew. *The Elusive Republic: Political Economy in Jeffersonian America*. New York: Norton, 1982.

McGehee, Minnie Lee. "The Rivanna Navigation Company." *Bulletin of the Fluvanna Historical Society* 5 (1967): 1–20.

McMurry, Sally. "Who Read the Agricultural Journals? Evidence from Chenango County, New York, 1839–1865." *Agricultural History* 63 (1989): 1–18.

McNeill, J. R., and Verena Winiwarter, eds. *Soils and Societies: Perspectives from Environmental History.* Isle of Harris, UK: White Horse, 2006.

Meikle, Jeffrey. "Leo Marx's *The Machine in the Garden.*" *Technology and Culture* 44 (2003): 147–159.

Meisel, Max. *A Bibliography of American Natural History: The Pioneer Century, 1769–1865.* New York: Premier Publishing, 1924–1929. 3 volumes.

Melhado, Evan. "Mineralogy and the Autonomy of Chemistry around 1800." *Lychnos* 55 (1990): 229–262.

Melhado, Evan, and Tore Frangsmyr, eds. *Enlightenment Science in the Romantic Era: The Chemistry of Berzelius and Its Cultural Setting.* Cambridge, UK: Cambridge University Press, 1992.

Merchant, Carolyn. *The Death of Nature: Women, Ecology, and the Scientific Revolution.* New York: Harper & Row, 1980.

Merchant, Carolyn. *Ecological Revolutions: Nature, Gender, and Science in New England.* Chapel Hill, NC: University of North Carolina Press, 1989.

Merchant, Carolyn. "Farms and Subsistence." In *Major Problems in American Environmental History,* ed. Carolyn Merchant, 147–154. New York: Houghton Mifflin, 2005. 2nd edition.

Merchant, Carolyn, ed. *The Columbia Guide to American Environmental History.* New York: Columbia University Press, 2002.

Merrill, George. *The First Hundred Years of American Geology.* New Haven, CT: Yale University Press, 1924.

Merrill, George, ed. *Contributions to a History of American State Geological and Natural History Surveys.* Washington, DC: Govt. Printing Office, 1920.

Michel, Gregg. "From Slavery to Freedom: Hickory Hill, 1850–80." In *The Edge of the South: Life in Nineteenth-Century Virginia,* ed. Edward Ayers and John C. Willis, 109–133. Charlottesville, VA: University Press of Virginia, 1991.

Milici, Robert, and C. R. Bruce Hobbs, Jr. "William Barton Rogers and the First Geological Survey of Virginia, 1835–1841." *Earth Science History* 6 (1987): 3–13.

Millbrooke, Anne. "State Geological Surveys of the Nineteenth Century." Ph.D. diss., University of Pennsylvania, 1981.

Miller, Angela. *The Empire of the Eye: Landscape Representation and American Cultural Politics, 1825–1875.* Ithaca, NY: Cornell University Press, 1993.

Miller, Charles. *Jefferson and Nature.* Baltimore: Johns Hopkins University Press, 1988.

Miller, Randall, ed. *"Dear Master": Letters of a Slave Family.* Athens, GA: University of Georgia Press, 1990.

Miller, Randall, and William Pencak, eds. *Pennsylvania: A History of the Commonwealth.* State College, PA: Pennsylvania State University Press, 2000.

Mills, William J. "Metaphorical Vision: Changes in Western Attitudes to the Environment." *Annals of the Association of American Geographers* 72 (1982): 237–253.

Minteer, Ben. *The Landscape of Reform: Civic Pragmatism and Environmental Thought in America*. Cambridge, MA: MIT Press, 2006.

Mitman, Gregg. *The State of Nature: Ecology, Community, and American Social Thought, 1900–1950*. Chicago: University of Chicago Press, 1992.

Mitman, Gregg. "When Nature *Is* the Zoo: Vision and Power in the Art and Science of Natural History." In "Science in the Field," ed. Henrika Kuklick and Robert Kohler. *Osiris* 11 (1996): 117–143.

Munday, Pat. "Liebig's Metamorphosis: From Organic Chemistry to the Chemistry of Agriculture." *Ambix* 38 (1991): 135–154.

Myers, Greg. "Fictionality, Demonstration, and a Forum for Popular Science: Jane Marcet's 'Conversations on Chemistry.'" In *Natural Eloquence: Women Reinscribe Science*, ed. Barbara T. Gates and Ann B. Shteir, 43–60. Madison: University of Wisconsin Press, 1997.

Nye, David. *America as Second Creation: Technology and Narratives of New Beginnings*. Cambridge, MA: MIT Press, 2003.

Novak, Barbara. *Nature and Culture: American Landscape and Painting, 1825–1875*. New York: Oxford University Press, 1980.

Numbers, Ronald, and Todd Savitt, eds. *Science and Medicine in the Old South*. Baton Rouge, LA: Louisiana State University Press, 1989.

O'Connell, Joseph. "Metrology: The Creation of Universality by the Circulation of Particulars." *Social Studies of Science* 23 (1993): 129–173.

Oleson, Alexandra, and Sanborn Brown, eds. *The Pursuit of Knowledge in the Early American Republic: American Scientific and Learned Societies from Colonial Times to the Civil War*. Baltimore: Johns Hopkins University Press, 1976.

Ophir, A., and Steve Shapin. "The Place of Knowledge: A Methodological Survey." *Science in Context* 4 (1991): 3–21.

Orwin, Christabel, and Edith Whetham. *History of British Agriculture, 1846–1914*. London: David and Charles, 1971.

Ostrander, Gilman. *Republic of Letters: The American Intellectual Community, 1776–1865*. Madison, WI: Madison House, 1999.

Outland, Robert. "Slavery, Work, and the Geography of the North Carolina Naval Stores Industry, 1835–1860." *Journal of Southern History* 62 (1996): 27–56.

Pauly, Philip. "Fighting the Hessian Fly: American and British Responses to Insect Invasion, 1776–1789." *Environmental History* 7 (2002): 485–507.

Peskin, Lawrence. "How the Republicans Learned to Love Manufacturing: The First Parties and the 'New Economy.'" *Journal of the Early Republic* 22 (2002): 235–262.

Peskin, Lawrence. *Manufacturing Revolution: The Intellectual Origins of Early American Industry*. Baltimore: Johns Hopkins University Press, 2003.

Philips, Sarah. "Antebellum Agricultural Reform, Republican Ideology, and Sectional Tension." *Agricultural History* 74 (2000): 799–822.

Phillips, Dana. *The Truth of Ecology: Nature, Culture, and Literature in America*. New York: Oxford University Press, 2003.

Pollan, Michael. *The Omnivore's Dilemma: A Natural History of Four Meals*. New York: Penguin, 2006.

Poovey, Mary. *A History of the Modern Fact: Problems and Knowledge in the Science of Wealth and Society*. Chicago: University of Chicago Press, 1998.
Porter, Theodore. "The Promotion of Mining and the Advancement of Science: The Chemical Revolution of Mineralogy." *Annals of Science* 38 (1981): 543–570.
Porter, Theodore. *Trust in Numbers: The Pursuit of Objectivity in Science and Public Life*. Princeton, NJ: Princeton University Press, 1995.
Rabbitt, Mary. *A Brief History of the U.S. Geological Survey*. Arlington, VA: U.S. Geological Survey, 1979.
Ragsdale, Bruce. *A Planter's Republic: The Search for Economic Independence in Revolutionary Virginia*. Madison, WI: Madison House, 1996.
Ramberg, Peter. "The Death of Vitalism and the Birth of Organic Chemistry: Wohler's Urea Synthesis and the Disciplinary Identity of Organic Chemistry." *Ambix* 47 (2000): 170–195.
Rasmussen, Wayne, and Robert Tilton. *Old Virginia: The Pursuit of a Pastoral Ideal*. Charlottesville, VA: Howell Press, 2003.
Regis, Pamela. *Describing Early America: Bartram, Jefferson, Crèvecoeur, and the Rhetoric of Natural History*. DeKalb, IL: Northern Illinois University Press, 1992.
Reingold, Nathan. *Science, American Style*. New Brunswick, NJ: Rutgers University Press, 1991.
Reingold, Nathan, ed. *The Sciences in the American Context: New Perspectives*. Washington, DC: Smithsonian Institution Press, 1979.
Richards, Paul. *Indigenous Agricultural Revolution: Ecology and Food Production in West Africa*. London: Hutchinson, 1985.
Richards, Robert. *The Romantic Conception of Life: Science and Philosophy in the Age of Goethe*. Chicago: University of Chicago Press, 2002.
Rigal, Laura. *The American Manufactory: Art, Labor, and the World of Things in the Early Republic*. Princeton, NJ: Princeton University Press, 1998.
Rogers, Everett. *Diffusion of Innovations*. New York: Free Press, 1995. 4th edition.
Rogers, Muriel. "John Hartwell Cocke (1780–1866): From Jeffersonian Palladianism to Romantic Colonial Revivalism in Antebellum Virginia." Ph.D. diss., Virginia Commonwealth University, 2003.
Rosenberg, Charles. *No Other Gods: On Science and American Social Thought*. Baltimore: Johns Hopkins University Press, 1976.
Rosenberg, Charles. "Woods or Trees? Ideas and Actors in the History of Science." *Isis* 79 (1988): 565–570.
Rossiter, Margaret. *The Emergence of Agricultural Science: Justus Liebig and the Americans, 1840–1880*. New Haven, CT: Yale University Press, 1975.
Rossiter, Margaret. "The Organization of Agricultural Improvement in the United States, 1785–1865." In *The Pursuit of Knowledge in the Early American Republic*, ed. Alexandra Oleson and Sanborn Brown, 279–298. Baltimore: Johns Hopkins University Press, 1974.
Rouse, Joseph. *Engaging Science: How to Understand Its Practices Philosophically*. Ithaca, NY: Cornell University Press, 1996.
Rubin, Julius. "The Limits of Agricultural Progress in the Nineteenth-Century South." *Agricultural History* 59 (1979): 362–373.

Russell, Edmund. *War and Nature: Fighting Insects with Chemicals, from World War I to Silent Spring*. New York: Cambridge University Press, 2001.
Ryden, Kent. *Landscape with Figures: Nature & Culture in New England*. Iowa City, IA: University of Iowa Press, 2001.
Sachs, Aaron. *The Humboldt Current: Nineteenth-Century Exploration and the Roots of American Environmentalism*. New York: Viking, 2006.
Sayre, Laura. "Cultivating the Georgic." In *Black Earth and Ivory Tower: New American Essays from Farm to Classroom*, ed. Zachary Michael Jack, 184–192. Columbia, SC: University of South Carolina Press, 2005.
Sayre, Laura. "Farming by the Book: British Georgic in Prose and Practice, 1697–1820." Ph.D. diss., Princeton University, 2002.
Scarborough, William. "Science on the Plantation." In *Science and Medicine in the Old South*, ed. Ronald Numbers and Todd Savvitt, 79–106. Baton Rouge, LA: Louisiana State University Press, 1989.
Schama, Simon. *Landscape and Memory*. New York: Knopf, 1996.
Scheese, Don. *Nature Writing: The Pastoral Impulse in America*. New York: Twayne, 1996.
Schlebecker, John, and Andrew W. Hopkins. *A History of Dairy Journalism in the United States, 1810–1950*. Madison WI: University of Wisconsin Press, 1957.
Schling-Brodersen, Uschi. "Liebig's Role in the Establishment of Agricultural Chemistry." *Ambix* 39 (1992): 21–31.
Schlotterbeck, John. "Plantation and Farm: Social and Economic Change in Orange and Greene Counties, Virginia, 1716–1860." Ph.D. diss., Johns Hopkins University, 1980.
Schneer, Cecil J., ed. *Two Hundred Years of Geology in America*. Hanover, NH: University Press of New England, 1979.
Schneider, Daniel. "Local Knowledge, Environmental Politics, and the Founding of Ecology in the United States: Stephen Forbes and 'The Lake as a Microcosm' (1887)." *Isis* 91 (2000): 681–705.
Scott, James. *Seeing Like a State: How Certain Schemes to Improve the Human Condition Have Failed*. New Haven, CT: Yale University Press, 1998.
Secord, Ann. "Science in the Pub: Artisan Botanists in Early Nineteenth-Century Lancashire." *History of Science* 32 (1994): 269–315.
Sellers, Charles. *The Market Revolution: Jacksonian America, 1815–1846* New York: Oxford University Press, 1991.
Shade, William. *Democratizing the Old Dominion: Virginia and the Second Party System*. Charlottesville, VA: University Press of Virginia, 1993.
Shalhope, Robert. *John Taylor of Caroline: Pastoral Republican*. Columbia, SC: University of South Carolina Press, 1980.
Shapin, Steven. "Cordelia's Love: Credibility and the Social Studies of Science." *Perspectives on Science* 3 (1995): 255–275.
Shapin, Steven. "The Invisible Technician." *American Scientist* 77 (1989): 554–563.
Shapin, Steven. 'Nibbling at the Teats of Science': Edinburgh and the Diffusion of Science in the 1830s," In *Metropolis and Province: Science in British Culture,*

1780–1850, ed. Ian Inkster and Jack Morrell, 151–178. Philadelphia: University of Pennsylvania Press, 1983.

Shapin, Steven. *The Social History of Truth*. Chicago: University of Chicago Press, 1995.

Shapin, Steven, and Simon Schaffer. *Leviathan and the Air Pump: Hobbes, Boyle, and the Experimental Life*. Princeton, NJ: Princeton University Press, 1985.

Sharrer, G. Terry. *A Kind of Fate: Agricultural Change in Virginia, 1861–1920*. Ames, IA: Iowa University Press, 2000.

Siegel, Frederick. *The Roots of Southern Distinctiveness: Tobacco and Society in Danville, VA, 1780–1865*. Chapel Hill, NC: University of North Carolina Press, 1987.

Silver, Timothy. *A New Face on the Countryside: Indians, Colonists, and Slaves in South Atlantic Forests, 1500–1800*. New York: Cambridge University Press, 1990.

Slotten, Hugh. *Patronage, Practice, and the Culture of American Science: Alexander Dallas Bache and the U.S. Coast Survey*. New York: Cambridge University Press, 1994.

Smith, Kimberly. *Wendell Berry and the Agrarian Tradition: A Common Grace*. Lawrence, KS: University Press of Kansas, 2003.

Smith, Merritt Roe, and George Clancey, eds. *Major Problems in the History of American Technology*. New York: Houghton Mifflin, 1998.

Smith, Michael. "The Conflict between 'Practical Utility' and Geology: Denison Olmstead, Elisha Mitchell, and the 1823 to 1828 Geological Surveys of North Carolina." *Southeastern Geology* 38 (1999): 145–154.

Socolow, Arthur, ed. *The State Geological Surveys: A History*. Tallahassee, FL: Association of American State Geologists, 1988.

Spanagel, David. "Chronicles of a Land Etched by God, Water, Fire, Time, and Ice." Ph.D. diss., Harvard University, 1996.

Spanagel, David. Review of *Nature's Management: Writings on Landscape and Reform, 1822–1859*, by Edmund Ruffin, ed. Jack Temple Kirby. *H-SHEAR*, H-Net Reviews, October, 2001. Accessed 26 January 2009 at http://www.h-net.msu.edu/reviews/showrev.cgi?path=274641002733673.

Stafford, Barbara. *Artful Science: Enlightenment Entertainment and the Eclipse of Visual Education*. Cambridge, MA: MIT Press, 1994.

Star, Susan Leigh, and James Greisemer. "Institutional Ecology, 'Translations,' and Boundary Objects: Amateurs and Professionals in Berkeley's Museum of Vertebrate Zoology, 1907–39." *Social Studies of Science* 19 (1989): 387–420.

Steinberg, Ted. "Down to Earth: Nature, Agency, and Power in History." *American Historical Review* 107 (2002): 798–820.

Steinberg, Ted. *Down to Earth: Nature's Role in American History*. Oxford: Oxford University Press, 2002.

Steinberg, Ted. *Nature Incorporated: Industrialization and the Waters of New England*. New York: Cambridge University Press, 1991.

Stephens, Lester. *Science, Race, and Religion in the American South: John Bachman and the Charleston Circle of Naturalists, 1815–1895*. Chapel Hill, NC: University of North Carolina Press, 2000.

Stewart, Mart. "Southern Environmental History." In *A Companion to the American South*, ed. John Boles, 411–423. Malden, MA: Blackwell, 2002.

Stewart, Mart. *"What Nature Suffers to Groe": Life, Labor, and Landscape on the Georgia Coast, 1680–1920*. Athens, GA: University of Georgia Press, 1996.

Stilgoe, John R. *Common Landscape of America, 1580–1845*. New Haven, CT: Yale University Press, 1982.

Stokes, Melvyn, and Stephen Conway, eds. *The Market Revolution in America: Social, Political, and Religious Expressions, 1800–1880*. Charlottesville, VA: University Press of Virginia, 1996.

Stoll, Steven. *Larding the Lean Earth: Soil and Society in Nineteenth-Century America*. New York: Hill and Wang, 2002.

Stowell, Robert. *A Thoreau Gazetteer*. Ed. William Howarth. Princeton, NJ: Princeton University Press, 1970.

Sweet, Timothy. *American Georgics: Economy and Environment in Early American Literature*. Philadelphia: University of Pennsylvania Press, 2002.

Taylor, George. *The Transportation Revolution, 1815–1860*. New York: Harper Torchbooks, 1968.

Taylor, Joseph. *Making Salmon: An Environmental History of the Northwest Fisheries Crisis*. Seattle: University of Washington Press, 1999.

Thompson, Paul. *The Spirit of the Soil: Agriculture and Environmental Ethics*. New York: Routledge, 1995.

Thompson, Paul, and Thomas Hilde, eds. *The Agrarian Roots of Pragmatism*. Nashville, TN: Vanderbilt University Press, 2000.

Throne, Mildred. "'Book Farming' in Iowa, 1840–1870." *Iowa Journal of History* 49 (1951): 117–142.

Tillman, James. "The Transcendental Georgic in *Walden*." *ESQ* 21 (1975): 137–141.

Tolley, Kim. *The Science Education of American Girls: A Historical Perspective*. New York: Routledge, 2003.

True, Rodney. "Early Days of the Albemarle Agricultural Society." *Annual Report of the American Historical Association for the Year 1918*, 222–240. Washington: Government Printing Office, 1921.

Turner, Charles. *Virginia's Green Revolution: Essays on the Nineteenth Century Virginia Agricultural Reform and Fairs*. Waynesboro, VA: Humphries Press, 1986.

Turner, Stephen. "The Survey in Nineteenth Century American Geology: The Evolution of a Form of Patronage." *Minerva* 25 (1987): 282–330.

University of Virginia Library. "Historical Census Browser." Electronic edition. Accessed 26 February 2008 at http://fisher.lib.virginia.edu/collections/stats/histcensus/.

Upton, Dell. "New Views of the Virginia Landscape." *The Virginia Magazine of History and Biography* 96 (1988): 403–470.

Valencius, Conevery Bolton. *The Health of the Country: How American Settlers Understood Themselves and Their Land*. New York: Basic Books, 2002.

Van der Ploeg, R. R., W. Bohm, and M. B. Kirkham. "History of Soil Science: On the Origin of the Theory of Mineral Nutrition of Plants and the Law of the Minimum." *Soil Science Society of America Journal* 63 (1999): 1055–1062.

Van Leer, David. "Nature's Book: The Language of Science in the American Renaissance." In *Romanticism and the Sciences,* ed. Andrew Cunningham and Nicholas Jardine, 307–321. Cambridge, UK: Cambridge University Press, 1990.

Van Noy, Rick. "Surveying the Sublime: Literary Cartographers and the Spirit of Place." In *The Greening of Literary Scholarship, Theory, and the Environment,* ed. Steven Rosendale, 181–206. Iowa City, IA: University of Iowa Press, 2002.

Virtual Library of Virginia. "Virginia Heritage Database." Accessed 26 January 2009 at http://www.lib.virginia.edu/small/vhp/index.html.

Wallenstein, Peter. *Cradle of America: Four Centuries of Virginia History.* Lawrence: University Press of Kansas, 2007.

Walls, Laura. *Seeing New Worlds: Henry David Thoreau and Nineteenth-Century Natural Science.* Madison, WI: University of Wisconsin Press, 1995.

Warner, Michael. *The Letters of the Republic: Publication and the Public Sphere in Eighteenth-Century America.* Cambridge, MA: Harvard University Press, 1990.

Wellman, Kathleen. "Materialism and Vitalism." In *The Oxford Companion to the History of Modern Science,* ed. John Heilbron, 490–491. Oxford: Oxford University Press, 2003.

White, Richard. "Are You an Environmentalist or Do You Work for a Living?" In *Uncommon Ground,* ed. William Cronon, 171–185. New York: Norton, 1995.

White, Richard. "Environmental History, Ecology, and Meaning." *Journal of American History.* 76 (1990): 1111–1116.

White, Richard. *The Organic Machine: The Remaking of the Columbia River.* New York: Hill and Wang, 1995.

Wilson, Leonard, ed. *Benjamin Silliman and His Circle: Studies on the Influence of Benjamin Silliman on Science in America.* New York: Science History Publications, 1979.

Wines, Richard A. *Fertilizer in America: From Waste Recycling to Resource Exploitation.* Philadelphia: Temple University Press, 1985.

Withers, Charles. "Improvement and Enlightenment: Agriculture and Natural History in the Work of the Rev. Dr. John Walker (1731–1803)." In *Philosophy and Science in the Enlightenment,* ed. Peter Jones, 102–116. Edinburgh: John Donald, 1988.

Withers, Charles. "A Neglected Scottish Agriculturist: The 'Georgical Lectures' and Agricultural Writings of the Rev Dr John Walker (1731–1803)." *Agricultural History Review* 34 (1985): 132–146.

Withers, Charles. "On Georgics and Geology: James Hutton's 'Elements of Agriculture' and Agricultural Science in Eighteenth-Century Scotland." *Agricultural History Review* 42 (1994): 38–48.

Withers, Charles. "William Cullen's Agricultural Lectures and Writings and the Development of Agricultural Science in Eighteenth-Century Scotland." *Agricultural History Review* 37 (1989): 144–156.

Worster, Donald. *Nature's Economy: A History of Ecological Ideas.* Cambridge, UK: Cambridge University Press, [1977] 1993. 2nd edition.

Worster, Donald. *A River Running West: The Life of John Wesley Powell.* New York: Oxford University Press, 2001.

Worster, Donald. "Transformations of the Earth: Toward an Agroecological Perspective in History." *Journal of American History* 76 (1990): 1087–1106.

Worster, Donald. *The Wealth of Nature: Environmental History and the Ecological Imagination*. New York: Oxford University Press, 1993.

Worster, Donald, ed. *American Environmentalism: The Formative Period, 1860–1915*. New York: John Wiley and Sons, 1973.

Worster, Donald, ed. *The Ends of the Earth: Perspectives on Modern Environmental History*. New York: Cambridge University Press, 1988.

Zakim, Michael. "Sartorial Ideologies: From Homespun to Ready-Made." *American Historical Review* 106 (2001): 1553–1586.

INDEX

Lecture Notes in Mathematics

Edited by A. Dold and B. Eckmann

584

C. Brezinski

Accélération de la Convergence en Analyse Numérique

Springer-Verlag
Berlin · Heidelberg · New York 1977

Author
Claude Brezinski
UER d'iEEA-informatique
Université de Lille I
B.P. 36
59650 Villeneuve d'Ascq/France

Library of Congress Cataloging in Publication Data

Brezinski, Claude, 1941-
Accélération de la convergence en analyse numérique.

(Lecture notes in mathematics ; 584)
Includes index.
1. Numerical analysis--Acceleration of convergence. 2. Series. 3. Fractions, Continued. I. I. Title. II. Series: Lecture notes in mathematics (Berlin) ; 584.
QA3.L28 no. 584 [QA297] 510'.8s [519.4] 77-6813

AMS Subject Classifications (1970): 65B05, 65B10, 65B15, 65B99, 65D15, 65F05, 65F10, 65F15, 65H10, 65L10

ISBN 3-540-08241-7 Springer-Verlag Berlin · Heidelberg · New York
ISBN 0-387-08241-7 Springer-Verlag New York · Heidelberg · Berlin

This work is subject to copyright. All rights are reserved, whether the whole or part of the material is concerned, specifically those of translation, re-printing, re-use of illustrations, broadcasting, reproduction by photocopying machine or similar means, and storage in data banks.

Under § 54 of the German Copyright Law where copies are made for other than private use, a fee is payable to the publisher, the amount of the fee to be determined by agreement with the publisher.

© by Springer-Verlag Berlin · Heidelberg 1977

Printed in Germany

Printing and binding: Beltz Offsetdruck, Hemsbach/Bergstr.
2141/3140-543210

Math
Sep.

PLAN

1575237

Numerical analysis is very much an experimental science.

P. Wynn

INTRODUCTION

Le but de ce livre est d'être une introduction aux méthodes d'accélération de la convergence en analyse numérique. L'accélération de la convergence est un domaine important de l'analyse numérique qui reste encore peu exploré à l'heure actuelle bien que des domaines voisins (approximants de Padé, fractions continues) fassent l'objet de nombreuses recherches.

Un grand nombre de méthodes utilisées en analyse numérique et en mathématiques appliquées sont des méthodes itératives. Il arrive malheureusement que, dans la pratique, ces méthodes convergent avec une telle lenteur que leur emploi effectif est à exclure. C'est pour cette raison que l'on utilise simultanément des méthodes d'accélération de la convergence. Ce livre est donc destiné aussi bien aux mathématiciens qui veulent étudier ce domaine qu'à tous ceux qui désirent utiliser les méthodes d'accélération de la convergence.

Dans ce qui suit, après de brefs rappels mathématiques, on s'attachera à l'étude d'un certain nombre d'algorithmes d'accélération de la convergence. On verra également que ces algorithmes débouchent sur des méthodes nouvelles en analyse numérique et qui n'ont qu'un rapport lointain avec le sujet initial : résolution des systèmes d'équations linéaires et non linéaires, calcul des valeurs propres d'une matrice, quadratures numériques, etc.

Bien qu'un certain nombre d'exemples numériques illustrent les théorèmes, ce livre est théorique. Un ouvrage pratique contenant de nombreuses applications ainsi que les programmes FORTRAN des algorithmes devrait bientôt paraître [35].

Je remercie particulièrement le rapporteur qui a lu mon texte et m'a suggéré de nombreuses améliorations ainsi que le Professeur A. DOLD qui a bien voulu en accepter la publication.

Ce livre est issu d'un cours de troisième cycle que j'enseigne à l'Université de Lille depuis 1973 ; de nombreuses personnes y ont donc contribué. Ma reconnaissance est acquise à F. CORDELLIER et B. GERMAIN BONNE pour leur aide précieuse et le temps qu'ils m'ont consacré ainsi qu'à Mademoiselle M. DRIESSENS pour sa parfaite dactylographie du texte.

Je tiens enfin à remercier le Professeur P. WYNN pour son soutien amical et ses conseils tout au long de ce travail.

CHAPITRE I

COMPARAISON DE SUITES CONVERGENTES

I - 1 Rappels

Les notions exposées dans ce chapitre font constamment appel aux relations de comparaison dont nous rappelons ici les définitions :

Soient $\{u_n\}$ et $\{v_n\}$ deux suites de nombres réels qui tendent vers zéro lorsque n tend vers l'infini :

Si $\exists$ N et C > 0 : $\forall n > N$ on a $|v_n| < C \; |u_n|$

alors on écrit : $v_n = O(u_n)$

Si $\forall \varepsilon > 0 \;\; \exists N : \forall n > N$ on a $|v_n| < \varepsilon \; |u_n|$ alors on écrit : $v_n = o(u_n)$.

En d'autres termes $\lim_{n\to\infty} v_n/u_n = 0$.

Les principales propriétés des relations de comparaison ainsi que les règles qui président à leur manipulation sont supposées connues. Pour un exposé général on pourra se reporter à [157] et, pour un exposé plus détaillé à [72] et [18]. On y trouvera aussi des notions sur les échelles de comparaison et les développements asymptotiques.

I - 2 Ordre d'une suite

Dans la suite du chapitre on ne considèrera que des suites de nombres réels positifs ou nuls qui convergent vers zéro. Ceci n'est pas restrictif : soit en effet (E,d) un espace métrique et $\{S_n\}$ une suite d'éléments de E qui converge vers S, les quantités $d(S_n,S)$ sont bien des nombres réels positifs ou nuls et la suite $d(S_n,S)$ converge bien vers zéro.

<u>Définition 1</u> : On dit que la suite $\{u_n\}$ est d'ordre r si :

$u_{n+1} = O(u_n^r)$ et si $u_n^r = O(u_{n+1})$

Si on utilise la définition de la notation O ceci revient à dire que

$\exists\ 0 < A \leqslant B < +\infty$ tels que :

$$A \leqslant \frac{u_{n+1}}{u_n^r} \leqslant B \qquad \forall n > N$$

Théorème 1 : S'il existe, r est unique.

démonstration : supposons qu'il existe $p \neq r$ tel que la suite soit aussi d'ordre p. On a alors :

$$C_2\, u_n^r \leqslant u_{n+1} \leqslant C_1\, u_n^r \text{ et } C_3\, u_n^p \leqslant u_{n+1} \leqslant C_4\, u_n^p$$

d'où

$$u_{n+1} \leqslant C_1\, u_n^{r-p} \quad u_n^p \leqslant \frac{C_1}{C_3}\, u_n^{r-p} \quad u_{n+1}$$

ce qui donne :

$$1 \leqslant \frac{C_1}{C_3}\, u_n^{r-p}$$

si $r > p$ alors u_n^{r-p} tend vers 0 quand n tend vers l'infini.

On a donc $r = p$.

Si $r < p$ on écrit $u_{n+1} \leqslant C_4\, u_n^{p-r}\, u_n^r$ et la suite de la démonstration est identique.

REMARQUES :

1) Dans de nombreux ouvrages on trouve l'ordre d'une suite défini uniquement par $u_{n+1} = O(u_n^r)$. Il faut alors remarquer que cette définition n'assure pas l'unicité de r. En effet si nous considèrons la suite $u_n = a^{b^n}$ avec $0 < a < 1$ et $b > 1$. Cette suite est d'ordre b d'après la définition 1 alors que :

$$u_{n+1} = u_n^b \leqslant u_n^c \text{ pour } 1 < c < b \text{ et } \forall n > N$$

d'où la nouvelle définition :

Définition 2 : si l'on a $u_{n+1} = O(u_n^r)$ on dira que la suite $\{u_n\}$ est d'ordre r au moins tandis que si l'on a $u_n^r = O(u_{n+1})$ on dira que la suite est d'ordre r au plus.

On a les propriétés évidentes suivantes :

propriété 1 : si $u_{n+1} = O(u_n^r)$ alors $u_{n+1} = o(u_n^p)$ si $p < r$

propriété 2 : si $u_n^r = O(u_{n+1})$ alors $u_n^p = o(u_{n+1})$ si $p > r$. r ne peut pas être inférieur à 1.

2) Dans l'exemple $u_n = a^{b^n}$ on voit que l'ordre b peut être un nombre réel positif. Si la suite $\{u_n\}$ est générée par $u_{n+1} = f(u_n)$ et si f est suffisamment différentiable au voisinage de zéro alors l'ordre r est égal au plus petit entier k tel que $f^{(i)}(0) = 0$ pour $i = 0, \ldots, k-1$ et $f^{(k)}(0) \neq 0$.

3) Dans certains ouvrages on rencontre souvent l'ordre d'une suite défini comme le plus petit réel positif r tel que :

$$\lim_{n \to \infty} \frac{u_{n+1}}{u_n^r} = C \neq 0 \text{ ou de } +\infty.$$

Cette limite peut ne pas exister mais $\{u_n\}$ peut cependant avoir un ordre au sens de la définition 1. Il n'y a qu'à considérer la suite $u_n = 1/n$ si n pair et $1/2n$ sinon.

4) Remarquons enfin que la définition 1 ne permet pas d'attribuer un ordre à n'importe quelle suite (par exemple $u_n = \lambda^{n^2}$ avec $0 < \lambda < 1$). On peut donc se poser la question de savoir si la définition 1 est insuffisante ou si l'on est réellement incapable de définir un ordre pour certaines suites.

Définition 3 : on appelle coefficient asymptotique d'erreur le nombre

$$C = \limsup_{n \to \infty} \frac{u_{n+1}}{u_n^r}$$

Les notions d'ordre et de coefficient asymptotique d'erreur ne sont pas des notions purement théoriques ; elles ont une relation étroite avec le nombre de chiffres exacts obtenu :

puisque la suite $\{u_n\}$ converge vers zéro u_n représente l'erreur absolue. Posons $e_n = -\log_{10} u_n$; e_n est le nombre de chiffres significatifs décimaux exacts de u_n (par exemple si $u_n = 10^{-3} = 0{,}001$ on a bien $e_n = 3$).

Pour n suffisamment grand on a :

$$e_{n+1} = r\, e_n + R \text{ avec } R = -\log_{10} C.$$

On voit donc que si r = 1 on ajoute environ R chiffres significatifs exacts en passant de u_n à u_{n+1} : par exemple si C = 0,999 alors $R = 4.10^{-4}$ et il faudra 2500 termes de plus pour gagner un seul chiffre significatif.

Par contre si r > 1 on multiplie environ par r le nombre de chiffres significatifs exacts en passant de u_n à u_{n+1}. On voit donc l'intérêt des suites d'ordre plus grand que un.

Propriété 3 : On a :

$$C = \limsup_{n\to\infty} \left[\frac{u_n}{u_0^{r^n}} \right]^p$$

avec p = 1/n si r = 1 et $p = \dfrac{r-1}{r^n-1}$ si r > 1.

La démonstration est laissée en exercice.

I - 3 Comparaison de deux suites

Soient maintenant $\{u_n\}$ et $\{v_n\}$ deux suites de nombres réels positifs qui convergent vers zéro. Nous allons donner un certain nombre de définitions qui permettent de comparer leurs "vitesses" de convergence.

Définition 4 :

On dit que $\{u_n\}$ converge comme $\{v_n\}$ si :

$$u_n = O(v_n) \text{ et } v_n = O(u_n)$$

on peut affiner cette définition en disant que $\{v_n\}$ converge mieux que $\{u_n\}$ si le nombre C donné par :

$$C = \limsup_{n\to\infty} \frac{v_n}{u_n}$$

est strictement inférieur à un.

<u>Théorème 2</u> : si $\lim_{n\to\infty} \frac{u_{n+1}}{u_n} = a < 1$ et $\lim_{n\to\infty} \frac{v_n}{u_n} = b$

alors $\{v_n\}$ converge mieux que $\{u_{n+k}\}$ $\forall k <$ Log b / Log a

démonstration :

$$\lim_{n\to\infty} \frac{v_n}{u_{n+k}} = \lim_{n\to\infty} \frac{v_n}{u_n} \cdot \lim_{n\to\infty} \frac{u_n}{u_{n+k}} = \frac{b}{a^k}$$

. Une condition suffisante pour que $\{v_n\}$ converge mieux que $\{u_{n+k}\}$ est que $0 \leqslant b / a^k < 1$. D'où Log b < k Log a ce qui donne la condition du théorème puisque Log a < 0. Il faut remarquer que si b < 1 alors k > 0 et que si b > 1 alors k < 0. Si b = 0 alors la proposition est vraie pour tout k positif.

<u>Définition 5</u> :

on dit que $\{v_n\}$ converge plus vite que $\{u_n\}$ si $v_n = o(u_n)$

Soit T une méthode qui permet de transformer la suite $\{u_n\}$ en une suite $\{v_n\}$ qui converge également vers zéro. Si $v_n = o(u_n)$ on dit que l'on a accéléré la convergence et que la méthode T est une méthode d'accélération de la convergence.

Il est bien évident que l'on peut définir l'accélération de la convergence de façon différente. Il nous arrivera quelquefois par la suite de dire que $\{v_n\}$ converge plus vite que $\{u_n\}$ si $v_{n+1} - v_n = o(u_{n+1} - u_n)$. Si c'est cette définition qui est utilisée nous le préciserons toujours.

On a les résultats suivants :

<u>Théorème 3</u> : hypothèses : 1 - $u_{n+1} = O(u_n)$

2 - $u_n = O(u_{n+1})$

3 - $u_{n+1} = o(u_n)$

4 - $v_n = O(u_n)$

5 - $v_n = o(u_n)$

alors 1 et 5 impliquent $v_n = o(u_{n-k})$ $\forall k \geqslant 0$

" 2 et 5 " $v_n = o(u_{n+k})$ "

alors 3 et 4 impliquent $v_n = o(u_{n-k})$ $\forall k \geq 0$

" 1 et 4 " $v_n = o(u_{n-k})$ "

" 2 et 4 " $v_n = O(u_{n+k})$ "

Ce théorème est très facile à établir. Le résultat fondamental auquel on aboutit est le suivant :

<u>Théorème 4</u> : Si $\{v_n\}$ est d'ordre r au moins, si $\{u_n\}$ est d'ordre p au plus et si $r > p$ alors $\{v_n\}$ converge plus vite que $\{u_n\}$.

Démonstration : on a par définition :

$$v_{n+1} \leq A\, v_n^r \quad \text{et} \quad u_n^p / B \leq u_{n+1}$$

et par récurrence :

$$v_{n+k} \leq A^{1/(1-r)} (v'_n)^{r^k} \quad \text{et} \quad (u'_n)^{p^k} / B^{1/(1-p)} \leq u_{n+k}$$

avec $$v'_n = A^{1/(r-1)} v_n \quad \text{et} \quad u'_n = u_n / B^{1/(p-1)}$$

d'où $$\frac{v_{n+k}}{u_{n+k}} \leq C \left(\frac{v'_n}{u'_n}\right)^{p^k} (v'_n)^{r^k - p^k} \quad \text{avec } C = A^{1/(1-r)}\, B^{1/(1-p)}$$

Soit n un indice tel que $v'_n < 1$ ce qui est toujours possible puisque v_n tend vers zéro. Alors on a :

$$w_k = \left(\frac{v'_n}{u'_n}\right)^{p^k} (v'_n)^{r^k - p^k}$$

$$\text{Log } w_k = p^k \left[\text{Log } \frac{v'_n}{u'_n} + \left(\left(\frac{r}{p}\right)^k - 1\right) \text{Log } v'_n\right]$$

or $\text{Log } v'_n < 0$ et $(r/p)^k - 1 > 0$. Donc $\lim_{k\to\infty} \text{Log } w_k = -$ car $p \geq 1$ et par conséquent $\lim_{k\to\infty} w_k = 0$ ce qui termine la démonstration.

Ce théorème montre que si deux suites ne sont pas du même ordre alors il suffit de comparer leurs ordres pour connaître celle qui converge plus vite que l'autre. La situation est beaucoup plus délicate si les deux suites ont le même ordre. Deux suites peuvent en effet avoir le même ordre et la même constante asymptotique d'erreur et cependant l'une peut converger plus vite que l'autre : par exemple considérons $u_n = 1/n$ et $v_n = 1/n^\alpha$ avec $\alpha > 1$. Ces deux suites sont d'ordre un et leur coefficient asymptotique d'erreur vaut un. Il est cependant clair que $\{v_n\}$ converge plus vite que $\{u_n\}$ d'une part et d'autre part on voit que $\{v_n\}$ converge comme $\{u_n^\alpha\}$ au sens de la définition 4. Les notions d'ordre et de coefficient asymptotique d'erreur ne sont donc pas des notions assez fines pour comparer les vitesses de convergence de deux suites ni pour chiffrer l'accélération de la convergence. On est donc amené à introduire une notion d'ordre dans la comparaison de la convergence de deux suites ; c'est la notion d'α-équivalence.

définition 6 : On dit que $\{v_n\}$ est α-équivalente à $\{u_n\}$ si $v_n = O(u_n^\alpha)$ et $u_n^\alpha = O(v_n)$. α est le coefficient d'équivalence de $\{v_n\}$ par rapport à $\{u_n\}$

Remarques :

1°) Cette définition englobe celle de l'ordre d'une suite en prenant $v_n = u_{n+1}$.

2°) De même qu'il n'est pas toujours possible de définir un ordre pour toute suite convergente, il n'est pas toujours possible de comparer deux suites à l'aide de l'α-équivalence (par exemple $u_n = 1/n^2$ et $v_n = 1/n$ si n pair et $1/n^3$ si n impair).

Comme pour l'ordre d'une suite on a le :

Théorème 5 : S'il existe, α est unique

démonstration : elle peut être calquée sur celle du théorème 1 ; nous la donnerons en utilisant les relations de comparaison sans utiliser leurs définitions.

Supposons qu'il existe β tel que $\{v_n\}$ soit β-équivalente à $\{u_n\}$. On a :

$$v_n = O(u_n^\alpha) \qquad u_n^\alpha = O(v_n)$$

$$v_n = O(u_n^\beta) \qquad u_n^\beta = O(v_n)$$

$v_n = O(u_n^\alpha) = O(u_n^{\alpha-\beta}\, u_n^\beta) = O(u_n^{\alpha-\beta}\, v_n)$ d'où

$$1 = O(u_n^{\alpha-\beta})$$

si $\alpha > \beta$ ceci est impossible car $u_n^{\alpha-\beta}$ tend vers 0 lorsque n tend vers l'infini.

Si $\alpha < \beta$ on écrit $v_n = O(u_n^{\beta}) = O(u_n^{\beta-\alpha}\ u_n^{\alpha})$ et la suite de la démonstration est identique. Par conséquent $\alpha = \beta$.

Théorème 6 : Si $\{v_n\}$ est α-équivalente à $\{u_n\}$ et si $\alpha \neq 0$ et est fini alors les deux suites ont le même ordre.

démonstration : Supposons que $\{u_n\}$ soit d'ordre p, on a :

$v_n = O(u_n^{\alpha});\ u_n^{\alpha} = O(v_n);\ u_{n+1} = O(u_n^{p})$ et $u_n^{p} = O(u_{n+1})$

en utilisant ces relations on obtient :

$$v_{n+1} = O(u_{n+1}^{\alpha}) = O(u_n^{p\alpha}) = O(v_n^{p})$$

$$v_n^{p} = O(u_n^{\alpha\, p}) = O(u_{n+1}^{\alpha}) = O(v_{n+1})$$

ce qui démontre que $\{v_n\}$ est d'ordre p.

Le concept d'α-équivalence est un cas particulier d'une notion plus générale introduite par Bourbaki [18] : la notion d'ordre d'une suite par rapport à une autre.

Définition 7 :

On dit que $\{v_n\}$ est d'ordre α par rapport à $\{u_n\}$ si :

$$\lim_{n\to\infty} \frac{\text{Log } v_n}{\text{Log } u_n} = \alpha$$

propriété 4 : si $\{v_n\}$ est α-équivalente à $\{u_n\}$ alors $\{v_n\}$ est d'ordre α par rapport à $\{u_n\}$.

Démonstration : la définition de l'α-équivalence signifie qu'il existe deux constantes A et B telles que

$$0 < A \leqslant \frac{v_n}{u_n^{\alpha}} \leqslant B < +\infty \qquad \forall n > N$$

prenons le logarithme de cette inégalité ; il vient :

$$\text{Log } A \leqslant \text{Log } v_n - \alpha \text{ Log } u_n \leqslant \text{Log } B$$

d'où $\lim\limits_{n\to\infty} \frac{\text{Log } v_n}{\text{Log } u_n} = \alpha$ puisque $\lim\limits_{n\to\infty} \text{Log } u_n = -\infty$

Ainsi un certain nombre de théorèmes démontrés pour l'ordre d'une suite par rapport à une autre se transposent immédiatement en termes d'α-équivalence. Les démonstrations de ces théorèmes sont laissées en exercices.

Théorème 7 :

Si $\{v_n\}$ est α-équivalente à $\{u_n\}$ alors

$v_n = o(u_n^{\alpha-a})$ et $u_n^{\alpha+a} = o(v_n)$ $\forall a > 0$

Théorème 8 :

Pour que $\{v_n\}$ soit $+\infty$-équivalente à $\{u_n\}$ il est nécessaire que $v_n = o(u_n^{\alpha})$ $\forall \alpha \geq 0$

Théorème 9 :

Pour que $\{v_n\}$ soit $-\infty$-équivalente à $\{u_n\}$ il est nécessaire que $u_n^{-\alpha} = o(v_n)$ $\forall \alpha \geq 0$

L'ordre d'une suite par rapport à une autre possède un certain nombre de propriétés rassemblées dans l'énoncé suivant dont la démonstration est également laissée en exercice :

propriété 5 : Si $\{v_n\}$ est α-équivalente à $\{u_n\}$ avec α fini cela n'implique pas que le rapport v_n / u_n^{α} ait une limite. Si $\{v_n\}$ est α-équivalente à $\{u_n\}$ avec α fini alors $\{v_n u_n^{-\alpha}\}$ est o-équivalente à $\{u_n\}$.
Si $\{v_n\}$ et $\{s_n\}$ sont respectivement α_1 et α_2 - équivalentes à $\{u_n\}$ et si $\alpha_1 + \alpha_2$ est défini alors $\{v_n s_n\}$ est $(\alpha_1 + \alpha_2)$-équivalente à $\{u_n\}$. Si $\{v_n\}$ est α-équivalente à $\{u_n\}$ alors $\{u_n\}$ est $1/\alpha$ -équivalente à $\{v_n\}$.

Bien que la notion d'α-équivalence soit un cas particulier de la notion d'ordre d'une suite par rapport à une autre, elle peut cependant être plus intéressante car elle permet d'obtenir les résultats supplémentaires suivants :

Théorème 10 :

la relation définie par $\{u_n\} \approx \{v_n\}$ si $\{u_n\}$ est 1-équivalente à $\{v_n\}$ est une relation d'équivalence sur l'ensemble des suites convergentes d'éléments de E (E est l'espace métrique introduit en I.2).

Démonstration : elle est évidente et laissée en exercice.

Définition 8 : on écrira $\{v_n\} << \{u_n\}$ s'il existe $\alpha > 1$ tel que $\{v_n\}$ soit α-équivalente à $\{u_n\}$.

On a le :

Théorème 11 :

La relation définie par $\{v_n\} \leq< \{u_n\}$ si $\{u_n\} = \{v_n\}$ ou si $\{v_n\} << \{u_n\}$ est une relation d'ordre sur l'ensemble des suites convergentes d'éléments de E.

Démonstration : reflexivité $\{u_n\} = \{u_n\}$

antisymétrie si $\{u_n\} \leq< \{v_n\}$ et si $\{v_n\} \leq< \{u_n\}$

alors $\{u_n\} = \{v_n\}$

La transitivité est évidente.

Le fait que cette relation ne soit pas une relation d'ordre total explique pourquoi la notion d'α-équivalence ne permet pas de comparer n'importe quelles suites.

Prenons maintenant $E = \mathbb{R}$. S'il est possible de choisir dans chaque classe d'équivalence une et une seule suite telle que l'ensemble G de ces suites vérifie :

1°) toute suite de G est positive dans un voisinage de $+\infty$

2°) toute suite de G (autre qu'une suite constante) tend vers zéro

3°) toute suite dont chaque terme est le produit des termes correspondants de deux suites appartenant à G, appartient elle-même à G. Il en est de même pour toute élévation à une puissance réelle (et en particulier le quotient de deux fonctions de G est dans G).

alors l'ensemble G forme une échelle de comparaison. Il est donc possible d'obtenir le développement asymptotique de certaines suites réelles convergentes suivant l'échelle G.

Par exemple l'ensemble des suites de la forme $u_n = \lambda^n$ avec $0 \leq \lambda \leq 1$ forme une échelle de comparaison. Il en est de même des suites $u_n = 1/n^\alpha$ avec $\alpha \geq 0$. Nous verrons plus loin que ces deux ensembles de suites jouent un rôle prépondérant dans certaines méthodes d'accélération de la convergence.

I - 4 Théorèmes sur la comparaison

Dans tout ce qui précède on a toujours implicitement supposé que l'on connaissait les limites respectives des suites utilisées. Bien souvent dans la pratique cette limite est précisément l'inconnue. Le but de ce paragraphe est de fournir un certain nombre de résultats qui permette d'éviter cet inconvénient. On supposera que E est complet.

Rappelons d'abord la définition de l'opérateur de différences Δ dont nous aurons constamment besoin dans toute la suite.

Définition 9 : Soit $\{f_n\}$ une suite d'éléments d'un ensemble E. On a :

$$\Delta^\circ f_n = f_n$$

$$\Delta^{k+1} f_n = \Delta^k f_{n+1} - \Delta^k f_n \qquad k = 0, 1, \ldots$$

Dans la suite de ce paragraphe nous supposerons toujours que les suites considérées sont des suites de nombres réels qui convergent vers zéro à moins qu'une autre condition ne soit explicitement spécifiée.

Théorème 12 :

Si $\exists N$ et $a < 1 < b$ tels que $\forall n > N$

$$\frac{v_{n+1}}{v_n} \notin [a,b] \text{ et si } \lim_{n\to\infty} \frac{u_n}{v_n} = c \text{ alors } \lim_{n\to\infty} \frac{\Delta u_n}{\Delta v_n} = c$$

démonstration : soit $\{z_n\}$ une suite qui converge vers c et $\{a_n\}$ une suite telle que $\exists N$ et $a < 1 < b$: $\forall n > N$ $\dfrac{a_{n+1}}{a_n} \notin [a,b]$.

Alors $\lim\limits_{n\to\infty} \dfrac{a_{n+1} z_{n+1} - a_n z_n}{a_{n+1} - a_n} = c$ puisque les trois conditions du théorème de Toeplitz sont vérifiées (pour ce théorème voir chapitre II, théorème 22). Posons

$w_n = a_n z_n$ alors $z_n = w_n / a_n$ et $\lim_{n\to\infty} \frac{w_n}{a_n} = c$ entraîne $\lim_{n\to\infty} \frac{\Delta w_n}{\Delta a_n} = c$ pour toute suite $\{w_n\}$

Une condition nécessaire pour que $\lim_{n\to\infty} a_{n+1} / a_n \neq 1$ est que $\{a_n\}$converge vers zéro. Il en est donc de même pour la suite $\{w_n\}$ ce qui termine la démonstration du théorème.

remarques :

1°) Le théorème de Toeplitz entraîne qu'il y a convergence pour toute suite $\{z_n\}$. Il peut cependant exister des suites $\{z_n\}$ telle que la propriété reste vraie même si la condition sur $\{a_n\}$ n'est pas vérifiée.

2°) Ce théorème est l'analogue pour les suites (ou les séries) de la règle de l'Hospital pour les fonctions.

3°) La réciproque de ce théorème n'est pas vraie. Prenons par exemple $u_n = 1/n$ et $v_n = (-1)^n/n$. On a $\frac{\Delta u_n}{\Delta v_n} = \frac{(-1)^n}{2n+1}$ d'où $\lim_{n\to\infty} \frac{\Delta u_n}{\Delta v_n} = 0$ et $\frac{u_n}{v_n} = (-1)^n$

Démontrons maintenant un résultat un peu plus général :

Théorème 13 :

Si $v_n = O(\Delta v_n)$ et si $u_n = O(v_n)$ alors $\Delta u_n = O(\Delta v_n)$

démonstration : On a $|\Delta u_n| \leq |u_{n+1}| + |u_n| = O(v_{n+1}) + O(v_n)$.

De plus $v_{n+1} = \Delta v_n + v_n$ et $|v_{n+1}| \leq |\Delta v_n| + |v_n| = O(\Delta v_n)$ ce qui démontre le résultat.

On a de même le :

Théorème 14 :

Si $v_n = O(\Delta v_n)$ et si $u_n = o(v_n)$ alors $\Delta u_n = o(\Delta v_n)$

la démonstration est analogue à celle du théorème précédent. Donnons maintenant un théorème qui permet de déduire la limite de u_n/v_n à partir de celle de $\Delta u_n/\Delta v_n$.

Théorème 15 :

Si $\{v_n\}$ est strictement monotone alors $\lim_{n\to\infty} \frac{\Delta u_n}{\Delta v_n} = a$ entraîne $\lim_{n\to\infty} \frac{u_n}{v_n} = a$ que a soit fini ou non.

démonstration : la démonstration dans le cas où $\{v_n\}$ est strictement décroissante a été donnée par Bromwich [51]; nous reproduisons ici sa démonstration. Le cas strictement croissant est analogue ; il a été étudié par Clark [63].

Supposons que $\lim\limits_{n\to\infty} \frac{\Delta u_n}{\Delta v_n} = a$; alors

$$\forall\varepsilon > 0 \; \exists N : \forall n > N \quad a - \varepsilon < \frac{\Delta u_n}{\Delta v_n} \quad a + \varepsilon$$

Or puisque $\Delta v_n < 0$ on a

$$(a - \varepsilon)(v_n - v_{n+1}) < u_n - u_{n+1} < (a + \varepsilon)(v_n - v_{n+1})$$

changeons n en n+1, n+2, ..., n+p-1 et ajoutons les inégalités ainsi obtenues :

$$(a - \varepsilon)(v_n - v_{n+p}) < u_n - u_{n+p} < (a+\varepsilon)(v_n - v_{n+p})$$

prenons la limite quand p tend vers l'infini ; il vient :

$$(a-\varepsilon)\, v_n \leq u_n \leq (a+\varepsilon)\, v_n$$

et par conséquent, puisque $v_n > 0$:

$$\left| \frac{u_n}{v_n} - a \right| \leq \varepsilon \qquad \forall n > N$$

ce qui démontre la première partie du théorème lorsque a est fini. Si a est infini alors $\forall A > 0 \;\; \exists N : \forall n > N$

$$\frac{\Delta u_n}{\Delta v_n} > A$$

d'où

$$u_n - u_{n+p} > A(v_n - v_{n+p})$$

et

$$u_n \geq A\, v_n$$

en faisant tendre p vers l'infini; on a donc $\frac{u_n}{v_n} \geq A \quad \forall\, n > N$.

Si $\{v_n\}$ n'est pas strictement monotone alors le théorème peut ne pas être vrai de même que la règle de L'Hospital pour les fonctions peut être fausse si la dérivée du dénominateur change de signe autant de fois que l'on veut lorsque la variable tend vers sa limite.

I - 5 L'indice de comparaison

Soit maintenant à comparer, du point de vue numérique, la rapidité de convergence de deux suites. Le matériel dont on dispose est fourni par l'indice d'efficacité introduit par Ostrowski [147]: considérons une suite d'ordre $r > 1$ telle que le calcul de chaque terme nécessite p opérations arithmétiques élémentaires (on appelle opération arithmétique élémentaire l'une des opérations x : + - prise comme base de mesure ; on saura par exemple qu'une multiplication vaut 1,8 additions et une division 2,2 additions).

Considérons maintenant une autre suite d'ordre r^2 telle que le calcul de chaque terme nécessite 2p opérations arithmétiques élémentaires. Du point de vue du temps de calcul nécessaire pour obtenir une certaine précision on n'aura rien gagné : la seconde suite converge deux fois plus vite mais elle nécessite deux fois plus de calculs. En effet pour multiplier par r^2 le nombre de chiffres significatifs exacts il faut un seul terme supplémentaire pour la seconde suite soit 2p opérations et il faut deux termes supplémentaires pour la première suite soit $p + p = 2p$ opérations. On dit que ces deux suites ont le même indice d'efficacité. Cet indice est défini par :

$$E(u_n) = r^{1/p} = E(r,p) \qquad \text{si } r > 1.$$

Il représente le facteur par lequel on multiplie le nombre de chiffres significatifs exacts par opération élémentaire. Plus l'indice d'efficacité est grand et plus la rapidité de convergence est élevée. Cet indice possède les propriétés évidentes suivantes :

propriété 6 :

$$E(nr,p) = n^{1/p}\, E(r,p)$$

$$E(r,np) = [E(r,p)]^{1/n}$$

$$E(r^n,np) = E(r,p)$$

$$E(r^n,p) = [E(r,p)]^n$$

$$E(r_1\, r_2,\, p_1+p_2) = [E(r_1,p_1)^{p_1}\; E(r_2,p_2)^{p_2}]^{\frac{1}{p_1+p_2}}$$

Puisque la notion d'ordre n'est pas une notion assez finie, la comparaison de leurs indices d'efficacité n'est pas suffisante pour comparer deux suites. On est donc amené à introduire un indice de comparaison :

Définition 10 : supposons que $\{u_n\}$ soit α-équivalente à $\{v_n\}$. On appelle indice de comparaison de $\{u_n\}$ par rapport à $\{v_n\}$ la quantité :

$$C(u_n, v_n) = \alpha \frac{E(u_n)}{E(v_n)}$$

par exemple en supposant qu'une division vaut une multiplication on trouve que :

$$C(\frac{1}{n^2}, \frac{1}{n}) = 2 \text{ alors que } E(\frac{1}{n^2}) = E(\frac{1}{n}) = 1$$

ce qui rend mieux compte de la réalité que la comparaison des indices d'efficacité. L'indice de comparaison possède les propriétés suivantes :

Propriété 7 :

$$C(u_n, u_n) = 1$$

$$C(u_n, v_n) . C(v_n, u_n) = 1$$

$$C(u_n, v_n) . C(v_n, w_n) = C(u_n, w_n)$$

si $C(u_n, v_n) > 1$ alors la vitesse de convergence de $\{u_n\}$ est supérieure à celle de $\{v_n\}$. Si $r = 1$ on pourrait définir l'indice d'efficacité par $E(R,p) = R/p$ avec $R = -\log C$ où C est le coefficient asymptotique d'erreur.

I - 6 Développement asymptotique d'une série

Dans ce paragraphe on va démontrer pour les séries à termes positifs un théorème analogue à celui que l'on connait pour la partie principale d'une primitive (voir par exemple [72]). En traduisant ce résultat en termes de suites on trouve un procédé d'accélération de la convergence très utilisé : le procédé Δ^2 d'Aitken qui est un cas particulier de l'ε-algorithme ; ces méthodes seront étudiées au chapitre III. On remarque aussi que le théorème sur la partie principale d'une primitive est un cas particulier de la première forme confluente de l'ε-algorithme qui sera étudiée au chapitre IV. Ce paragraphe apparait donc comme le lien entre les résultats que nous venons d'énoncer sur les suites et certains procédés d'accélération de la convergence.

Etablissons maintenant ce résultat :

Théorème 16 :

Soit u_n une série à termes positifs au voisinage de $+\infty$ et telle que Δu_n ne change pas de signe au voisinage de $+\infty$.

Posons $h_n = u_n / \Delta u_n$ et supposons que $\Delta h_n = o(1)$.

Alors :

- si $\Delta u_n > 0$ au voisinage de $+\infty$, la série diverge et l'on a au voisinage de $+\infty$

$$\sum_{n=0}^{k} u_n \sim \frac{u_k \, u_{k+1}}{\Delta u_k}$$

- si $\Delta u_n < 0$ au voisinage de $+\infty$ alors la série converge, et l'on a :

$$\sum_{n=k}^{\infty} u_n \sim - \frac{u_k^2}{\Delta u_k}$$

(le symbole $f_n \sim g_n$ signifie que $\lim_{n \to \infty} f_n / g_n = 1$ ou, en d'autres termes que $f_n - g_n = o(f_n) = o(g_n)$).

démonstration : si $\Delta u_n > 0$ alors $u_{n+1} > u_n > 0$ donc la série diverge. On a :

$$\sum_{n=0}^{k} u_n = \sum_{n=0}^{k} h_n \, \Delta u_n = h_k \, u_{k+1} - h_o \, u_o - \sum_{n=1}^{k} u_n \, \Delta h_{n-1} \text{ d'où}$$

$$u_o + \sum_{n=1}^{k} u_n (1 + \Delta h_{n-1}) = h_k \, u_{k+1} - h_o \, u_o$$

or $u_n (1 + \Delta h_{n-1}) \sim u_n$ car $\Delta h_n = o(1)$ donc

$$u_o + \sum_{n=1}^{k} u_n \, (1 + \Delta h_{n-1}) \sim \sum_{n=0}^{k} u_n \sim h_k \, u_{k+1} - h_o \, u_o$$

si $\sum_{n=0}^{\infty} u_n$ diverge alors $\sum_{n=0}^{k} u_n$ est prépondérant sur la constante $h_o u_o$ d'où la première partie du théorème.

Si $\Delta u_n < 0$ alors $|h_n - h_0| = |\sum_{k=0}^{n-1} \Delta h_k| = O\{\sum_{k=0}^{n-1} |\Delta h_k|\} = o(n)$ puisque la série $\sum_{k=0}^{\infty} 1$ diverge [72]. Par conséquent $\lim_{n\to\infty} n(1-u_{n+1}/u_n) = \infty$ et la série converge d'après le critère de Raabe [18]. On a :

$$\sum_{n=k}^{\infty} u_n = \sum_{n=k}^{\infty} h_n \Delta u_n = -h_k u_k - \sum_{n=k+1}^{\infty} u_n \Delta h_{n-1} \text{ d'où } u_k + \sum_{n=k+1}^{\infty} u_n(1+\Delta h_{n-1}) = -h_k u_k$$

or $u_n(1 + \Delta h_{n-1}) \sim u_n$ et donc $\sum_{n=k}^{\infty} u_n \sim -h_k u_k$.

On voit que, dans la démonstration de la seconde partie du théorème, on ne s'est servi ni de l'hypothèse de convergence de la série, ni du fait que la série est à termes positifs et que $\Delta u_n < 0$.

Le résultat $\sum_{n=k}^{\infty} u_n \sim -u_k^2 / \Delta u_k$ peut donc être démontré avec la seule hypothèse que $\Delta h_n = o(1)$. On voit que c'est un résultat très général.

En posant $S_n = \sum_{k=0}^{n} u_k$ et $S = \lim_{n\to\infty} S_n$

on peut transposer le théorème précédent pour l'appliquer aux suites de nombres réels ou complexes $\{S_n\}$ qui converge vers S, d'où le :

<u>Théorème 17</u> :

Soit $\{S_n\}$ une suite de nombres réels ou complexes qui converge vers S avec la propriété que :

$$\lim_{n\to\infty} \frac{\Delta S_{n+1}}{\Delta^2 S_{n+1}} - \frac{\Delta S_n}{\Delta^2 S_n} = 0$$

alors la suite notée $\{\varepsilon_2^{(n)}\}$ et définie par :

$$\varepsilon_2^{(n)} = S_n - \frac{(\Delta S_n)^2}{\Delta^2 S_n}$$

converge vers S et cela plus vite que $\{S_n\}$. De plus on a :

$$\varepsilon_2^{(n)} - S = o(\varepsilon_2^{(n)} - S_n)$$

Démonstration : c'est une simple transposition du théorème 16. La dernière partie se démontre en utilisant le fait que si $f_n \sim g_n$ alors $f_n - g_n = o(f_n) = o(g_n)$.

On verra au chapitre IV que l'on retrouve par ailleurs cette estimation de l'erreur $\varepsilon_2^{(n)} - S = o(\varepsilon_2^{(n)} - S_n)$.

La quantité $S_n - \frac{(\Delta S_n)^2}{\Delta^2 S_n}$ n'est autre qu'un procédé bien connu d'accélération de la convergence qui s'appelle le procédé Δ^2 d'Aitken. C'est un cas particulier d'un algorithme beaucoup plus puissant d'accélération de la convergence : l'ε-algorithme. C'est pour respecter les notations de cet algorithme que nous avons posé ici

$$\varepsilon_2^{(n)} = S_n - \frac{(\Delta S_n)^2}{\Delta^2 S_n}$$

Cette méthode sera étudiée très en détail au chapitre III.

On verra également au chapitre III que l'ε-algorithme utilise beaucoup la connexion qui existe entre suite et série de puissances. Nous terminerons ce chapitre en utilisant cette relation pour expliquer différemment ce qu'est accélérer la convergence.

Commençons par rappeler la :

<u>définition 11</u> : considérons la série de puissances $\sum_{k=0}^{\infty} a_k x^k$. On appelle rayon de convergence de cette série la quantité R définie par :

$$1 / R = \limsup_{k\to\infty} |a_k|^{1/k}$$

on sait que la série converge pour tout x tel que $|x| < R$.

Considérons maintenant les séries $\sum_{k=0}^{\infty} a_k x^k$ et $\sum_{k=0}^{\infty} b_k x^k$ dont les rayons de convergence respectifs sont R_1 et R_2. On a :

$$\frac{R_1}{R_2} = \frac{\limsup_{k\to\infty} |b_k|^{1/k}}{\limsup_{k\to\infty} |a_k|^{1/k}}$$

Supposons que $\limsup_{k\to\infty} |a_k|^{1/k} = \lim_{k\to\infty} |a_k|^{1/k}$. On a alors :

$$\frac{R_1}{R_2} = \limsup_{k\to\infty} \left|\frac{b_k}{a_k}\right|^{1/k}$$ ce qui signifie que $\frac{R_2}{R_1}$ est le rayon de convergence de la série $\sum_{k=0}^{\infty} \frac{b_k}{a_k} x^k$.

Donc si $R_2 / R_1 > 1$ alors la série $\sum_{k=0}^{\infty} b_k / a_k$ est convergente et par conséquent on a : $\lim_{k\to\infty} \frac{b_k}{a_k} = 0$

En transposant en termes de suites on obtient donc le :

Théorème 18 :

Soient $\{S_n\}$ et $\{V_n\}$ deux suites convergentes telles que $\lim_{n\to\infty} |\Delta S_n|^{1/n} = 1/R_1$ et $\limsup_{n\to\infty} |\Delta V_n|^{1/n} = 1/R_2$ avec $R_1 < R_2$. Alors $\{V_n\}$ converge plus vite que $\{S_n\}$ en ce sens que :

$$\lim_{n\to\infty} \frac{\Delta V_n}{\Delta S_n} = 0$$

On voit donc que, dans ce cas, accélérer la convergence n'est autre qu'augmenter le rayon de convergence.

REMARQUE : on peut trouver un exposé des résultats de ce chapitre dans les références [32,33].

CHAPITRE II

LES PROCEDES DE SOMMATION

II - 1 Formulation générale du problème

Soit (c) l'espace des suites convergentes de nombres réels et soit $S = \{S_n\}$ un élément de (c). On muni c de la norme :

$$||S|| = \sup_n |S_n|$$

(c) est alors un espace de Banach (voir par exemple [244] p. 325).

Le dual topologique de (c) est ℓ^1, espace des suites de nombres réels telles que $\sum_{n=0}^{\infty} |S'_n|$ converge. On muni ℓ^1 de la norme $||S'|| = \sum_{n=0}^{\infty} |S'_n|$. C'est alors un espace de Banach (voir par exemple [243]). On désignera par $<.,.>$ la dualité entre (c) et ℓ^1.

Le problème est le suivant : étant donné $S \in (c)$ on veut transformer S en une autre suite $V = \{V_n\}$ à l'aide d'une application linéaire T de (c) dans l'espace des suites. On dira que T est un procédé de sommation et l'on veut naturellement que la suite $\{V_n\}$ converge et que $\lim_{n\to\infty} V_n = \lim_{n\to\infty} S_n$ quelquesoit $S = \{S_n\} \in (c)$.

L'application T est définie par une matrice infinie $A = (a_{nk})$ et il nous faut donc chercher les conditions à imposer à A pour que les deux conditions précédentes soient satisfaites. C'est ce problème que nous allons maintenant examiner.

Considérons la suite des formes linéaires continues sur (c) qui à $S = \{S_n\} \in (c)$ fait correspondre V_n $n^{\text{ième}}$ terme de la suite transformée $V = \{V_n\}$ par l'application linéaire T. On est donc ramené au problème de la convergence faible d'une suite de formes linéaires continues sur (c) ou, en d'autres termes, à la convergence faible d'une suite d'éléments de ℓ^1.

Pour ce faire nous disposons de deux théorèmes : le théorème de la borne uniforme (quelquefois appelé théorème de Banach-Steinhaus, voir [120]) et un théorème dérivé du principe de prolongement des identités (voir [71]).

- Soit E un espace de Banach et E' son dual topologique.
- Soit $\{e'_n\}$ une suite de E' qui converge faiblement vers $e'_\infty \in E'$ c'est-à-dire telle que $\forall x \in E$ on ait $\lim_{n\to\infty} \langle x, e'_n\rangle = \langle x, e'_\infty\rangle$.
- Soit $\{e_n\}$ un système total d'éléments de E c'est-à-dire dont les combinaisons linéaires finies engendrent un sous-espace D partout dense dans E.
- Soit $\{x'_n\}$ une suite d'éléments de E'.

On a les résultats suivants :

Théorème 19 :

Une condition nécessaire et suffisante pour que $\{x'_n\}$ converge faiblement vers $x'_\infty \in E'$ est que :

1°) $||x'_n|| < M \quad \forall n$

2°) $\lim_{n\to\infty} \langle e_k, x'_n - x'_\infty\rangle = 0 \quad \forall k$

Théorème 20 :

Soit $\{x'_n\}$ une suite d'éléments de E' qui converge faiblement vers $x'_\infty \in E'$. Si

$\lim_{n\to\infty} \langle x, x'_n\rangle = \lim_{n\to\infty} \langle x, e'_n\rangle$

$\forall x \in D$ alors $\lim_{n\to\infty} \langle x, x'_n\rangle = \lim_{n\to\infty} \langle x, e'_n\rangle \quad \forall x \in E$.

Démonstration : on a $\overline{D} = E$; donc $\forall x \in E$

il existe une suite $\{x_n\}$ d'éléments de D telle que $\lim_{n\to\infty} x_n = x$.

Or $\langle x_n, x'_\infty\rangle = \langle x_n, e'_\infty\rangle$ $\forall n$ et par conséquent $\langle x, x'_\infty\rangle = \langle x, e'_\infty\rangle$ $\forall x \in E$ à cause de la continuité de x'_∞ et de e'_∞.

Nous allons appliquer les résultats de ces deux théorèmes à notre problème.

Pour cela nous prendrons E = (c) et e'_n sera la forme linéaire continue qui à $x = \{x_n\} \in c$ fait correspondre sa $n^{ième}$ composante x_n.

Le sous-espace D engendré par le système libre $e_o = (1, 1, \ldots)$, $e_1 = (1, 0, 0, \ldots)$, $e_2 = (0, 1, 0, \ldots)$ est partout dense dans (c) (voir par exemple [244]).

La $n^{ième}$ forme linéaire x'_n est définie par la $n^{ième}$ ligne de la matrice A et les conditions du théorème 19 s'écrivent :

1°) $||x'_n|| = \sum_{k=1}^{\infty} |a_{nk}| < M \quad \forall n$

2°) $\lim_{n\to\infty} \langle e_k, x'_n \rangle = \lim_{n\to\infty} a_{nk} = b_k \quad \forall k > 0$

$$\lim_{n\to\infty} \langle e_o, x'_n \rangle = \lim_{n\to\infty} \sum_{k=1}^{\infty} a_{nk} = b_o$$

d'où le théorème suivant :

<u>Théorème 21</u> :

Une condition nécessaire et suffisante pour que la matrice A définisse un endomorphisme de (c) est que :

1°) $\sum_{k=1}^{\infty} |a_{nk}| < M \quad \forall n$

2°) $\lim_{n\to\infty} a_{nk} = b_k \quad k = 1, 2, \ldots$

3°) $\lim_{n\to\infty} \sum_{k=1}^{\infty} a_{nk} = b_o$

Si maintenant $b_k = 0$ pour $k = 1, 2, \ldots$ et si $b_o = 1$ alors les conditions du théorème 20 sont satisfaites car $\lim_{n\to\infty} \langle e_k, e'_n \rangle = 0$ pour $k = 1, 2, \ldots$ et $\lim_{n\to\infty} \langle e_o, e'_n \rangle = 1$. D'où le :

<u>Théorème 22</u> : (Théorème de Toeplitz) :

Soit $\{S_n\}$ une suite convergente et soit $\{V_n\}$ la suite déduite de $\{S_n\}$ par :

$$\begin{pmatrix} V_o \\ V_1 \\ \cdot \\ \cdot \\ \cdot \end{pmatrix} = A \begin{pmatrix} S_o \\ S_1 \\ \cdot \\ \cdot \\ \cdot \end{pmatrix}$$

où $A = (a_{nk})$ est une matrice infinie. Une condition nécessaire et suffisante pour que la suite $\{V_n\}$ converge vers la même limite que $\{S_n\}$ et ceci quelquesoit $\{S_n\}$ est que les trois conditions suivantes soient vérifiées :

1°) $\sum_{k=1}^{\infty} |a_{nk}| < M \qquad \forall n$

2°) $\lim_{n\to\infty} a_{nk} = 0 \qquad k = 1, 2, \ldots$

3°) $\lim_{n\to\infty} \sum_{k=1}^{\infty} a_{nk} = 1$

Dans ce cas on dit que la matrice A définit un procédé régulier de sommation.

REMARQUE :

Le théorème de Toeplitz assure que $\{V_n\}$ converge vers la même limite que $\{S_n\}$ quelquesoit la suite $\{S_n\}$ convergente. Ce "quelquesoit $\{S_n\}$" provient de la convergence faible de $\{x'_n\}$ et assure toute sa généralité au théorème.

Il peut exister des procédés de sommation qui ne vérifient pas les conditions du théorème de Toeplitz et pour lesquels on a cependant $\lim_{n\to\infty} V_n = \lim_{n\to\infty} S_n$ mais seulement pour des suites d'un type bien particulier et non pour toute suite convergente $\{S_n\}$. Ce ne sont pas alors des procédés réguliers de sommation. D'où la :

<u>Définition 12</u> : Soit un procédé de sommation défini par une matrice A. Si, pour toute suite convergente $\{S_n\}$, la suite déduite $\{V_n\}$ à même limite alors on dit que le procédé de sommation est régulier.

II - 2 Etude de quelques procédés

Il existe trois procédés de sommation importants :

1°) Le procédé de Hölder

La matrice A est donnée par

$$\left.\begin{array}{ll} a_{ij} = 1/i & j = 1, \ldots, i \\ a_{ij} = 0 & j > i \end{array}\right\} \quad i = 1, \ldots, \infty$$

Cette matrice A définit la méthode notée (H, 1). La méthode (H, k) k entier est définie par la matrice A^k.

2°) Le procédé de Césaro :

La matrice A définissant le procédé noté (C, k) est donnée par $A = DHL^{k-1}$ où H est la matrice du procédé de Hölder (H, 1), où L est une matrice infinie triangulaire inférieure dont tous les termes sont égaux à 1 et où D est une matrice diagonale telle que la somme des éléments de chaque ligne de A soit égale à 1.

3°) Les méthodes d'Euler :

Elles sont définies par une matrice A dépendant d'un paramètre positif q :

$$\left.\begin{array}{ll} a_{ij} = \dfrac{q^{i-j}}{(q+1)^{i+1}} \dbinom{i+1}{j+1} & j = 1, \ldots, i \\ a_{ij} = 0 & j > i \end{array}\right\} \quad i = 1, \ldots$$

La plus utilisée des méthodes d'Euler est celle avec q = 1. Il existe de nombreuses autres méthodes de sommation. Elles ont été étudiées en détail par Hardy [107] et Peyerimhoff [155]. Les procédés de sommation ne sont pas des méthodes très intéressantes en ce qui concerne l'accélération de la convergence. Considérons par exemple la méthode définie par :

$$V_n = a\,S_n + bS_{n+1}$$

avec a + b = 1. Il est évident que cette méthode vérifie le théorème de Toeplitz.

On a :

$$V_n - S = a(S_n - S) + b(S_{n+1} - S)$$

et par conséquent une condition nécessaire et suffisante pour que $\{V_n\}$ converge plus vite que $\{S_n\}$ est que :

$$\lim_{n\to\infty} \frac{S_{n+1} - S}{S_n - S} = -\frac{a}{b}$$

ce qui restreint singulièrement l'ensemble des suites qui peuvent être accélérées par ce procédé. Pour qu'un procédé soit efficacement utilisable il faudrait qu'il soit capable d'accélérer la convergence de toutes les suites telles que

$$\lim_{n\to\infty} \frac{S_{n+1} - S}{S_n - S} = a \neq 1$$

Nous avons déjà rencontré à la fin du premier chapitre un procédé qui possède cette propriété : le procédé Δ^2 d'Aitken. Nous reviendrons en détail sur cette méthode ainsi que sur l'ε-algorithme au chapitre III.
Cependant avant de quitter définitivement les procédés de sommation il reste à parler en détail du procédé d'extrapolation de Richardson qui peut se rattacher à cette catégorie de méthodes et qui possède un certain nombre de propriétés très intéressantes. Le chapitre se terminera par une interprétation de certains procédés de sommation qui montre que ce sont des cas particuliers de l'ε-algorithme.

II - 3 Le procédé d'extrapolation de Richardson

Soit $\{S_n\}$ une suite qui converge vers S et soit $\{x_n\}$ une suite de paramètres qui converge vers zéro et telle qu'il n'existe pas deux indices distincts k et p tels que $x_k = x_p$ et que $x_n \neq 0 \ \forall n$.
Soit $T_k^{(n)}$ la valeur en x = 0 du polynôme d'interpolation de degré k qui passe par les k+1 couples :
$(x_n, S_n), (x_{n+1}, S_{n+1}), \ldots, (x_{n+k}, S_{n+k})$. A partir de la suite initiale $T_o^{(n)} = S_n$

on peut ainsi générer tout un ensemble de suites en faisant varier n et k : c'est le procédé d'extrapolation de Richardson[161]. Les quantités $T_k^{(n)}$ sont construites à partir du schéma de Neville-Aitken de construction du polynôme d'interpolation :

$$T_o^{(n)} = S_n$$

$$T_{k+1}^{(n)} = \frac{x_n\, T_k^{(n+1)} - x_{n+k+1}\, T_k^{(n)}}{x_n - x_{n+k+1}} \qquad n, k = 0, 1, \ldots$$

on place ces quantités dans un tableau à double entrée. L'indice inférieur k représente une colonne et l'indice supérieur représente une diagonale descendante. A partir des valeurs initiales $T_o^{(n)}$ on progresse de la gauche vers la droite ; les trois quantités $T_k^{(n)}$, $T_k^{(n+1)}$ et $T_{k+1}^{(n)}$ liées par la relation précédente sont situées au sommet d'un triangle dans lequel les deux quantités les plus à gauche servent à calculer celle située à droite comme cela est indiqué par les flèches dans le tableau suivant :

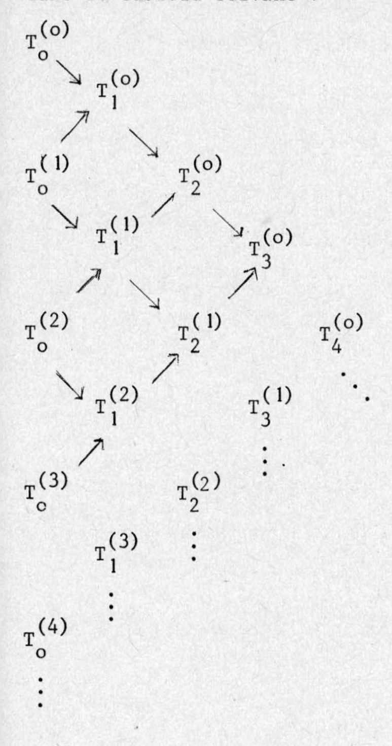

Puisque l'algorithme précédent est bâti à partir du polynôme d'interpolation on a bien évidemment le résultat fondamental suivant :

Théorème 23 :

Si $S_n = S + \sum_{i=1}^{k} a_i x_n^i \quad \forall n > N$ alors $T_k^{(n)} = S \quad \forall n > N$

On voit que l'application qui fait passer de la suite $\{T_k^{(n)}\}$ où k est fixé à la suite $\{T_{k+1}^{(n)}\}$ k fixé est un procédé de sommation. La matrice correspondante est donnée par :

$$a_{ii} = - \frac{x_{i+k+1}}{x_i - x_{i+k+1}}$$

$$a_{i,i+1} = \frac{x_i}{x_i - x_{i+k+1}}$$

et tous les autres termes sont nuls. On peut donc appliquer le théorème de Toeplitz à ce procédé et l'on obtient le :

Théorème 24 :

Une condition nécessaire et suffisante pour que $\lim_{n\to\infty} T_k^{(n)} = \lim_{n\to\infty} S_n$ pour toute suite convergente $\{S_n\}$ est que $\exists \alpha < 1 < \beta$:

$$\frac{x_{n+p+1}}{x_n} \notin [\alpha,\beta] \ \forall n \text{ et pour } p = 0, \ldots, k-1$$

Démonstration : pour que $\lim_{n\to\infty} T_k^{(n)} = \lim_{n\to\infty} S_n$ il faut et il suffit que les transformations linéaires de suite à suite $\{T_p^{(n)}\} \to \{T_{p+1}^{(n)}\}$ soient des procédés réguliers de sommation pour p = 0, ..., k-1. Les deux dernières conditions du théorème de Toeplitz sont automatiquement vérifiées puisque la matrice est bidiagonale et que $\sum_{j=1}^{\infty} a_{ij} = 1$ pour tout i. La première condition s'écrit :

$$\left|\frac{x_n}{x_n - x_{n+p+1}}\right| + \left|\frac{x_{n+p+1}}{x_n - x_{n+p+1}}\right| < M_p \quad \forall n$$

on doit avoir $K_p < |1 - \frac{x_{n+p+1}}{x_n}| \; \forall n$

Il doit donc exister α_p et β_p : $\alpha_p < 1 < \beta_p$ tels que $x_{n+p+1} / x_n \notin [\alpha_p, \beta_p] \; \forall n$,

d'où le théorème en prenant $\alpha = \inf_p \alpha_p$ et $\beta = \sup_p \beta_p$.

REMARQUE : on voit que ce résultat nécessite que $\lim_{n\to\infty} x_n = 0$ ou $l'\infty$

Etudions maintenant la convergence des suites $\{T_k^{(n)}\}$ pour n fixé. Le résultat suivant, appelé condition (α), a été démontré par Laurent [128] en partant du théorème 19 et en explicitant les coefficients du polynôme d'interpolation. La démonstration est laissée à titre d'exercice.

Théorème 25 : (condition (α)) :

Soit $\{x_n\}$ une suite de nombres positifs strictement décroissants et tendant vers zéro quand n tend vers l'infini.

Une condition nécessaire et suffisante pour que $\lim_{k\to\infty} T_k^{(n)} = \lim_{k\to\infty} S_k \; \forall n$ et ceci pour toute suite convergente $\{S_n\}$ est qu'il existe $\alpha > 1$ tel que $\frac{x_n}{x_{n+1}} \geq \alpha$.

REMARQUE : les théorèmes 24 et 25 entraînent la convergence pour toute suite convergente $\{S_n\}$. Cependant dans des cas particuliers il est possible de se passer de conditions sur $\{x_n\}$; on trouvera de tels théorèmes dans [41].

Etudions maintenant les conditions d'accélération de la convergence. On a le :

Théorème 26 :

Supposons que la condition (α) soit vérifiée. Une condition nécessaire et suffisante pour que $\{T_{k+1}^{(n)}\}$ k fixé converge vers S plus vite que $\{T_k^{(n)}\}$k fixé est que :

$$\lim_{n\to\infty} \frac{T_k^{(n+1)} - S}{T_k^{(n)} - S} = \lim_{n\to\infty} \frac{x_{n+k+1}}{x_n}$$

démonstration : supposons que $\{T_{k+1}^{(n)}\}$ k fixé converge plus vite que $\{T_k^{(n)}\}$, on a :

$$\lim_{n} \frac{T_{k+1}^{(n)} - S}{T_k^{(n)} - S} = 0 = \lim_{n\to\infty} \frac{x_n\,(T_k^{(n+1)} - S) - x_{n+k+1}\,(T_k^{(n)} - S)}{(T_k^{(n)} - S)(x_n - x_{n+k+1})}$$

$$= \lim_{n\to\infty} \frac{\dfrac{T_k^{(n+1)} - S}{T_k^{(n)} - S} - \dfrac{x_{n+k+1}}{x_n}}{1 - \dfrac{x_{n+k+1}}{x_n}}$$

puisque la condition (α) est vérifiée alors $\lim\limits_{n\to\infty} \dfrac{x_{n+k+1}}{x_n} \neq 1\ \forall k$, ce qui démontre

que la condition est nécessaire.

Réciproquement si la condition du théorème est vérifiée alors $\{T_{k+1}^{(n)}\}$ k fixé converge plus vite que $\{T_k^{(n)}\}$ k fixé.

Une application très importante du procédé d'extrapolation de Richardson est celui des quadratures numériques à l'aide de la méthode des trapèzes : soit à calculer

$$I = \int_a^b f(x)\,dx$$

la formule des trapèzes avec un pas $h = (b-a)/n$ nous donne une valeur approchée $\bar{I}$ de I :

$$\bar{I} = \frac{h}{2}\left(f(x_o) + 2\sum_{i=1}^{n-1} f(x_i) + f(x_n)\right)$$

avec $x_i = a + ih$ pour $i = 0, \ldots, n$.

On sait que l'on a, pour une fonction f suffisamment différentiable :

$$\bar{I} = I + a_1 h^2 + a_2 h^4 + \ldots$$

Appliquons la méthode des trapèzes au calcul de I avec les pas $h_o = H$, $h_1 = h_o/\alpha$, $h_2 = h_1/\alpha$ avec $\alpha > 1$. On obtient ainsi une suite de valeurs approchées de I à l'aide de la méthode des trapèzes que l'on notera S_o, S_1, ...

On peut utiliser le procédé de Richardson pour accélérer cette suite en prenant $x_n = h_n^2$ puisque l'erreur est un polynôme pair en h. On a alors :

$$T_k^{(n)} = I + a_k^{(k)} h_n^{2k+2} + \dots$$

d'où

$$\lim_{n\to\infty} \frac{T_k^{(n+1)} - I}{T_k^{(n)} - I} = \lim_{n\to\infty} \left(\frac{h_{n+1}}{h_n}\right)^{2k+2} = \frac{1}{\alpha^{2k+2}}$$

d'autre part on a :

$$\frac{h_{n+k+1}}{h_n} = \frac{\alpha^{2n}}{\alpha^{2n+2k+2}} = \frac{1}{\alpha^{2k+2}}$$

ce qui démontre que $\forall k$ $\{T_{k+1}^{(n)}\}$ k fixé converge plus vite que $\{T_k^{(n)}\}$ k fixé car la condition du théorème 26 est satisfaite.

Quand $\alpha = 2$ ce procédé est plus connu sous le nom de méthode de Romberg. Elle s'écrit :

$$T_o^{(n)} = S_n$$

$$T_{k+1}^{(n)} = \frac{2^{2k+2}\, T_k^{(n+1)} - T_k^{(n)}}{2^{2k+2} - 1}$$

Donnons un exemple numérique. Soit à calculer

$I = \int_o^1 \frac{dx}{x+0.01} = 4.615120517$. Avec $H = 1/3$ et $\alpha=2$ on obtient :

$T_o^{(o)} = 18.29$ $\quad T_o^{(1)} = 10.61$ $\quad T_o^{(2)} = 7.06$ $\quad T_o^{(3)} = 5.51$

$T_o^{(4)} = 4.91$ $\quad T_o^{(5)} = 4.69$ $\quad T_o^{(6)} = 4.63$ $\quad T_o^{(7)} = 4.62$

$T_o^{(8)} = 4.616$

A partir de ces neuf valeurs dont la meilleure a trois chiffres exacts on trouve :

$T_2^{(o)} = 5.72$

$T_2^{(1)} = 4.94$ $\quad T_4^{(c)} = 4.67$

$T_2^{(2)} = 4.68$ $\quad T_4^{(1)} = 4.6234$ $\quad T_6^{(o)} = 4.61570$

$T_2^{(3)} = 4.62$ $\quad T_4^{(2)} = 4.6157$ $\quad T_6^{(1)} = 4.61514$ $\quad T_8^{(o)} = 4.615120793$

$T_2^{(4)} = 4.6159$ $\quad T_4^{(3)} = 4.61514$ $\quad T_6^{(2)} = 4.615120794$

$T_2^{(5)} = 4.615155$ $\quad T_4^{(4)} = 4.6151208$

$T_2^{(6)} = 4.6151214$

c'est-à-dire que l'on obtient 7 chiffres exacts.

On a vu dans l'application du procédé de Richardson à la méthodes des trapèzes que l'on avait été amené à effectuer le changement de variable $x = h^2$. Dans certains cas il peut être intéressant d'effectuer d'autres changements de variables suivant la nature de la suite que l'on a à accélérer. Pour ces questions on pourra se reporter à [34].

Si l'on ne connait pas explicitement la dépendance de S_n par rapport au paramètre x_n on peut songer à prendre $x_n = \Delta S_n$. On obtient alors l'algorithme :

$$T_o^{(n)} = S_n$$

$$T_{k+1}^{(n)} = \frac{\Delta S_n . T_k^{(n+1)} - \Delta S_{n+k+1} . T_k^{(n)}}{\Delta S_n - \Delta S_{n+k+1}}$$

qui a été obtenu et étudié par Germain-Bonne [87]. Pour cette méthode on a le :

<u>Théorème 27</u> :

Pour l'algorithme précédent une condition nécessaire et suffisante pour que $T_k^{(n)} = S$ $\forall n > N$ est que la suite $\{S_n\}$ vérifie :

$$a_o(S_n - S) + a_1 \Delta S_n + \ldots + a_k(\Delta S_n)^k = 0 \quad \forall n > N$$

démonstration : elle est évidente à partir de la définition du polynôme d'interpolation comme un rapport de deux déterminants. On a :

$$T_k^{(n)} = \frac{\begin{vmatrix} S_n & \ldots\ldots & S_{n+k} \\ x_n & \ldots\ldots & x_{n+k} \\ \ldots & \ldots & \ldots \\ x_n^k & \ldots\ldots & x_{n+k}^k \end{vmatrix}}{\begin{vmatrix} 1 & \ldots\ldots & 1 \\ x_n & \ldots\ldots & x_{n+k} \\ \ldots & \ldots & \ldots \\ x_n^k & \ldots\ldots & x_{n+k}^k \end{vmatrix}}$$

La condition est nécessaire car on doit avoir :

$$\begin{vmatrix} S_n - S & \ldots\ldots & S_{n+k} - S \\ \Delta S_n & \ldots\ldots & \Delta S_{n+k} \\ \text{-----} & \text{-----} & \text{-----} \\ (\Delta S_n)^k & \ldots\ldots & (\Delta S_{n+k})^k \end{vmatrix} = 0$$

qui a lieu s'il existe $a_o, \ldots, a_k$ non tous nuls tels que

$$a_o(S_n - S) + a_1 \Delta S_n + \ldots + a_k(\Delta S_n)^k = 0 \quad \forall n > N$$

Réciproquement il est évident que cette condition est suffisante.

Remarques : 1°) si k = 0 on obtient

$$T_1^{(n)} = \frac{S_n S_{n+2} - S_{n+1}^2}{\Delta^2 S_n} = S_n - \frac{(\Delta S_n)^2}{\Delta^2 S_n}$$

qui n'est autre que le procédé Δ^2 d'Aitken.

2°) On voit que $T_k^{(n)}$ est continu et différentiable par rapport aux variables $S_n, \ldots, S_{n+k}$.

II - 4 Interprétation des procédés totaux

Nous allons maintenant donner une interprétation de certains procédés de sommation : les procédés totaux.

<u>définition 13</u> : Soit un procédé régulier de sommation défini par une matrice A. Appliquons ce procédé à une suite constante $S_n = S \quad \forall n$. Si la suite transformée $\{V_n\}$ est telle que $V_n = S \quad \forall n$ on dit que le procédé de sommation est total.

On a immédiatement le :

<u>théorème 28</u> :

Une condition nécessaire et suffisante pour que le procédé régulier défini par la matrice A soit total est que :

$$\sum_{k=1}^{\infty} a_{nk} = 1 \qquad n = 1, 2, \ldots$$

Jusqu'à présent nous avions appliqué le procédé de sommation à partir du premier terme de la suite $\{S_n\}$:

$$V_n = \sum_{p=1}^{\infty} a_{np} S_p \qquad n = 1, 2, \ldots$$

En appliquant ce procédé à partir d'un terme quelconque de $\{S_n\}$ on obtient tout un ensemble de suites :

$$V_k^{(n)} = \sum_{p=1}^{\infty} a_{kp} S_{p+n} \qquad n = 0, 1, \ldots$$
$$k = 1, 2, \ldots$$

k étant fixé étudions la convergence de la suite $\{V_k^{(n)}\}$.

On voit immédiatement que :

<u>Théorème 29</u> :

Une condition nécessaire et suffisante pour que $\lim_{n\to\infty} V_k^{(n)} = \lim_{n\to\infty} S_n$ est que :

$$\sum_{p=1}^{\infty} a_{kp} = 1$$

Cherchons maintenant quelle doit être la forme exacte de la suite $\{S_n\}$ pour que $V_k^{(n)} = S \quad \forall n$. Nous supposerons que $\forall k$ il existe $p(k)$ tel que $\forall p > p(k)$ $a_{kp} = 0$.

Pour k fixé $V_k^{(n)}$ sera égal à S $\forall n$ si et seulement si l'équation aux différences :

$$S = \sum_{p=1}^{p(k)} a_{kp} S_{p+n} \quad \text{est vérifiée quelquesoit n.}$$

La solution générale de cette équation aux différences est connue (voir par exemple [106]) d'où le résultat :

Théorème 30 :

Soit un procédé total de sommation défini par la matrice infinie A telle que

$\forall k \quad \exists p(k) : \forall p > p(k)$ on ait $a_{kp} = 0$. Une condition nécessaire et suffisante pour que $\sum_{p=1}^{p(k)} a_{kp} S_{p+n} = S \quad \forall n$ est que :

$$1°)\ S_n = S + \sum_{i=1}^{p} A_i(n)\, r_i^n + \sum_{i=p+1}^{q} [B_i(n) \cos b_i n + C_i(n) \sin b_i n]\, e^{w_i n}$$

$$+ \sum_{i=1}^{m} c_i\, \delta_{in}$$

où A_i, B_i et C_i sont des polynômes en n tels que si d_i est égal au degré de A_i plus un pour i = 1, ..., p et au plus grand des degrés de B_i et de C_i plus un pour i = p+1, ..., q on ait :

$$m + \sum_{i=1}^{p} d_i + 2 \sum_{i=p+1}^{q} d_i = p(k) - 1$$ avec m=o s'il n'y a aucun terme en δ_{in}

2°) le polynôme :

$$\sum_{p=1}^{p(k)} a_{kp}\, t^{p-1}$$

admette les p racines réelles distinctes $r_i \quad i = 1, \ldots, p$

les 2(q-p) racines complexes distinctes données par $e^{w_i} (\cos b_i \pm j \sin b_i)$

i = p+1, ..., q et m racines nulles ($j = \sqrt{-1}$).

Les conséquences de ce théorème sont importantes. On peut interpréter un procédé total comme une extrapolation pour n infini par une somme d'exponentielles de la forme donnée au 1°) du théorème 30.

Les r_i, les w_i et les b_i sont fixés et spécifiques du procédé de sommation utilisé. Les seules inconnues sont les coefficients des polynômes A_i, B_i et C_i ainsi que les c_i et S, ce qui explique pourquoi les procédés de sommation sont linéaires. La détermination de ces inconnues nécessite la connaissance de p(k) termes de la suite initiale. Tout procédé total de sommation est un cas particulier d'un procédé où les r_i, les w_i et les b_i ne seraient pas fixés mais seraient également des inconnues. Un tel procédé existe : c'est l'ε-algorithme. Sur ces questions et sur la formalisation de l'interprétation des procédés de sommation comme des extrapolations on pourra consulter [25].

Les procédés réguliers de sommation sont, par définition, convergents pour toute suite convergente. Cependant, comme nous le montre le théorème 30 (et nous le reverrons au paragraphe IV-8) ils ne sont capables d'accélérer la convergence que pour des suites bien particulières parce qu'ils ne font aucune hypothèse sur la suite à transformer. Etant donnée une suite, il faut, pour accélérer sa convergence, supposer qu'elle a une loi de formation assez régulière. La connaissance, même approximative, de cette loi permettra alors de bâtir une transformation de suite capable d'accélérer la convergence. L'étude de telles transformations fait l'objet des chapitres suivants. Il est bien évident qu'il faudra construire des transformations capables d'accélérer la convergence de classes de suites les plus vastes possibles. Tous les procédés que nous étudierons sont des procédés simultanés d'accélération de la convergence c'est-à-dire qu'ils construisent, en parallèle à la suite à transformer et sans la modifier, une seconde suite qui doit converger plus rapidement vers la même limite. Ces méthodes sont non linéaires.

CHAPITRE III

L'ε-ALGORITHME

L'ε-algorithme est sans doute l'algorithme d'accélération de la convergence le plus puissant connu à l'heure actuelle. D'autre part c'est pour cet algorithme que l'on dispose du plus de résultats théoriques. C'est pour ces deux raisons qu'un chapitre entier est consacré à son étude. D'autres algorithmes non linéaires d'accélération de la convergence seront étudiés au chapitre IV.

III - 1 Le procédé Δ^2 d'Aitken

Comme nous l'avons déjà vu le procédé Δ^2 d'Aitken consiste à transformer la suite $\{S_n\}$ en une suite $\{\varepsilon_2^{(n)}\}$ donnée par :

$$\varepsilon_2^{(n)} = \frac{S_{n+2}\, S_n - S_{n+1}^2}{S_{n+2} - 2S_{n+1} + S_n} = S_{n+1} - \frac{\Delta S_n . \Delta S_{n+1}}{\Delta^2 S_n}$$

Ce procédé a été trouvé par Aitken [1]. On a encore :

$$\varepsilon_2^{(n)} = S_{n+1} - \frac{\Delta S_{n+1}}{\dfrac{\Delta S_{n+1}}{\Delta S_n} - 1}$$

cette dernière forme nous permet d'énoncer immédiatement le :

Théorème 31 :

Si $\exists\ \alpha < 1 < \beta$ tels que $\Delta S_{n+1} / \Delta S_n \notin [\alpha,\beta]\ \forall n > N$ alors $\lim\limits_{n\to\infty} \varepsilon_2^{(n)} = \lim\limits_{n\to\infty} S_n$.

La démonstration de ce théorème est évidente.

Etant donnée une suite $\{S_n\}$ il se peut que certains termes de la suite $\{\varepsilon_2^{(n)}\}$ ne puissent pas être calculés si $\Delta^2 S_n = 0$. Un théorème de convergence comme le théorème précédent ne s'applique donc qu'aux termes de la suite $\{\varepsilon_2^{(n)}\}$qui peuvent effectivement être calculés.

Avec les conditions du théorème 31 tous les termes $\varepsilon_2^{(n)}$ peuvent être calculés pour $n > N$ car la condition $\Delta S_{n+1} / \Delta S_n \notin [\alpha,\beta]$ $\forall n > N$ impose que $\Delta S_{n+1} \neq \Delta S_n$ et donc que $\Delta^2 S_n \neq 0$ pour tout $n > N$.

En ce qui concerne l'accélération de la convergence on a le :

<u>Théorème 32</u> :

Si on applique le procédé Δ^2 d'Aitken à une suite $\{S_n\}$ qui converge vers S et si :

$$\lim_{n\to\infty} \frac{S_{n+1} - S}{S_n - S} = \lim_{n\to\infty} \frac{\Delta S_{n+1}}{\Delta S_n} = \rho \neq 1 \text{ alors}$$

$\{\varepsilon_2^{(n)}\}$ converge vers S plus vite que $\{S_{n+1}\}$

démonstration : en utilisant la définition de l'accélération de la convergence on voit immédiatement que $\{\varepsilon_2^{(n)}\}$ converge plus vite que $\{S_{n+1}\}$ si

$$\lim_{n\to\infty} \frac{\dfrac{S_{n+2} - S}{S_{n+1} - S} - 1}{\dfrac{\Delta S_{n+1}}{\Delta S_n} - 1} = 1$$

si la condition du théorème est satisfaite alors, puisque $\rho \neq 1$, $\{\varepsilon_2^{(n)}\}$ converge vers S d'une part et, d'autre part, $\{\varepsilon_2^{(n)}\}$ converge plus vite que $\{S_{n+1}\}$

REMARQUE : on voit qu'il est nécessaire d'introduire des hypothèses supplémentaires sur la suite $\{S_n\}$ pour avoir accélération de la convergence au lieu d'avoir simplement la convergence.

Nous allons maintenant étudier quelles sont les conditions que doivent vérifier la suite $\{S_n\}$ pour que $\varepsilon_2^{(n)} = S$ $\forall n > N$.

On a vu que :

$$\varepsilon_2^{(n)} = \frac{S_{n+2}\, S_n - S_{n+1}^2}{\Delta^2 S_n}$$

ce qui peut encore s'écrire :

$$\varepsilon_2^{(n)} = \frac{\begin{vmatrix} S_n & S_{n+1} \\ \Delta S_n & \Delta S_{n+1} \end{vmatrix}}{\begin{vmatrix} 1 & 1 \\ \Delta S_n & \Delta S_{n+1} \end{vmatrix}}$$

On veut donc que :

$$\frac{\begin{vmatrix} S_n & S_{n+1} \\ \Delta S_n & \Delta S_{n+1} \end{vmatrix}}{\begin{vmatrix} 1 & 1 \\ \Delta S_n & \Delta S_{n+1} \end{vmatrix}} = S \quad \forall n > N$$

d'où

$$\begin{vmatrix} S_n - S & S_{n+1} - S \\ \Delta S_n & \Delta S_{n+1} \end{vmatrix} = \begin{vmatrix} S_n - S & S_{n+1} - S \\ S_{n+1} - S & S_{n+2} - S \end{vmatrix} = 0 \qquad \forall n > N$$

par conséquent, pour que ce déterminant soit nul, il faut et il suffit qu'il existe a_o et a_1 non tous les deux nuls tels que :

$$a_o(S_n - S) + a_1(S_{n+1} - S) = 0 \quad \forall n > N$$

Si $a_o + a_1 = 0$ alors on voit que $S_n = S_{n+1}$ $\forall n$ et on ne peut pas alors appliquer le procédé Δ^2 d'Aitken à la suite $\{S_n\}$.

Inversement si $a_o(S_n - S) + a_1(S_{n+1} - S) = 0$ $\forall n > N$ avec $a_o + a_1 \neq 0$ alors $\varepsilon_2^{(n)} = S$ $\forall n > N$, d'où le :

Théorème 33 :

Une condition nécessaire et suffisante pour que $\varepsilon_2^{(n)} = S$ $\forall n > N$ est que la suite $\{S_n\}$ vérifie :

$$a_o(S_n - S) + a_1(S_{n+1} - S) = 0 \quad \forall n > N$$

avec $a_o + a_1 \neq 0$.

Résolvons maintenant l'équation aux différences du théorème 33. On trouve immédiatement que $S_n = S + \alpha\lambda^n$ $\forall n > N$. La condition $a_o + a_1 \neq 0$ se traduit par $\lambda \neq 1$. D'où le :

Théorème 34 :

Une condition nécessaire et suffisante pour que $\varepsilon_2^{(n)} = S$ $\forall n > N$ est que la suite $\{S_n\}$ soit telle :

$$S_n = S + \alpha\lambda^n \quad \forall n > N$$

avec $\lambda \neq 1$.

REMARQUES :

1°) On voit que $\varepsilon_2^{(n)} = S$ $\forall n > N$ même si $|\lambda| > 1$ c'est-à-dire même si la suite $\{S_n\}$ est divergente. Si $|\lambda| < 1$ S est alors la limite de $\{S_n\}$. Le procédé Δ^2 d'Aitken donne donc la limite d'une suite convergente dont le reste est une progression géométrique.

2°) Le procédé Δ^2 d'Aitken peut donc être interprété comme une extrapolation par une exponentielle ; étant donnés trois termes consécutifs S_n, S_{n+1} et S_{n+2} d'une suite, on détermine α , λ et S tels que :

$$S_n = S + \alpha\lambda^n$$
$$S_{n+1} = S + \alpha\lambda^{n+1}$$
$$S_{n+2} = S + \alpha\lambda^{n+2}$$

puis on prend $\varepsilon_2^{(n)} = S$. On recommence ensuite avec S_{n+1}, S_{n+2} et S_{n+3} pour obtenir $\varepsilon_2^{(n+1)}$.

III - 2 La transformation de Shanks et l'ε-algorithme

Shanks [170] a introduit et étudié une transformation non linéaire de suite à suite qui généralise le procédé Δ^2 d'Aitken en ce sens que cette transformation appelée $e_k(S_n)$ est construite de sorte que $e_k(S_n) = S$ $\forall n > N$ pour toute suite $\{S_n\}$ telle que :

$$\sum_{i=0}^{k} a_i (S_{n+i} - S) = 0 \quad \forall n > N$$

avec $\sum_{i=0}^{k} a_i \neq 0$.

On doit donc avoir :

$$\begin{vmatrix} S_n - S & \ldots\ldots & S_{n+k} - S \\ S_{n+1} - S & \ldots & S_{n+k+1} - S \\ \text{------} & & \text{------} \\ S_{n+k} - S & \ldots\ldots & S_{n+2k} - S \end{vmatrix} = 0 \qquad \forall n > N$$

Remplaçons la seconde ligne par sa différence avec la première, la troisième par sa différence avec la seconde et ainsi de suite. On obtient ainsi :

$$\begin{vmatrix} S_n & \ldots\ldots & S_{n+k} \\ \Delta S_n & \ldots\ldots & \Delta S_{n+k} \\ \text{------} & & \text{------} \\ \Delta S_{n+k-1} & \ldots & \Delta S_{n+2k-1} \end{vmatrix} = S \begin{vmatrix} 1 & \ldots\ldots & 1 \\ \Delta S_n & \ldots\ldots & \Delta S_{n+k} \\ \text{------} & & \text{------} \\ \Delta S_{n+k-1} & \ldots & \Delta S_{n+2k-1} \end{vmatrix}$$

D'où la transformation $e_k(S_n)$ de Shanks qui, en faisant de nouveau des combinaisons de lignes et de colonnes, s'écrit :

$$e_k(S_n) = \frac{\begin{vmatrix} S_n & \ldots\ldots & S_{n+k} \\ S_{n+1} & \ldots\ldots & S_{n+k+1} \\ \text{------} & & \text{------} \\ S_{n+k} & \ldots\ldots & S_{n+2k} \end{vmatrix}}{\begin{vmatrix} \Delta^2 S_n & \ldots\ldots & \Delta^2 S_{n+k-1} \\ \text{------} & & \text{------} \\ \Delta^2 S_{n+k-1} & \ldots & \Delta^2 S_{n+2k-2} \end{vmatrix}}$$

Cette formule a été en réalité découverte par Jacobi [115] puis reprise par Schmidt [168].

Par construction même de cette transformation on a le :

<u>Théorème 35</u> :

une condition nécessaire et suffisante pour que $e_k(S_n) = S \ \forall n > N$ est que la suite $\{S_n\}$

vérifie

$$\sum_{i=0}^{k} a_i \, (S_{n+i} - S) = 0 \quad \forall n > N$$

avec $\sum_{i=0}^{k} a_i \neq 0.$

REMARQUE :

Si l'on prend k = 1 on voit que $e_1(S_n) = \varepsilon_2^{(n)} \quad \forall n$

Du point de vue pratique on voit que la mise en oeuvre de la transformation $e_k(S_n)$ de Shanks est difficile dès que k atteint 4 ou 5 car elle nécessite l'évaluation de déterminants.

L'ε-algorithme est un algorithme récursif dû à P. Wynn [223] pour éviter le calcul effectif de ces déterminants. Les règles de cet algorithme sont les suivantes :

$$\varepsilon_{-1}^{(n)} = 0 \quad \varepsilon_0^{(n)} = S_n \quad n = 0, 1, \ldots$$

$$\varepsilon_{k+1}^{(n)} = \varepsilon_{k-1}^{(n+1)} + \frac{1}{\varepsilon_k^{(n+1)} - \varepsilon_k^{(n)}} \quad k, n = 0, 1, \ldots$$

On place ces quantités dans un tableau à double entrée : le tableau ε. L'indice inférieur k représente une colonne et l'indice supérieur n une diagonale descendante. La relation de l'ε-algorithme relie, dans ce tableau, des quantités situées aux quatre sommets d'un losange. La quantité située la plus à droite est calculée à partir des trois autres comme l'indiquent les flèches dans le schéma suivant :

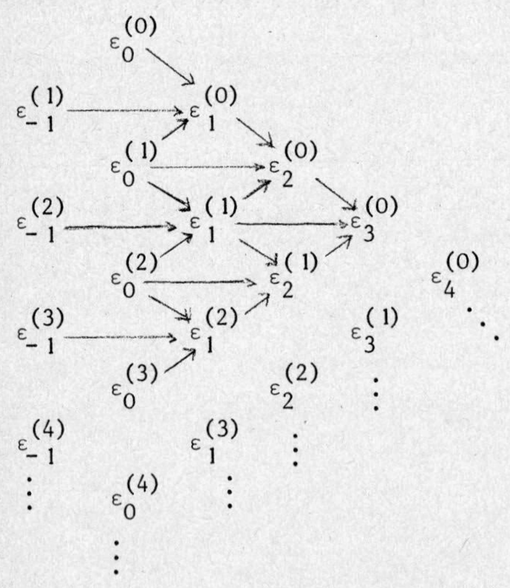

A partir des conditions initiales on progresse donc de gauche à droite dans ce tableau.

Le calcul de $\varepsilon_{2k}^{(n)}$ nécessite la connaissance de $S_n, S_{n+1}, \ldots, S_{n+2k}$.

Voyons maintenant la façon dont l'ε-algorithme est relié à la transformation de Shanks ; auparavant on a la :

<u>définition 14</u> : soit $\{u_n\}$ une suite de nombres. On appelle déterminants de Hankel, les déterminants définis par :

$$H_o^{(n)}(u_n) = 1 \qquad n = 0, 1, \ldots$$

$$H_k^{(n)}(u_n) = \begin{vmatrix} u_n & \cdots\cdots & u_{n+k-1} \\ u_{n+1} & \cdots\cdots & u_{n+k} \\ \multicolumn{3}{c}{\text{- - - - - - - -}} \\ u_{n+k-1} & \cdots & u_{n+2k-2} \end{vmatrix} \qquad \begin{array}{l} n = 0, 1, \ldots \\ k = 1, 2, \ldots \end{array}$$

<u>Propriété 8</u> : on a :

$$e_k(S_n) = \frac{H_{k+1}^{(n)}(S_n)}{H_k^{(n)}(\Delta^2 S_n)}$$

<u>Propriété 9</u> : les déterminants de Hankel d'une suite $\{u_n\}$ vérifient la relation :

$$H_k^{(n+1)}(u_{n+1}) \cdot H_k^{(n-1)}(u_{n-1}) - [H_k^{(n)}(u_n)]^2 = H_{k+1}^{(n-1)}(u_{n-1}) \cdot H_{k-1}^{(n+1)}(u_{n+1})$$

pour n, k = 1, 2, ...

<u>Propriété 10</u> : On appelle développement de Schweins de quotients de déterminants la relation :

$$\frac{\begin{vmatrix} b_1 & a_{12} & \cdots & a_{1n} \\ \multicolumn{4}{c}{\text{- - - - - -}} \\ b_n & a_{n2} & \cdots & a_{nn} \end{vmatrix}}{\begin{vmatrix} a_{11} & a_{12} & \cdots & a_{1n} \\ \multicolumn{4}{c}{\text{- - - - - -}} \\ a_{n1} & a_{n2} & \cdots & a_{nn} \end{vmatrix}} - \frac{\begin{vmatrix} b_1 & a_{12} & \cdots\cdots & a_{1,n-1} \\ \multicolumn{4}{c}{\text{- - - - - -}} \\ b_{n-1} & a_{n-1,2} & \cdots & a_{n-1,n-1} \end{vmatrix}}{\begin{vmatrix} a_{11} & a_{12} & \cdots\cdots & a_{1,n-1} \\ \multicolumn{4}{c}{\text{- - - - - -}} \\ a_{n-1,1} & a_{n-1,2} & \cdots & a_{n-1,n-1} \end{vmatrix}} = \frac{\begin{vmatrix} b_1 & a_{11} & \cdots\cdots & a_{1,n-1} \\ \multicolumn{4}{c}{\text{- - - - - -}} \\ b_n & a_{n1} & \cdots\cdots & a_{n,n-1} \end{vmatrix} \begin{vmatrix} a_{12} & \cdots\cdots & a_{1n} \\ \multicolumn{3}{c}{\text{- - - - - -}} \\ a_{n-1,2} & \cdots & a_{n-1,n} \end{vmatrix}}{\begin{vmatrix} a_{11} & a_{12} & \cdots\cdots & a_{1,n-1} \\ \multicolumn{4}{c}{\text{- - - - - -}} \\ a_{n-1,1} & a_{n-1,2} & \cdots & a_{n-1,n-1} \end{vmatrix} \begin{vmatrix} a_{11} & a_{12} & \cdots & a_{1n} \\ \multicolumn{4}{c}{\text{- - - - - -}} \\ a_{n1} & a_{n2} & \cdots & a_{nn} \end{vmatrix}}$$

Sur ce développement on pourra consulter [2].

Nous pouvons maintenant établir le résultat fondamental qui relie la transformation Shanks et l'ε-algorithme [223] :

<u>Théorème 36</u> :

$$\varepsilon_{2k}^{(n)} = e_k(S_n) \qquad \varepsilon_{2k+1}^{(n)} = 1/e_k\,(\Delta S_n)$$

démonstration : il est facile de voir immédiatement que :

$$\varepsilon_1^{(n)} = 1/e_o\,(\Delta S_n) = 1/\Delta S_n$$

$$\varepsilon_2^{(n)} = e_1\,(S_n) = \frac{S_{n+2}\,S_n - S_{n+1}^2}{\Delta^2\,S_n}$$

supposons avoir démontré la propriété jusqu'à la colonne 2k. Démontrons qu'elle est encore vraie pour la colonne 2k+1. En utilisant la propriété 8 la relation :

$$\varepsilon_{2k+1}^{(n)} = \varepsilon_{2k-1}^{(n+1)} + 1\,/\,(\varepsilon_{2k}^{(n+1)} - \varepsilon_{2k}^{(n)})$$

devient :

$$\frac{\begin{vmatrix} 1 & \cdots & 1 \\ \Delta^2 S_n & \cdots & \Delta^2 S_{n+k} \\ \vdots & & \vdots \\ \Delta^2 S_{n+k-1} & \cdots & \Delta^2 S_{n+2k-1} \end{vmatrix}}{\begin{vmatrix} \Delta S_n & \cdots & \Delta S_{n+k} \\ \vdots & & \vdots \\ \Delta S_{n+k} & \cdots & \Delta S_{n+2k} \end{vmatrix}} - \frac{\begin{vmatrix} 1 & \cdots & 1 \\ \Delta^2 S_{n+1} & \cdots & \Delta^2 S_{n+k} \\ \vdots & & \vdots \\ \Delta^2 S_{n+k-1} & \cdots & \Delta^2 S_{n+2k-2} \end{vmatrix}}{\begin{vmatrix} \Delta S_{n+1} & \cdots & \Delta S_{n+k} \\ \vdots & & \vdots \\ \Delta S_{n+k} & \cdots & \Delta S_{n+2k-1} \end{vmatrix}} =$$

$$\frac{1}{\dfrac{\begin{vmatrix} S_{n+1} & \cdots & S_{n+k+1} \\ \Delta S_{n+1} & \cdots & \Delta S_{n+k+1} \\ \vdots & & \vdots \\ \Delta S_{n+k} & \cdots & \Delta S_{n+2k} \end{vmatrix}}{\begin{vmatrix} 1 & \cdots & 1 \\ \Delta S_{n+1} & \cdots & \Delta S_{n+k+1} \\ \vdots & & \vdots \\ \Delta S_{n+k} & \cdots & \Delta S_{n+2k} \end{vmatrix}} - \dfrac{\begin{vmatrix} S_n & \cdots & S_{n+k} \\ \Delta S_n & \cdots & \Delta S_{n+k} \\ \vdots & & \vdots \\ \Delta S_{n+k-1} & \cdots & \Delta S_{n+2k-1} \end{vmatrix}}{\begin{vmatrix} 1 & \cdots & 1 \\ \Delta S_n & \cdots & \Delta S_{n+k} \\ \vdots & & \vdots \\ \Delta S_{n+k-1} & \cdots & \Delta S_{n+2k-1} \end{vmatrix}}}$$

en réarrangeant les lignes et les colonnes, le membre de gauche de cette relation peut s'écrire :

$$\frac{\begin{vmatrix} 1 & \cdots & 1 & 1 \\ \Delta^2 S_{n+1} & \cdots & \Delta^2 S_{n+k} & \Delta^2 S_n \\ \vdots & & & \vdots \\ \Delta^2 S_{n+k} & \cdots & \Delta^2 S_{n+2k-1} & \Delta^2 S_{n+k-1} \end{vmatrix}}{\begin{vmatrix} \Delta S_{n+1} & \cdots & \Delta S_{n+k} & \Delta S_n \\ \Delta^2 S_{n+1} & \cdots & \Delta^2 S_{n+k} & \Delta^2 S_n \\ \vdots & & & \vdots \\ \Delta^2 S_{n+k} & \cdots & \Delta^2 S_{n+2k-1} & \Delta^2 S_{n+k-1} \end{vmatrix}} - \frac{\begin{vmatrix} 1 & \cdots & 1 \\ \Delta^2 S_{n+1} & & \Delta^2 S_{n+k} \\ \vdots & & \vdots \\ \Delta^2 S_{n+k-1} & \cdots & \Delta^2 S_{n+2k-2} \end{vmatrix}}{\begin{vmatrix} \Delta S_{n+1} & \cdots & \Delta S_{n+k} \\ \Delta^2 S_{n+1} & \cdots & \Delta^2 S_{n+k} \\ \vdots & & \vdots \\ \Delta^2 S_{n+k-1} & \cdots & \Delta^2 S_{n+2k-2} \end{vmatrix}}$$

D'où en utilisant un développement de Schweins et en inversant les rapports :

$$\frac{\begin{vmatrix} 1 & \cdots & 1 & 1 \\ \Delta S_{n+1} & \cdots & \Delta S_{n+k} & \Delta S_n \\ \Delta^2 S_{n+1} & \cdots & \Delta^2 S_{n+k} & \Delta^2 S_n \\ \vdots & & & \vdots \\ \Delta^2 S_{n+k-1} & \cdots & \Delta^2 S_{n+2k-2} & \Delta^2 S_{n+k-2} \end{vmatrix} \cdot \begin{vmatrix} \Delta^2 S_{n+1} & \cdots & \Delta^2 S_{n+k} \\ \vdots & & \vdots \\ \Delta^2 S_{n+k} & \cdots & \Delta^2 S_{n+2k-1} \end{vmatrix}}{\begin{vmatrix} \Delta S_{n+1} & \cdots & \Delta S_{n+k} & \Delta S_n \\ \Delta^2 S_{n+1} & \cdots & \Delta^2 S_{n+k} & \Delta^2 S_n \\ \vdots & & & \vdots \\ \Delta^2 S_{n+k} & \cdots & \Delta^2 S_{n+2k-1} & \Delta^2 S_{n+k-1} \end{vmatrix} \cdot \begin{vmatrix} \Delta S_{n+1} & \cdots & \Delta S_{n+k} \\ \Delta^2 S_{n+1} & \cdots & \Delta^2 S_{n+k} \\ \vdots & & \vdots \\ \Delta^2 S_{n+k-1} & \cdots & \Delta^2 S_{n+2k-2} \end{vmatrix}} =$$

$$\frac{\begin{vmatrix} 1 & \cdots & 1 \\ \Delta S_n & \cdots & \Delta S_{n+k} \\ \vdots & & \vdots \\ \Delta S_{n+k-1} & \cdots & \Delta S_{n+2k-1} \end{vmatrix} \cdot \begin{vmatrix} 1 & \cdots & 1 \\ \Delta S_{n+1} & \cdots & \Delta S_{n+k+1} \\ \vdots & & \vdots \\ \Delta S_{n+k} & \cdots & \Delta S_{n+2k} \end{vmatrix}}{\begin{vmatrix} \Delta S_n & \cdots & \Delta S_{n+k} \\ \vdots & & \vdots \\ \Delta S_{n+k} & \cdots & \Delta S_{n+2k} \end{vmatrix} \cdot \begin{vmatrix} \Delta S_{n+1} & \cdots & \Delta S_{n+k} \\ \vdots & & \vdots \\ \Delta S_{n+k} & \cdots & \Delta S_{n+2k-1} \end{vmatrix}}$$

Quant au membre de droite de la première relation il peut s'écrire :

$$(-1)^k \frac{\begin{vmatrix} 1 & \cdots\cdots & 1 \\ \Delta S_n & \cdots\cdots & \Delta S_{n+k} \\ \cdots & \cdots & \cdots \\ \Delta S_{n+k-1} & \cdots & \Delta S_{n+2k-1} \end{vmatrix} \cdot \begin{vmatrix} 1 & \cdots\cdots & 1 \\ \Delta S_{n+1} & \cdots\cdots & \Delta S_{n+k+1} \\ \cdots & \cdots & \cdots \\ \Delta S_{n+k} & \cdots\cdots & \Delta S_{n+2k} \end{vmatrix}}{D}$$

avec

$$D = \begin{vmatrix} \Delta S_{n+1} & \cdots\cdots & \Delta S_{n+k+1} \\ \cdots & \cdots & \cdots \\ \Delta S_{n+k} & \cdots\cdots & \Delta S_{n+2k} \\ S_{n+1} & \cdots\cdots & S_{n+k+1} \end{vmatrix} \cdot \begin{vmatrix} \Delta S_{n+1} & \cdots\cdots & \Delta S_{n+k} & \Delta S_n \\ \cdots & \cdots & \cdots & \cdots \\ \Delta S_{n+k} & \cdots\cdots & \Delta S_{n+2k-1} & \Delta S_{n+k-1} \\ 1 & \cdots\cdots & 1 & 1 \end{vmatrix}$$

$$- \begin{vmatrix} \Delta S_{n+1} & \cdots\cdots & \Delta S_{n+k} & \Delta S_n \\ \cdots & \cdots & \cdots & \cdots \\ \Delta S_{n+k} & \cdots\cdots & \Delta S_{n+2k-1} & \Delta S_{n+k-1} \\ S_{n+1} & \cdots\cdots & S_{n+k} & S_n \end{vmatrix} \cdot \begin{vmatrix} \Delta S_{n+1} & \cdots\cdots & \Delta S_{n+k+1} \\ \cdots & \cdots & \cdots \\ \Delta S_{n+k} & \cdots\cdots & \Delta S_{n+2k} \\ 1 & \cdots\cdots & 1 \end{vmatrix}$$

en utilisant une identité dérivée de [2] :

$$\begin{vmatrix} a_1 & a_2 \\ b_1 & b_2 \end{vmatrix} = \Big| |a_1| \; |b_2| - |b_1| \; |a_2| \Big|$$

On trouve que :

$$D = \begin{vmatrix} \Delta S_{n+1} & \cdots\cdots & \Delta S_{n+k+1} & \Delta S_n \\ \cdots & \cdots & \cdots & \cdots \\ \Delta S_{n+k} & \cdots\cdots & \Delta S_{n+2k} & \Delta S_{n+k-1} \\ S_{n+1} & \cdots\cdots & S_{n+k+1} & S_n \\ 1 & \cdots\cdots & 1 & 1 \end{vmatrix} \cdot \begin{vmatrix} \Delta S_{n+1} & \cdots\cdots & \Delta S_{n+k} \\ \cdots & \cdots & \cdots \\ \Delta S_{n+k} & \cdots\cdots & \Delta S_{n+2k-1} \end{vmatrix}$$

ce qui démontre l'égalité des deux membres de la première égalité. On démontre de la même façon que la propriété énoncée dans le théorème reste vraie pour la colonne 2k+2, ce qui termine la démonstration.

REMARQUE : on voit que seules les quantités d'indice inférieur pair sont intéressantes. Les autres ne sont que des calculs intermédiaires.

La conséquence immédiate de ce théorème est donc la :

Propriété 10 :

$$\varepsilon_{2k}^{(n)} = \frac{H_{k+1}^{(n)}(S_n)}{H_k^{(n)}(\Delta^2 S_n)} \quad \text{et} \quad \varepsilon_{2k+1}^{(n)} = \frac{H_k^{(n)}(\Delta^3 S_n)}{H_{k+1}^{(n)}(\Delta S_n)}$$

La démonstration du théorème 36 prouve également la relation suivante entre les déterminants de Hankel :

Propriété 11 :

$$H_k^{(n)}(\Delta^2 u_n) \, . \, H_k^{(n)}(u_n) - [H_k^{(n)}(\Delta u_n)]^2 = H_{k+1}^{(n)}(u_n) . \, H_{k-1}^{(n)}(\Delta^2 u_n)$$

III - 3 Propriétés de l' ε-algorithme

Si l'on écrit la définition de $\varepsilon_{2k+2}^{(n)}$ et de $\varepsilon_{2k}^{(n)}$ à partir des déterminants de Hankel et si l'on effectue un développement de Schweins de $\varepsilon_{2k+2}^{(n)} - \varepsilon_{2k}^{(n)}$ on trouve la propriété suivante :

Propriété 12 :

$$\varepsilon_{2k+2}^{(n)} - \varepsilon_{2k}^{(n)} = - \frac{[H_{k+1}^{(n)}(\Delta S_n)]^2}{H_{k+1}^{(n)}(\Delta^2 S_n) . \, H_k^{(n)}(\Delta^2 S_n)}$$

La démonstration est laissée en exercice.

On a vu que le calcul de $\varepsilon_{2k}^{(n)}$ nécessitait la connaissance de $S_n, S_{n+1}, \ldots, S_{n+2k}$. Supposons que nous inversions la numérotation de ces 2k+1 termes on a le résultat :

Propriété 13 : Soit $\varepsilon_{2k}^{(n)}$ la valeur obtenue en appliquant l'ε-algorithme à $S_n, S_{n+1}, \ldots$ S_{n+2k}. Si on applique l'ε-algorithme à $u_n = S_{n+2k}, u_{n+1} = S_{n+2k-1}, \ldots, u_{n+2k} = S_n$ on obtient la même quantité $\varepsilon_{2k}^{(n)}$ n, k = 0, 1, ...

démonstration : elle est évidente en utilisant la propriété 10. Cela revient à intervertir des lignes et des colonnes dans les déterminants de Hankel. Ce résultat a été obtenu par Gilewicz [90] qui en donne une démonstration faisant intervenir la connexion entre l'ε-algorithme et la table de Padé.

L'ε-algorithme est une transformation non linéaire de suite à suite, c'est-à-dire que si on l'applique à la somme terme à terme de deux suites les quantités $\varepsilon_k^{(n)}$ que l'on obtient ne sont pas les sommes des quantités obtenues en appliquant l'ε-algorithme séparément à chacune des suites. Cependant on a la :

Propriété 14 : si l'application de l'ε-algorithme à $\{S_n\}$ et à $\{aS_n + b\}$ fournit respectivement les quantités $\varepsilon_k^{(n)}$ et $\bar{\varepsilon}_k^{(n)}$ alors :

$$\bar{\varepsilon}_{2k}^{(n)} = a\ \varepsilon_{2k}^{(n)} + b \qquad\qquad \bar{\varepsilon}_{2k+1}^{(n)} = \varepsilon_{2k+1}^{(n)} / a$$

démonstration : elle est évidente à partir de la définition des $\varepsilon_k^{(n)}$ à l'aide des déterminants de Hankel. Elle est laissée en exercice.

Wynn a également démontré le résultat suivant [242] :

Propriété 15 : si on applique l'ε-algorithme à une suite $\{S_n\}$ qui vérifie $\sum_{i=0}^{k} a_i S_{n+i} =$

pour tout n et si les racines $\lambda_1, \ldots, \lambda_k$ du polynôme $\sum_{i=0}^{k} a_i \lambda^i$ sont réelles, distincte et telles que :

$|\lambda_1| > |\lambda_2| > \ldots > |\lambda_j| > 1 > |\lambda_{j+1}| > \ldots > |\lambda_k|$ alors

$$\lim_{n\to\infty} \frac{\varepsilon_{2i-1}^{(n+1)}}{\varepsilon_{2i-1}^{(n)}} = \frac{1}{\lambda_i} \qquad i = 1, \ldots, k$$

et

$$\lim_{n\to\infty} \frac{\varepsilon_{2i-2}^{(n+1)}}{\varepsilon_{2i-2}^{(n)}} = \begin{cases} \lambda_i & i = 1, \ldots, j \\ 1 & i = j+1, \ldots, k \end{cases}$$

si $a \neq 0$ et λ_i pour $i = 1,\ldots,k$ si $a = 0$

la démonstration de cette propriété est liée au :

Théorème 37 :

Une condition nécessaire et suffisante pour que $\varepsilon_{2k}^{(n)} = S$ $\forall n > N$ est que :

$$S_n = S + \sum_{i=1}^{p} A_i(n)\, r_i^n + \sum_{i=p+1}^{q} [B_i(n) \cos b_i n + C_i(n) \sin b_i n]\ e^{w_i n}$$

$$+ \sum_{i=0}^{m} c_i \, \delta_{in} \qquad \forall n > N$$

avec $r_i \neq 1$ pour $i = 1,\ldots,p$

A_i, B_i et C_i sont des polynômes en n tels que si d_i est égal au degré de A_i plus un pour $i = 1, \ldots, p$ et au plus grand des degrés de B_i et de C_i plus un pour $i = p+1, \ldots, q$ on ait :

$$m + \sum_{i=1}^{p} d_i + 2 \sum_{i=p+1}^{q} d_i = k-1$$

avec la convention que $m = -1$ s'il n'y a aucun terme en δ_{in}.

Démonstration : de même que le théorème 34 était une conséquence immédiate du théorème 33, ce théorème est la conséquence du théorème 35. Il n'y a aucune démonstration à faire, on écrit seulement la solution générale de l'équation aux différences du théorème 35. La condition $\sum_{i=0}^{k} a_i \neq 0$ impose simplement que $r_i \neq 1$ $i = 1, \ldots, p$

On remarquera que ce théorème est une généralisation du théorème 34 obtenu pour le procédé Δ^2 d'Aitken ainsi que du théorème 30 pour les procédés de sommation totaux.

On peut donc interpréter l'ε-algorithme comme une extrapolation par une somme de telles exponentielles où les coefficients de A_i, B_i et C_i ainsi que les r_i, b_i w_i et c_i sont des inconnues. Sur ce théorème on pourra consulter [47]. Wynn [195] a également démontré le résultat suivant :

<u>Propriété 16</u> : si on applique l'ε-algorithme à une suite $\{S_n\}$ telle que :

$$\sum_{i=0}^{k} a_i \, S_{n+i} = 0 \qquad \forall n > N$$

avec $\sum_{i=0}^{k} a_i \neq 0$ alors :

$$\varepsilon_0^{(n)} \varepsilon_1^{(n)} - \varepsilon_1^{(n)} \varepsilon_2^{(n)} + \varepsilon_2^{(n)} \varepsilon_3^{(n)} - \ldots + \varepsilon_{2k-2}^{(n)} \varepsilon_{2k-1}^{(n)} = - \sum_{i=1}^{k} i \, a_i \Big/ \sum_{i=0}^{k} a_i \qquad \forall n > N$$

Démonstration : la relation de l'ε-algorithme peut s'écrire :

$$\varepsilon_{i+1}^{(n)}\,\varepsilon_{i}^{(n+1)} - \varepsilon_{i-1}^{(n+1)}\,\varepsilon_{i}^{(n+1)} - \varepsilon_{i}^{(n)}\,\varepsilon_{i+1}^{(n)} + \varepsilon_{i-1}^{(n+1)}\,\varepsilon_{i}^{(n)} = 1$$

faisons i = 0, ..., 2k-1 et effectuons une somme alternée des équations ainsi obtenue c'est-à-dire la première moins la seconde plus la troisième et ainsi de suite. Si l'on pose :

$$B(n) = \varepsilon_{0}^{(n)}\,\varepsilon_{1}^{(n)} - \varepsilon_{1}^{(n)}\,\varepsilon_{2}^{(n)} + \ldots + \varepsilon_{2k-2}^{(n)}\,\varepsilon_{2k-1}^{(n)}$$

on trouve, en utilisant le théorème 35 et le fait que $\varepsilon_{2k}^{(n)} = 0\ \forall n$, que la combinaison précédente s'écrit :

$$B(n) - B(n+1) = 0 \qquad \forall n$$

ce qui démontre que B(n) = constante ∀n. Cette première partie de la démonstration a été obtenue par Bauer [15]. La valeur de la constante à laquelle est égale B(n) a été obtenue par Wynn. La démonstration est trop longue et trop technique pour être donnée ici.

En effectuant des éliminations dans la relation de l'ε-algorithme, Wynn [212] a obtenu la :

Propriété 17 :

$$[\varepsilon_{k+2}^{(n-1)} - \varepsilon_{k}^{(n)}]^{-1} - [\varepsilon_{k}^{(n)} - \varepsilon_{k-2}^{(n+1)}]^{-1} = [\varepsilon_{k}^{(n+1)} - \varepsilon_{k}^{(n)}]^{-1} - [\varepsilon_{k}^{(n)} - \varepsilon_{k}^{(n-1)}]^{-1}$$

démonstration : on a :

$$\varepsilon_{k+1}^{(n-1)} - \varepsilon_{k-1}^{(n)} = [\varepsilon_{k}^{(n)} - \varepsilon_{k}^{(n-1)}]^{-1}$$

$$\varepsilon_{k+1}^{(n)} - \varepsilon_{k-1}^{(n+1)} = [\varepsilon_{k}^{(n+1)} - \varepsilon_{k}^{(n)}]^{-1}$$

soustrayons et réarrangeons les termes à gauche du signe égal :

$$\varepsilon_{k+1}^{(n)} - \varepsilon_{k+1}^{(n-1)} - [\varepsilon_{k-1}^{(n+1)} - \varepsilon_{k-1}^{(n)}] = [\varepsilon_{k}^{(n+1)} - \varepsilon_{k}^{(n)}]^{-1} - [\varepsilon_{k}^{(n)} - \varepsilon_{k}^{(n-1)}]^{-1}$$

d'où la relation cherchée puisque

$$\varepsilon_{k+2}^{(n-1)} - \varepsilon_{k}^{(n)} = [\varepsilon_{k+1}^{(n)} - \varepsilon_{k+1}^{(n-1)}]^{-1}$$

$$\varepsilon_{k}^{(n)} - \varepsilon_{k-2}^{(n+1)} = [\varepsilon_{k-1}^{(n+1)} - \varepsilon_{k-1}^{(n)}]^{-1}$$

comme on le voit la relation de la propriété 17 permet de passer de colonnes paires à colonnes paires dans le tableau ε. On relie ainsi cinq quantités situées aux sommets et au centre d'un losange :

$$\begin{array}{ccc} & N & \\ W & C & E \\ & S & \end{array}$$

la relation s'écrit donc (règle de la croix) :

$$(C-N)^{-1} + (C-S)^{-1} = (C-W)^{-1} + (C-E)^{-1}$$

En partant de là, Cordellier [65] a donné une interprétation géométrique de l'ε-algorithme. Cette relation est également le point de départ de l'obtention des règles particulières de l'ε-algorithme comme nous le verrons au paragraphe IV.8.

Soit G_k l'ensemble des suites finies de 2k+1 termes telles que si $x = (S_n, \ldots, S_{n+2k}) \in G_k$ alors $H_k^{(n)}(\Delta^2 S_n) \neq 0$. Soit E_k l'application qui à $x = (S_n, \ldots, S_{n+2k}) \in G_k$ fait correspondre $\varepsilon_{2k}^{(n)} = H_{k+1}^{(n)}(S_n) / H_k^{(n)}(\Delta^2 S_n)$. Par conséquent puisque $H_k^{(n)}(\Delta^2 S_n) \neq 0$ l'application E_k est continue dans G_k. D'où le :

<u>Théorème 38</u> :

L'ε-algorithme est continu dans G_k.

$H_{k+1}^{(n)}(S_n)$ et $H_k^{(n)}(\Delta^2 S_n)$ sont des applications différentiables de G_k dans $\mathbb{R}$ puisque un déterminant est différentiable par rapport à chacun de ses éléments ; d'où le :

<u>Théorème 39</u> :

L'ε-algorithme est différentiable dans G_k.

III - 4 Interprétation de l'ε-algorithme

Nous allons maintenant montrer que le calcul de $\varepsilon_{2k}^{(n)}$ revient à résoudre le système suivant :

$$a_o S_n + a_1 S_{n+1} + \ldots + a_k S_{n+k} = C$$

$$a_o S_{n+1} + a_1 S_{n+2} + \ldots + a_k S_{n+k+1} = C$$

$$a_o S_{n+k} + a_1 S_{n+k+1} + \ldots + a_k S_{n+2k} = C$$

où C est une constante arbitraire non nulle puis à calculer :

$$\varepsilon_{2k}^{(n)} = C \,/ \sum_{i=0}^{k} a_i$$

Dans ce système remplaçons la seconde ligne par sa différence avec la première, la troisième par sa différence avec la seconde et ainsi de suite ; on obtient :

$$a_o S_n + \ldots\ldots + a_k S_{n+k} = C$$

$$a_o \Delta S_n + \ldots\ldots + a_k \Delta S_{n+k} = 0$$

$$a_o \Delta S_{n+k-1} + \ldots + a_k \Delta S_{n+2k-1} = 0$$

Effectuons maintenant la même opération sur les colonnes :

$$b_o S_n + b_1 \Delta S_n + \ldots\ldots + b_k \Delta S_{n+k-1} = C$$

$$b_o \Delta S_n + b_1 \Delta^2 S_n + \ldots\ldots + b_k \Delta^2 S_{n+k-1} = 0$$

$$b_o \Delta S_{n+k-1} + b_1 \Delta^2 S_{n+k-1} + \ldots + b_k \Delta^2 S_{n+2k-2} = 0$$

avec $b_j = \sum_{i=j}^{k} a_i$ pour $j = 0, \ldots, k$

d'où immédiatement :

$$b_o = \sum_{i=0}^{k} a_i = C \, \frac{H_k^{(n)}(\Delta^2 S_n)}{H_{k+1}^{(n)}(S_n)}$$

ce qui montre que :

$$C \,/ \sum_{i=0}^{k} a_i = \varepsilon_{2k}^{(n)}$$

On voit que $H_k^{(n)}(\Delta^2 S_n) \neq 0$ entraîne $\sum_{i=0}^{k} a_i \neq 0$ et réciproquement. De plus si l'on remplace C par aC où a est une constante non nulle alors les a_i sont remplacés par $a\, a_i$ et par conséquent $\varepsilon_{2k}^{(n)} = C \,/ \sum_{i=0}^{k} a_i$ demeure inchangé.

On peut également démontrer l'identité entre ces deux façons de procéder en posant $C = \varepsilon_{2k}^{(n)} \sum_{i=0}^{k} a_i$, d'où :

$$a_o S_n + \ldots\ldots + a_k S_{n+k} = \varepsilon_{2k}^{(n)} \sum_{i=0}^{k} a_i$$

$$\text{---------------------------------}$$

$$a_o S_{n+k} + \ldots + a_k S_{n+2k} = \varepsilon_{2k}^{(n)} \sum_{i=0}^{k} a_i$$

ou encore

$$a_o(S_n - \varepsilon_{2k}^{(n)}) + \ldots + a_k(S_{n+k} - \varepsilon_{2k}^{(n)}) = 0$$

$$\text{---------------------------------}$$

$$a_o(S_{n+k} - \varepsilon_{2k}^{(n)}) + \ldots + a_k(S_{n+2k} - \varepsilon_{2k}^{(n)}) = 0$$

d'où, pour avoir une solution différente de la solution nulle :

$$\begin{vmatrix} S_n - \varepsilon_{2k}^{(n)} \ldots\ldots S_{n+k} - \varepsilon_{2k}^{(n)} \\ \text{------------------------} \\ S_{n+k} - \varepsilon_{2k}^{(n)} \ldots S_{n+2k} - \varepsilon_{2k}^{(n)} \end{vmatrix} = 0$$

ce qui démontre, d'après le théorème 35, que l'on a bien :

$$\varepsilon_{2k}^{(n)} = H_{k+1}^{(n)}(S_n) \,/\, H_k^{(n)}(\Delta^2 Sn).$$

On peut également, comme l'a fait Greville [105], poser $\sum_{i=0}^{k} a_i = 1$ ce qui donne $C = \varepsilon_{2k}^{(n)}$ d'où le nouveau système de k+2 équations à k+2 inconnues :

$$a_o + \ldots\ldots + a_k = 1$$

$$-\varepsilon_{2k}^{(n)} + a_o S_n + \ldots + a_k S_{n+k} = 0$$

$$\text{---------------------------------}$$

$$-\varepsilon_{2k}^{(n)} + a_o S_{n+k} + \ldots + a_k S_{n+2k} = 0$$

A l'aide de cette interprétation algébrique, Germain-Bonne [88] a démontré qu'il y avait identité entre l'application de l'ε-algorithme et la méthode des moments telle qu'elle est décrite, par exemple, dans le livre de Vorobyev [184].

Considérons les vecteurs :

$$x_i = \begin{pmatrix} S_i \\ S_{i+1} \\ \vdots \\ S_{i+k-1} \end{pmatrix} \quad \text{pour } i = 0, \ldots, k+1$$

La méthode des moments revient à chercher la matrice carrée A d'ordre k+1 et le vecteur $b \in \mathbb{R}^{k+1}$ tels que :

$$x_1 = Ax_0 + b$$

$$-----------$$

$$x_{k+1} = Ax_k + b$$

On a donc

$$\Delta x_1 = A.\Delta x_0$$

$$---------$$

$$\Delta x_k = A.\Delta x_{k-1}$$

La matrice A est de la forme :

$$\begin{pmatrix} 0 & 1 & \dots\dots & & 0 \\ 0 & 0 & 1 & \dots & 0 \\ - & - & - & - & - \\ 0 & 0 & 0 & \dots & 1 \\ a_1 & a_2 & a_3 & \dots & a_k \end{pmatrix}$$

La dernière ligne de chacun des systèmes précédents détermine les a_i. Toutes les composantes de b sont nulles sauf la dernière qui est égale à $S_k - a_1 S_0 - \dots - a_k S_{k-1}$. Toutes les composantes du vecteur x tel que x = Ax + b sont égales à la quantité $\varepsilon_{2k}^{(0)}$ obtenue par application de l'ε-algorithme à $S_0, \dots, S_{2k}$.

Il semble que la liaison entre l'ε-algorithme et la méthode des moments soit beaucoup plus profonde qu'une simple liaison algébrique et qu'elle passe par la théorie des polynômes orthogonaux. Cette question est actuellement à l'étude.

Revenons au début de l'interprétation algébrique. Nous avons vu que l'application de l'ε-algorithme à $S_n, \dots, S_{n+2k}$ revenait à résoudre le système :

$$\begin{pmatrix} S_n & \dots\dots & S_{n+k} \\ - & - & - \\ S_{n+k} & \dots\dots & S_{n+2k} \end{pmatrix} \begin{pmatrix} a_0 \\ \cdot \\ \cdot \\ \cdot \\ a_k \end{pmatrix} = \begin{pmatrix} 1 \\ \cdot \\ \cdot \\ \cdot \\ 1 \end{pmatrix}$$

puis à calculer $\varepsilon_{2k}^{(n)} = 1 / \sum_{i=0}^{k} a_i$. Si nous appelons $A_{k+1}^{(n)}$ la matrice de ce système, on voit que $\varepsilon_{2k}^{(n)}$ est égal à la somme des éléments de $[A_{k+1}^{(n)}]^{-1}$. on peut donc calculer $\varepsilon_{2k}^{(n)}$ si l'on possède une méthode simple pour obtenir $[A_{k+1}^{(n)}]^{-1}$. Une telle méthode existe et est classique en analyse numérique : c'est la méthode de bordage qui permet de calculer $[A_{k+1}^{(n)}]^{-1}$ si l'on connaît $[A_k^{(n)}]^{-1}$ (voir par exemple [77]). Si l'on tient compte de la structure particulière des matrices de Hankel $A_k^{(n)}$ alors une variante de la méthode de bordage due à Trench [175] nous permet un calcul simple de $\{\varepsilon_{2k}^{(n)}\}$ pour n fixé et k = 0,1,... A première vue, cette méthode est beaucoup plus compliquée que l'ε-algorithme ; cependant elle nécessite environ trois fois moins d'opérations arithmétiques que l'ε-algorithme. Cette méthode est la suivante :

initialisations : $\gamma_{-1} = 0 \quad \lambda_{-1} = 1 \quad d_{-1} = 1 \quad d_{-2} = 0 \quad \lambda_0 = S_n$

$u_i^{(-2)} = 0 \quad \forall\, i$

$u_0^{(-1)} = 1$ et $u_i^{(-1)} = 0 \quad \forall\, i \neq 0$

$u_p^{(q)} = 0 \quad \forall\, q > p+1$ et $\forall\, q < 0$

$u_{p+1}^{(p)} = 1 \quad \forall\, p \geq 0$

$\varepsilon_0^{(n)} = S_n$

Calcul de $\varepsilon_{2k+2}^{(n)}$ pour $n \geq 0$ fixé et pour k = 0,1,... par :

$$\gamma_k = \sum_{i=0}^{k} S_{n+k+i+1}\, u_i^{(k-1)}$$

$$u_i^{(k)} = (\gamma_{k-1}\lambda_{k-1}^{-1} - \gamma_k \lambda_k^{-1})\, u_i^{(k-1)} + u_{i-1}^{(k-1)} - \lambda_k \lambda_{k-1}^{-1} u_i^{(k-2)} \text{ pour } i=0,\ldots,k \text{ et } u_{k+1}^{(k)} = 1.$$

$$d_k = (\gamma_{k-1}\lambda_{k-1}^{-1} - \gamma_k \lambda_k^{-1} + 1)\, d_{k-1} - \lambda_k \lambda_{k-1}^{-1} d_{k-2}$$

$$\lambda_{k+1} = \sum_{i=0}^{k+1} S_{n+k+i+1}\, u_i^{(k)}$$

et enfin $[\varepsilon_{2k+2}^{(n)}]^{-1} = [\varepsilon_{2k}^{(n)}]^{-1} + d_k^2 \lambda_{k+1}^{-1}$

Une étude complète de cette méthode se trouve dans [27].

III - 5 L'ε-algorithme et la table de Padé

L'importance de ce paragraphe est primordiale. Beaucoup de physiciens sont intéressés par l'approximation de Padé car ils obtiennent souvent la solution de leurs problèmes sous forme de série de puissances ; chaque terme de la série étant très difficile à calculer ils emploient l'approximation de Padé pour essayer d'améliorer leurs résultats et d'avoir le moins possible de termes à calculer. Nous allons montrer l'identité entre l'ε-algorithme et la moitié supérieure de la table de Padé. Etant donnée une série de puissances $f(x) = \sum_{i=0}^{\infty} c_i x^i$ il est possible de construire, sous certaines conditions, une suite double de fonctions rationnelles de x qui seront notées $[p/q]$, avec :

$$[p / q] = \frac{\sum_{i=0}^{p} a_i x^i}{\sum_{i=0}^{q} b_i x^i}$$

Cette fonction rationnelle $[p / q]$ possède la propriété fondamentale que son développement en puissances croissantes de x coïncide avec celui de f(x) jusqu'au terme de degré p+q compris.

En d'autres termes on a :

$$f(x) - [p / q] = 0 (x^{p+q+1})$$

La fraction rationnelle $[p / q]$ s'appelle approximant de Padé de f(x). La table à double entrée obtenue en plaçant $[p / q]$ à l'intersection de la ligne q et de la colonne p pour p, q = 0, 1, ... s'appelle la table de Padé [149]. De nombreux ouvrages ont été écrits sur la table de Padé : les deux plus importants sont ceux de Perron [152] et de Wall [185]. On pourra également consulter [9,96,97] qui sont plus récents.

Le calcul des coefficients a_i et b_i de $[p / q]$ s'effectue en écrivant que :

$$[p/q] = P(x) / Q(x) \text{ et } f(x) Q(x) - P(x) = 0(x^{p+q+1})$$

ou en d'autres termes :

$$(c_o + c_1 x + c_2 x^2 + \dots)(b_o + b_1 x + \dots + b_q x^q) - (a_o + a_1 x + \dots + a_p x^p) = 0(x^{p+q+1})$$

ou encore :

$$(c_o + c_1 x + \ldots)(b_o + \ldots + b_q x^q) = a_o + \ldots + a_p x^p + 0.x^{p+1} + \ldots + 0.x^{p+q}$$

$$+ O(x^{p+q+1})$$

En identifiant de part et d'autre du signe égal les coefficients des termes de même degré en x on obtient les coefficients b_i comme solution du système de q équations à q+1 inconnues :

$$c_{p+1} b_o + c_p b_1 + \ldots + c_{p-q+1} b_q = 0$$

$$c_{p+q} b_o + c_{p+q-1} b_1 + \ldots + c_p b_q = 0$$

en prenant $b_o = 1$ puis les a_i sont calculés en utilisant les p+1 relations :

$$c_o b_o = a_o$$

$$c_1 b_o + c_o b_1 = a_1$$

$$c_2 b_o + c_1 b_1 + c_o b_2 = a_2$$

$$c_p b_o + c_{p-1} b_1 + \ldots + c_{p-q} b_q = a_p$$

avec la convention que tous les c_i avec i négatif sont nuls.

On a le :

Théorème 40 :

une condition nécessaire et suffisante pour que [p / q] existe est que :

$$H_q^{(p-q+1)} (c_{p-q+1}) \neq 0$$

démonstration : c'est la condition pour que le système linéaire donnant les b_i admette une solution. Cette solution est d'ailleurs unique d'où le :

Théorème 41 :

s'il existe [p / q] est unique

Longman [133] a proposé une méthode pour calculer les coefficients a_i et b_i d'un approximant de Padé en fonction des coefficients des approximants adjacents : dénotons par

a'_i et b'_i les coefficients de [p / q-1] , par a_i et b_i ceux de [p-1,q] et par a^*_i et b^*_i ceux de [p-1 / q-1] ; on a les relations suivantes :

$$b'_i = b_i - \frac{b^*_{i-1}\, b_q}{b^*_{q-1}} \qquad i = 1, \ldots, q-1$$

$$b'_o = 1$$

$$a_i = a'_i - \frac{a^*_{i-1}\, a'_p}{a^*_{p-1}} \qquad i = 1, \ldots, p-1$$

$$a_o = c_o$$

On trouvera des méthodes similaires dans [27,61,227].
Les approximants de Padé possèdent un certain nombre de propriétés ; pour énoncer ces propriétés il sera plus facile de dénoter par $[p/q]_f(x)$ l'approximant de Padé [p/q] correspondant à la série f(x). On a :

Propriété 18 :

si $f(o) \neq o$ alors $[p/q]_{1/f}(x) = 1 / [q/p]_f(x)$

Propriété 19 :

si $f(o) = o$ alors $[p-1/q]_{f/x}(x) = [p/q]_f(x)/x$

Propriété 20 :

soit $R_k(x)$ un polynôme de degré k en x

si $p \geqslant q+k$ alors :

$$[p/q]_{f+R_k}(x) = [p/q]_f(x) + R_k(x)$$

On a également le résultat suivant :

Propriété 21 :

$$[p/q] = \frac{\begin{vmatrix} \sum_{i=0}^{p-q} c_i\, x^{q+i} & \cdots\cdots & \sum_{i=0}^{p} c_i\, x^i \\ c_{p-q+1} & \cdots\cdots & c_{p+1} \\ \hline c_p & \cdots\cdots & c_{p+q} \end{vmatrix}}{\begin{vmatrix} x^q & \cdots\cdots & 1 \\ c_{p-q+1} & \cdots\cdots & c_{p+1} \\ \hline c_p & \cdots\cdots & c_{p+q} \end{vmatrix}}$$

Supposons que l'approximant de Padé [p/q] soit en fait, après simplification par un facteur commun, le rapport d'un polynôme de degré exactement égal à p-k sur un polynôme de degré exactement égal à q-k. Cette fraction rationnelle coïncide avec f jusqu'au terme de degré p + q - 2k inclus puisqu'elle coïncide en réalité avec f jusqu'au terme de degré p + q inclus. C'est, par conséquent, un approximant de Padé et, en raison de leur unicité $[p/q] \equiv [p-k/q-k]$.

Multiplions maintenant le numérateur et le dénominateur de [p-k / q-k] par un même polynôme de degré $n \leq k$. On obtient le rapport d'un polynôme de degré p-k+n sur un polynôme de degré q-k+n qui coïncide avec f jusqu'au terme de degré p+q-2k+2n (puisqu'il coïncide en fait avec f jusqu'au terme de degré p+q inclus) ; c'est donc un approximant de Padé et, en raison de leur unicité, on a :

$$[p - n / q - n] \equiv [p/q] \qquad \text{pour } n = 0,\ldots,k$$

Montrons maintenant que si deux approximants adjacents sur une diagonale, c'est-à-dire [p/q] et [p-1 / q-1], sont identiques alors, en fait, on a :

$$[p/q] \equiv [p-1/q] \equiv [p/q-1] \equiv [p-1/q-1]$$

Les coefficients des dénominateurs de [p/q] et [p-1/q-1] sont identiques ; on a donc :

$$\begin{pmatrix} c_{p-1} & \cdots\cdots & c_{p-q+1} \\ - - - & - - - & - - - \\ c_{p+q-3} & \cdots\cdots & c_{p-1} \end{pmatrix} \begin{pmatrix} b_1 \\ \cdot \\ \cdot \\ \cdot \\ b_{q-1} \end{pmatrix} = - \begin{pmatrix} c_p \\ \cdot \\ \cdot \\ \cdot \\ c_{p+q-2} \end{pmatrix}$$

$$\begin{pmatrix} c_p & \cdots & c_{p-1} & \cdots & c_{p-q+1} \\ - - - & & - - - & & - - - \\ c_{p+q-2} & & c_{p+q-3} & \cdot\cdot & c_{p-1} \\ c_{p+q-1} & & c_{p+q-2} & \cdot\cdot & c_p \end{pmatrix} \begin{pmatrix} b_1 \\ \cdot \\ \cdot \\ \cdot \\ b_{q-1} \\ b_q \end{pmatrix} = - \begin{pmatrix} c_{p+1} \\ \cdot \\ \cdot \\ \cdot \\ c_{p+q-1} \\ c_{p+q} \end{pmatrix}$$

On a évidemment $b_q = 0$ puisque le polynôme est en fait de degré q-1.

Considérons maintenant [p/q-1]. Ses coefficients b'_i sont donnés par :

$$\begin{pmatrix} c_p & \cdots\cdots\cdots & c_{p-q+2} \\ - & - - - - - - - - - & - \\ c_{p+q-2} & \cdots\cdots\cdots & c_p \end{pmatrix} \begin{pmatrix} b'_1 \\ \vdots \\ b'_{q-1} \end{pmatrix} = - \begin{pmatrix} c_{p+1} \\ \vdots \\ c_{p+q-1} \end{pmatrix}$$

En comparant avec le système précédent, on voit que $b'_i = b_i$ pour $i = 1,\ldots,q-1$ et, par conséquent, que $a'_i = a_i \ \forall\, i$. Considérons maintenant [p-1/q]. Ses coefficients b''_i sont donnés par :

$$\begin{pmatrix} c_{p-1} & \cdots\cdots\cdots & c_{p-q} \\ - & - - - - - - - - - & - \\ c_{p+q-2} & \cdots\cdots\cdots & c_{p-1} \end{pmatrix} \begin{pmatrix} b''_1 \\ \vdots \\ b''_q \end{pmatrix} = - \begin{pmatrix} c_p \\ \vdots \\ c_{p+q-1} \end{pmatrix}$$

En comparant avec le premier système, on voit que $b''_i = b_i$ pour $i = 1,\ldots,q-1$ et que $b''_q = 0$. On a aussi $a''_i = a_i \ \forall\, i$. Les quatre approximants de Padé sont donc identiques. Si nous appliquons cette propriété au cas où $[p-n/q-n] \equiv [p/q]$ pour $n = 0,\ldots,k$ alors on voit que tous les approximants [p-n/q-m] sont identiques pour $n,m = 0,\ldots,k$. C'est ce qu'on appelle la structure en blocs (carrés) de la table de Padé puisque tous ces approximants remplissent un bloc carré de la table de Padé dont les quatre angles sont [p-k/q-k], [p-k/q], [p/q-k] et [p/q]. C'est Padé [149] qui, le premier, a étudié en détail cette structure en blocs. Sur cette question on pourra consulter également [93,185].

La connexion qui existe entre la table de Padé et l'ε-algorithme a été mise en lumière par Shanks [170] et Wynn [213] :

Théorème 42 :

Soit $f(x) = \sum_{i=0}^{\infty} c_i x^i$ une série de puissances. Si on applique l'ε-algorithme aux sommes partielles de cette série : $S_n = \sum_{i=o}^{n} c_i x^i$ alors :

$$\varepsilon_{2k}^{(n)} = [n + k/k]$$

démonstration : on a vu au début de ce paragraphe que

$$\varepsilon_{2k}^{(n)} = \frac{\begin{vmatrix} S_n & \ldots\ldots & S_{n+k} \\ \Delta S_n & \ldots\ldots & \Delta S_{n+k} \\ \hline \Delta S_{n+k-1} & \ldots & \Delta S_{n+2k-1} \end{vmatrix}}{\begin{vmatrix} 1 & \ldots\ldots & 1 \\ \Delta S_n & \ldots\ldots & \Delta S_{n+k} \\ \hline \Delta S_{n+k-1} & \ldots & \Delta S_{n+2k-1} \end{vmatrix}}$$

on a $\Delta S_n = c_{n+1} x^{n+1}$ d'où

$$\varepsilon_{2k}^{(n)} = \frac{\begin{vmatrix} \sum_{i=0}^{n} c_i x_i & \ldots\ldots & \sum_{i=0}^{n+k} c_i x^i \\ c_{n+1} x^{n+1} & \ldots\ldots & c_{n+k+1} x^{n+k+1} \\ \hline c_{n+k} x^{n+k} & \ldots\ldots & c_{n+2k} x^{n+2k} \end{vmatrix}}{\begin{vmatrix} 1 & \ldots\ldots\ldots & 1 \\ c_{n+1} x^{n+1} & \ldots\ldots & c_{n+k+1} x^{n+k+1} \\ \hline c_{n+k} x^{n+k} & \ldots\ldots & c_{n+2k} x^{n+2k} \end{vmatrix}}$$

Multiplions la première colonne du numérateur et du dénominateur par x^k, les secondes par x^{k-1}, etc. et les dernières par 1. On obtient :

$$\varepsilon_{2k}^{(n)} = \frac{\begin{vmatrix} \sum_{i=0}^{n} c_i x^{k+i} & \cdots\cdots & \sum_{i=0}^{n+k} c_i x^i \\ c_{n+1} x^{n+k+1} & \cdots & c_{n+k+1} x^{n+k+1} \\ \text{-----} & & \text{-----} \\ c_{n+k} x^{n+2k} & \cdots & c_{n+2k} x^{n+2k} \end{vmatrix}}{\begin{vmatrix} x^k & \cdots\cdots & 1 \\ c_{n+1} x^{n+k+1} & \cdots & c_{n+k+1} x^{n+k+1} \\ \text{-----} & & \text{-----} \\ c_{n+k} x^{n+2k} & \cdots & c_{n+2k} x^{n+2k} \end{vmatrix}}$$

divisons maintenant les secondes lignes du numérateur et du dénominateur par x^{n+k+1}, les troisièmes par x^{n+k+2}, etc. et les dernières par x^{n+2k}. On trouve que $\varepsilon_{2k}^{(n)} = [n+k/k]$ d'après la propriété 21.

L'ε-algorithme permet donc de construire la moitié de la table de Padé. L'autre moitié de la table de Padé peut être obtenue en utilisant la relation de la propriété 17 de la façon suivante :

on part des conditions aux limites extérieures $[-1/q] = 0$ et $[p/-1] = \infty$ et des conditions aux limites intérieures $[p/0] = \sum_{i=0}^{p} c_i x^i$ pour $p = 0, 1, \ldots$ et $[0/q] = \left(\sum_{i=0}^{q} d_i x^i\right)^{-1}$ où la série $\sum_{i=0}^{\infty} d_i x^i$ est telle que :

$$\left(\sum_{i=0}^{\infty} c_i x^i\right)\left(\sum_{i=0}^{\infty} d_i x^i\right) = 1.$$

La propriété 17 s'écrit :

$$([p/q+1] - [p/q])^{-1} - ([p/q] - [p/q-1])^{-1} = ([p+1/q] - [p/q])^{-1} - ([p/q] - [p-1/q])^{-1}$$

On obtient par conséquent ainsi toute la table de Padé :

	∞	∞	--------	∞
0	[0/0]	[1/0]	--------	[p/0]
0	[0/1]	[1/1]	--------	[p/1]
⋮	⋮	⋮		⋮
0	[0/q]	[1/q]	--------	[p/q]

Remarque 1 : Une colonne du tableau ε correspond à une ligne de la table de Padé.

Remarque 2 : La méthode de bordage du paragraphe précédent permet de calculer les coefficients des approximants de Padé.

Remarque 3 : Si f(x) est une fraction rationnelle de degré N sur M et si on applique l'ε-algorithme aux sommes partielles de f alors, pour $N \geq M$, on a :

$$\varepsilon_{2M}^{(n)} = f(x) \qquad \forall\, n \geq N - M$$

Ce résultat n'est pas en contradiction avec les théorèmes 35 et 37. Supposons en effet que :

$$f(x) = \sum_{i=0}^{\infty} c_i\, x^i = \sum_{i=0}^{N} a_i\, x^i \,/ \sum_{i=0}^{M} b_i\, x^i \qquad \text{avec} \quad N \geq M$$

alors, par identification des termes de même degré en x, on trouve que :

$$b_0\, c_n + b_1\, c_{n-1} + \ldots + b_M\, c_{n-M} = 0 \qquad \text{pour} \quad n = N+1,\, N+2,\, \ldots$$

Si l'on multiplie cette égalité par x^n alors on a :

$$b_0\, c_n\, x^n + b_1\, x\, c_{n-1}\, x^{n-1} + \ldots + b_M\, x^M\, c_{n-M}\, x^{n-M} = 0 \qquad \text{pour} \quad n = N+1, \ldots$$

Si nous sommons ces égalités à partir de $n = p \geq N+1$, alors on obtient :

$$b_0 \sum_{n=p}^{\infty} c_n\, x^n + b_1\, x \sum_{n=p}^{\infty} c_{n-1}\, x^{n-1} + \ldots + b_M\, x^M \sum_{n=p}^{\infty} c_{n-M}\, x^{n-M} = 0$$

pour p = N+1, N+2,... fixé. Ce qui peut s'écrire :

$$b_0(f(x) - S_{p-1}) + b_1 x(f(x) - S_{p-2}) + \ldots + b_M\, x^M(f(x) - S_{p-M-1}) = 0$$

pour p = N+1, N+2,... en posant $S_k = \sum_{i=0}^{k} c_i\, x^i$ pour tout k.

En posant $f(x) = S$, $e_M = b_0$, $e_{M-1} = b_1 x$, ..., $e_0 = b_M x^M$, cette égalité peut s'écrire pour x fixé :

$$\sum_{i=0}^{M} e_i (S_{n+i} - S) = 0 \quad \text{pour} \quad n = N-M, N-M+1, \ldots$$

ce qui est autre que la relation du théorème 35 et par conséquent :

$$\varepsilon_{2M}^{(n)} = S = f(x) \quad \forall\, n \geq N-M$$

Inversement, considérons une suite $\{S_n\}$ qui vérifie :

$$\sum_{i=0}^{M} e_i (S_{n+i} - S) = 0 \quad \forall\, n \geq N-M$$

posons $c_0 = S_0$ et $c_i = \Delta S_{i-1}$ pour $i = 1,2,\ldots$

et
$$f(x) = \sum_{i=0}^{\infty} c_i x^i.$$

Alors $f(1) = S$ et S_n est égal à la $n^{\text{ième}}$ somme partielle de $f(1)$. L'égalité précédente s'écrit donc :

$$\sum_{i=0}^{M} e_i \sum_{j=n+1}^{\infty} c_{i+j} = 0 \quad \text{pour tout} \quad n \geq N-M$$

D'où encore :

$$\sum_{j=n+1}^{\infty} \sum_{i=0}^{M} e_i c_{i+j} = 0 \quad \text{pour tout} \quad n \geq N-M.$$

En écrivant cette égalité pour n et n+1 et en soustrayant, on trouve que :

$$\sum_{i=0}^{M} e_i c_{n+i+1} = 0 \quad \text{pour tout} \quad n \geq N-M$$

ce qui démontre que $f(x)$ est le développement en puissances croissantes de x d'une fraction rationnelle dont le numérateur est un polynôme de degré N en x et dont le dénominateur est de degré M avec $N \geq M$ puisque $n \geq 0$. Il y a donc équivalence totale entre les suites vérifiant les théorèmes 35 et 37 et les sommes partielles de fractions rationnelles.

Terminons ce paragraphe par la liaison entre les approximants de Padé et les polynômes orthogonaux. Etant donnée une suite $\{c_n\}$ on se définit la fonctionnelle c sur l'espace des polynômes réels par :

$$c(x^n) = c_n \quad \text{pour} \quad n = 0, 1, \ldots$$

Il est tout à fait classique [3] de construire une famille de polynômes orthogonaux $\{P_n\}$ par rapport à cette fonctionnelle c, c'est-à-dire que :

$$c(P_k\ P_n) = 0 \quad \text{si} \quad k \neq n.$$

On peut définir également des polynômes de seconde espèce $\{Q_n\}$ par :

$$Q_n(t) = c\left(\frac{P_n(x) - P_n(t)}{x - t}\right)$$

où t est un paramètre et où c agit sur la variable x. P_n est de degré n et Q_n est de degré n-1. Posons :

$$\tilde{P}_n(x) = x^n\ P_n(x^{-1})$$

$$\tilde{Q}_n(x) = x^{n-1}\ Q_n(x^{-1})$$

Considérons la série $f(x) = \sum_{i=0}^{\infty} c_i\ x^i$; alors on a :

$$[n-1/n]_f(x) = \tilde{Q}_n(x)\ /\ \tilde{P}_n(x).$$

Cette connexion entre la théorie des polynômes orthogonaux et les approximants de Padé est très intéressante car elle permet de bâtir une théorie très cohérente des approximants de Padé, des fractions continues et de certaines méthodes d'accélération de la convergence. Elle permet également de rattacher entre autre les approximants de Padé aux formules de quadrature de Gauss, à la méthode des moments, à celle de Lanczos, à l'approximation d'opérateurs et à la méthode du gradient conjugué. Ce point de vue est actuellement en plein développement. Sur ce sujet, on pourra consulter [4,5, 94, 179, 180, 181, 182].

On trouvera les développements récents sur la table de Padé dans [60,206].

III - 6 Théorèmes de convergence

Avant de donner des théorèmes de convergence pour l'ε-algorithme il faut précis un point important. On voit que lorsqu'on applique l'ε-algorithme à une suite $\{S_n\}$ il se peut que deux quantités $\varepsilon_k^{(n+1)}$ et $\varepsilon_k^{(n)}$ deviennent égales pour une certaine valeur de k et de n. Il est alors impossible de continuer à construire le tableau ε car il y aurait une division par zéro. Dans la suite la convergence devra toujours être comprise avec la restriction énoncée par Wynn [235] : "bien que des conditions spéciales puissent être imposées à la suite $\{S_n\}$ pour éviter cette division par zéro, dans l'exposition d'une théorie générale où l'on impose aucune condition sur la suite initiale, les résultats énoncés ne concernent que les nombres qui peuvent être calculés. Une autre remarque importante est que les théorèmes de convergence concernant la table de Padé donnent des théorèmes de convergence pour l'ε-algorithme, inversement les théorèmes de convergence établis pour l'ε-algorithme fournissent des théorèmes de convergence ponctuelle pour la table de Padé. Le premier théorème que nous allons énoncer a été démontré par Montessus de Ballore [144] pour la table de Padé. Nous le donnons ici en termes d'ε-algorithme et de suite et sans démonstration :

Théorème 43 :

Soit $\{S_n\}$ une suite qui converge vers S et qui est telle que $\lim\sup_{n\to\infty} |\Delta S_n|^{1/n} = 1$.
Soit f(z) la série associée :

$$f(z) = \sum_{i=0}^{\infty} c_i z^i \text{ avec } c_o = S_o \text{ et } c_k = \Delta S_{k-1} \text{ pour } k = 1, 2, \ldots$$

Supposons que f(z) possède k pôles comptés avec leurs multiplicités sur le cercle $|z| = 1$ et pas d'autres singularités. Si on applique l'ε-algorithme à la suite $\{S_n\}$ alors la suite $\{\varepsilon_{2k}^{(n)}\}$ converge vers S lorsque n tend vers l'infini.

Il existe de nombreux théorèmes de convergence pour la table de Padé. En général ce sont des théorèmes de convergence uniforme, en mesure ou en capacité.

Sur ces questions on pourra consulter par exemple les références [10,97] ainsi que les articles de Wynn [207,208,209,210,211]et ceux de Basdevant [13, 14]. Pour être complet il faudrait également citer les articles de Zinn-Justin, Bessis, Nuttall, etc ... On trouvera les références à ces articles dans ceux qui viennent d'être cités et dans [37].

Si nous ne nous intéressons qu'aux suites alors des théorèmes de convergence simple sont suffisants. Ces théorèmes se divisent en plusieurs groupes suivant les hypothèses faites :

- théorèmes de convergence pour des suites de forme bien déterminée : par exemple $S_n = S + \sum_{i=1}^{\infty} \alpha_i \lambda_i^n$
- théorèmes de convergence pour des classes de suites présentant certaines propriétés particulières comme, par exemple, les suites totalement monotones.
- théorèmes de convergence pour des suites dont on connait la loi de formation : par exemple $S_{n+1} = f(S_n, S_{n-1}, \ldots, S_{n-k})$
- théorèmes de convergence pour la colonne 2k du tableau ε quand on connait certaines propriétés des colonnes précédentes.

Pour le premier groupe, on a les résultats suivants qui ont été obtenus par Wynn [234] et que nous donnons ici sans démonstration.

<u>Théorème 44</u> :

Si on applique l'ε-algorithme à une suite $\{S_n\}$ telle que :

$$S_n \sim S + \sum_{i=1}^{\infty} a_i (n+b)^{-i} \qquad a_1 \neq 0$$

alors pour k fixé :

$$\varepsilon_{2k}^{(n)} \sim S + \frac{a_1}{(k+1)(n+b)}$$

<u>Théorème 45</u> :

si on applique l'ε-algorithme à une suite $\{S_n\}$ telle que :

$$S_n \sim S + (-1)^n \sum_{i=1}^{\infty} a_i (n+b)^{-i} \qquad a_1 \neq 0$$

alors pour k fixé :

$$\varepsilon_{2k}^{(n)} \sim S + \frac{(-1)^n (k!)^2 a_1}{2^{2k} (n+b)^{2k+1}}$$

Théorème 46 :

Si on applique l'ε-algorithme à une suite $\{S_n\}$ telle que :

$$S_n \sim S + \sum_{i=1}^{\infty} a_i \lambda_i^n \quad \text{avec} \quad 1 > \lambda_1 > \lambda_2 > \ldots > 0$$

alors pour k fixé :

$$\varepsilon_{2k}^{(n)} \sim S + \frac{a_{k+1} (\lambda_{k+1} - \lambda_1)^2 \ldots\ldots (\lambda_{k+1} - \lambda_k)^2 \lambda_{k+1}^n}{(1 - \lambda_1)^2 \ldots\ldots (1 - \lambda_k)^2}$$

Théorème 47 :

si on applique l'ε-algorithme à une suite $\{S_n\}$ telle que :

$$S_n \sim S + (-1)^n \sum^{\infty} a_i \lambda_i^n \quad \text{avec} \quad 1 > \lambda_1 > \lambda_2 > \ldots > 0$$

alors pour k fixé :

$$\varepsilon_{2k}^{(n)} \sim S + (-1)^n \frac{a_{k+1} (\lambda_{k+1} - \lambda_1)^2 \ldots\ldots (\lambda_{k+1} - \lambda_k)^2 \lambda_{k+1}^n}{(1 + \lambda_1)^2 \ldots\ldots (1 + \lambda_k)^2}$$

On voit que l'utilisation de ces quatre théorèmes nécessite d'avoir beaucoup d'informations sur la suite $\{S_n\}$. Leur emploi est donc restreint. Il vaut donc mieux s'orienter vers des théorèmes de convergence pour des classes de suites. Une classe de suites très importante en analyse numérique est la classe des suites totalement monotones.

Définition 15 : on dit que la suite $\{S_n\}$ est totalement monotone si $(-1)^k \Delta^k S_n \geq 0$ pour n, k = 0, 1, ... On écrira $\{S_n\} \in TM$.

Wynn [196] a montré que de nombreuses suites déduites d'une suite totalement monotone sont, elles aussi, totalement monotones :

Théorème 48 :

Soit $\{S_n\} \in TM$ alors :

1°) $\{(1 - S_n)^{-1}\} \in TM$ si $S_o < 1$

2°) $\{\prod_{i=0}^{n-1} S_i^{-1}\} \in TM$ si $\lim_{n\to\infty} S_n \geq 1$

3°) $\{\prod_{i=0}^{n-1} (1-S_i)\} \in TM$ si $S_o \leq 1$

4°) $\{a^{(-1)^{k+1}} \Delta^k S_n\} \in TM$ si $0 < a \leq 1$ et $k \geq 0$ entier fini fixé

5°) $\{a^{\sum_{i=0}^{n-1} S_i}\} \in TM$ si $0 \leq a \leq 1$

Dans le même ordre d'idée, Brezinski [36] a démontré le :

Théorème 48 bis : Soit $f(t) = \sum_{k=0}^{\infty} c_k x^k$ une série de puissances de rayon de convergence R et telle que $c_k \geq 0$ pour tout k et soit $\{S_n\} \in TM$. Si $S_o < R$ alors $\{f(S_n)\} \in TM$.

Démonstration : rappelons d'abord, ce qui est trivial à démontrer, que si $\{u_n\} \in TM$ et $\{v_n\} \in TM$ alors $\{u_n v_n\} \in TM$ et $\{au_n + bv_n\} \in TM$ si $a,b \geq 0$. Posons $f_k(t) = c_o + \ldots + c_k t^k$. Pour k fixé $\{f_k(S_n)\} \in TM$. Puisque $0 \leq \ldots \leq S_1 \leq S_o < R$ alors $f(S_n) = \lim_{k\to\infty} f_k(S_n)$ existe pour tout n. $\{f(S_n)\}$ est donc la limite, pour k tendant vers l'infini, d'une suite de suites TM. C'est donc aussi une suite TM puisque

$$(-1)^p \Delta^p f_k(S_n) \geq 0 \qquad n,p,k = 0,1,\ldots$$

$$\lim_{k\to\infty} (-1)^p \Delta^p f_k(S_n) = (-1)^p \Delta^p f(S_n) \geq 0 \qquad p,n = 0,1,\ldots$$

Remarque 1 : Si la série converge pour t = R alors le résultat précédent reste valable si $S_o = R$.

Remarque 2 : Si $(-1)^k c_k \geq 0$ pour tout k alors $\{f(-S_n)\} \in TM$ si $\{S_n\} \in TM$ et $S_o < R$.

Donnons maintenant quelques exemples pour illustrer ces résultats. Soit $\{S_n\} \in TM$; alors les suites suivantes sont également TM si les conditions indiquées sont satisfaites.

suites	conditions	
$a^r - (a - S_n)^r$	$0 \leq r \leq 1$	$S_o \leq a$
$(a - S_n)^{-r}$	$r \geq 0$	$S_o < a$
$\mathrm{tg}\ (S_n)$	$S_o < \Pi/2$	
a^{S_n}	$a > 1$	
$-\ \mathrm{Log}\ (1-S_n)$	$S_o < 1$	
$\mathrm{Arcsin}\ (S_n)$	$S_o < 1$	
$\mathrm{Arcos}\ (S_n)$	$S_o < 1$	
$\mathrm{sh}\ (S_n)$	pas de condition	
$\mathrm{ch}\ (S_n)$	pas de condition	

Cette liste n'est évidemment pas limitative ; on peut composer les fonctions précédentes pour obtenir de nouvelles suites TM. On peut également utiliser le fait que sommes et produits de suites TM sont aussi TM ou encore faire appel au lemme 1 (voir plus loin).

Wynn [236] a également montré que les sommes partielles de certaines séries d'opérateurs utilisées en analyse numérique étaient totalement monotones ; c'est le cas de la série d'interpolation de Newton, de la série de Newton pour la dérivation, de la série d'intégration de Newton-Gregory et de la formule d'Euler-Maclaurin pour certaines fonctions. Wynn a donné aussi des exemples qui montrent le gain appréciable que procure l'ε-algorithme quand on l'applique à des suites totalement monotones [237].

Avant de démontrer la convergence de l'ε-algorithme pour des suites totalement monotones nous avons besoin d'établir un certain nombre de lemmes [234].

Lemme 1 : Si $\{S_n\} \in TM$ alors $\{(-1)^k \Delta^k S_n\} \in TM$ $k = 1, 2, \ldots$

la démonstration est laissée en exercice.

Lemme 2 : Toute suite totalement monotone est convergente.

La démonstration est évidente.

Une des plus importante caractéristiques des suites totalement monotones est d'être reliées à la théorie des moments. On a notamment le résultat suivant que nous énoncerons sans démonstration. L'ouvrage fondamental sur ces questions est le livre de Widder [188].

<u>Lemme 3</u> : une condition nécessaire et suffisante pour que $\{S_n\} \in TM$ est que :

$S_n = \int_0^1 x^n \, dg(x)$ pour $n = 0, 1, \ldots$ où g est une fonction bornée non décroissante dans $[0, 1]$.

Les suites TM possèdent un certain nombre de propriétés remarquables. Ainsi, en utilisant l'inégalité de Hölder, on trouve [89] que si $\{S_n\} \in TM$ alors :

$$S_{r(n+k)}^{1/r} \leq S_{np}^{1/p} \, S_{kq}^{1/q}$$

avec $1/p + 1/q = 1/r$, $r(n+k) \in \mathbb{N}$, $np \in \mathbb{N}$ et $kq \in \mathbb{N}$.

Pour $p = q = 2$ et $r = 1$ on a donc :

$$S_{n+k}^2 \leq S_{2n} \, S_{2k} \qquad n,k = 0,1,\ldots$$

Pour $n = 0$, $k = 1$ et $q = r+1$ on trouve que :

$$0 \leq \frac{S_1}{S_o} \leq \left(\frac{S_2}{S_o}\right)^{1/2} \leq \ldots \leq \left(\frac{S_n}{S_o}\right)^{1/n} \leq \left(\frac{S_{n+1}}{S_o}\right)^{1/(n+1)} \leq \ldots \leq \frac{1}{R} \leq 1$$

où R est le rayon de convergence de la série $f(x) = \sum_{i=0}^{\infty} S_i \, x^i$.

Pour $k = 0$ et $r = 1$ on obtient :

$$S_n^p \leq S_{np} \, S_o^{p-1} \qquad n,p = 0,1,\ldots$$

On déduit de ces inégalités un certain nombre de conséquences pour la convergence des suites TM : Soit $\{S_n\} \in TM$ et soit S sa limite. Si $S_n \neq S \; \forall \, n$ alors $\exists \, \lambda \in \,]0,1[$ tel que $\lambda^n = O(S_n - S)$. Inversement soit $\{S_n\}$ une suite de limite S telle que $S_n > S \; \forall \, n$; si $\forall \, \lambda \in \,]0,1[\; S_n - S = o(\lambda^n)$ alors $\{S_n - S\} \notin TM$.

Le résultat fondamental pour la suite a été démontré par Wynn [234] :

<u>Lemme 4</u> : si $\{S_n\} \in TM$ alors $H_k^{(n)}(S_n) \geq 0$ pour $n, k = 0, 1, \ldots$

démonstration : l'idée de la démonstration est la suivante : à partir du lemme 3 on a $S_n = \int_0^1 x^n \, dg(x)$. A partir de la théorie des moments de Stieltjes on montre que $H_k^{(o)}(S_o) \geq 0$ et $H_k^{(1)}(S_1) \geq 0$ si et seulement si $S_n = \int_0^\infty x^n \, d\bar{g}(x)$ pour n = 0, 1, ... où $\bar{g}$ est une fonction bornée non décroissante sur $[0, +\infty)$. Dans ce dernier cas on a :

$$S_{n+p} = \int_0^\infty x^n \, d\bar{g}^{(p)}(x) \quad n, p = 0, 1, \ldots$$

avec $d\bar{g}^{(p)}(x) = x^p \, d\bar{g}(x)$. $\bar{g}^{(p)}(x)$ est également bornée et non décroissante sur $[0, +\infty)$. Ainsi on a $H_k^{(n)}(S_n) \geq 0$ pour k, n = 0, 1, ... On obtient ensuite le résultat du lemme en posant $\bar{g}(x) = g(x)$ pour $x \in [0,1]$ et $\bar{g}(x) = g(1)$ pour $x \in [1, +\infty)$.

A partir des lemmes 1 et 4 on a donc immédiatement le :

Lemme 5 : $H_k^{(n)}(\Delta^{2p} S_n) \geq 0$ et $(-1)^k H_k^{(n)}(\Delta^{2p+1} S_n) \geq 0$

d'où en utilisant ce lemme et la propriété 10 qui relie l'ε-algorithme aux déterminants de Hankel :

Lemme 6 : si on applique l'ε-algorithme à $\{S_n\} \in TM$ alors :

$$\varepsilon_{2k}^{(n)} \geq 0 \text{ et } \varepsilon_{2k+1}^{(n)} \leq 0 \text{ pour } n, k = 0, 1, \ldots$$

Utilisons maintenant la propriété 12 :

$$\varepsilon_{2k+2}^{(n)} - \varepsilon_{2k}^{(n)} = - \frac{[H_{k+1}^{(n)}(\Delta S_n)]^2}{H_{k+1}^{(n)}(\Delta^2 S_n) . H_k^{(n)}(\Delta^2 S_n)}$$

et les lemmes 5 et 6. On obtient immédiatement le :

Lemme 7 : Si on applique l'ε-algorithme à $\{S_n\} \in TM$ alors :

$$0 \leq \varepsilon_{2k+2}^{(n)} \leq \varepsilon_{2k}^{(n)} \quad n, k = 0, 1, \ldots$$

Remarque : en utilisant un résultat peu connu sur les matrices définies positives [72, problème 17, p.51] on montre que si $\{S_n\} \in TM$ et $\{V_n\} \in TM$ alors $e_k(S_n + V_n) \geq e_k(S_n) + e_k(V_n) \geq 0 \;\; \forall \, n,k$.

Démontrons maintenant le résultat fondamental [24] :

Théorème 49 : Si on applique l'ε-algorithme à une suite $\{S_n\}$ qui converge vers S et s'il existe deux constantes $a \neq 0$ et b telles que $\{a \, S_n + b\} \in TM$ alors :

$$\lim_{n\to\infty} \varepsilon_{2k}^{(n)} = S \quad \text{pour } k = 0, 1, \ldots$$

démonstration : supposons que $\{S_n\} \in TM$ et soit S sa limite. Alors $\{S_n - S\} \in TM$. Si on applique l'ε-algorithme à $\{S_n - S\}$ alors on obtient des quantités $\varepsilon_{2k}^{(n)}$ qui vérifient l'inégalité du lemme 7. Si nous faisons k = 0 dans cette inégalité on voit que $\lim_{n\to\infty} \varepsilon_2^{(n)} = 0$ puisque $\{S_n - S\}$ converge vers zéro. On a donc $\lim_{n\to\infty} \varepsilon_{2k}^{(n)} = 0 \; \forall k$. Le reste de la démonstration provient tout simplement de la propriété 14 de l'ε-algorithme.

Nous allons maintenant étudier la convergence des diagonales du tableau ε pour les suites totalement monotones. Auparavant on a :

Lemme 8 : si on applique l'ε-algorithme à une suite $\{S_n\} \in TM$ alors

$$\varepsilon_{2k+1}^{(n)} \leq \varepsilon_{2k-1}^{(n)} \leq 0 \qquad k, n = 0, 1, \ldots$$

et

$$\lim_{n\to\infty} \varepsilon_{2k+1}^{(n)} = -\infty \qquad k = 0, 1, \ldots$$

démonstration : Puisque $\varepsilon_{2k+1}^{(n)} = 1/e_k(\Delta s_n)$ on a, d'après la propriété 12 et le lemme 5 :

$$\frac{1}{\varepsilon_{2k+1}^{(n)}} - \frac{1}{\varepsilon_{2k-1}^{(n)}} = -\frac{[H_k^{(n)}(\Delta^2 S_n)]^2}{H_{k-1}^{(n)}(\Delta^3 S_n).H_k^{(n)}(\Delta^3 S_n)} \geq 0$$

et par conséquent en utilisant le lemme 6 :

$$\varepsilon_{2k+1}^{(n)} \leq \varepsilon_{2k-1}^{(n)} \leq 0$$

De plus $\varepsilon_1^{(n)} = 1 / \Delta S_n$ d'où $\lim_{n\to\infty} \varepsilon_1^{(n)} = -\infty$ ce qui termine la démonstration du lemme.

Lemme 9 : Si on applique l'ε-algorithme à une suite $\{S_n\} \in TM$ alors

$$\Delta\varepsilon_k^{(n)} \leq 0 \qquad n, k = 0, 1, \ldots$$

démonstration : on a :

$\varepsilon_{2k+1}^{(n+1)} - \varepsilon_{2k-1}^{(n+1)} = \varepsilon_{2k+1}^{(n)} - \varepsilon_{2k-1}^{(n+1)} + \Delta\varepsilon_{2k+1}^{(n)} \leq 0$ d'après le lemme 8 ; d'où :

$$\varepsilon_{2k+1}^{(n)} \leq \varepsilon_{2k-1}^{(n+1)} - \Delta\varepsilon_{2k+1}^{(n)}$$

ou encore :

$$\varepsilon_{2k-1}^{(n+1)} + \frac{1}{\Delta\varepsilon_{2k}^{(n)}} \leq \varepsilon_{2k-1}^{(n+1)} - \Delta\varepsilon_{2k+1}^{(n)}$$ ce qui donne :

$$\frac{1}{\Delta\varepsilon_{2k}^{(n)}} \leq -\Delta\varepsilon_{2k+1}^{(n)} \qquad (1)$$

d'autre part on a :

$$0 \leq \varepsilon_{2k+2}^{(n)} = \varepsilon_{2k}^{(n+1)} + \frac{1}{\Delta\varepsilon_{2k+1}^{(n)}} \leq \varepsilon_{2k}^{(n)}$$

ou encore

$$\Delta\varepsilon_{2k}^{(n)} \leq -\frac{1}{\Delta\ \varepsilon_{2k+1}^{(n)}} \qquad (2)$$

si $\Delta\varepsilon_{2k}^{(n)} \geq 0$ alors (1) devient :

$$1 \leq -\Delta\varepsilon_{2k}^{(n)}\ \Delta\varepsilon_{2k+1}^{(n)}$$

et (2) s'écrit :

$$1 \leq -1/\Delta\varepsilon_{2k}^{(n)}\ \Delta\varepsilon_{2k+1}^{(n)}$$

Ces deux inégalités sont incompatibles et par conséquent :

$$\Delta\varepsilon_{2k}^{(n)} \leq 0 \qquad \forall\ n,k.$$

De même si $\Delta\varepsilon_{2k+1}^{(n)} \geq 0$ alors (1) s'écrit :

$$1/\Delta\varepsilon_{2k}^{(n)}\ \Delta\varepsilon_{2k+1}^{(n)} \leq -1$$

et (2) devient :

$$\Delta\varepsilon_{2k}^{(n)} \ \Delta\varepsilon_{2k+1}^{(n)} \leq -1$$

Ces deux inégalités sont incompatibles et donc $\Delta\varepsilon_{2k+1}^{(n)} \leq 0 \quad \forall\, n,k.$

Une conséquence des lemmes 7 et 9 s'obtient immédiatement à partir de la règle de la croix (propriété 17) :

Lemme 9bis : Si on applique l'ε-algorithme à une suite TM alors

$$0 \leq \varepsilon_{2k+2}^{(n)} \leq \varepsilon_{2k}^{(n+2)} \qquad n,k = 0,1,\ldots$$

Pour la démonstration, voir celle du lemme 17bis.

On peut maintenant démontrer la convergence des diagonales du tableau ε [21] :

Théorème 50 :

Si on applique l'ε-algorithme à une suite $\{S_n\}$ qui converge vers S et s'il existe deux constantes $a \neq 0$ et b telles que $\{a\, S_n + b\} \in$ TM alors :

$$\lim_{k\to\infty} \varepsilon_{2k}^{(n)} = S \qquad \text{pour } n = 0,\, 1,\, \ldots$$

démonstration : d'après le lemme 7 on a $0 \leq \varepsilon_{2k+2}^{(n)} \leq \varepsilon_{2k}^{(n)}$.

Pour n fixé la suite $\{\varepsilon_{2k}^{(n)}\}$ est décroissante et bornée inférieurement. Elle est donc convergente. Appelons $T^{(n)}$ sa limite :

$$T^{(n)} = \lim_{k\to\infty} \varepsilon_{2k}^{(n)}.$$

Nous allons montrer que $T^{(n)} = S\ \forall n$. On a $0 \leq T^{(n)} \leq \varepsilon_{2k}^{(n)}\ \forall k$. Puisque la suite $\{S_n - S\} \in$ TM alors :

$$0 \leq T^{(n)} - S \leq \varepsilon_{2k}^{(n)} - S$$

d'où $\lim_{n\to\infty} T^{(n)} = S$ puisque $\lim_{n\to\infty} \varepsilon_{2k}^{(n)} = S\ \forall k$ d'après le théorème 49. D'un autre côté on a :

$$0 \leqslant \varepsilon_{2k+2}^{(n)} = \varepsilon_{2k}^{(n+1)} + \frac{1}{\Delta\varepsilon_{2k+1}^{(n)}} \leqslant \varepsilon_{2k}^{(n+1)}$$

puisque $\Delta\varepsilon_{2k+1}^{(n)} \leqslant 0$ d'après le lemme 9. D'où, en passant à la limite :

$$0 \leqslant S \leqslant T^{(n)} \leqslant T^{(n+1)} \leqslant \ldots$$

et par conséquent $T^{(n)} = S \ \forall n$ puisque $\lim_{n\to\infty} T^{(n)} = S$. Le reste de la démonstration provient de la propriété 14 de l'ε-algorithme. On voit donc que, pour les suites totalement monotones aux constantes multiplicatives et additives a et b près, on peut démontrer la convergence vers S des colonnes et des diagonales du tableau ε. On obtient de plus des inégalités entre les termes de ce tableau :

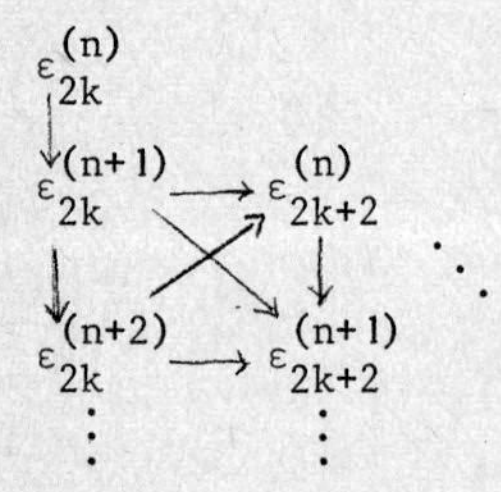

Les flèches allant de la quantité la plus grande à la quantité la plus petite (on a supposé que S = 0).

Nous allons maintenant étudier la convergence de l'ε-algorithme pour une autre classe de suites : les suites totalement oscillantes.

<u>définition 16</u> : on dit que la suite $\{S_n\}$ est totalement oscillante si la suite $\{(-1)^n S_n\}$ est totalement monotone. On écrira $\{S_n\} \in TO$.

Le résultat suivant est évident :

<u>Lemme 10</u> : si $\{S_n\} \in TO$ alors $\{(-1)^k \Delta^k S_n\} \in TO$

<u>Lemme 11</u> : Toute suite totalement oscillante convergente converge vers zéro.

démonstration : en posant $u_n = (-1)^n S_n$ on a d'après la définition 16 :

$$\Delta u_n = (-1)^n(- S_{n+1} - S_n) \leqslant 0 \text{ d'où}$$

$$0 \geqslant - S_{2n+2} \geqslant S_{2n+1} \geqslant - S_{2n} \geqslant S_{2n-1}$$

ce qui démontre la convergence de $\{u_n\}$. Supposons que $\{S_n\}$ converge vers $S \neq 0$. On a donc :

$$\forall \varepsilon > 0 \quad \exists N : \forall n > N \quad |S_n - S| < \varepsilon$$

or $S_{2n} \geq 0$ et $S_{2n+1} \leq 0$, donc si $S > 0$ alors

$\forall n \quad |S_{2n+1} - S| > S$ d'où si $0 < \varepsilon < S$

$$|S_p - S| < S < |S_{2n+1} - S| \quad \forall p > N \text{ et } \forall n$$

ce qui est impossible, donc $S = 0$.

Si $S < 0$ on fait le même raisonnement avec S_{2n}. On a le résultat fondamental suivant [234]:

Lemme 12 : Si $\{S_n\} \in TO$ alors $(-1)^{nk}\, H_k^{(n)}\,(S_n) \geq 0$

Démonstration : elle est évidente car, dans les déterminants de Hankel, k colonnes sont multipliées par $(-1)^n$. De même on a immédiatement le :

Lemme 13 : Si $\{S_n\} \in TO$ alors

$(-1)^{kn}\, H_k^{(n)}\,(\Delta^{2p}\, S_n) \geq 0$ et $(-1)^{k(n+1)}\, H_k^{(n)}\,(\Delta^{2p+1}\, S_n) \geq 0$

La propriété 10 de l'ε-algorithme ainsi que ces deux derniers lemmes nous donnent donc le :

Lemme 14 : Si on applique l'ε-algorithme à $\{S_n\} \in TO$ alors :

$$(-1)^n\, \varepsilon_{2k}^{(n)} \geq 0 \text{ et } (-1)^n\, \varepsilon_{2k+1}^{(n)} \leq 0 \text{ pour } n, k = 0, 1, \ldots$$

En utilisant la propriété 12 et les lemmes 13 et 14 on a :

Lemme 15 : Si on applique l'ε-algorithme à $\{S_n\} \in TO$ alors :

$$0 \leq \varepsilon_{2k+2}^{(2n)} \leq \varepsilon_{2k}^{(2n)} \text{ et } \varepsilon_{2k}^{(2n+1)} \leq \varepsilon_{2k+2}^{(2n+1)} \leq 0 \qquad n, k = 0, 1, \ldots$$

d'où le résultat suivant [24] dont la démonstration est analogue à celle du théorème 49 :

Théorème 51 :

Si on applique l'ε-algorithme à une suite $\{S_n\}$ qui converge vers S et s'il existe deux constantes $a \neq 0$ et b telles que $\{a\, S_n + b\} \in TO$ alors :

$$\lim_{n\to\infty} \varepsilon_{2k}^{(n)} = S \qquad \text{pour } k = 0, 1, \ldots$$

Il est possible de démontrer, pour les suites totalement oscillantes, la convergence vers S des diagonales du tableau ε.

Nous avons auparavant besoin des lemmes suivants [21] :

Lemme 16 : Si on applique l'ε-algorithme à une suite $\{S_n\} \in TO$ alors :

$$(-1)^n \varepsilon_{2k+1}^{(n)} \leq (-1)^n \varepsilon_{2k-1}^{(n)} \leq 0 \qquad n, k = 0, 1, \ldots$$

et

$$\lim_{n \to \infty} \varepsilon_{2k+1}^{(2n)} = -\infty \qquad k = 0, 1, \ldots$$

$$\lim_{n \to \infty} \varepsilon_{2k+1}^{(2n+1)} = +\infty$$

démonstration : elle peut être effectuée comme celle du lemme 8. Cependant il est plus facile d'utiliser le fait que :

$$\varepsilon_{2k+1}^{(n)} = 1 / e_k (\Delta S_n)$$

et le fait que $\{-\Delta S_n\} \in TO$. D'après le lemme 15 on a donc :

$$0 \geq (-1)^n e_k(\Delta S_n) \geq (-1)^n e_{k-1} (\Delta S_n) \text{ d'où}$$

$$0 \geq \frac{(-1)^n}{\varepsilon_{2k+1}^{(n)}} \geq \frac{(-1)^n}{\varepsilon_{2k-1}^{(n)}}$$

ce qui donne bien :

$$(-1)^n \varepsilon_{2k+1}^{(n)} \leq (-1)^n \varepsilon_{2k-1}^{(n)} \leq 0$$

et $\lim_{n\to\infty} \varepsilon_{2k+1}^{(2n)} = -\infty$, $\lim_{n\to\infty} \varepsilon_{2k+1}^{(2n+1)} = +\infty$ puisque $\lim_{n\to\infty} \varepsilon_1^{(2n)} = -\infty$ et $\lim_{n\to\infty} \varepsilon_1^{(2n+1)} = +\infty$

Lemme 17 : Si on applique l'ε-algorithme à une suite $\{S_n\} \in TO$ alors :

$$(-1)^n \Delta\varepsilon_{2k}^{(n)} \leq (-1)^n \Delta\varepsilon_{2k+2}^{(n)} \leq 0$$

$$(-1)^n \Delta\varepsilon_{2k+1}^{(n)} \geq (-1)^n \Delta\varepsilon_{2k-1}^{(n)} \geq 0 \qquad \text{pour } n, k = 0, 1, \ldots$$

et

$$\lim_{n\to\infty} \Delta\varepsilon_{2k+1}^{(2n)} = +\infty$$

$$\lim_{n\to\infty} \Delta\varepsilon_{2k+1}^{(2n+1)} = -\infty \qquad \text{pour } k = 0, 1, \ldots$$

Démonstration : elle est évidente à partir des lemmes 15 et 16. Ecrivons par exemple le lemme 15 :

$$\varepsilon_{2k}^{(2n+1)} \leq \varepsilon_{2k+2}^{(2n+1)} \leq 0$$

$$-\varepsilon_{2k}^{(2n)} \leq -\varepsilon_{2k+2}^{(2n)} \leq 0$$

d'où en ajoutant ces deux inégalités :

$$\Delta\varepsilon_{2k}^{(2n)} \leq \Delta\varepsilon_{2k+2}^{(2n)} \leq 0$$

On aurait de même :

$$0 \leq -\varepsilon_{2k+2}^{(2n+1)} \leq -\varepsilon_{2k}^{(2n+1)}$$

$$0 \leq \varepsilon_{2k+2}^{(2n+2)} \leq \varepsilon_{2k}^{(2n+2)}$$

d'où en ajoutant

$$0 \leq \Delta\varepsilon_{2k+2}^{(2n+1)} \leq \Delta\varepsilon_{2k}^{(2n+1)}$$

Les inégalités sur les quantités d'indices inférieurs impairs s'obtiennent de façon analogue à partir du lemme 16.

Les limites proviennent de la convergence vers zéro de $\varepsilon_{2k}^{(n)}$ lorsque n tend vers l'infini.

Il est possible de démontrer également l'inégalité suivante qui est plus complète :

$$-\frac{(-1)^n}{\Delta\varepsilon_{2k-1}^{(n)}} \leq (-1)^n \,\Delta\varepsilon_{2k}^{(n)} \leq -\frac{(-1)^n}{\Delta\varepsilon_{2k+1}^{(n)}} \leq (-1)^n \,\Delta\varepsilon_{2k+2}^{(n)} \leq 0$$

REMARQUE : les théorèmes 49 et 51 assurent seulement la convergence des suites $\{\varepsilon_{2k}^{(n)}\}$ pour k fixé. Ils n'assurent pas une convergence de $\{\varepsilon_{2k+2}^{(n)}\}$ plus rapide que celle de $\{\varepsilon_{2k}^{(n)}\}$ pour k fixé. Prenons par exemple $S_n = 1 + 1 / (n+1)$. On a $\{S_n - 1\} \in TM$. On trouve que :

$$\varepsilon_{2k}^{(n)} = \frac{S_{n+k} + k}{k+1} \quad \text{d'où} \quad \lim_{n\to\infty} \frac{\varepsilon_{2k+2}^{(n)} - 1}{\varepsilon_{2k}^{(n)} - 1} = \frac{k+1}{k+2} \quad \forall k.$$

Lemme 17 bis : Si on applique l'ε-algorithme à une suite $\{S_n\} \in TO$ alors :

$$(-1)^n \varepsilon_{2k+2}^{(n)} \leq (-1)^n \varepsilon_{2k}^{(n+2)}$$

Démonstration : la règle de la croix (propriété 17) s'écrit :

$$(N-C)^{-1} + (S-C)^{-1} = (W-C)^{-1} + (E-C)^{-1}$$

ou encore :

$$(W-N)(E-C)(S-C) = (S-E)(N-C)(W-C).$$

Les différences N-C, W-C, E-C et S-C sont toutes les quatre de même signe. Il en est donc de même de W-N et de S-E. Or pour k=0, on a :

$$(-1)^n (\varepsilon_2^{(n)} - S_{n+2}) = (-1)^{n+1} (\Delta S_{n+1})^2/\Delta^2 S_n \leq 0$$

ce qui démontre le lemme.

Théorème 51 bis : Si on applique l'ε-algorithme à une suite $\{S_n\}$ qui converge vers S et s'il existe deux constantes $a \neq 0$ et b telles que $\{aS_n + b\} \in TO$ alors :

$$\lim_{k\to\infty} \varepsilon_{2k}^{(n)} = S \quad \text{pour} \quad n=0,1,\ldots$$

Démonstration : Supposons que $\{S_n\} \in TO$; pour n fixé la suite $\{(-1)^n \varepsilon_{2k}^{(n)}\}$ est décroissante et bornée inférieurement d'après le lemme 15. Elle converge donc. Posons $T^{(n)} = \lim_{k\to\infty} \varepsilon_{2k}^{(n)}$. On a $(-1)^n(T^{(n)} - \varepsilon_{2k}^{(n)}) \leq 0$ et donc $\lim_{n\to\infty} T^{(n)} = 0$.

D'autre part, d'après le lemme 17 bis et en faisant tendre k vers l'infini $0 \leq (-1)^n T^{(n)} \leq (-1)^n T^{(n+2)}$. Ceci n'est possible que si $T^{(n)} = 0$ pour tout n. La propriété 14 complète immédiatement la démonstration de ce théorème lorsque c'est la suite $\{aS_n + b\}$ qui est TO.

Nous allons maintenant démontrer un résultat d'accélération de la convergence. Nous ne ferons la démonstration que pour des suites TM ; le cas des suites TO se traite de façon analogue. Si $\{S_n\} \in TM$ alors $H_2^{(n)}(S_n - S) \geq 0$. Par conséquent si l'on suppose que $S_n \neq S$, $\forall$ n (dans le cas contraire on ne pourrait pas appliquer l'ε-algorithme) alors :

$$0 < \frac{S_1 - S_0}{S_0 - S} \leq \frac{S_2 - S}{S_1 - S} \leq \ldots \leq 1$$

Donc $\exists$ a $\in$]0,1[tel que :

$$S_{n+1} - S = (a+e_n)(S_n - S)$$

avec $\lim_{n\to\infty} e_n = 0$. Si l'on suppose que a $\neq$ 1 alors, d'après le théorème 32 [110] :

$$\varepsilon_2^{(n)} - S = o(S_{n+2} - S) \quad \text{pour} \quad n \to \infty$$

Par conséquent, d'après le lemme 9 bis, on a :

$$0 \leq \varepsilon_{2k}^{(n)} - S \leq \varepsilon_{2k-2}^{(n+2)} - S \leq \ldots \leq \varepsilon_2^{(n+2k-2)} - S \leq S_{n+2k} - S$$

D'où le :

<u>Théorème 51 ter</u> : Si on applique l'ε-algorithme à une suite $\{S_n\}$ qui converge vers S, si $S_n \neq S$ $\forall$ n, s'il existe deux constante a $\neq$ 0 et b telles que $\{aS_n+b\} \in$ TM ou TO et si $\lim_{n\to\infty} (S_{n+1}-S)/(S_n-S) \neq 1$ (toujours vrai dans le cas TO) alors :

$$\varepsilon_{2k}^{(n)} - S = o\,(S_{n+2k} - S) \quad \text{pour k fixé et } n \to \infty$$

$$\varepsilon_{2k}^{(n)} - S = o\,(S_{n+2k} - S) \quad \text{pour n fixé et } k \to \infty.$$

Toutes ces inégalités et ces théorèmes de convergence et d'accélération de la convergence peuvent s'exprimer en termes d'approximants de Padé. Ils sont à rapprocher des résultats obtenus par Wynn [208] pour les séries de Stieltjes de la forme :

$$f(x) = \sum_{i=0}^{\infty} (-1)^i c_i x^i$$

où $\{c_n\} \in$ TM. Soit R le rayon de convergence de cette série. On a alors :

$$f(x) = \int_0^{1/R} \frac{dg(t)}{1+xt} \geq 0 \quad \text{si} \quad c_n = \int_0^{1/R} t^n \, dg(t)$$

Les résultats précédents se transposent facilement en terme de série et de table de Padé et l'on a :

$$\forall\, x \in [0,R[\text{ et pour } n,k = 0,1,\ldots$$

$$0 \leq (-1)^n \left([n+k+1/k+1]_f(x) - f(x)\right) \leq (-1)^n \left([n+k/k]_f(x) - f(x)\right)$$

$$0 \leq (-1)^n \left([n+k+1/k+1]_f(x) - f(x)\right) \leq (-1)^n \left([n+k+2/k]_f(x) - f(x)\right)$$

$$\forall\, x \in [0,R[\text{ et pour } k = 0,1,\ldots \text{ et } n = -1,0,1,\ldots$$

$$0 \leq (-1)^n \left(f(x) - [k+1/n+k+2]_f(x)\right) \leq (-1)^n \left(f(x) - [k/n+k+1]_f(x)\right)$$

$$0 \leq (-1)^n \left(f(x) - [k+1/n+k+2]_f(x)\right) \leq (-1)^n \left(f(x) - [k/n+k+3]_f(x)\right)$$

$$\forall\, x \in [0,R[\text{ et pour } k = 0,1,\ldots$$

$$[k/k]_f(x) \geq f(x) \geq [k-1/k]_f(x)$$

$$[k/k+1]_f(x) \geq [k+1/k]_f(x)$$

$$\forall\, x \in]-R,0] \text{ et pour } n,k = 0,1,\ldots$$

$$0 \leq f(x) - [n+k+1/k+1]_f(x) \leq f(x) - [n+k/k]_f(x)$$

$$0 \leq f(x) - [n+k+1/k+1]_f(x) \leq f(x) - [n+k+2/k]_f(x)$$

$$0 \leq f(x) - [n+k+1/k]_f(x) \leq f(x) - [n+k/k]_f(x)$$

$$0 \leq f(x) - [n+k+1/k+1]_f(x) \leq f(x) - [n+k+1/k]_f(x)$$

$$0 \leq f(x) - [k+1/n+k+2]_f(x) \leq f(x) - [k/n+k+1]_f(x)$$

$$0 \leq f(x) - [k+1/n+k+1]_f(x) \leq f(x) - [k/n+k+2]_f(x)$$

$$0 \leq f(x) - [k/n+k+2]_f(x) \leq f(x) - [k/n+k+1]_f(x)$$

$$0 \le f(x) - [k+1/n+k+2]_f(x) \le f(x) - [k/n+k+2]_f(x)$$

$\forall\, x \in]-R,0]$ et pour $k = 0,1,\dots$

$$0 \le f(x) - [k/k+1]_f(x) \le f(x) - [k/k]_f(x)$$

$$0 \le f(x) - [k+1/k+1]_f(x) \le f(x) - [k/k+1]_f(x)$$

$$0 \le f(x) - [k+1/k+1]_f(x) \le f(x) - [k/k+2]_f(x)$$

$$0 \le f(x) - [k+1/k]_f(x) \le f(x) - [k/k+1]_f(x)$$

En plus $\forall\, x \in]-R,R[$ et pour $n,k = 0,1,\dots$ on a :

$$[n+k/k]_f(x) \ge 0$$

$$[k/n+k]_f(x) \ge 0$$

On démontre également que les pôles de $[n+k/k]_f(x)$ appartiennent à $(-\infty,-R]$ pour $n = -1,0,1,\dots$ et $k = 0,1,\dots$ et que les pôles de $[k/n+k]_f(x)$ n'appartiennent pas à $]-R,R[$ pour $n=2,3,\dots$ et $k=0,1,\dots$

$\forall\, x \in]-R,R[$ on a :

$$\lim_{n\to\infty} [n+k/k]_f(x) = \lim_{n\to\infty} [k/n+k]_f(x) = f(x) \qquad k=0,1,\dots$$

$$\lim_{k\to\infty} [n+k/k]_f(x) = \lim_{k\to\infty} [k/n+k]_f(x) = f(x) \qquad n=0,1,\dots$$

L'étude des approximants de Padé pour les séries de Stieltjes est fortement liée à la théorie des polynômes orthogonaux [6]. Sur ces questions on pourra également consulter [205]. On trouvera d'autres résultats de convergence dans [117,209].

Etudions maintenant les théorèmes de convergence qui portent sur la loi de formation des termes de la suite [41]. On a :

Théorème 52 :

- si $S_n = f(S_{n-1}, \ldots, S_{n-k})$ pour $n = 0, 1, \ldots$ et $S_{-1}, \ldots, S_{-k}$ étant donnés
- si $\lim_{n\to\infty} S_n = S$
- si f est une fois dérivable par rapport à chacun de ses arguments et si la somme de ces dérivées partielles en $(S, \ldots, S) \in \mathbb{R}^k$ est différente de un alors :

$$\lim_{n\to\infty} \varepsilon_{2k}^{(n)} = S$$

De plus si $S_n - S = O(S_{n+1} - S)$ alors :

$$\lim_{n\to\infty} \frac{\varepsilon_{2k}^{(n)} - S}{S_{n+2k} - S} = 0$$

Démonstration : en utilisant le développement de Taylor d'une fonction de plusieurs variables et le fait que $S = f(S, \ldots, S)$, on a :

$$S_n - S = a_1(S_{n-1} - S) + \ldots + a_k(S_{n-k} - S) + R_n \qquad (1)$$

où a_i est la dérivée partielle de f par rapport à sa $i^{\text{ème}}$ variable en $(S, \ldots, S) \in \mathbb{R}^k$ et avec $\lim_{n\to\infty} R_n = 0$. D'après l'interprétation de l'ε-algorithme donnée au paragraphe III - 4 on sait qu'il revient à chercher $\varepsilon_{2k}^{(n-k)}$ et $b_1^{(n)}, \ldots, b_k^{(n)}$ tels que :

$$S_p - \varepsilon_{2k}^{(n-k)} = b_1^{(n)} (S_{p-1} - \varepsilon_{2k}^{(n-k)}) + \ldots + b_k^{(n)} (S_{p-k} - \varepsilon_{2k}^{(n-k)}) \qquad (2)$$

pour $p = n, \ldots, n+k$. On sait également que l'application

$$(S_{n-k}, \ldots, S_{n+k}) \to (b_1^{(n)}, \ldots, b_k^{(n)})$$

est continue. Par conséquent :

$$\forall \varepsilon > 0 \ \exists N : \forall n > N \quad |b_i^{(n)} - a_i| < \varepsilon \quad \text{pour } i = 1, \ldots, k.$$

Dans (2) faisons $p = n$ et soustrayons (1) de (2) ; il vient :

$$S_n - S - (\varepsilon_{2k}^{(n-k)} - S) = \sum_{i=1}^{k} b_i^{(n)} \; [S_{n-i} - S - (\varepsilon_{2k}^{(n-k)} - S)]$$

$$S_n - S = \sum_{i=1}^{k} a_i (S_{n-i} - S) + R_n$$

$$S - \varepsilon_{2k}^{(n-k)} = \sum_{i=1}^{k} (b_i^{(n)} - a_i)(S_{n-i} - S) - (\varepsilon_{2k}^{(n-k)} - S) \sum_{i=1}^{k} b_i^{(n)} - R_n$$

d'où :

$$(S - \varepsilon_{2k}^{(n-k)})(1 - \sum_{i=1}^{k} b_i^{(n)}) = \sum_{i=1}^{k} (b_i^{(n)} - a_i)(S_{n-i} - S) - R_n$$

donc puisque $\sum_{i=1}^{k} a_i \neq 1$ on a $\lim_{n\to\infty} \varepsilon_{2k}^{(n-k)} = S$ ce qui démontre la première partie du théorème.

De plus si $S_{n-k} - S = O(S_p - S)$ pour $p = n-k, \ldots, n+k$ on a :

$$R_n = \phi(S_{n-1} - S, \ldots, S_{n-k} - S) \sum_{i=1}^{k} |S_{n-i} - S|$$

d'où $\varepsilon_{2k}^{(n-k)} - S = o(S_{n+k} - S)$ lorsque n tend vers l'infini ce qui termine la démonstration.

REMARQUE : si f est une fonction affine de chacun de ses arguments alors $R_n = 0 \ \forall n$ et l'on est ramené au théorème 35 : $\varepsilon_{2k}^{(n)} = S \ \forall n$. Ce théorème est donc une généralisation du théorème 35 ainsi que d'un théorème démontré par Henrici[110] pour le procédé Δ^2 d'Aitken. On retrouve le résultat d'Henrici en faisant k = 1 dans le théorème 52.

Exemple :

$$S_{-2} = 1, \ S_{-1} = 0 \text{ et}$$

$$S_n = \exp\left[-(S_{n-1} + S_{n-2}) / 2\right]$$

On a $S = \lim_{n\to\infty} S_n = 0.56714329 \ldots$

On est dans les conditions d'application du théorème précédent avec k = 2. On obtient :

$\varepsilon_0^{(n)}$	$\varepsilon_2^{(n)}$	$\varepsilon_4^{(n)}$
1		
0	0,377...	
0,606...	0,775...	0,571...
0,738...	0,654...	0,570...
0,510...	0,533...	0,5678...

$\varepsilon_0^{(n)}$	$\varepsilon_2^{(n)}$	$\varepsilon_4^{(n)}$
0,535...	0,490...	0,5673...
0,592...	0,575...	0,5672...
0,568...	0,553...	0,56715...
0,559...	0,564...	0,56715...
0,568...	0,568...	0,567144...
0,568...	0,568...	0,5671436...
0,566...	0,566...	
0,567...		

On voit que les suites $\{\varepsilon_2^{(n)}\}$ et $\{\varepsilon_4^{(n)}\}$ convergent bien vers S mais que seule $\{\varepsilon_4^{(n)}\}$ converge plus vite que $\{S_{n+4}\}$.

Théorème 53 :

- si $\sum_{i=0}^{k} a_i(n)\ (S_{n+i} - S) = x_n$ pour $n = 0, 1, \ldots$

- si $\lim_{n\to\infty} a_i(n) = a_i$ pour $i = 0, \ldots, k$

- si $\sum_{i=0}^{k} a_i \neq 0$

- si $\lim_{n\to\infty} x_n = 0$

alors $\lim_{n\to\infty} \varepsilon_{2k}^{(n)} = S$

Démonstration : le raisonnement est analogue à celui du théorème précédent. On a :

$$\sum_{i=0}^{k} b_i^{(n)}\ (S_{p+i} - \varepsilon_{2k}^{(n)}) = 0 \qquad \text{pour } p = n, \ldots, n+k$$

ou encore, pour $p = n$:

$$\sum_{i=0}^{k} b_i^{(n)}\ [S_{n+i} - S - (\varepsilon_{2k}^{(n)} - S)] = 0$$

d'où :

$$(\varepsilon_{2k}^{(n)} - S) \sum_{i=0}^{k} b_i^{(n)} = x_n + \sum_{i=0}^{k} (b_i^{(n)} - a_i(n))(S_{n+i} - S)$$

Or $\lim_{n\to\infty} b_i^{(n)} = a_i$ à cause de la continuité de l'application

$$(S_n, \ldots, S_{n+2k}) \to (b_1^{(n)}, \ldots, b_k^{(n)})$$

Puisque $\sum_{i=0}^{k} a_i \neq 0$ on a bien $\lim_{n\to\infty} \varepsilon_{2k}^{(n)} = S$.

Exemple : $S_1 = 1$ et $S_n = e^{-n} + S_{n-1} / (n-1)$ pour $n = 2, \ldots$

On est dans les conditions d'application du théorème précédent avec $k = 1$ et $S = 0$.

On obtient

$\varepsilon_0^{(n)}$	$\varepsilon_2^{(n)}$
1	
1.13	1.02
0.61	-1.01
0.22	$-0.49\ 10^{-1}$
$0.62\ 10^{-1}$	$-0.50\ 10^{-2}$
$0.15\ 10^{-1}$	$-0.31\ 10^{-3}$
$0.34\ 10^{-2}$	0;78 10^{-4}
$0.82\ 10^{-3}$	

Nous terminerons ce paragraphe par deux théorèmes de convergence des colonnes de l'ε-algorithme qui nécessitent la connaissance du comportement des colonnes précédentes. Le premier théorème a été donné dans [22]. Il apparait comme une généralisation d'un résultat dû à Marx [138] et à Tucker [176].

<u>Théorème 54</u> :

Si $\lim_{n\to\infty} \varepsilon_{2k}^{(n)} = S$, si $\Delta\varepsilon_{2k-1}^{(n)} \leq 0$, $\forall\, n > N$

et si $\Delta\varepsilon_{2k}^{(n)} \leq \Delta\varepsilon_{2k}^{(n+1)} \leq 0$, $\forall\, n > N$ et pour k fixé alors il existe une sous-suite de $\{\varepsilon_{2k+2}^{(n)}\}$ qui converge vers S.

Démonstration :

On considère la série de terme général u_n qui est équivalente à la suite $\{\varepsilon_{2k}^{(n)}\}$ pour k fixé et qui est donnée par :

$$\varepsilon_{2k}^{(n)} = \sum_{i=0}^{n} u_i$$

avec $u_0 = \varepsilon_{2k}^{(0)}$ et $u_{p+1} = \Delta\varepsilon_{2k}^{(p)}$ pour $p = 0,1,\ldots$

Cette série est convergente d'après la première hypothèse. La troisième condition du théorème s'écrit :

$$u_n \leq u_{n+1} \leq 0 \quad \forall\, n > N$$

Nous allons montrer que cette condition entraîne que la suite

$$\left\{ \frac{1}{u_{n+1}} - \frac{1}{u_n} \right\} \qquad n \geq M \geq N$$

n'est pas bornée inférieurement pour tout $M \geq N$.

Pour cela posons $v_n = 1/u_n$ et supposons $\exists A$ fini tel que

$$A < v_{n+1} - v_n \leq 0 \qquad \forall\, n \geq M \geq N$$

On a :

$$A + v_n < v_{n+1} \leq 0$$

ou encore $$A + v_n < \frac{1}{u_{n+1}} \leq 0$$

$$u_{n+1} < \frac{1}{A+v_n} \leq 0$$

et de même $$A + v_{n+1} < v_{n+2} \leq 0$$

$$2A + v_n < A + v_{n+1}$$

d'où $$u_{n+2} < \frac{1}{2A+v_n} \leq 0$$

donc $\sum_{k=1}^{\infty} u_{n+k} < \sum_{k=1}^{\infty} \frac{1}{kA+v_n} \leq 0$, $\forall\, n \geq M \geq N$

Puisque la série $\sum_{k=1}^{\infty} 1/(kA+v_n)$ diverge il en est de même de $\sum_{k=1}^{\infty} u_{n+k}$ ce qui est contraire aux hypothèses. Par conséquent $A = -\infty$. $\forall\, M$.

Or $\Delta\varepsilon_{2k+1}^{(n)} = \Delta\varepsilon_{2k-1}^{(n+1)} + \frac{1}{\Delta\varepsilon_{2k}^{(n+1)}} - \frac{1}{\Delta\varepsilon_{2k}^{(n)}} \leq 0.$

La suite $\{\Delta\varepsilon_{2k+1}^{(n)}\}$ pour $n \geq M$ n'est donc pas bornée inférieurement $\forall\, M$ et, par conséquent, il existe une sous-suite de $\{\varepsilon_{2k+2}^{(n)}\}$ qui converge vers S.

Remarque : si $\lim_{n\to\infty} \varepsilon_{2k-2}^{(n)} = S$ alors $\lim_{n\to\infty} \Delta\varepsilon_{2k-1}^{(n)} = -\infty$ et donc, sous les mêmes hypothèses, $\lim_{n\to\infty} \Delta\varepsilon_{2k+1}^{(n)} = -\infty$ et $\lim_{n\to\infty} \varepsilon_{2k+2}^{(n)} = S$.

Donnons maintenant un dernier résultat qui généralise un théorème dû à Pennacchi [150].

Si l'on tient compte du théorème 36 alors la relation de l'ε-algorithme peut s'écrire :

$$e_k(S_n) = e_{k-1}(S_{n+1}) + \frac{e_{k-1}(\Delta S_n)\, e_{k-1}(\Delta S_{n+1})}{e_{k-1}(\Delta S_n) - e_{k-1}(\Delta S_{n+1})}$$

d'où immédiatement le :

Théorème 55 :

Si on applique l'ε-algorithme à une suite $\{S_n\}$ qui converge vers S, si $\lim_{n\to\infty} \varepsilon_{2k}^{(n)} = S$, si $\lim_{n\to\infty} 1/\varepsilon_{2k+1}^{(n)} = 0$ et si $\lim_{n\to\infty} \varepsilon_{2k+1}^{(n+1)} / \varepsilon_{2k+1}^{(n)} = a \neq 1$ alors $\lim_{n\to\infty} \varepsilon_{2k+2}^{(n)} = S$.

Démonstration : elle est évidente et laissée en exercice. Elle peut se démontrer directement à partir de la relation de l'ε-algorithme.

Pour terminer ce chapitre nous allons donner un exemple numérique pour illustrer la puissance de l'ε-algorithme. Considérons la série :

$$\text{Log }(1+x) = x - \frac{x^2}{2} + \frac{x^3}{3} - \ldots$$

Elle converge pour $-1 < x \leq 1$. Appliquons l'ε-algorithme à la suite des sommes partielles de cette série pour x fixé. On obtient :

x = 1 Log 2 = 0,6931471805599453

n	$\{S_n\}$	$\{\varepsilon_n^{(0)}\}$
2	0,83	0,7
4	0,783	0,6933
6	0,759	0,693152
8	0,745	0,69314733
10	0,736	0,6931471849
12	0,730	0,69314718068
14	0,725	0,693147180563
16	0,721	0,69314718056000
18	0,718	0,6931471805599485

x = 2 Log 3 = 1,09861228868110

n	$\{S_n\}$	$\{\varepsilon_n^{(0)}\}$
2	$0,26\ 10^1$	1,14
4	$0,506\ 10^1$	1,101
6	$0,126\ 10^2$	1,0988
8	$0,375\ 10^2$	1,098625
10	$0,121\ 10^3$	1,0986132
12	$0,410\ 10^3$	1,09861235

ce qui montre que l'ε-algorithme peut être utilisé pour sommer des séries divergentes et induire la convergence de procédés itératifs divergents.

III - 7 Application à la quadrature numérique

Une application importante de l'ε-algorithme et de certains théorèmes qui viennent d'être énoncés est l'accélération de la convergence des méthodes de quadrature numérique sur un intervalle fini. La rédaction de ce paragraphe fait largement appel au travail de Genz [84]. Considérons des intégrales de la forme :

$$I = \int_0^1 f(x)\, dx$$

ainsi qu'une méthode de quadrature approchée :

$$\bar{I} = \sum_{i=1}^{n} w_i\, f(x_i) \text{ avec } \sum_{i=1}^{n} w_i = 1$$

où les x_i et les w_i peuvent dépendre de n.

A l'aide de cette formule il est classique de construire des méthodes composites d'intégration de la façon suivante ; on écrit que :

$$I = \int_0^{x_1} f(x)\, dx + \int_{x_1}^{x_2} f(x)\, dx + \ldots + \int_{x_{m-1}}^{1} f(x)\, dx$$

puis chacune de ces intégrales est calculée de façon approchée en utilisant la formule de $\bar{I}$ correctement modifiée. Pour simplifier nous supposerons que tous les intervalles $[x_i, x_{i+1}]$ sont égaux. La méthode des trapèzes appartient à cette classe de formules composites lorsque l'on prend :

$$\bar{I} = \frac{1}{2}\,(f(0) + f(1)).$$

Nous appelerons I_m la valeur approchée de I obtenue à l'aide d'une méthode composite comprenant m sous-intervalles. On a :

$$I_m = h \sum_{k=0}^{m-1} \sum_{i=1}^{n} w_i \, f((x_i+k)h) \qquad \text{avec } h = 1/m.$$

On démontre (voir par exemple [69]) que, sous des conditions assez faibles sur f :

$$\lim_{m\to\infty} I_m = I$$

De plus si f est analytique dans [0,1] alors une simple généralisation de la formule d'Euler-Madaurin nous donne [136] :

$$I_m = I + a_1 h + a_2 h^2 + \ldots + a_k h^k + 0(h^{k+1})$$

Il est évident que si l'on peut éliminer les termes de plus bas degré en h dans ce développement limité de I_m alors on obtiendra des approximations de I qui seront meilleures. C'est ce que fait implicitement la méthode de Romberg qui a été étudiée et illustrée au paragraphe II-3.

A la place de la méthode de Romberg, on peut appliquer l'ε-algorithme à la suite $\{S_n = I_{2^n}\}$ [56,57,58]. on aura alors [84] :

$$S_n = I + a_1\lambda_1^n + a_2\lambda_2^n + \ldots + a_k\lambda_k^n + 0(\lambda_{k+1}^n)$$

avec $\lambda_i = 2^{-i}$. On voit que si l'on ne tient pas compte du terme en $0(\lambda_{k+1}^n)$ alors on est dans les conditions d'applications des théorèmes 35 et 37 ; $\varepsilon_{2k}^{(n)}$ sera donc vraisemblablement une très bonne approximation de I ($\forall$ n). Si k est grand et si l'on ne peut pas disposer des 2k+1 premiers éléments de la suite $\{S_n\}$ nécessaires au calcul de $\varepsilon_{2k}^{(0)}$ alors on se contentera d'une valeur intermédiaire $\varepsilon_{2i}^{(0)}$ $(i < k)$ qui devrait cependant fournir une bonne approximation de I d'après le théorème 46. On s'aperçoit dans la pratique [59] que la méthode de Romberg fournit de meilleurs résultats que l'ε-algorithme lorsque f est continue dans [0,1]. On verra au paragraphe IV-3 une méthode qui donne de meilleurs résultats que la méthode de Romberg. Sur les procédés linéaires d'extrapolations on pourra consulter l'article de synthèse de Joyce [118].

Venons-en maintenant au cas où f présente, à l'une des bornes de l'intervalle d'intégration, une singularité de la forme :

$$f(x) = x^{\alpha}(1-x)^{\beta}\ g(x)$$

avec g analytique dans [0,1] et α et β non entiers.
Alors on sait que [136] :

$$I_m = I + a_1 h^{1+\alpha} + a_2 h^{2+\alpha} + \ldots + a_k h^{k+\alpha} + 0(h^{k+\alpha+1})$$

$$+ b_1 h^{1+\beta} + b_2 h^{2+\beta} + \ldots + b_k h^{k+\beta} + 0(h^{k+\beta+1})$$

Si l'on veut appliquer la méthode de Romberg il faut traiter séparément les termes en α et ceux en β [80].

Par contre si nous considérons la suite $\{S_n = I_{2^n}\}$ alors, en négligeant les termes en 0 :

$$S_n = I + a_1\lambda_1^n + \ldots + a_k\lambda_k^n + b_1\delta_1^n + \ldots + b_k\delta_k^n$$

(avec $\lambda_i = 2^{-(\alpha+i)}$ et $\delta_i = 2^{-(\beta+i)}$). L'application de l'$\varepsilon$-algorithme à cette suite $\{S_n\}$ fournira donc, comme précédemment, de très bonnes approximations de I. Pour des singularités logarithmiques on obtient des développements limités analogues et l'ε-algorithme est également très efficace [119].

Si maintenant f possède une singularité au milieu de l'intervalle avec, par exemple :

$$f(x) = (x-a)^{\alpha}\ g(x) \qquad\qquad 0 < a < 1$$

et g analytique dans [0,1] alors on a [136] :

$$I_m = I + a_1 h + \ldots + a_k h^k + 0(h^{k+1})$$

$$+ b_1^{(m)}\ h^{\alpha+1} + \ldots + b_k^{(m)}\ h^{\alpha+k} + 0(h^{k+\alpha+1})$$

avec, cette fois-ci, des coefficients $b_i^{(m)}$ qui dépendent de m de sorte que :

$$b_i^{(m)} = y_i(ma) \text{ et } y_i(ma) = y_i(ma+1)$$

Si a est un nombre rationnel il aura un développement binaire périodique ; soit p cette période, alors :

$$b_i^{(2^n)} = b_i^{(2^{n+p})}$$

Par conséquent la suite $\{S_n = I_{2^n}\}$ vérifiera :

$$S_n = I + \sum_{i=1}^{k} a_i \lambda_i^n + \sum_{i=1}^{k} c_i^{(n)} \delta_i^n$$

avec $\lambda_i = 2^{-i}$, $c_i^{(n)} = b_i^{(2^n)}$ et $\delta_i = 2^{-(\alpha+i)}$.

Genz [84] a démontré que si on applique l'ε-algorithme à une telle suite alors $\varepsilon_{2kp}^{(n)} = I \ \forall\, n$. Les quantités $\varepsilon_{2kp}^{(n)}$ seront donc de bonnes approximations de I puisque nous avons négligé les termes en 0.

Signalons, pour être complets, que la méthode de Romberg peut également s'appliquer au calcul des valeurs principales de Cauchy [114, 156].

Soit, par exemple, à calculer :

$$I = \int_0^1 \frac{dx}{x^2-0,01} = -\ 1,0033534\ \ldots$$

En utilisant pour $\bar{I}$ une formule de quadrature de Gauss-Legendre avec huit points et en appliquant l'ε-algorithme à la suite $\{S_n = I_{2^n}\}$ on obtient :

$\{\varepsilon_0^{(n)}\}$	$\{\varepsilon_2^{(n)}\}$	$\{\varepsilon_4^{(n)}\}$	$\{\varepsilon_6^{(n)}\}$
331,0			
15,2	86,7		
107,5	54,5	155,7	
-17,2	-371,5	-1,0033535	-1,0033534
-109,5	-56,5	-1,0033534	
15,2	369,4		
107,5			

Des méthodes pour calculer les intégrales avec une borne infinie ou non seront étudiées au paragraphe IV-8.

Les méthodes non linéaires d'accélération de la convergence semblent donc être très performantes pour le calcul approché des intégrales. De nombreuses études théoriques restent encore à faire sur ce sujet.

CHAPITRE IV

ETUDE DE DIVERS ALGORITHMES D'ACCELERATION DE LA CONVERGENCE

Le but de ce chapitre est d'étudier un certain nombre d'algorithmes d'accélération de la convergence. On verra que l'on est ensuite conduit, pour les algorithmes d'un certain type, à un algorithme qui en est une généralisation naturelle. Le chapitre se terminera par une brève revue des essais actuels de formalisation des méthodes d'accélération de la convergence.

IV - 1 Le procédé d'Overholt

Le but du procédé d'Overholt [148] est de fournir des approximations d'ordre de plus en plus élevé de la limite S des suites $\{S_n\}$ telles que :

$$S_{n+1} - S = \sum_k a_k (S_n - S)^k \qquad \forall n$$

où un nombre fini ou infini de coefficients a_k peuvent être nuls.

Posons $d_n = S_n - S$. Supposons que $S_o = S + d_o$ et que $S_1 = S + a_1 d_o + a_2 d_o^2 + \dots$; on obtient :

$$\frac{S_1 - a_1 S_o}{1 - a_1} = S + \frac{a_2 d_o^2}{1 - a_1} + \dots$$

qui est une approximation d'ordre 2 de S. En général a_1 est inconnu ; on en détermine une approximation du premier ordre $\bar{a}_1$ en utilisant S_2 :

$$S_2 = S + a_1 d_1 + a_2 d_1^2 + \dots$$

ce qui donne :

$$\bar{a}_1 = \frac{S_2 - S_1}{S_1 - S_o} = a_1 + (1 + a_1) a_2 d_o + \dots$$

Cette approximation est d'un ordre suffisant pour que la quantité

$$V_1^{(o)} = \frac{S_1 - \bar{a}_1 S_o}{1 - \bar{a}_1}$$

Soit une approximation du second ordre de S :

$$V_1^{(o)} = S - \frac{a_1 a_2}{1 - a_1} d_o^2 + \ldots$$

De même toute autre approximation du premier ordre de a_1 conduirait à une formule du second ordre. Cette remarque est à la base du procédé d'Overholt qui apparait comme une extension du procédé Δ^2 d'Aitken puisque $V_1^{(o)} = \varepsilon_2^{(o)}$. Cette extension est analogue à celle qui permet de passer dans la formule de Neville-Aitken pour l'extrapolation polynômiale d'un polynôme de degré k à un polynôme de degré k+1. Après avoir éliminé les termes du premier ordre en d_n on peut éliminer ceux du deuxième ordre et ainsi de suite. On obtient ainsi le procédé d'Overholt :

$$V_o^{(n)} = S_n \qquad n = 0, 1, \ldots$$

$$V_{k+1}^{(n)} = \frac{(\Delta S_{n+k})^{k+1} V_k^{(n+1)} - (\Delta S_{n+k+1})^{k+1} V_k^{(n)}}{(\Delta S_{n+k})^{k+1} - (\Delta S_{n+k+1})^{k+1}} \qquad n,k = 0,1,\ldots$$

REMARQUES :

1°) nous avons modifié les notations d'Overholt par souci d'homogénéitéavec les autres algorithmes.

2°) Cette méthode est à rapprocher de celle obtenue par Germain-Bonne [87] en prenant $x_n = \Delta S_n$ dans le procédé d'extrapolation de Richardson (voir paragraphe. II - 3)

Nous allons suivre la démarche inverse de celle qui a conduit Overholt aux règles de l'algorithme c'est-à-dire que nous allons supposer que $V_k^{(n)}$ est une approximation d'ordre k+1 de S et, en utilisant les règles de l'algorithme, nous allons montrer que $V_{k+1}^{(n)}$ est une approximation d'ordre k+2 de S.

Supposons donc que :

$$V_k^{(n)} = S + a_{kk}\, d_n^{k+1} + a_{k,\,k+1}\, d_n^{k+2} + \ldots$$

Portons dans l'algorithme ; on trouve immédiatement que :

$$V_{k+1}^{(n)} = a_{k+1,k+1}\, d_n^{k+2} + \ldots$$

avec :

$$a_{k+1,k+1} = \frac{a_1^{k+1}}{1-a_1^{k+1}} \left[(a_1 - 1)\, a_{k,k+1} - (k+1)\, a_2\, a_{kk}\right]$$

Pour ce procédé d'Overholt on a les résultats théoriques suivants [43] :

Théorème 56 :

Si $V_k^{(n)} = S + a_k(S_{n-1} - S)^{k+1}$ et si $\Delta S_{n+k} = b_k(S_n - S)$ avec $b_k \neq 0\ \forall n$ alors :

$$V_{k+1}^{(n)} = S \quad \forall n$$

La démonstration de ce résultat est évidente et est laissée en exercice.

Théorème 57 :

Une condition nécessaire et suffisante pour que $\lim_{n\to\infty} V_k^{(n)} = S\ \forall k$ pour toute suite $\{S_n\}$ qui converge vers S est qu'il existe $\alpha < 1 < \beta$ tels que :

$$\frac{\Delta S_{n+1}}{\Delta S_n} \notin [\alpha,\beta] \quad \forall n$$

démonstration : on voit que la transformation $\{V_k^{(n)}\} \to \{V_{k+1}^{(n)}\}$ est une transformation linéaire de suite à suite. On peut donc lui appliquer le théorème de Toeplitz (théorème 22, paragraphe II - 1).

Les deux dernières conditions du théorème de Toeplitz sont automatiquement vérifiées car le procédé est total et la matrice associée est bidiagonale. La première condition s'écrit :

$$\left| \frac{(\Delta S_{n+k-1})^{k+1}}{(\Delta S_{n+k-1})^{k+1} - (\Delta S_{n+k})^{k+1}} \right| + \left| \frac{(\Delta S_{n+k})^{k+1}}{(\Delta S_{n+k-1})^{k+1} - (\Delta S_{n+k})^{k+1}} \right| < M_k$$

et ceci ∀n ; d'où :

$$N_k < \left| 1 - \left(\frac{\Delta S_{n+k}}{\Delta S_{n+k-1}} \right)^{k+1} \right|$$

donc il doit exister α_k et β_k avec $\alpha_k < 1 < \beta_k$ tels que

$$\left(\frac{\Delta S_{n+k}}{\Delta S_{n+k-1}} \right)^{k+1} \notin [\alpha_k, \beta_k]$$

et par conséquent il existe $\alpha < 1 < \beta$ tels que :

$$\frac{\Delta S_{n+1}}{\Delta S_n} \notin [\alpha, \beta]$$

Le procédé d'Overholt est particulièrement bien adapté à la résolution d'une équation non linéaire. Soit en effet à résoudre $x = F(x)$ où $F : \mathbb{R} \to \mathbb{R}$. Si l'on suppose F suffisamment différentiable au voisinage de la racine x et si l'on effectue les itérations $x_{n+1} = F(x_n)$ alors on a :

$$d_{n+1} = a_1 d_n + a_2 d_n^2 + \ldots$$

avec $d_n = x_n - x$ et $a_k = f^{(k)}(x) / k\,!$

Le procédé d'Overholt permet ainsi de construire une méthode itérative d'ordre quelconque k de la façon suivante :

x_o donné

$(n+1)^{\text{ième}}$ itération $u_o = x_n$

$$u_{p+1} = F(u_p) \qquad p = 0, \ldots, k-1$$

application du procédé d'Overholt à $V_o^{(o)} = u_o, \ldots, V_o^{(k)} = u_k$

$$x_{n+1} = V_{k-1}^{(o)}$$

On peut également se contenter d'accélérer les itérations $x_{n+1} = F(x_n)$ en construisant la suite $\{V_k^{(o)}\}$.

Soit par exemple à chercher la racine unique $x = 0.5671432904\ldots$ de $x = e^{-x}$. On obtient :

$V_o^{(n)}$	$V_n^{(o)}$	
0	0.576	
0.5	0.56715	
0.566	0.5671432904	10 chiffres exacts au lieu
0.56714316		de 6

Le procédé d'Overholt n'a pas encore été étudié plus à fond. En particulier il n'y a pas de théorème de convergence de $\{V_k^{(n)}\}$ pour n fixé. D'autre part les expériences numériques avec cet algorithme sont encore rares et des applications restent à trouver.

IV - 2 Les procédés p et q

On peut considérer les procédés p et q [41] comme des modifications de la transformation de Shanks. Le procédé p est défini par le rapport de deux déterminants :

$$P_k(x_n, S_n) = - \frac{\begin{vmatrix} x_n & S_n & \cdots & S_{n+k} \\ \hline x_{n+k+1} & S_{n+k+1} & \cdot & S_{n+2k+1} \end{vmatrix}}{\begin{vmatrix} \Delta x_n & \Delta^2 S_n & \cdots & \Delta^2 S_{n+k-1} \\ \hline \Delta x_{n+k} & \Delta^2 S_{n+k} & \cdots & \Delta^2 S_{n+2k-1} \end{vmatrix}}$$

Théorème 58 :

Une condition nécessaire et suffisante pour que $P_k(x_n, S_n) = S \quad \forall n > N$ est que la suite $\{S_n\}$ vérifie :

$$\sum_{i=0}^{k} a_i (S_{n+i} - S) = x_n \quad \text{avec} \quad \sum_{i=0}^{k} a_i \neq 0 \quad \forall n > N$$

La démonstration est évidente. Elle est analogue à celle du théorème 35 pour la transformation de Shanks. Elle est laissée en exercice.

On a les propriétés suivantes :

Propriété 22 : $P_k(\Delta S_{n-1}, S_n) = \varepsilon_{2k+2}^{(n-1)}$

$$P_k(\Delta S_{n+k}, S_n) = \varepsilon_{2k+2}^{(n)}$$

$$P_k(\Delta S_n, S_{n+1}) = P_k(\Delta S_{n+k}, S_n)$$

$$P_k(ax_n \quad , cS_n + d) = cP_k(x_n, S_n) + d$$

D'autre part il est possible de trouver pour ce procédé une interprétation analogue à celle établie pour l'ε-algorithme en III - 4. On peut également démontrer un théorème analogue au théorème 53 pour l'ε-algorithme :

Théorème 59 :

- Si $\sum_{i=0}^{k} a_i(n)\,(S_{n+i} - S) = x_n$ pour $n = 0, 1, \ldots$

- Si $\lim_{n\to\infty} a_i(n) = a_i$ pour $i = 0, \ldots, k$

- Si $\sum_{i=0}^{k} a_i \neq 0$

alors $\lim_{n\to\infty} P_k(x_n, S_n) = S$

Pour le procédé p il n'existe pas encore d'algorithme permettant d'éviter le calcul des déterminants mis en jeu comme c'est le cas pour la transformation de Shanks à l'aide de l'ε-algorithme, ce qui en restreint les applications.

Le procédé q est défini, lui aussi, comme un rapport de deux déterminants :

$$q_k(x_n, S_n) = \frac{\begin{vmatrix} S_n & x_n & \cdots\cdots & x_{n+k} \\ \multicolumn{4}{c}{\text{------------------------}} \\ S_{n+k+1} & x_{n+k+1} & \cdots & x_{n+2k+1} \end{vmatrix}}{\begin{vmatrix} \Delta x_n & \cdots\cdots\cdots & \Delta x_{n+k} \\ \multicolumn{3}{c}{\text{--------------------}} \\ \Delta x_{n+k} & \cdots\cdots\cdots & \Delta x_{n+2k} \end{vmatrix}}$$

Théorème 60 :

Une condition nécessaire et suffisante pour que $q_k(x_n, S_n) = S \ \forall n > N$ est que la suite $\{S_n\}$ vérifie :

$$S_n - S = \sum_{i=0}^{k} a_i \, x_{n+i} \qquad \forall n > N$$

La démonstration est laissée en exercice.

On a la :

Propriété 23 : $q_0(x_n, S_n) = p_0(x_n, S_n)$

$$q_k(ax_n \quad , cS_n + d) = cq_k(x_n, S_n) + d$$

$$q_k(\Delta S_n, S_n) = \varepsilon_{2k+2}^{(n)}$$

Théorème 61 :

- Si $S_n - S = \sum_{i=0}^{k} a_i(n) \, x_{n+i}$ pour $n = 0, 1, \ldots$

- Si $\lim_{n\to\infty} a_i(n) = a_i$ pour $i = 0, \ldots, k$

alors $\lim_{n\to\infty} q_k(x_n, S_n) = S$

Le procédé q est à rapprocher d'un procédé d'accélération de la convergence dont nous ne parlerons pas ici : la transformation G [104]. En effet soit $\{S_n\}$ une suite telle que $S_n = f(y_n)$ avec $\lim_{n\to\infty} y_n = \infty$. Si dans le procédé q on prend $x_n = f'(y_n)$ on retrouve alors exactement la transformation G. Un algorithme pour éviter le calcul des déterminants qui interviennent dans le procédé q peut se déduire de celui de la transformation G [159] et [153]. Il existe également une méthode due à P. Barrucand pour mettre en oeuvre le procédé p.

IV - 3 Le ρ-algorithme

Nous avons vu au paragraphe II - 3 que le procédé de Richardson consistait à faire passer un polynôme d'interpolation de degré k par les k+1 couples $(x_n, S_n), \ldots, (x_{n+k}, S_{n+k})$ à l'aide de la formule de Neville-Aitken puis à calculer la valeur de ce polynôme en $x = 0$. Le ρ-algorithme consiste à faire passer une fraction rationnelle d'interpolation dont numérateur et dénominateur sont des polynômes de degré k par les $2k+1$ couples de points $(x_n, S_n), \ldots, (x_{n+2k}, S_{n+2k})$, à l'aide de la formule d'interpolation de Thiele (voir [141,142] par exemple) puis à calculer la valeur de cette fraction rationnelle en $x = \infty$. Définissons d'abord ce que sont les différences réciproques d'une fonction. Soit une fonction dont on connait la valeur S_n en un certain nombre de points x_n pour $n = 0, 1, \ldots$

Définition 17 : On appelle différences réciproques les quantités :

$$\rho_o^{(n)} = S_n$$

$$\rho_1^{(n)} = \frac{x_n - x_{n+1}}{\rho_o^{(n)} - \rho_o^{(n+1)}}$$

$$\rho_2^{(n)} = \frac{x_n - x_{n+2}}{\rho_1^{(n)} - \rho_1^{(n+1)}} + \rho_o^{(n+1)}$$

$$\rho_k^{(n)} = \frac{x_n - x_{n+k}}{\rho_{k-1}^{(n)} - \rho_{k-1}^{(n+1)}} + \rho_{k-2}^{(n+1)}$$

On démontre que la fraction rationnelle $R(x)$ dont numérateur et dénominateur sont des polynômes de degré k et telle que

$$R(x_p) = S_p \quad \text{pour } p = n, \ldots, n+2k$$

se met sous la forme :

$$R(x) = \frac{\rho_{2k}^{(n)} x^k + \ldots}{x^k + \ldots}$$

Par conséquent on a $\lim_{x \to \infty} R(x) = \rho_{2k}^{(n)}$ ce qui donne l'idée de prendre cette quantité $\rho_{2k}^{(n)}$ comme approximation de la limite de la suite $\{S_n\}$ lorsque n tend vers l'infini. Le calcul de $\rho_{2k}^{(n)}$ s'effectue à l'aide de la forme étendue du ρ-algorithme qui n'est autre que le calcul des différences réciproques.

$$\rho_{-1}^{(n)} = 0 \qquad \rho_o^{(n)} = S_n \qquad \text{pour } n = 0, 1, \ldots$$

$$\rho_{k+1}^{(n)} = \rho_{k-1}^{(n+1)} + \frac{x_{k+n+1} - x_n}{\rho_k^{(n+1)} - \rho_k^{(n)}} \qquad n, k = 0, 1, \ldots$$

On voit que la structure de cet algorithme est analogue à celle de l'ε-algorithme et que l'on peut construire un tableau identique au tableau ε. Les quantités d'indice inférieur impair ne sont que des calculs intermédiaires et n'ont aucune signification. On a le

Théorème 62 :

Si on applique le ρ-algorithme à une suite $\{S_n\}$ telle que :

$$S_n = \frac{Sx_n^k + a_1 x_n^{k-1} + \ldots + a_k}{x_n^k + b_1 x_n^{k-1} + \ldots + b_k}$$

alors $\rho_{2k}^{(n)} = S \ \forall n$

Les propriétés du ρ-algorithme ressemblent à celles de l'ε-algorithme. On a d'abord la :

Propriété 25 :

$$\rho_{2k}^{(n)} = \frac{\begin{vmatrix} 1 & S_n & x_n & x_n S_n & \ldots & x_n^{k-1} & x_n^{k-1} S_n & x_n^k S_n \\ \hline 1 & S_{n+2k} & x_{n+2k} & x_{n+2k} S_{n+2k} & \ldots & & & x_{n+2k}^k S_{n+2k} \end{vmatrix}}{\begin{vmatrix} 1 & S_n & x_n & x_n S_n & \ldots & x_n^{k-1} & x_n^{k-1} S_n & x_n^k \\ \hline 1 & S_{n+2k} & x_{n+2k} & x_{n+2k} S_{n+2k} & \ldots & & & x_{n+2k}^k \end{vmatrix}}$$

Les propriétés algébriques du ρ-algorithme sont les suivantes :

Propriétés 26 : Si l'application du ρ-algorithme à $\{S_n\}$ et à $\{a S_n + b\}$ fournit respectivement les quantités $\rho_k^{(n)}$ et $\bar{\rho}_k^{(n)}$ alors :

$$\bar{\rho}_{2k}^{(n)} = a\, \rho_{2k}^{(n)} + b \qquad \bar{\rho}_{2k+1}^{(n)} = \rho_{2k+1}^{(n)} / a$$

Si l'application du ρ-algorithme à $\{S_n\}$ et à $\{\frac{aS_n + b}{cS_n + d}\}$ fournit respectivement les quantités $\rho_k^{(n)}$ et $\bar{\rho}_k^{(n)}$ alors :

$$\bar{\rho}_{2k}^{(n)} = \frac{a\,\rho_{2k}^{(n)} + b}{c\,\rho_{2k}^{(n)} + d}$$

REMARQUE : Les ε et ρ-algorithmes font partie de la classe des algorithmes de losange car leur relation relie des quantités situées au quatre sommets d'un losange dans le tableau construit. Les propriétés générales de ces algorithmes ont été étudiées par Bauer [16,17] . On montre ainsi qu'il existe une liaison entre l'ε-algorithme et l'algorithme qd de Rutishauser [164]. Un résumé de ceci se trouve au chapitre VII.

Wynn [221] a montré que l'on pouvait aussi considérer les algorithmes de losange comme des approximations par les différences finies du premier ordre de systèmes d'équations aux dérivées partielles du premier ordre par rapport à deux variables indépendantes Des propriétés des solutions de ces équations aux dérivées partielles ont été obtenues par Wynn [202,204] mais, pour l'instant, aucune application n'a pu leur être trouvée. Le ρ-algorithme a été utilisé pour la première fois comme transformation de suite à suite et pour accélérer la convergence par Wynn [241] mais en se restreignant au choix $x_n = n\ \forall n$.

Dans certains cas il est préférable d'utiliser un autre choix pour les abscisses x_n. Ainsi, si l'on applique le ρ-algorithme à la suite des approximations de la valeur d'une intégrale à l'aide de la formule des trapèzes avec des pas $h_n = H/2^n$, il semble judicieux de prendre $x_n = 1/h_n^2$. On obtient ainsi une méthode que l'on peut mettre en parallèle avec la méthode de Romberg :

$$\rho_{k+1}^{(n)} = \rho_{k-1}^{(n+1)} + \frac{2^{2n}\,(2^{2k+2} - 1)}{\rho_k^{(n+1)} - \rho_k^{(n)}} \qquad n, k = 0, 1, \ldots$$

cet algorithme a été proposé par Brezinski [23]. En utilisant un résultat de Gragg [91] on montre que l'erreur est identique à celle faite par la méthode de Romberg mais cependant avec un certain avantage pour le ρ-algorithme. Reprenons l'exemple traité en II - 3, on obtient :

$\rho_2^{(0)} = 5.58$

$\rho_2^{(1)} = 4.89$ $\quad \rho_4^{(0)} = 4.65$

$\rho_2^{(2)} = 4.67$ $\quad \rho_4^{(1)} = 4.6199$ $\quad \rho_6^{(0)} = 4.61537$

$\rho_2^{(3)} = 4.62$ $\quad \rho_4^{(2)} = 4.6154$ $\quad \rho_6^{(1)} = 4.6151273$ $\quad \rho_8^{(0)} = 4.615120586$

$\rho_2^{(4)} = 4.6157$ $\quad \rho_4^{(3)} = 4.6151293$ $\quad \rho_6^{(2)} = 4.615120593$

$\rho_2^{(5)} = 4.615155$ $\quad \rho_4^{(4)} = 4.6151206$

$\rho_2^{(6)} = 4.6151212$

On voit que les résultats obtenus sont meilleurs que ceux donnés par la méthode de Romberg. On trouvera d'autres exemples numériques dans [22]. L'application de méthodes d'accélération de la convergence a des formules de quadrature est un sujet qui a suscité de nombreuses études (paragraphe III-7).

Wynn [224,239] a proposé de nouveaux algorithmes d'accélération de la convergence spécialement adaptés au calcul d'intégrales. Nous les étudierons au paragraphe VI-8.

REMARQUES :

1°) Le ρ-algorithme est un algorithme d'extrapolation par une fraction rationnelle dont numérateur et dénominateur ont même degré. On peut le considérer comme un cas particulie d'une méthode due à Bulirsch et Stoer [50] où les degrés du numérateur et du dénominateu sont quelconques. Ces auteurs ont également appliqué leur méthode à la quadrature numérique [51].

2°) Sur l'extrapolation rationnelle on pourra aussi consulter les références [127,189].

3°) Le ρ-algorithme est peu utilisé en pratique sans doute à cause du manque de théorème de convergence le concernant.

IV - 4 Généralisations de l'ε-algorithme

La seule différence entre les règles de l'ε-algorithme et du ρ-algorithme est l'introduction d'une suite de paramètres $\{x_n\}$. Il est donc tentant d'essayer d'introduire également des paramètres dans l'ε-algorithme. Ceci a été effectué de deux façons qui aboutissent à deux généralisations de l'ε-algorithme [41].

La première généralisation est la suivante :

$$\varepsilon_{-1}^{(n)} = 0 \qquad \varepsilon_0^{(n)} = S_n \qquad \text{pour } n = 0, 1, \ldots$$

$$\varepsilon_{k+1}^{(n)} = \varepsilon_{k-1}^{(n+1)} + \frac{\Delta x_n}{\varepsilon_k^{(n+1)} - \varepsilon_k^{(n)}} \qquad n, k = 0, 1, \ldots$$

Nous allons donner quelques résultats théoriques sur cet algorithme.

Définition 18 : Soit f une fonction linéaire de $S_{p_1}, \ldots, S_{p_n}$ et posons $r_n = \Delta S_n / \Delta x_n$.

L'opérateur R est défini par :

$$Rf\,(S_{p_1}, \ldots, S_{p_n}) = f(r_{p_1}, \ldots, r_{p_n})$$

On voit que R est une généralisation de l'opérateur Δ que l'on retrouve si $\Delta x_n = 1$.

On peut par conséquent définir les puissances successives de l'opérateur R. On a :

Propriété 27 :

$$Rc = 0 \qquad c = \text{constante}$$

$$R(af + bg) = a\,Rf + b\,Rg$$

posons $v_n = S_n \,.\, \Delta x_n$

$$R^{k+1} v_n = \Delta\left(\frac{R^k v_n}{\Delta x_n}\right) = R^k \quad \left[\Delta\left(\frac{v_n}{\Delta x_n}\right)\right]$$

La démonstration de ces propriétés est laissée en exercice. On peut démontrer que les quantités $\varepsilon_{2k}^{(n)}$ que nous poserons égales à $e_k(S_n)$, comme pour la transformation de Shanks, sont égales à un rapport de deux déterminants ainsi que les quantités $\varepsilon_{2k+1}^{(n)}$. La démonstration est calquée sur celle de Wynn pour montrer l'identité entre l'ε-algorithme et la transformation de Shanks (théorème 36). Nous n'en donnerons que les grandes lignes :

Théorème 63 :

On a :

$$e_k(S_n)=\varepsilon_{2k}^{(n)} = \frac{\begin{vmatrix} v_n & \cdots\cdots & v_{n+k} \\ Rv_n & \cdots\cdots & Rv_{n+k} \\ \cdots & \cdots & \cdots \\ R^k v_n & \cdots\cdots & R^k v_{n+k} \end{vmatrix}}{\begin{vmatrix} \Delta x_n & \cdots\cdots & \Delta x_{n+k} \\ Rv_n & \cdots\cdots & Rv_{n+k} \\ \cdots & \cdots & \cdots \\ R^k v_n & \cdots\cdots & R^k v_{n+k} \end{vmatrix}}$$

et

$$\varepsilon_{2k+1}^{(n)} = \frac{1}{e_k(r_n)} = \frac{\begin{vmatrix} \Delta x_n & \cdots\cdots & \Delta x_{n+k} \\ R^2 v_n & \cdots\cdots & R^2 v_{n+k} \\ \cdots & \cdots & \cdots \\ R^{k+1} v_n & \cdots & R^{k+1} v_{n+k} \end{vmatrix}}{\begin{vmatrix} Rv_n & \cdots\cdots & Rv_{n+k} \\ \cdots & \cdots & \cdots \\ R^{k+1} v_n & \cdots & R^{k+1} v_{n+k} \end{vmatrix}}$$

démonstration : montrons que la relation :

$$\varepsilon_{2k+1}^{(n)} = \varepsilon_{2k-1}^{(n+1)} + \frac{\Delta x_n}{\varepsilon_{2k}^{(n+1)} - \varepsilon_{2k}^{(n)}}$$

est vérifiée. On a :

$$\varepsilon_{2k+1}^{(n)} = \frac{\begin{vmatrix} 1 & \cdots\cdots\cdots & 1 \\ R^2 v_n/\Delta x_n & \cdots & R^2 v_{n+k}/\Delta x_{n+k} \\ \cdots & \cdots & \cdots \\ R^{k+1} v_n/\Delta x_n & \cdots & R^{k+1} v_{n+k}/\Delta x_{n+k} \end{vmatrix}}{\begin{vmatrix} Rv_n/\Delta x_n & \cdots & Rv_{n+k}/\Delta x_{n+k} \\ \cdots & \cdots & \cdots \\ R^{k+1} v_n/\Delta x_n & \cdots & R^{k+1} v_{n+k}/\Delta x_{n+k} \end{vmatrix}}$$

D'où en utilisant un développement de Schweins :

$$\frac{\varepsilon_{2k+1}^{(n)} - \varepsilon_{2k-1}^{(n+1)}}{\Delta x_n} = \frac{\begin{vmatrix} 1 & \cdots & 1 \\ Rv_n/\Delta x_n & \cdots & Rv_{n+k}/\Delta x_{n+k} \\ \vdots & & \vdots \\ R^k v_n/\Delta x_n & \cdots & R^k v_{n+k}/\Delta x_{n+k} \end{vmatrix} \begin{vmatrix} R^2 v_{n+1} & \cdots & R^2 v_{n+k} \\ \vdots & & \vdots \\ R^{k+1} v_{n+1} & \cdots & R^{k+1} v_{n+k} \end{vmatrix}}{\begin{vmatrix} Rv_n & \cdots & Rv_{n+k} \\ \vdots & & \vdots \\ R^{k+1} v_n & \cdots & R^{k+1} v_{n+k} \end{vmatrix} \begin{vmatrix} Rv_{n+1}/\Delta x_{n+1} & \cdots & Rv_{n+k}/\Delta x_{n+k} \\ \vdots & & \vdots \\ R^k v_{n+1}/\Delta x_{n+1} & \cdots & R^k v_{n+k}/\Delta x_{n+k} \end{vmatrix}}$$

D'autre part on a :

$$\frac{(-1)^k}{\varepsilon_{2k}^{(n+1)} - \varepsilon_{2k}^{(n)}} = \frac{\begin{vmatrix} 1 & \cdots & 1 \\ Rv_{n+1}/\Delta x_{n+1} & \cdots & Rv_{n+k+1}/\Delta x_{n+k+1} \\ \vdots & & \vdots \\ R^k v_{n+1}/\Delta x_{n+1} & \cdots & R^k v_{n+k+1}/\Delta x_{n+k+1} \end{vmatrix} \begin{vmatrix} 1 & \cdots & 1 \\ Rv_n/\Delta x_n & \cdots & Rv_{n+k}/\Delta x_{n+k} \\ \vdots & & \vdots \\ R^k v_n/\Delta x_n & \cdots & R^k v_{n+k}/\Delta x_{n+k} \end{vmatrix}}{D}$$

avec

$$D = \begin{vmatrix} Rv_{n+1}/\Delta x_{n+1} & \cdots & Rv_{n+k+1}/\Delta x_{n+k+1} \\ \vdots & & \vdots \\ R^k v_{n+1}/\Delta x_{n+1} & \cdots & R^k v_{n+k+1}/\Delta x_{n+k+1} \\ v_{n+1}/\Delta x_{n+1} & \cdots & v_{n+k+1}/\Delta x_{n+k+1} \end{vmatrix} \begin{vmatrix} Rv_{n+1}/\Delta x_{n+1} & \cdots & Rv_{n+k}/\Delta x_{n+k} & Rv_n/\Delta x_n \\ \vdots & & \vdots & \vdots \\ R^k v_{n+1}/\Delta x_{n+1} & \cdots & R^k v_{n+k}/\Delta x_{n+k} & R^k v_n/\Delta x_n \\ 1 & \cdots & 1 & 1 \end{vmatrix}$$

$$- \begin{vmatrix} Rv_{n+1}/\Delta x_{n+1} & \cdots & Rv_{n+k}/\Delta x_{n+k} & Rv_n/\Delta x_n \\ \vdots & & \vdots & \vdots \\ R^k v_{n+1}/\Delta x_{n+1} & \cdots & R^k v_{n+k}/\Delta x_{n+k} & R^k v_n/\Delta x_n \\ v_{n+1}/\Delta x_{n+1} & \cdots & v_{n+k}/\Delta x_{n+k} & v_n/\Delta x_n \end{vmatrix} \begin{vmatrix} Rv_{n+1}/\Delta x_{n+1} & \cdots & Rv_{n+k+1}/\Delta x_{n+k+1} \\ \vdots & & \vdots \\ R^k v_{n+1}/\Delta x_{n+1} & \cdots & R^k v_{n+k+1}/\Delta x_{n+k+1} \\ 1 & \cdots & 1 \end{vmatrix}$$

d'où :

$$\frac{\Delta x_n}{\varepsilon_{2k}^{(n+1)} - \varepsilon_{2k}^{(n)}} =$$

$$\Delta x_n \frac{\begin{vmatrix} 1 & \cdots & 1 \\ Rv_{n+1}/\Delta x_{n+1} & \cdots & Rv_{n+k+1}/\Delta x_{n+k+1} \\ \cdots & & \cdots \\ R^k v_{n+1}/\Delta x_{n+1} & \cdots & R^k v_{n+k+1}/\Delta x_{n+k+1} \end{vmatrix} \begin{vmatrix} 1 & \cdots & 1 \\ Rv_n/\Delta x_n & \cdots & Rv_{n+k}/\Delta x_{n+k} \\ \cdots & & \cdots \\ R^k v_n/\Delta x_n & \cdots & R^k v_{n+k}/\Delta x_{n+k} \end{vmatrix}}{\begin{vmatrix} 1 & \cdots & 1 \\ v_n/\Delta x_n & \cdots & v_{n+k+1}/\Delta x_{n+k+1} \\ \cdots & & \cdots \\ R^k v_n/\Delta x_n & \cdots & R^k v_{n+k+1}/\Delta x_{n+k+1} \end{vmatrix} \begin{vmatrix} Rv_{n+1}/\Delta x_{n+1} & \cdots & Rv_{n+k}/\Delta x_{n+k} \\ \cdots & & \cdots \\ R^k v_{n+1}/\Delta x_{n+1} & \cdots & R^k v_{n+k}/\Delta x_{n+k} \end{vmatrix}}$$

qui est égal à $\varepsilon_{2k+1}^{(n)} - \varepsilon_{2k-1}^{(n+1)}$ puisque :

$$\begin{vmatrix} 1 & \cdots & 1 \\ Rv_{n+1}/\Delta x_{n+1} & \cdots & Rv_{n+k+1}/\Delta x_{n+k+1} \\ \cdots & & \cdots \\ R^k v_{n+1}/\Delta x_{n+1} & \cdots & R^k v_{n+k+1}/\Delta x_{n+k+1} \end{vmatrix} = \begin{vmatrix} R^2 v_{n+1} & \cdots & R^2 v_{n+k} \\ \cdots & & \cdots \\ R^{k+1} v_{n+1} & \cdots & R^{k+1} v_{n+k} \end{vmatrix}$$

et que :

$$\begin{vmatrix} 1 & \cdots & 1 \\ v_n/\Delta x_n & \cdots & v_{n+k+1}/\Delta x_{n+k+1} \\ \cdots & & \cdots \\ R^k v_n/\Delta x_n & \cdots & R^k v_{n+k+1}/\Delta x_{n+k+1} \end{vmatrix} = \begin{vmatrix} Rv_n & \cdots & Rv_{n+k} \\ \cdots & & \cdots \\ R^{k+1} v_n & \cdots & R^{k+1} v_{n+k} \end{vmatrix}$$

en remplaçant chaque colonne par sa différence avec la précédente. On démontrerait de même que :

$$\varepsilon_{2k+2}^{(n)} = \varepsilon_{2k}^{(n+1)} + \frac{\Delta x_n}{\varepsilon_{2k+1}^{(n+1)} - \varepsilon_{2k+1}^{(n)}}$$

En utilisant les relations précédentes on voit immédiatement :

Propriété 28 : Si l'application de la première généralisation de l'ε-algorithme à $\{S_n\}$ et à $\{aS_n + b\}$ fournit respectivement les quantités $\varepsilon_k^{(n)}$ et $\bar{\varepsilon}_k^{(n)}$ alors

$$\bar{\varepsilon}_{2k}^{(n)} = a\,\varepsilon_{2k}^{(n)} + b \qquad \bar{\varepsilon}_{2k+1}^{(n)} = \varepsilon_{2k+1}^{(n)} / a$$

En d'autres termes on a :

$$e_k\,(aS_n + b) = a\,e_k(S_n) + b$$

$$\frac{1}{e_k(ar_n+b)} = \frac{1}{a\,e_k(r_n)}$$

Démontrons maintenant un théorème analogue au théorème 35 de l'ε-algorithme :

<u>Théorème 64</u> :

Une condition nécessaire et suffisante pour que $\varepsilon_{2k}^{(n)} = S\ \forall n > N$ est qu'il existe $a_o, \ldots, a_k$ non tous nuls tels que :

$$a_o(v_n - S.\,\Delta x_n) + \sum_{i=1}^{k} a_i\,.\,R^i v_n = 0 \qquad \forall n > N$$

Démonstration : écrivons que $\varepsilon_{2k}^{(n)} = S\ \forall n > N$. D'après le théorème précédent on a :

$$\begin{vmatrix} v_n & \ldots & v_{n+k} \\ Rv_n & \ldots & Rv_{n+k} \\ \text{-----} & & \text{-----} \\ R^k v_n & \ldots & R^k v_{n+k} \end{vmatrix} = S \begin{vmatrix} \Delta x_n & \ldots & \Delta x_{n+k} \\ Rv_n & \ldots & Rv_{n+k} \\ \text{-----} & & \text{-----} \\ R^k v_n & \ldots & R^k v_{n+k} \end{vmatrix}$$

d'où :

$$\begin{vmatrix} v_n - S\,\Delta x_n & \ldots & v_{n+k} - S\,\Delta x_{n+k} \\ Rv_n & \ldots & Rv_{n+k} \\ \text{-----} & & \text{-----} \\ R^k v_n & \ldots & R^k v_{n+k} \end{vmatrix} = 0$$

Une condition nécessaire et suffisante pour que ce déterminant soit nul est qu'il existe $a_o, \ldots, a_k$ non tous nuls tels que :

$$a_o(v_n - S\,\Delta x_n) + \sum_{i=1}^{k} a_i\,R^i v_n = 0 \qquad \forall n > N$$

L'équation aux différences que nous venons d'obtenir est difficilement résoluble pour k quelconque comme cela était le cas pour l'ε-algorithme. Cependant on obtient les théorèmes suivants :

<u>Théorème 65</u> :

Si on applique la première généralisation de l'ε-algorithme à une suite $\{S_n\}$ telle que :

$$S_n = S + \sum_{i=0}^{p+k-1} c_i \, \delta_{in}$$

alors $\varepsilon_{2k}^{(p)} = S$

démonstration : on a pour une telle suite $R^i v_{n+k} = 0 \ \forall i \geq 1$

d'où :

$$\varepsilon_{2k}^{(p)} = \frac{\begin{vmatrix} v_p & \ldots & v_{p+k-1} & S.\,\Delta x_{p+k} \\ Rv_p & \ldots & Rv_{p+k-1} & 0 \\ \multicolumn{4}{c}{\text{-----------------}} \\ R^k v_p & \ldots & R^k v_{p+k-1} & 0 \end{vmatrix}}{\begin{vmatrix} \Delta x_p & \ldots & \Delta x_{p+k-1} & x_{p+k} \\ Rv_p & \ldots & Rv_{p+k-1} & 0 \\ \multicolumn{4}{c}{\text{-----------------}} \\ R^k v_p & \ldots & R^k v_{p+k-1} & 0 \end{vmatrix}} = S$$

<u>Théorème 66</u> :

Une condition nécessaire et suffisante pour que $\varepsilon_2^{(n)} = S \ \forall n > N$ est que $S_n = S + a \prod_{i=0}^{n-1} \lambda$

$\forall n > N$ avec $\lambda_i = 1 + c.\ \Delta x_i$

démonstration : on a $Rv_n = \Delta S_n$, d'où, d'après le théorème 64 :

$$a_o (S_n - S)\ \Delta x_n + a_1\ \Delta S_n = 0$$

posons $d_n = S_n - S$; il vient :

$$a_o\ d_n\ \Delta x_n + a_1\ \Delta d_n = 0$$

ce qui donne, puisque $a_1 \neq 0$

$$d_{n+1} = \left(1 - \frac{a_o}{a_1} \Delta x_n\right) d_n = d_o \prod_{i=0}^{n} \left(1 - \frac{a_o}{a_1} \Delta x_i\right)$$

ce qui démontre que la condition est nécessaire. La condition suffisante se démontre aussi facilement.

On voit que ce théorème est une généralisation du théorème 34 pour le procédé Δ^2 d'Aitke

On retrouve le procédé d'Aitken et le théorème 34 si $\Delta x_n = b \ \forall n$. Dans ce cas on a

$\lambda_i = \lambda = 1 + bc \ \forall i$ et $S_n = S + a\,\lambda^n$. $\forall n$.

Les expériences numériques ainsi que les résultats théoriques font encore défaut pour cette première généralisation de l'ε-algorithme. Voici un exemple où elle permet d'obtenir de meilleurs résultats que l'ε-algorithme. Considérons la suite d'ordre 1.1 donnée par :

$$S_n = 1 + 3e^{-1.4\,x_n} \quad \text{avec } x_n = 1.1^{(n-1)}$$

on obtient :

$S_o = 1.73979 \quad S_1 = 1.64314 \quad S_4 = 1.38631$

avec l'ε-algorithme on trouve :

$$\varepsilon_4^{(0)} = 1.73799$$

tandis que sa première généralisation fournit :

$$\varepsilon_4^{(0)} = 1.00272$$

La seconde généralisation de l'ε-algorithme est donnée par :

$$\varepsilon_{-1}^{(n)} = 0 \qquad \varepsilon_0^{(n)} = S_n \qquad n = 0, 1, \ldots$$

$$\varepsilon_{k+1}^{(n)} = \varepsilon_{k-1}^{(n+1)} + \frac{\Delta x_{n+k}}{\varepsilon_k^{(n+1)} - \varepsilon_k^{(n)}} \qquad n, k = 0, 1, \ldots$$

Il n'a pas été possible, pour cette seconde généralisation d'obtenir des résultats analogues aux théorèmes 63, 64 et 65. Cependant on a le :

<u>Théorème 67</u> :

Une condition nécessaire et suffisante pour que $\varepsilon_2^{(n)} = S \ \forall n > N$ est que $S_n = S + a \prod_{i=0}^{n-1} \lambda_i$ $\forall n > N$ avec $\lambda_i = \dfrac{1}{1 + c.\Delta x_i}$

démonstration : $\varepsilon_1^{(n)} = \Delta x_n \,/\, \Delta S_n$ d'où

$$\varepsilon_2^{(n)} = S = S_n + \frac{\Delta x_{n+1}}{\dfrac{\Delta x_{n+1}}{\Delta S_{n+1}} - \dfrac{\Delta x_n}{\Delta S_n}}$$

posons $d_n = S_n - S$; il vient :

$$d_{n+1} = \frac{\Delta x_{n+1}\, \Delta d_n\, \Delta d_{n+1}}{\Delta x_n\, \Delta d_{n+1} - \Delta x_{n+1}\, \Delta d_n}$$

ou encore :

$$\frac{d_{n+1}}{\Delta x_{n+1}\, \Delta d_n} = \frac{\Delta\, d_{n+1}}{\Delta x_n\, \Delta d_{n+1} - \Delta x_{n+1}\, \Delta d_n} = \frac{d_{n+2}}{\Delta x_n\, \Delta d_{n+1}}$$

ce qui donne

$$\frac{d_{n+1}\, \Delta x_n}{\Delta d_n} = \frac{d_{n+2}\, \Delta x_{n+1}}{\Delta d_{n+1}} = b \quad \forall n > N$$

on doit donc avoir :

$$d_{n+1}\, \Delta x_n = b\, \Delta d_n$$

$$d_{n+1}\, (b - \Delta x_n) = b\, d_n$$

$$d_{n+1} = \frac{b}{b - \Delta x_n}\, d_n \text{ et } d_n = d_o \prod_{i=0}^{n-1} \frac{b}{b - \Delta x_i}$$

La condition suffisante est immédiate.

Si l'on reprend l'exemple numérique précédent on obtient, avec cette seconde généralisation de l'ε-algorithme :

$$\varepsilon_4^{(0)} = 1.00224$$

En pratique ces deux généralisations de l'ε-algorithme semblent interessantes. De nombreux problèmes restent à régler et en particulier celui du choix de la suite $\{x_n\}$ des paramètres Donnons encore deux exemples :

1°) $S_n = 1 + 1 / n$ $\quad x_n = \text{Log}(1+n)$ pour $n = 1, \ldots$

on obtient :

$S_{37} = 1.027$ $\quad \varepsilon_{36}^{(0)} = 1.004$ avec l'ε-algorithme

$\varepsilon_{36}^{(0)} = 1.000008$ avec la première généralisation

$\varepsilon_{36}^{(0)} = 1.00000002$ avec la seconde généralisation

2°) $S_n = n \sin \frac{1}{n}$ et $x_n = \text{Log}\,(1+n)$

$S_{37} = 0.99987$ $\varepsilon_{36}^{(0)} = 0.9999938$ avec l'ε-algorithme

$\varepsilon_{36}^{(0)} = 1.0000000027$ avec la première généralisation

$\varepsilon_{36}^{(0)} = 0.99999999976$ avec la seconde généralisation

En ce qui concerne la convergence de ces deux méthodes on a seulement les résultats suivants :

Théorème 68 :

Pour les première et seconde généralisations de l'ε-algorithme, si

$$\lim_{n\to\infty} \frac{\Delta x_{n+1}}{\Delta x_n} \neq \lim_{n\to\infty} \frac{\Delta S_{n+1}}{\Delta S_n} \quad \text{alors}$$

$$\lim_{n\to\infty} \varepsilon_2^{(n)} = \lim_{n\to\infty} S_n$$

La démonstration est laissée en exercice.

On voit que l'on peut donc traiter avec ces généralisations les suites dites à convergence logarithmique c'est-à-dire telles que :

$$\lim_{n\to\infty} \frac{S_{n+1} - S}{S_n - S} = 1 \quad \text{ou telles que} \lim_{n\to\infty} \frac{\Delta S_{n+1}}{\Delta S_n} = 1.$$

Ces suites sont en général difficiles à accélérer et le procédé Δ^2 d'Aitken peut même ne pas converger dans ce cas. Pour de telles suites on est alors amené à étudier des algorithmes spéciaux [38] ; nous en verrons un exemple au paragraphe suivant. Sur ce sujet on pourra également consulter [87].

IV - 5 Le problème de l'accélération de la convergence

L'ε-algorithme et ses deux généralisations ainsi que le ρ-algorithme sont tous de la forme :

$$\theta_{-1}^{(n)} = 0 \quad \theta_0^{(n)} = S_n \qquad n = 0, 1, \ldots$$

$$\theta_{k+1}^{(n)} = \theta_{k-1}^{(n+1)} + D_k^{(n)} \qquad n, k = 0, 1, \ldots$$

avec :

$D_k^{(n)} = 1 / (\theta_k^{(n+1)} - \theta_k^{(n)})$ pour l'ε-algorithme

$D_k^{(n)} = \Delta x_n / (\theta_k^{(n+1)} - \theta_k^{(n)})$ pour sa première généralisation

$D_k^{(n)} = \Delta x_{n+k} / (\theta_k^{(n+1)} - \theta_k^{(n)})$ pour sa seconde généralisation

$D_k^{(n)} = (x_{n+k+1} - x_n) / (\theta_k^{(n+1)} - \theta_k^{(n)})$ pour le ρ-algorithme

Le but de ce paragraphe est d'étudier de façon globale le problème de l'accélération de la convergence à l'aide de ces quatre algorithmes. Il est bien évident que l'on englobe aussi dans cette théorie tout algorithme de cette forme.

Dans ce paragraphe nous parlerons d'accélération de la convergence au sens suivant : soient $\{V_n\}$ et $\{S_n\}$ deux suites convergentes ; on dira que $\{V_n\}$ converge plus vite que $\{S_n\}$ si :

$$\lim_{n\to\infty} \frac{\Delta V_n}{\Delta S_n} = 0$$

Théorème 69 :

Supposons que $\lim_{n\to\infty} \theta_{2k+2}^{(n)} = \lim_{n\to\infty} \theta_{2k}^{(n)}$. Une condition nécessaire et suffisante pour que $\{\theta_{2k+2}^{(n)}\}$ converge plus vite que $\{\theta_{2k}^{(n+1)}\}$ pour k fixé, est que :

$$\lim_{n\to\infty} \frac{\Delta D_{2k+1}^{(n)}}{\Delta \theta_{2k}^{(n+1)}} = -1$$

où l'opérateur Δ porte sur les indices supérieurs .

démonstration : on a $\Delta\theta_{2k+2}^{(n)} = \Delta\theta_{2k}^{(n+1)} + \Delta D_{2k+1}^{(n)}$.

on voit donc immédiatement que $\lim_{n\to\infty} \frac{\Delta\theta_{2k+2}^{(n)}}{\Delta\theta_{2k}^{(n+1)}} = 0$ entraîne la condition donnée dans le théorème. La condition suffisante est évidente [24].

Si la condition du théorème 69 n'est pas vérifiée alors il n'y aura pas accélération de la convergence en passant de la colonne 2k à la colonne 2k+2. Afin d'accélérer la convergence on va introduire dans l'algorithme un facteur d'accélération w_k comme on le fait dans la méthode de surrelaxation pour résoudre les systèmes d'équations linéaires.

L'algorithme deviendra donc :

$$\theta_{2k+1}^{(n)} = \theta_{2k-1}^{(n+1)} + D_{2k}^{(n)}$$

$$\theta_{2k+2}^{(n)} = \theta_{2k}^{(n+1)} + w_k\, D_{2k+1}^{(n)}$$

Le choix optimal de w_k est caractérisé par le résultat suivant :

Théorème 70 :

Supposons que $\lim_{n\to\infty} \theta_{2k+2}^{(n)} = \lim_{n\to\infty} \theta_{2k}^{(n)}$. Une condition nécessaire et suffisante pour que $\{\theta_{2k+2}^{(n)}\}$ converge plus vite que $\{\theta_{2k}^{(n+1)}\}$ est de prendre :

$$w_k = - \lim_{n\to\infty} \frac{\Delta\theta_{2k}^{(n+1)}}{\Delta D_{2k+1}^{(n)}}$$

Démonstration : on a $\Delta\theta_{2k+2}^{(n)} = \Delta\theta_{2k}^{(n+1)} + w_k\, \Delta D_{2k+1}^{(n)}$.

$$\text{D'où } \lim_{n\to\infty} \frac{\Delta\theta_{2k+2}^{(n)}}{\Delta\theta_{2k}^{(n+1)}} = 0 = 1 + w_k \lim_{n\to\infty} \frac{\Delta D_{2k+1}^{(n)}}{\Delta\theta_{2k}^{(n+1)}}$$

le reste de la démonstration est évident.

Il est possible de donner une interprétation fort simple de ce paramètre w_k ainsi que des résultats que nous venons d'énoncer en considérant les quantités $w_k\, D_{2k+1}^{(n)}$ comme les termes successifs d'un développement asymptotique. Soit $\{S_n\}$ une suite qui converge vers S et soit G l'ensemble des suites $\{D_{2k+1}^{(n)}\}$ pour k fixé. Supposons que $D_{2k+1}^{(n)} = o\,(D_{2k-1}^{(n+1)})$ quelquesoit k fixé. Alors, s'il vérifie cette propriété, l'ensemble G est une échelle de comparaison. Supposons que $S - S_{n+k}$ possède un développement asymptotique jusqu'à l'ordre k au voisinage de +∞ par rapport à G et que ce développement puisse s'écrire :

$$S - S_{n+k} = \sum_{i=1}^{k} w_{i-1}\, D_{2i-1}^{(n+k-i)} + o(D_{2k-1}^{(n)})$$

Le problème est de trouver les coefficients w_i de ce développement asymptotique. On a :

$$\theta_{2k}^{(n)} = S_{n+k} + \sum_{i=1}^{k} w_{i-1}\, D_{2i-1}^{(n+k-i)}$$

d'où :

$$S = \theta_{2k}^{(n)} + o(D_{2k-1}^{(n)})$$

$$= \theta_{2k-2}^{(n+1)} + w_{k-1} D_{2k-1}^{(n)} + o(D_{2k-1}^{(n)})$$

On choisit w_{k-1} de façon que :

$$0 = \Delta\theta_{2k-2}^{(n+1)} + w_{k-1} \Delta D_{2k-1}^{(n)} + o(\Delta D_{2k-1}^{(n)})$$

ce qui entrainera que $S = \theta_{2k}^{(n)} + O(D_{2k-1}^{(n)})$ puisque les séries $\sum_{i=n}^{\infty} \Delta\theta_{2k-2}^{(i)}$ et $\sum_{i=n}^{\infty} D_{2k-1}^{(i)}$ sont convergentes $\forall n$. On a donc :

$$w_{k-1} \Delta D_{2k-1}^{(n)} = -\Delta\theta_{2k-2}^{(n+1)} + o(\Delta D_{2k-1}^{(n)}) \text{ d'où}$$

$$w_{k-1} = -\lim_{n\to\infty} \frac{\Delta\theta_{2k-2}^{(n+1)}}{\Delta D_{2k-1}^{(n)}}$$

ce qui n'est autre que la condition du théorème 70. Le théorème 69 apparait ainsi comme une condition nécessaire et suffisante pour que $w_k = 1$. Le fait que le choix de w_k donné par le théorème 70 fournisse l'algorithme optimal signifie simplement que w_k est le coefficient de $D_{2k+1}^{(n)}$ dans le développement asymptotique de $S - S_{n+k}$. Ce choix de w_k donne le seul algorithme pour lequel on ait :

$$S - \theta_{2k}^{(n)} = o(D_{2k-1}^{(n)})$$

Cette relation peut s'écrire :

$$S - \theta_{2k-2}^{(n+1)} = w_{k-1} D_{2k-1}^{(n)} + o(D_{2k-1}^{(n)})$$

d'où

$$S - \theta_{2k-2}^{(n+1)} \sim w_{k-1} D_{2k-1}^{(n)}$$

et par conséquent

$$S - \theta_{2k-2}^{(n+1)} = O(D_{2k-1}^{(n)})$$

On voit que l'on a ainsi généralisé un résultat obtenu au théorème 17.

Ce paramètre w_k apparait aussi comme le lien entre les ε et ρ-algorithmes : en effet prenons $D_k^{(n)} = 1 / (\theta_k^{(n+1)} - \theta_k^{(n)})$ c'est-à-dire l'ε-algorithme ; considérons la suite $S_n = 1 + \frac{1}{n+1}$ on trouve $w_o = 2$ ce qui n'est autre que la forme simplifiée du ρ-algorithme avec $x_n = n$. Inversement considérons la suite $S_n = S + ab^n$ on trouve $w_o = 1$.

Pour cet algorithme avec paramètre d'accélération on ne connait pas de théorème analogue au théorème 35 pour l'ε-algorithme. Considérons par exemple :

$$\theta_2^{(n)} = S_{n+1} - w_o \frac{\Delta S_n \, \Delta S_{n+1}}{\Delta^2 S_n}$$

Nous voulons trouver la condition que doit vérifier $\{S_n\}$ pour que $\theta_2^{(n)} = S \; \forall n$. C'est un problème de sommation de fonction dont on ne connait pas de solution générale [106]. Cela provient du fait que l'équation aux différences $\Delta f(x) = 1/x$ n'a pas de fonctions élémentaires comme solutions.

Nous allons montrer, qu'à l'aide de ce paramètre w_k, il est possible d'accélérer des suites à convergence logarithmique.

Prenons par exemple une suite telle que :

$$d_{n+1} = d_n + a_1 \, d_n^p + \ldots$$

avec $p > 1$ et $d_n = S_n - S$. Pour une telle suite on a :

$$D_1^{(n)} = - \frac{\Delta S_n \, \Delta S_{n+1}}{\Delta^2 S_n} = - \frac{d_n + pa_1 \, d_n^p + \ldots}{p + \ldots}$$

on a donc $\lim_{n \to \infty} D_1^{(n)} = 0$ et par conséquent $\lim_{n \to \infty} \theta_2^{(n)} = S$.

En utilisant le fait que $d_{n+1}^p = d_n^p + pa_1 \, d_n^{2p-1} + \ldots$ on trouve facilement que $w_o = p$ [38]. Si p est connu il est alors possible d'accélérer la convergence de la suite $\{S_n\}$. Ainsi pour la suite $S_o = 0.938$, $S_{n+1} = S_n - 0.005 \, S_n^2$ on obtient :

n	S_n	$\theta_2^{(n)}$
0	0.938	$0.219 \; 10^{-2}$
5	0.916	$0.209 \; 10^{-2}$
15	0.876	$0.191 \; 10^{-2}$
25	0.839	$0.176 \; 10^{-2}$

Considérons maintenant le cas des séries convergentes :

$$S = a_0 + a_1 + \dots$$

Appliquons cet algorithme aux sommes partielles de la série $S_n = \sum_{i=0}^{n} a_i$.

On trouve alors que si

$\lim_{n\to\infty} a_{n+1} / a_n = a \neq 1$ alors $w_o = 1$ d'où le :

Théorème 71 :

Si on applique l'ε-algorithme aux sommes partielles d'une série telle que :

$$\lim_{n\to\infty} \frac{a_{n+1}}{a_n} = a \neq 1$$

alors $\{\varepsilon_2^{(n)}\}$ converge vers S plus vite que $\{S_{n+1}\}$.

La démonstration de ce théorème est laissée en exercice.

Considérons par exemple la série

$$\text{Log } 2 = 0.693147 \dots = \sum_{i=0}^{\infty} \frac{(-1)^n}{n+1}$$

pour laquelle $a = -1$. On obtient :

n	S_n	$\varepsilon_2^{(n)}$	$\rho_2^{(n)}$
0	1	0.7	0.9
5	0.616667	0.692857	0.626190
15	0.662871	0.693124	0.664552
25	0.674286	0.693141	0.674959

Si $a = 1$ il est possible dans certains cas de calculer w_o. Ainsi pour la série de terme général $a_n = 1/n^\alpha$ avec $\alpha > 1$ on trouve $w_o = \alpha/(\alpha-1)$. Si $\alpha = 2$ on a $w_o = 2$ et c'est par conséquent le ρ-algorithme simplifié qui accélère la convergence. On obtient dans ce cas ($S = 1.64493$) les résultats suivants :

n	S_n	$\varepsilon_2^{(n)}$	$\rho_2^{(n)}$
0	1	1.45	1.65
5	1.49139	1.57846	1.64513
15	1.58435	1.61638	1.64495
25	1.60720	1.62676	1.64494

IV - 6 Le θ-algorithme

Que la suite soit à convergence logarithmique ou non il est bien évident que le paramètre w_k introduit dans les algorithmes au paragraphe précédent est difficile à calculer en pratique puisqu'il fait intervenir une limite.

D'où l'idée immédiate de remplacer

$$w_k = - \lim_{n\to\infty} \frac{\Delta\theta_{2k}^{(n+1)}}{\Delta D_{2k+1}^{(n)}} \quad \text{par} \quad w_k^{(n)} = - \frac{\Delta\theta_{2k}^{(n+1)}}{\Delta D_{2k+1}^{(n)}}$$

C'est ce nouvel algorithme que nous appelerons le θ-algorithme.

C'est une généralisation d'un algorithme obtenu par Germain-Bonne [87] en effectuant une extrapolation linéaire à partir de $\{S_n\}$ et de $\{\varepsilon_2^{(n)}\}$. Les règles du θ-algorithme sont les suivantes [38] :

$$\theta_{-1}^{(n)} = 0 \qquad \theta_0^{(n)} = S_n \qquad \text{pour } n = 0, 1, \ldots$$

$$\theta_{2k+1}^{(n)} = \theta_{2k-1}^{(n+1)} + D_{2k}^{(n)}$$

$$\theta_{2k+2}^{(n)} = \frac{D_{2k+1}^{(n+1)} \theta_{2k}^{(n+1)} - D_{2k+1}^{(n)} \theta_{2k}^{(n+2)}}{D_{2k+1}^{(n+1)} - D_{2k+1}^{(n)}}$$

En prenant $D_k^{(n)} = 1 / (\theta_k^{(n+1)} - \theta_k^{(n)})$ on trouve que :

$$\theta_{2k+2}^{(n)} = \frac{\theta_{2k}^{(n+2)} \Delta\theta_{2k+1}^{(n+1)} - \theta_{2k}^{(n+1)} \Delta\theta_{2k+1}^{(n)}}{\Delta^2 \theta_{2k+1}^{(n)}}$$

On voit que ce θ-algorithme n'est plus un algorithme de losange ; par exemple le calcul de $\theta_2^{(n)}$ nécessite la connaissance de S_n, S_{n+1}, S_{n+2} et S_{n+3} alors que ce dernier terme n'intervenait pas dans le calcul de $\varepsilon_2^{(n)}$ ou $\rho_2^{(n)}$.

Pour cet algorithme on a le :

Théorème 72 :

Supposons que $\lim_{n\to\infty} \theta_{2k}^{(n)} = S$. Une condition nécessaire et suffisante pour que $\lim_{n\to\infty} \theta_{2k+2}^{(n)} = S$

est qu' il existe $\alpha < 1 < \beta$ tels que :

$$\frac{D_{2k+1}^{(n+1)}}{D_{2k+1}^{(n)}} \notin [\alpha,\beta] \qquad \forall n$$

démonstration : on peut considérer la transformation $\{\theta_{2k}^{(n)}\} \to \{\theta_{2k+2}^{(n)}\}$ pour k fixé comme un procédé de sommation ; on lui appliquera par conséquent le théorème de Toeplitz (théorème 22). Les deux dernières conditions de ce théorème sont automatiquement vérifiée puisque le procédé est total et que la matrice associée est bidiagonale. La première condition s'écrit :

$$\left| \frac{D_{2k+1}^{(n)}}{\Delta D_{2k+1}^{(n)}} \right| + \left| \frac{D_{2k+1}^{(n+1)}}{\Delta D_{2k+1}^{(n)}} \right| < M_k \qquad \forall n$$

on doit donc avoir :

$$N_k < \left| 1 - \frac{D_{2k+1}^{(n+1)}}{D_{2k+1}^{(n)}} \right| \qquad \forall n$$

Par conséquent il doit exister $\alpha < 1 < \beta$ tels que :

$$D_{2k+1}^{(n+1)} / D_{2k+1}^{(n)} \notin [\alpha,\beta].$$

On a également le résultat suivant :

Théorème 73 :

Supposons que $\lim_{n\to\infty} \theta_{2k}^{(n)} = \lim_{n\to\infty} \theta_{2k+2}^{(n)} = S$. Si $\lim_{n\to\infty} \frac{\theta_{2k}^{(n+1)} - S}{D_{2k+1}^{(n)}}$ et $\lim_{n\to\infty} \frac{\Delta\theta_{2k}^{(n+1)}}{\Delta D_{2k+1}^{(n)}}$ existent, sont finies et sont égales alors $\{\theta_{2k+2}^{(n)}\}$ converge plus vite que $\{\theta_{2k}^{(n+1)}\}$ en ce sens que

$$\lim_{n\to\infty} \frac{\theta_{2k+2}^{(n)} - S}{\theta_{2k}^{(n+1)} - S} = 0$$

démonstration : elle est immédiate puisque $\theta_{2k+2}^{(n)} - S = \theta_{2k}^{(n+1)} - S + w_k^{(n)} D_{2k+1}^{(n)}$

et que $w_k^{(n)} = - \Delta\theta_{2k}^{(n+1)} / \Delta D_{2k+1}^{(n)}$.

Reprenons les trois exemples numériques du paragraphe précédent. Avec $k = 0$, c'est-à-dire en considérant la suite $\{\theta_2^{(n)}\}$ on obtient :

n	ex. 1	ex. 2	ex. 3
0	0.933	0.655555	1.13888
5	$-0.211\ 10^{-2}$	0.693118	1.64461
15	$-0.193\ 10^{-2}$	0.693146	1.64491
25	$-0.177\ 10^{-2}$	0.693147	1.64493

Les résultats numériques ainsi que les applications et des théorèmes de convergence manquent encore pour le θ-algorithme.

Du point de vue numérique le θ-algorithme semble très intéressant. De nombreux exemples ont été testés [75]. Ils montrent que, étant donné une suite à accélérer, on obtient de bons résultats soit avec l'ε-algorithme, soit avec le ρ-algorithme mais pratiquement jamais avec les deux algorithmes à la fois. D'autre part les exemples montrent que le θ-algorithme se comporte pratiquement toujours comme celui qui donne les meilleurs résultats et parfois même mieux que les deux. Ceci tient au fait que, comme nous l'avons vu au paragraphe IV-5, le θ-algorithme peut être considéré comme le lien entre l'ε-algorithme et le ρ-algorithme. Il n'y a d'ailleurs aucune difficulté à démontrer que si la suite $\{S_n\}$ est de l'une des deux formes :

$$S_n = S + ab^n$$

ou

$$S_n = S + a / (n+1)$$

alors $\theta_2^{(n)} = S$ pour tout n.

Cette propriété est également vraie pour d'autres types de suites [67].

Les résultats théoriques pour le θ-algorithme manquent encore. Ils sont difficiles à obtenir parce que l'on ne possède, comme base de travail, que de la règle de l'algorithme. Il n'y a pas, dans l'état actuel de nos connaissances, de rapports de déterminants comme c'est le cas pour l'ε-algorithme. Des variantes et des généralisations du θ-algorithme sont actuellement à l'étude [154].

IV - 7 Les transformations de Levin

Revenons au procédé Δ^2 d'Aitken. Si la suite est de la forme $S_n - S = a\,\Delta S_n$ alors $\varepsilon_2^{(n)} = S$ pour tout n. Levin [130] a eu l'idée de généraliser cela au cas où la suite vérifie une relation de la forme :

$$S_n - S = \Delta\, S_n\, p_{k-1}(n) \;/\; g(n)$$

où p_{k-1} est un polynôme de degré k-1 de n dont les coefficients sont inconnus et où g est une fonction connue de n. On a alors :

$$\Delta^k(g(n)(S_n - S) \,/\Delta\, S_n) = \Delta^k\, p_{k-1}(n) = 0$$

d'où

$$S = \Delta^k(g(n)\, S_n \,/\, \Delta S_n) \;/\; \Delta^k(g(n) \,/\, \Delta S_n)$$

Si la suite $\{S_n\}$ n'est pas de la forme précédente on pourra cependant lui appliquer cette transformation. Nous noterons :

$$W_k^{(n)} = \Delta^k(g(n)\, S_n \,/\, \Delta S_n) \;/\; \Delta^k(g(n) \,/\, \Delta S_n)$$

Suivant ce que l'on prend pour g on retrouve les méthodes T, U et V décrites dans l'article de Levin [130]. De plus pour g(n) = 1 et k = 1 on a $W_1^{(n)} = \varepsilon_2^{(n)}$ et pour k = 2 on obtient $W_2^{(n)} = \theta_2^{(n)}$. Cette dernière propriété permet d'ailleurs de retrouver les résultats de Cordellier sur la première étape du Θ-algorithme [67]. Si $g(x) = \Delta x_n$ et si k = 1 alors $\{W_1^{(n)}\}$ est la première colonne paire $\{\varepsilon_2^{(n)}\}$ fournie par la première généralisation de l'ε-algorithme (paragraphe IV-4).

Il existe de nombreux exemples où cette transformation donne de meilleurs résultats que l'ε-algorithme ; c'est le cas pour la suite :

$$S_n = \sum_{i=0}^{n} (-1)^i / (2i + 1)$$

On a alors :

$$\Delta S_n = \Delta^2 S_n \, p_1(n) / g(n)$$

avec $\quad g(n) = 4n + 8 \quad$ et $\quad p_1(n) = - 2n - 5$

Ce procédé peut s'appliquer au calcul des intégrales impropres et à l'inversion de la transformée de Laplace [131]. Soit à calculer : $I = \int_0^\infty f(t)\, dt$.

Posons : $\quad F(x) = \int_0^x f(t)\, dt.$

Supposons, de façon analogue au cas des suites, que :

$$F(x) - I = f(x) \, p_{k-1}(x) / g(x).$$

On aura :

$$I = \Delta^k(g(x) F(x) / f(x)) / \Delta^k(g(x) / f(x))$$

où Δ est défini par $\Delta u(x) = u(x + h) - u(x)$.

De nombreuses études restent encore à faire sur ce procédé. A. Sidi (Université de Tel-Aviv) s'occupe actuellement de la généralisation de la méthode au cas où :

$$g(n)(S_n - S) = \Delta S_n \, p_{k-1}(n) + \Delta^2 S_n \, p_k(n) + \ldots$$

$$g(x)(F(x)-I) = f(x) \, p_{k-1}(x) + f''(x) \, p_k(x) + \ldots$$

cela permettrait de calculer des intégrales de la forme :

$$I = \int_o^\infty x^{-a} \cos bx \, J_o(cx)\, dx$$

où f vérifie une équation différentielle d'ordre 4.

IV - 8 Formalisation des procédés d'accélération de la convergence

Dans ce qui précède nous avons étudié un certain nombre d'algorithmes d'accélération de la convergence. Il est évidemment tentant d'essayer de mettre en lumière les propriétés qui doivent être vérifiées par un tel algorithme pour qu'il accélère la convergence. On pourrait ainsi, d'une part, unifier l'étude de tels algorithmes, et d'autre part, construire de nouveaux algorithmes d'accélération de la convergence une fois le formalisme bien établi. La première formalisation de ces méthodes a été donnée par Pennacchi [150]: ce sont les transformations rationnelles de suites. Ces transformations sont des cas particuliers d'un formalisme plus général dû à Germain-Bonne [85 ,86]. Nous exposerons d'abord les résultats de Pennacchi.

Définition 19 : on appelle transformation rationnelle d'ordre p et de degré m l'application $T_{p,m}$ qui à la suite $\{S_n\}$ fait correspondre la suite $\{V_{p,m}(n)\}$ pour p et m fixés donnée par :

$$V_{p,m}(n) = S_n + \frac{P_m(\Delta S_n, \ldots, \Delta S_{n+p-1})}{Q_{m-1}(\Delta S_n, \ldots, \Delta S_{n+p-1})}$$

où P_m et Q_{m-1} sont des polynômes homogènes de degrés respectifs m et m-1 des p variables $\Delta S_n, \ldots, \Delta S_{n+p-1}$. On posera $R_m = P_m / Q_{m-1}$ et $R_m \equiv 0$ si $\Delta S_n = \ldots = \Delta S_{n+p-1} = 0$ pour $m > 1$.

Les propriétés suivantes sont évidentes ; elles sont laissées en exercice.

Propriété 29 :

- $T_{p,m}[\{S\}] = \{S\}$ où $\{S\}$ est une suite constante
- $T_{p,m}[\{aS_n + b\}] = a\, T_{p,m}[\{S_n\}] + b$
- une condition nécessaire et suffisante pour que $V_{p,m}(n) = S \; \forall n > N$ est que $\{S_n\}$ vérifie $(S_n - S)\, Q_{m-1} + P_m = 0 \; \forall n > N$.
- La transformation rationnelle d'une progression arithmétique ou géométrique est une progression de même nature et de même raison.
- Les puissances successives d'une transformation rationnelle ne sont généralement pas des transformations rationnelles. La puissance d'une transformation rationnelle étant définie par :

$$T^2_{p,m}\ [\{S_n\}] = T_{p,m}\ [\{V_{p,m}(n)\}] = X_{p,m}(n)$$

$$T^3_{p,m}\ [\{S_n\}] = T_{p,m}\ [\{X_{p,m}(n)\}] \quad \text{etc} \ldots$$

Définition 20 :

On dit que $\{S_n\}$ est régulière si :

- $\lim_{n\to\infty} S_n = S$ existe et est finie

- $\exists N : \forall n > N \quad \Delta S_n \neq 0$

- $\lim_{n\to\infty} \frac{\Delta S_{n+1}}{\Delta S_n} = \rho$ existe, est finie et inférieure à 1.

Théorème 74 :

Si $\{S_n\}$ est régulière et si

$Q_{m-1}\ (1, \rho, \ldots, \rho^{p-1}) \neq 0$

alors $\lim_{n\to\infty} V_{p,m}(n) = S$

démonstration : posons $\rho_n = \Delta S_{n+1} / \Delta S_n$. On a :

$$\Delta S_{n+i} / \Delta S_n = \rho_n\ \rho_{n+1} \ldots \rho_{n+i-1}$$

posons $\sigma_{n,0} = 1$

$$\sigma_{n,i} = \rho_n \ldots \rho_{n+i-1} \quad \text{pour } i = 1, \ldots, p-1$$

on a $\lim_{n\to\infty} \sigma_{n,i} = \rho^i$ et

$$V_{p,m}(n) = S_n + \Delta S_n .\ R_m(1, \rho_n, \rho_n\ \rho_{n+1}, \ldots, \rho_n \ldots \rho_{n+p-2})$$

$$= S_n + \Delta S_n .\ R_m(\sigma_{n0}, \sigma_{n1}, \ldots, \sigma_{n,p-1})$$

Théorème 75 :

Une condition nécessaire et suffisante pour que $T_{p,m}$ accélère la convergence de toute suite régulière $\{S_n\}$ est que :

$$R_m(1, \rho, \ldots, \rho^{p-1}) = \frac{1}{1-\rho} \quad \forall \rho$$

démonstration :

$$V_{p,m}(n) - S = S_n - S + \Delta S_n . R_m(\sigma_{n0}, \ldots, \sigma_{n,p-1})$$

$$\frac{V_{p,m}(n) - S}{S_n - S} = 1 - \frac{\Delta S_n}{\Delta S_n + \Delta S_{n+1} + \ldots}\ R_m$$

car $S_n - S = -(\Delta S_n + \Delta S_{n+1} + \ldots)$; d'où

$$\frac{V_{p,m}(n) - S}{S_n - S} = 1 - \frac{1}{\sigma_{n0} + \sigma_{n1} + \ldots} R_m$$

$$\lim_{n\to\infty} \frac{V_{p,m}(n) - S}{S_n - S} = 0 = 1 - (1 - \rho)\, R_m(1, \rho, \ldots, \rho^{p-1})$$

ce qui termine la démonstration.

Cette condition peut encore s'écrire :

$$Q_{m-1}(1, \rho, \ldots, \rho^{p-1}) - (1 - \rho)\, P_m(1, \rho, \ldots, \rho^{p-1}) \equiv 0$$

Elle exprime l'annulation identique d'un polynôme de degré $m(p-1) + 1$ dont les $m(p-1)+2$ coefficients sont fonctions linéaires des $\binom{p+m-1}{p-1}$ coefficients de P_m et des $\binom{p+m-2}{p-1}$ coefficients de Q_{m-1}.

Etant donné qu'une fraction rationnelle n'est définie qu'à un facteur multiplicatif prè la relation du théorème précédent se traduit par un système non homogène de $m(p-1)+2$ équations linéaires avec un nombre d'inconnues égal à :

$$\binom{p+m-1}{p-1} + \binom{p+m-2}{p-1} - 1 = \binom{p+m-2}{p-1} \frac{p+2m-1}{m} - 1$$

Par conséquent le nombre de coefficients qu'il est possible de fixer arbitrairement dar une transformation rationnelle pour qu'elle accélère la convergence est :

$$\nu(p,m) = \binom{p+m-2}{p-1} \frac{p+2m-1}{m} - m(p-1) - 3$$

On trouve que $\nu(1,m) = \nu(p,1) = -1$, d'où le résultat

<u>Théorème 76</u> :

$T_{1,m}$ et $T_{p,1}$ ne peuvent pas accélérer la convergence de toute suite régulière.

Si $m = 1$ on voit que la transformation $\{S_n\} \to \{V_{p,1}(n)\}$ est une transformation linéaire

Par conséquent les procédés de sommation de ce type sont incapables d'accélérer la convergence de toutes les suites régulières $\{S_n\}$.

Le fait que $\nu(2,2) = 0$ entraîne le :

<u>Théorème 77</u> :

Il existe une et une seule transformation $T_{2,2}$ qui accélère la convergence de toute suite régulière. Cette transformation est donnée par :

$$V_{2,2}(n) = \frac{S_{n+2}\, S_n - S_{n+1}^2}{\Delta^2 S_n} = \varepsilon_2^{(n)}$$

Une fois de plus on voit que l'on retrouve le procédé Δ^2 d'Aitken.

On montre également que ce procédé Δ^2 d'Aitken est optimal. En effet, donnons d'abord la :

<u>Définition 21</u> : on dit que $T_{p,m}$ et $T_{q,k}$ sont équivalentes si :

$$T_{p,m}\,[\{S_n\}] = T_{q,k}\,[\{S_n\}]$$

on peut alors montrer que :

<u>Théorème 78</u> :

Pour $m > 2$, toute transformation $T_{2,m}$ qui accélère la convergence est toujours équivalente à $T_{2,2}$

<u>Théorème 79</u> :

Il existe une transformation unique d'ordre 2 qui accélère la convergence : c'est le procédé Δ^2 d'Aitken.

Les transformations de suites introduites par Germain-Bonne sont plus générales en ce sens qu'on ne suppose pas qu'elles mettent en jeu des polynômes et qu'elles font intervenir une suite de paramètres $\{x_n\}$. Nous allons maintenant donner les principaux de ces résultats.

Soit $G : \mathbb{R}^{k+1} \times \mathbb{R}^{k+1} \to \mathbb{R}$ telle que :

- G soit continue séparément par rapport à ses 2(k+1) variables
- $G(a\, y_0, \ldots, a\, y_k \;;\; a\, x_0, \ldots, a\, x_k) = a\, G(y_0, \ldots, y_k \;;\; x_0, \ldots, x_k)\ \forall a \in \mathbb{R}$
- $G(y_0 + b, \ldots, y_k + b \;;\; x_0, \ldots, x_k) = G(y_0, \ldots, y_k \;;\; x_0, \ldots, x_k) + b\ \ \forall b \in \mathbb{R}$

Etant donnée une suite $\{S_n\}$ qui converge vers S et une suite $\{x_n\}$ de paramètres convergente et de limite connue on veut étudier les conditions que doivent vérifier G, $\{x_n\}$ et $\{S_n\}$ pour que la suite $\{T_n\}$ donnée par :

$$T_n = G(S_n, \ldots, S_{n+k} \;;\; x_n, \ldots, x_{n+k})$$

converge vers S et cela plus vite que $\{S_n\}$.

On a :

<u>propriété 30</u> : $G(y, y, \ldots y ; 0, 0, \ldots, 0) = y \quad \forall y \in \mathbb{R}$

<u>Théorème 80</u> :

Soit D_{k+1} le sous-ensemble de $\mathbb{R}^{k+1}$ constitué des vecteurs ayant toutes leurs composantes différentes de zéro. Toute fonction G définie et continue sur $\mathbb{R}^{k+1} \times D_{k+1}$ peut se mettre sous la forme :

$$G(y_0, \ldots, y_k ; x_0, \ldots, x_k) = y_0 + x_0\, g\left(\frac{\Delta y_0}{x_0}, \ldots, \frac{\Delta y_{k-1}}{x_{k-1}} ; \frac{x_1}{x_0}, \ldots, \frac{x_k}{x_{k-1}}\right)$$

Les démonstrations de ces résultats sont laissées en exercices. On obtient les théorèmes suivants de convergence et d'accélération de la convergence :

<u>Théorème 81</u> :

Soit $\{S_n\}$ une suite qui converge vers S et $\{x_n\}$ une suite qui converge vers 0. Pour toute transformation G définie sur $\mathbb{R}^{k+1} \times \mathbb{R}^{k+1}$ on a :

$$\lim_{n\to\infty} G(S_n, \ldots, S_{n+k} ; x_n, \ldots, x_{n+k}) = S$$

démonstration : elle est immédiate puisque G est continue et que $G(S, \ldots, S ; 0, \ldots, 0)=S$

<u>Théorème 82</u> :

- Si $\{x_n\}$ converge vers zéro

- Si $x_{n+1} / x_n \neq 0$ et borné $\forall n > N$

- Si $\{S_n\}$ converge vers S

- Si $\Delta S_n / x_n$ borné $\forall n > N$

alors $\lim_{n\to\infty} G(S_n, \ldots, S_{n+k} ; x_n, \ldots, x_{n+k}) = S$ pour toute transformation G définie sur $\mathbb{R}^{k+1} \times D_{k+1}$.

démonstration : on a :

$$T_n = S_n + x_n\, g\left(\frac{\Delta S_n}{x_n}, \ldots, \frac{\Delta S_{n+k-1}}{x_{n+k-1}} ; \frac{x_{n+1}}{x_n}, \ldots, \frac{x_{n+k}}{x_{n+k-1}}\right)$$

d'où le théorème puisque g reste bornée lorsque n tend vers l'infini.

On a enfin le résultat fondamental suivant :

Théorème 83 :

- Si $\{x_n\}$ converge vers zéro

- si $\lim_{n\to\infty} x_{n+1} / x_n = a \neq 0$ ou 1

- si $\{S_n\}$ converge vers S

- si $\lim_{n\to\infty} (S_n - S) / x_n = \lim_{n\to\infty} \Delta S_n / \Delta x_n = b \neq 0$

alors une condition nécessaire et suffisante pour que la transformation G accélère la convergence de toute suite $\{S_n\}$ qui vérifie les propriétés précédentes est que la fonction g associée vérifie :

$$g(y, \ldots, y \,;\, x, \ldots, x) = \frac{y}{1-x}$$

démonstration : la condition d'accélération de la convergence s'écrit :

$$\lim_{n\to\infty} \frac{T_n - S}{S_n - S} = 0 = \lim_{n\to\infty} \left[1 + \frac{x_n}{S_n - S}\, g\left(\frac{\Delta S_n}{x_n}, \ldots \,;\, \frac{x_{n+1}}{x_n}, \ldots\right)\right]$$

$$0 = 1 + \frac{1}{b}\, g(b(a-1), \ldots \,;\, a, \ldots)$$

d'où la condition du théorème en posant $a = x$ et $b(a-1) = y$.

La réciproque est évidente.

REMARQUES :

1°) la condition $a \neq 0$ est nécessaire car les fonctions g ne sont définies que si $\mathbb{R}^k \times D_k$.

La condition $a \neq 1$ est nécessaire pour exprimer la condition d'accélération.

2°) On retrouve le procédé Δ^2 d'Aitken en prenant :

$$G(y_0, y_1, y_2 \,;\, x_0, x_1, x_2) = y_0 + \Delta y_0 \frac{1}{1 - \frac{\Delta y_1}{\Delta y_0}}$$

on retrouve $\rho_2^{(n)}$ en prenant

$$G(y_0, y_1, y_2 \,;\, x_0, x_1, x_2) = y_1 + \frac{x_2 - x_0}{\frac{\Delta x_1}{\Delta y_1} - \frac{\Delta x_0}{\Delta y_0}}$$

3°) La fonction G ayant le moins de variables conduit à un procédé linéaire :

$$G(y_0, y_1 \,;\, x_0, x_1) = y_0 - \frac{\Delta y_0}{\Delta x_0}\, x_0$$

4°) On retrouve les transformations rationnelles de Pennacchi en prenant :

$$G(y_0, \ldots, y_p \,;\, x_0, \ldots, x_p) = y_0 + \frac{P_m(\Delta y_0, \ldots, \Delta y_{p-1})}{Q_{m-1}(\Delta y_0, \ldots, \Delta y_{p-1})}$$

IV - 9 Mise en oeuvre des algorithmes

Avant de terminer ce chapitre nous parlerons de la mise en oeuvre pratique des algorithmes de la forme :

$$\theta_{-1}^{(n)} = 0 \quad \theta_0^{(n)} = S_n \qquad n = 0, 1, \ldots$$

$$\theta_{k+1}^{(n)} = \theta_{k-1}^{(n+1)} + D_k^{(n)} \qquad n, k = 0, 1, \ldots$$

avec

$$D_k^{(n)} = w_k^{(n)} \;/\; \Delta\theta_k^{(n)}$$

Nous commencerons par étudier les règles particulières. Supposons que les quantités $\theta_{k-2}^{(n+1)}$ et $\theta_{k-2}^{(n+2)}$ deviennent toutes les deux égales à b ; alors $\theta_{k-1}^{(n+1)}$ devient infini, $\theta_k^{(n)}$ et $\theta_k^{(n+1)}$ deviennent eux aussi égaux à b et $\theta_{k+1}^{(n)}$ est indéterminé. La situation peut se résumer ainsi :

$$\begin{array}{ccccc}
 & & \theta_{k-1}^{(n)} & & \\
 & \theta_{k-2}^{(n+1)} = b & & \theta_k^{(n)} = b & \\
\theta_{k-3}^{(n+2)} & & \theta_{k-1}^{(n+1)} = \infty & & \theta_{k+1}^{(n)} = ? \\
 & \theta_{k-2}^{(n+2)} = b & & \theta_k^{(n+1)} = b & \\
 & & \theta_{k-1}^{(n+2)} & &
\end{array}$$

On est donc, dans ce cas, obligé d'appliquer, à la place de la règle habituelle de l'algorithme, des règles particulières. De même si $\theta_{k-2}^{(n+1)}$ et $\theta_{k-2}^{(n+2)}$ sont très voisins il y a une importante perte de précision due à la troncature. Alors $\theta_{k-1}^{(n+1)}$ est mal détermi[né] et cette imprécision se répercute dans la suite de l'application de l'algorithme.

Dans ce cas on emploie encore des règles particulières pour éviter cette perte de précision. Les règles particulières que nous présentons ici sont des généralisations de celles données par Wynn [231]. On trouvera le détail des calculs dans [22].

On a :

$$\theta_{k+1}^{(n)} = \theta_{k-1}^{(n+1)} + \frac{a_k^{(n)}}{\theta_{k-2}^{(n+2)} + \frac{a_{k-1}^{(n+1)}}{\Delta\theta_{k-1}^{(n+1)}} - \theta_{k-2}^{(n+1)} - \frac{a_{k-1}^{(n)}}{\Delta\theta_{k-1}^{(n)}}}$$

or $\theta_{k-2}^{(n+2)} - \theta_{k-2}^{(n+1)} = \dfrac{a_{k-2}^{(n+1)}}{\theta_{k-1}^{(n+1)} - \theta_{k-3}^{(n+2)}}$

en posant :

$$a = (a_{k-1}^{(n+1)} + a_{k-1}^{(n)} - a_{k-2}^{(n+1)} - a_k^{(n)})\ \theta_{k-1}^{(n+1)} + a_{k-1}^{(n+1)}\ \theta_{k-1}^{(n+2)}$$

$$(1 - \theta_{k-1}^{(n+2)} / \theta_{k-1}^{(n+1)})^{-1} + a_{k-1}^{(n)}\ (1 - \theta_{k-1}^{(n)} / \theta_{k-1}^{(n+1)})^{-1}\ \theta_{k-1}^{(n)}$$

$$- a_{k-2}^{(n+1)}\ \theta_{k-3}^{(n+2)}\ (1 - \theta_{k-3}^{(n+2)} / \theta_{k-1}^{(n+1)})^{-1}$$

On obtient :

$$\theta_{k+1}^{(n)} = a(a_k^{(n)} + a / \theta_{k-1}^{(n+1)})^{-1}$$

Si $a_k^{(n)} = 1\ \forall n, k$ on retrouve les règles particulières de l'ε-algorithme. Dans ce cas si $\varepsilon_{k-2}^{(n+1)}$ et $\varepsilon_{k-2}^{(n+2)}$ sont égaux on obtient :

$$\varepsilon_{k+1}^{(n)} = \varepsilon_{k-1}^{(n+2)} + \varepsilon_{k-1}^{(n)} - \varepsilon_{k-3}^{(n+2)}$$

c'est-à-dire que si l'on revient aux notations du paragraphe III - 3 :

N

W C E

S

cette dernière règle particulière s'écrit :

N + S = W + E

Cordellier [89] a généralisé ces règles particulières au cas où il y a plus de quatre quantités égales dans le tableau ε, c'est-à-dire quand on est dans la situation :

$$\begin{array}{lll}
\theta_{k-2}^{(n+1)} = b & \theta_k^{(n)} = b & \theta_{k+2}^{(n-1)} = b \ldots \\
\theta_{k-2}^{(n+2)} = b & \theta_k^{(n+1)} = b & \theta_{k+2}^{(n)} = b \ldots \\
\theta_{k-2}^{(n+3)} = b & \theta_k^{(n+2)} = b & \theta_{k+2}^{(n+1)} = b \ldots \\
\vdots & \vdots & \vdots
\end{array}$$

où le carré contenant des quantités égales est de dimension quelconque.

Wynn a également étudié la stabilité [234,232] et la propagation des erreurs [233] pour l'ε-algorithme. Nous ne développerons pas cette question ici. Disons seulement que si une erreur $\delta_k^{(n)}$ est faite sur $\varepsilon_k^{(n)}$ elle se propage suivant le schéma suivant :

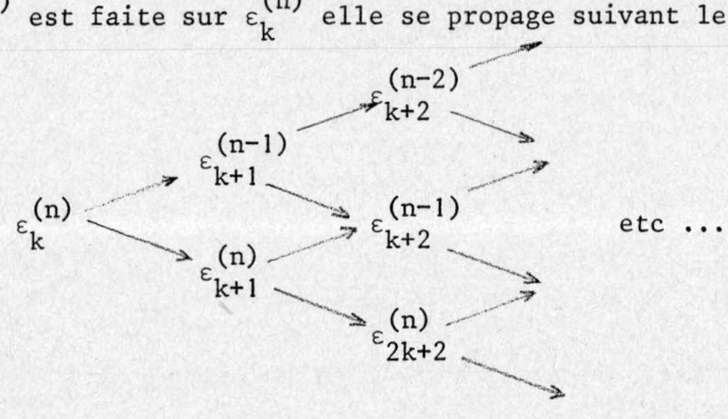

Voyons maintenant la façon dont il faut utiliser ces algorithmes. La première méthode est une utilisation a postériori c'est-à-dire que connaissant $S_n, \ldots, S_{n+2k}$ on calcule $\theta_{2k}^{(n)}$ et l'on arrête les calculs. Dans ce cas les programmes sont simples à écrire [158]. Une autre procédure consiste à utiliser les algorithmes en parallèle avec le calcul des termes de la suite : connaissant S_0, S_1 et S_2 on calcule $\theta_2^{(0)}$ on estime la précision obtenue par $\theta_2^{(0)} - S = O(D_1^{(0)})$.
Si cette précision est insuffisante on calcule S_3 ce qui permet de calculer $\theta_2^{(1)}$; on estime la précision par $\theta_2^{(1)} - S = O(D_1^{(1)})$; si cette précision est insuffisante on calc S_4 puis $\theta_4^{(0)}$ et ainsi de suite jusqu'à ce que la précision désirée soit atteinte. On emploie pour programmer cette utilisation des algorithmes, la technique du losange, donnée par Wynn [225,219]. On trouvera en [35] les programmes FORTRAN correspondants.

Au point de vue volume des calculs on montre que si l'on part des 2k+1 quantités $S_n, \ldots, S_{n+2k}$ alors le calcul de $\theta_{2k}^{(n)}$ nécessite l'évaluation de k(2k+1) quantités $\theta_p^{(q)}$ soit 2k(2k+1) additions ou soustractions et k(2k+1) divisions pour l'ε-algorithme.

Signalons enfin qu'une fois construit un tableau θ il est possible, en choisissant dans ce tableau de nouvelles quantités $\theta_0^{(n)}$, de recommencer une nouvelle application de l'algorithme et ainsi de suite. Il existe trois façons principales de procéder :

- l'application répétée associée
- l'application répétée correspondante
- l'application itérée

On peut les symboliser par le schéma :

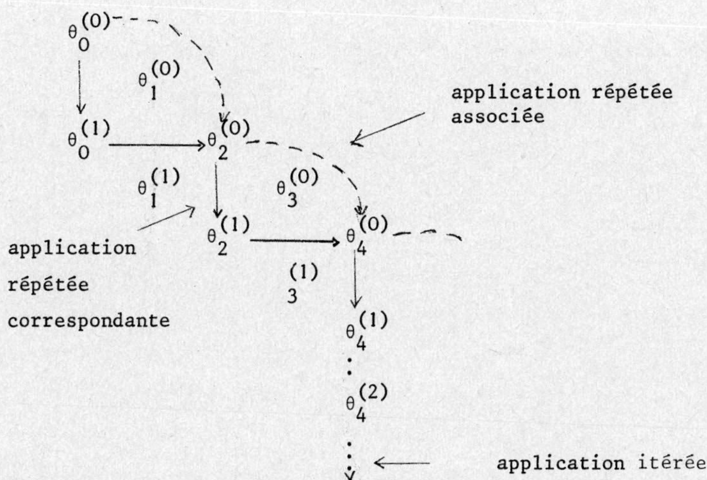

On peut obtenir, de cette façon, des résultats bien meilleurs que ceux fournis par une seule application de l'algorithme.

Des études théoriques sur ce problème sont nécessaires. Les applications répétées associées et correspondantes sont liées aux fractions continues et correspondantes dans la théorie de l'ε-algorithme [226]. Nous n'avons pas du tout parlé ici de la connexion entre l'ε-algorithme et les fractions continues. Cette question sera traitée au chapitre VII.

Il est essentiel, si l'on veut espérer obtenir de bons résultats numériques avec des algorithmes d'accélération de la convergence, de les programmer de façon extrêmement soigneuse. La propagation des erreurs dues à l'ordinateur est quelquefois catastrophique. Il est donc nécessaire d'utiliser les règles particulières et même parfois de corriger l'arithmétique de l'ordinateur. On trouvera dans [19,22,35] des programmes FORTRAN et dans [158,219,226] des programmes ALGOL.

CHAPITRE V

TRANSFORMATION DE SUITES NON SCALAIRES

Dans tout ce qui précède les suites que nous avons transformées à l'aide des algorithmes étaient des suites de nombres réels ou complexes. Dans ce chapitre nous allons étudier l'accélération de la convergence de suites plus générales que des suites de nombres : suites de matrices carrées, suites d'éléments d'un espace de Banach et surtout suites de vecteurs.

Nous effectuerons toujours ces transformations à l'aide de modifications appropriées de l'ε-algorithme et cela pour deux raisons :

- l'ε-algorithme est le plus puissant de tous les algorithmes que nous avons étudié pour des suites de nombres.
- Il n'existe de résultats théoriques sur les suites non scalaires que pour l'ε-algorithme.

Signalons que la théorie mathématique complète qui est sous-jacente à ces questions a été bâtie par Wynn [200,201,215,216,217,228].

V - 1 L'ε-algorithme matriciel

Il n'y a aucune difficulté à définir un ε-algorithme qui s'applique à des suites de matrices carrées. En effet la règle de l'ε-algorithme scalaire peut s'écrire ;

$$\varepsilon_{k+1}^{(n)} = \varepsilon_{k-1}^{(n+1)} + (\Delta\varepsilon_k^{(n)})^{-1}$$

Pour des matrices la puissance -1 désignera simplement l'inverse d'une matrice [242]. Soit donc $\{S_n\}$ une suite de matrices carrées ; l'ε-algorithme matriciel sera donc :

$$\varepsilon_{-1}^{(n)} = 0 \qquad \varepsilon_0^{(n)} = S_n \qquad \text{pour } n = 0, 1, \ldots$$

$$\varepsilon_{k+1}^{(n)} = \varepsilon_{k-1}^{(n+1)} + (\Delta\varepsilon_k^{(n)})^{-1} \qquad n, k = 0, 1, \ldots$$

où $\varepsilon_{-1}^{(n)} = 0$ est une matrice carrée dont tous les éléments sont nuls.

Pour cet algorithme on a les résultats suivants [30] :

Théorème 84 :

Soit A une matrice carrée inversible telle que I-A soit inversible.

Si on applique l'ε-algorithme matriciel à la suite :

$$S_n = \sum_{k=0}^{n} A^k \qquad n = 0, 1, \ldots$$

alors $\varepsilon_2^{(n)} = (I - A)^{-1} \quad n = 0, 1, \ldots$

et ceci indépendamment de la dimension de la matrice.

Démonstration : on a $\Delta S_n = A^{n+1}$ d'où $\varepsilon_1^{(n)} = A^{-n-1}$ puisque A^{-1} existe. D'où

$$\varepsilon_2^{(n)} = S_{n+1} + (A^{-n-2} - A^{-n-1})^{-1}$$

$$= I + A + \ldots + A^{n+1} + (A^{-n-2}(I - A))^{-1}$$

Or, puisque I-A est inversible, on a formellement :

$$(I - A)^{-1} = I + A + A^2 + \ldots$$

d'où $\varepsilon_2^{(n)} = I + A + \ldots + A^{n+1} + (I + A + A^2 + \ldots)\, A^{n+2}$

$$= I + A + \ldots + A^{n+1} + A^{n+2} + \ldots = (I - A)^{-1}$$

On voit que la dimension de la matrice n'intervient pas dans la démonstration. Ce théorème généralise un résultat donné par Householder [112] dans le cas de l'ε-algorithme scalaire.

Théorème 85 :

Soit $\{S_n\}$ une suite de matrices carrées telles que $S_{n+1} - S = (A + E_n)(S_n - S)$ où A et E_n sont des matrices carrées telles que le rayon spectral de A soit strictement inférieur à un et que $\lim_{n\to\infty} E_n = 0$. Si on applique l'ε-algorithme matriciel à la suite $\{S_n\}$ alors :

$$\lim_{n\to\infty} \varepsilon_2^{(n)} = S$$

et

$$\lim_{n\to\infty} (\varepsilon_2^{(n)} - S)(S_n - S)^{-1} = 0$$

démonstration : remarquons d'abord que le fait que A ait un rayon spectral strictement inférieur à un entraîne que I - A est inversible et que $\{S_n\}$ converge vers S. On a :

$$\varepsilon_2^{(n)} = S_{n+1} - \Delta_n (\Delta_n^2)^{-1} \Delta_{n+1}$$

où les matrices Δ_n et Δ_n^2 sont définies par :

$$\Delta_n = S_{n+1} - S_n \qquad \Delta_n^2 = \Delta_{n+1} - \Delta_n \qquad \text{et où } D_n = S_n - S$$

on voit immédiatement que

$$\Delta_n = (A - I + E_n) D_n$$

$$\Delta_n^2 = [(A - I)^2 + E'_n] D_n \text{ avec } E'_n = AE_n + E_{n+1} A - 2E_n + E_{n+1} E_n$$

et par conséquent $\lim_{n\to\infty} E'_n = 0$; d'où

$$\varepsilon_2^{(n)} - S = (A + E_n) D_n - (A - I + E_n) [(A - I)^2 + E'_n]^{-1}[(A - I)(A + E_n) + E_{n+1}(A+E_n)]D_n$$

donc $\lim_{n\to\infty} \varepsilon_2^{(n)} = S.$

On voit aussi que $\lim_{n\to\infty}(\varepsilon_2^{(n)} - S) D_n^{-1} = 0$

Ce théorème est une généralisation d'un résultat obtenu par Henrici [110] pour l'ε-algorithme scalaire.

On voit également que si $E_n = 0$ $\forall n$ alors $\varepsilon_2^{(n)} = S$ $\forall n$. On peut démontrer le même théorème que le théorème 86 si la suite $\{S_n\}$ est telle que :

$$S_{n+1} - S = (S_n - S)(A + E_n)$$

Par contre il est impossible de démontrer des résultats analogues dans les cas où :

$$S_{n+1} - S = A(S_n - S) + (S_n - S) E_n$$

$$S_{n+1} - S = (S_n - S) A + E_n(S_n - S)$$

Sur les applications de l'ε-algorithme matriciel on pourra consulter [242]. Wynn a égaleme étudié l'application de l'ε-algorithme à des suites de matrices rectangulaires [214] en utilisant la notion de pseudo-inverse définie par Moore [145] et Penrose [151]. Concernant ces applications il a formulé un certain nombre de conjectures ; des contre exemples pour quelques unes d'entre elles ont été trouvées par Greville [105]. Sur l'ε-algorithme matriciel signalons également le travail de Pyle [160].

La théorie de l'ε-algorithme matriciel est exposé dans [215] ; Wynn montre que le calcul des approximants de Padé matriciels peut se faire avec l'ε-algorithme matriciel.

V - 2 Transformation de suites dans un espace de Banach

Soit E un espace de Banach réel de norme notée $||.||$. Soit $\{S_n\}$ une suite d'éléments de E qui converge vers $S \in E$. On peut transformer cette suite à l'aide de l'ε-algorithme modifié de la façon suivante [40] :

$$\varepsilon_{-1}^{(n)} = 0 \in E \qquad \varepsilon_0^{(n)} = S_n \qquad n = 0, 1, \ldots$$

$$\varepsilon_{k+1}^{(n)} = \varepsilon_{k-1}^{(n+1)} + \frac{\Delta\varepsilon_k^{(n)}}{||\Delta\varepsilon_k^{(n)}||^2} \qquad n, k = 0, 1, \ldots$$

avec comme d'habitude $\Delta\varepsilon_k^{(n)} = \varepsilon_k^{(n+1)} - \varepsilon_k^{(n)}$.

Cette règle de l'ε-algorithme n'est pas le fruit du hasard. D'abord lorsque $E = \mathbb{R}$ on retrouve l'ε-algorithme scalaire et ensuite cette règle généralise celle de l'ε-algorithme vectoriel que nous étudierons au paragraphe suivant. On peut démontrer les résultats suivants :

<u>Propriété 31</u> : $\varepsilon_k^{(n)} \in E \quad n, k = 0, 1, \ldots$

c'est évident.

<u>Propriété 32</u> : Si l'application de l'ε-algorithme à $\{S_n\}$ et à $\{a S_n + b\}$ fournit respectivement les quantités $\varepsilon_k^{(n)}$ et $\bar{\varepsilon}_k^{(n)}$ alors :

$$\bar{\varepsilon}_{2k}^{(n)} = a\,\varepsilon_{2k}^{(n)} + b \qquad \bar{\varepsilon}_{2k+1}^{(n)} = \varepsilon_{2k+1}^{(n)} / a$$

Propriété 33 :

$$\frac{\varepsilon_{k+2}^{(n-1)}-\varepsilon_k^{(n)}}{||\varepsilon_{k+2}^{(n-1)}-\varepsilon_k^{(n)}||^2} - \frac{\varepsilon_k^{(n)}-\varepsilon_{k-2}^{(n+1)}}{||\varepsilon_k^{(n)}-\varepsilon_{k-2}^{(n+1)}||^2} = \frac{\varepsilon_k^{(n+1)}-\varepsilon_k^{(n)}}{||\varepsilon_k^{(n+1)}-\varepsilon_k^{(n)}||^2} - \frac{\varepsilon_k^{(n)}-\varepsilon_k^{(n-1)}}{||\varepsilon_k^{(n)}-\varepsilon_k^{(n-1)}||^2}$$

Les démonstrations de ces propriétés sont faciles et laissées en exercice.

On voit d'ailleurs que dans cet ε-algorithme la quantité $\varepsilon_k^{(n)} / ||\Delta\varepsilon_k^{(n)}||^2$ joue le rô de $1 / \Delta\varepsilon_k^{(n)}$ dans l'ε-algorithme scalaire. Cela suggère de définir l'inverse y^{-1} de $y \in$ par :

Définition 22 :

Soit $y \in E$; on appelle inverse de y l'élément $y^{-1} \in E$ défini par $y^{-1} = \frac{y}{||y||^2}$

Propriété 34 : $(y^{-1})^{-1} = y$

La démonstration est évidente.

On peut définir l'ε-algorithme sur des ensembles plus généraux que les espaces de Banac à condition de pouvoir définir dans ces ensembles l'inverse y^{-1} d'un élément de telle sorte que $(y^{-1})^{-1} = y$. La propriété 33 reste alors vérifiée. L'étude de cette question est abordée au paragraphe V-6.

Donnons maintenant un certain nombre de résultats. Appelons H.1 l'ensemble des hypothès suivantes :

- Soit $\{S_n\}$ une suite d'éléments de E
- Soit $\{a_n\}$ une suite de nombres réels tels qu'il existe $\alpha < 1 < \beta$ tels que $a_n \notin [\alpha,\beta]$ $\forall n$ et que $a_n \neq 0$ $\forall n$
- $S_{n+1} - S = a_n(S_n - S)$ $\forall n$ avec $S \in E$

Théorème 86 :

Si on applique l'ε-algorithme à une suite $\{S_n\}$ qui vérifie les hypothèses H1 et si $a_n = a \neq 1$ $\forall n > N$ alors :

$$\varepsilon_2^{(n)} = S \quad \forall n > N$$

Démonstration : on a :

$$\Delta S_n = (S_{n+1} - S) - (S_n - S) = \frac{a_n - 1}{a_n} (S_{n+1} - S)$$

$$\varepsilon_1^{(n)} = \frac{a_n}{a_n - 1} \frac{S_{n+1} - S}{||S_{n+1} - S||^2} = \frac{a_n}{a_n - 1} (S_{n+1} - S)^{-1}$$

en utilisant la définition de l'inverse donnée précédemment. On trouve :

$$\Delta\varepsilon_1^{(n)} = \left(\frac{1}{a_{n+1} - 1} - \frac{a_n}{a_n - 1}\right)(S_{n+1} - S)^{-1}$$

$$(\Delta\varepsilon_1^{(n)})^{-1} = \left(\frac{1}{a_{n+1} - 1} - \frac{a_n}{a_n - 1}\right)^{-1} (S_{n+1} - S)$$

Si $a_n = a \neq 1 \ \forall n > N$ on obtient :

$$(\Delta\varepsilon_1^{(n)})^{-1} = S - S_{n+1}$$

d'où $\qquad \varepsilon_2^{(n)} = S_{n+1} + S - S_{n+1} = S \quad \forall n > N$

Ce théorème a été démontré par Greville [105] dans le cas d'un ensemble avec un inverse vérifiant la propriété 34.

<u>Théorème 87</u> :

Si on applique l'ε-algorithme à une suite $\{S_n\}$ qui vérifie les hypothèses H1 et qui converge vers S alors $\lim_{n\to\infty} \varepsilon_2^{(n)} = S$.

Démonstration : reprenons la démonstration précédente. Puisque $a_n \notin [\alpha,\beta]$ alors $\left(\frac{1}{a_{n+1} - 1} - \frac{a_n}{a_n - 1}\right)^{-1}$ est borné supérieurement en valeur absolue. Soit M cette borne. De plus on a :

$$\varepsilon_2^{(n)} - S = \left[1 + \left(\frac{1}{a_{n+1} - 1} - \frac{a_n}{a_n - 1}\right)^{-1}\right] (S_{n+1} - S) \text{ d'où } ||\varepsilon_2^{(n)} - S|| \leqslant (1 + M)\, ||S_{n+1} - S||$$

ce qui démontre le théorème.

<u>Théorème 88</u> :

Si on applique l'ε-algorithme à une suite $\{S_n\}$ qui vérifie les hypothèses H1 et qui converge vers S et si la suite $\{a_n\}$ converge vers a alors :

$$||\varepsilon_2^{(n)} - S|| = o\,(||S_{n+1} - S||)$$

démonstration : reprenons la démonstration du théorème 86. Puisque $a_n \notin [\alpha,\beta]$ avec $\alpha < 1 < \beta$ alors $a \neq 1$; d'où

$$\lim_{n\to\infty} \frac{1}{a_{n+1} - 1} - \frac{a_n}{a_n - 1} = -1$$

ce qui démontre le théorème.

Ce théorème est une généralisation d'un théorème d'Henrici [110] déjà cité et établi pour le procédé Δ^2 d'Aitken.

Appelons maintenant H2 l'ensemble suivant d'hypothèses :

- Soit $\{S_n\}$ une suite d'éléments de E
- Soit $\{e_n\}$ une suite d'éléments de E qui converge vers zéro
- Soit $y \in E$ avec $y \neq 0$
- Soit $\lambda \in \mathbb{R}$ avec $0 < \lambda < 1$
- $S_n - S = \lambda^n (y + e_n)$

Théorème 89 :

Si on applique l'ε-algorithme à une suite $\{S_n\}$ qui vérifie les hypothèses H2 alors :

$$\lim_{n\to\infty} \varepsilon_2^{(n)} = S$$

et $$||\varepsilon_2^{(n)} - S|| = o\,(||S_{n+k} - S||) \quad \forall k \geq 0$$

démonstration : nous ne donnerons pas le détail des calculs. On trouve que :

$$(\Delta\varepsilon_1^{(n)})^{-1} = \lambda^n v_n \quad \text{où} \quad v_n \in E \text{ et } \lim_{n\to\infty} v_n = -\lambda y$$

d'où

$$\lim_{n\to\infty} \frac{||\varepsilon_2^{(n)} - S||}{\lambda^{n+1}} = 0$$

ce qui démontre le théorème.

Une application de ce théorème est l'accélération des suites de vecteurs produits par relaxation :

$$S_{n+1} = AS_n + b$$

où S_n, $b \in \mathbb{R}^p$ et où A est une matrice carrée.

Supposons que le rayon spectral de A soit strictement inférieur à un ; alors I - A est inversible et la suite $\{S_n\}$ converge vers l'unique solution S du système (I - A)S = b. Supposons de plus que :

$$1 > \rho(A) = |\lambda_1| > |\lambda_2| \geq |\lambda_3| \geq \ldots \geq |\lambda_n|$$

où les λ_i sont les valeurs propres de A. On a alors

$$S_n - S = \lambda_1^n y_1 + \lambda_1^n e_n \text{ avec } \lim_{n\to\infty} e_n = 0 \in \mathbb{R}^p$$

$$\text{et } e_n = \left(\frac{\lambda_2}{\lambda_1}\right)^n y_1 + v_n \quad \text{avec } \lim_{n\to\infty} v_n = 0.$$

Les hypothèses H2 sont donc vérifiées et l'on a d'après le théorème 89 :

$$\lim_{n\to\infty} \frac{||S_{n+1} - S||}{||S_n - S||} = |\lambda_1|$$

$$\lim_{n\to\infty} \frac{||\varepsilon_2^{(n)} - S||}{||S_{n+k} - S||} = 0 \qquad \forall k \geq 0$$

Donnons enfin un dernier résultat :

Théorème 90 :

Si on applique l'ε-algorithme à une suite $\{S_n\}$ d'éléments de E qui converge vers S et telle que :

$$||\Delta S_{n+1}|| = a_n \; ||\Delta S_n||$$

avec $a_n \notin [\alpha,\beta]$ $\forall n$ et $\alpha < 1 < \beta$ alors :

$$\lim_{n\to\infty} \varepsilon_2^{(n)} = S$$

démonstration :

$$\varepsilon_2^{(n)} - S = S_{n+1} - S + (\Delta\varepsilon_1^{(n)})^{-1}$$

$$||\varepsilon_2^{(n)} - S|| \leq ||S_{n+1} - S|| + \frac{1}{||\Delta\varepsilon_1^{(n)}||^2}$$

$$\text{or } \Delta\varepsilon_1^{(n)} = \frac{\Delta S_{n+1}}{||\Delta S_{n+1}||^2} - \frac{\Delta S_n}{||\Delta S_n||^2}$$

$$\frac{1}{||\Delta\varepsilon_1^{(n)}||} \leq \frac{1}{\left| \frac{1}{||\Delta S_{n+1}||} - \frac{1}{||\Delta S_n||} \right|} = \frac{||\Delta S_n||}{\left|\frac{1}{a_n} - 1\right|}$$

ce qui termine la démonstration.

Des études théoriques et numériques plus poussées sont nécessaires pour cet algorithme.

Prenons $E = \mathbb{R}^2$ et munissons le de la norme du **max**. Considérons la suite de vecteurs produits par la relation :

$$S_{n+2} = S_{n+1} + S_n \text{ avec } S_0 = \binom{0}{1} \text{ et } S_1 = \binom{1}{0}$$

on obtient :

$$\varepsilon_4^{(0)} = \binom{7}{-9/2}$$

Cet exemple prouve que le théorème 35 démontré pour l'ε-algorithme scalaire ne peut pas être étendu à tout espace de Banach.

Si nous prenons maintenant $E = \mathbb{R}^p$ et la norme euclidienne définie par $||x||^2 = (x,x)$ où (x,x) désigne le produit scalaire de $\mathbb{R}^p$ alors on retrouve les règles de l'ε-algorithme vectoriel définit par Wynn [242] pour des vecteurs de $\mathbb{C}^p$. Nous allons maintenant étudier très en détail cet algorithme.

V - 3 L'ε-algorithme vectoriel

Soit $\{S_n\}$ une suite de vecteurs de $\mathbb{C}^p$. On peut transformer cette suite de vecteurs à l'aide de l'ε-algorithme proposé par Wynn [242]. Les règles de cet algorithme sont les suivantes :

$$\varepsilon_{-1}^{(n)} = 0 \qquad \varepsilon_0^{(n)} = S_n \qquad n = 0, 1, \ldots$$

$$\varepsilon_{k+1}^{(n)} = \varepsilon_{k-1}^{(n+1)} + (\Delta\varepsilon_k^{(n)})^{-1} \qquad n, k = 0, 1, \ldots$$

où l'inverse y^{-1} d'un vecteur $y \in \mathbb{C}^p$ est défini par :

$$y^{-1} = \frac{\bar{y}}{(y,y)}$$

$\bar{y}$ dénote le vecteur complexe conjugué du vecteur y et (y,y) est le produit scalaire dans $\mathbb{C}^p$: $(y,y) = \sum_{i=1}^{p} y_i \bar{y}_i$. On peut aussi définir l'inverse par $y/(y,y)$.

Il est bien évident que cet ε-algorithme vectoriel possède les propriétés énoncées au paragraphe précédent puisque $(y,y)^{1/2}$ est la norme euclidienne dans $\mathbb{C}^p$. Les théorèmes démontrés au paragraphe précédent restent également valables dans le cas qui nous intéresse maintenant.

Le théorème fondamental sur lequel repose la théorie algébrique de l'ε-algorithme vectoriel a été conjecturé par Wynn [214] et démontré par Mc Leod [139]. Ce résultat est le suivant :

Théorème 91 :

Si on applique l'ε-algorithme vectoriel à une suite $\{S_n\}$ qui vérifie

$$\sum_{i=0}^{k} a_i (S_{n+i} - S) = 0 \qquad \forall n > N$$

où les a_i sont des nombres réels avec $a_k \neq 0$ alors :

$$\varepsilon_{2k}^{(n)} = S \qquad \forall n > N \qquad \text{si } \sum_{i=0}^{k} a_i \neq 0$$

$$\text{et} \qquad \varepsilon_{2k}^{(n)} = 0 \qquad \forall n > N \qquad \text{sinon}$$

Nous ne donnerons pas la démonstration de ce théorème car elle est complexe ; disons simplement qu'elle fait intervenir un isomorphisme entre vecteurs de $\mathbb{C}^p$ et matrices $2^p \times 2^p$.

Il faut remarquer que ce théorème est vérifié par ce que la norme de vecteurs correspondante au produit scalaire est la norme euclidienne. C'est ce que montre le contre exemple donné à la fin du paragraphe précédent où la norme du max est utilisée et où $\varepsilon_{2k}^{(n)} \neq S$ bien qu'une relation du type $\sum_{i=0}^{k} a_i (S_{n+i} - S) = 0$ soit vérifiée entre les vecteurs.

On remarquera que la condition du théorème 91 est une condition suffisante alors que le théorème 35 pour l'ε-algorithme scalaire montrait que cette condition est nécessaire et suffisante. Peut-être une démonstration du théorème 91 différente de celle de Mc Leod et plus directement liée à l'algorithme permettrait-elle de montrer que cette condition est nécessaire. On remarquera qu'on ne possède pas pour l'ε-algorithme vectoriel de définition des quantités $\varepsilon_k^{(n)}$ à l'aide de déterminants.

On remarquera également que la condition du théorème 91 impose aux a_i d'être des nombres réels. Il est vraisemblable que le théorème reste vrai pour des a_i complexes mais cela n'a pas pu être démontré alors que tous les exemples numériques le prouvent. Il serait très intéressant de démontrer le même résultat avec des a_i complexes.

On a la :

<u>Propriété 35</u> : Si l'application de l'ε-algorithme vectoriel à $\{S_n\}$ et à $\{a\, S_n + b\}$ où $a \in \mathbb{R}$ et où b est un vecteur de même dimension que S_n fournit respectivement les vecteurs $\varepsilon_k^{(n)}$ et $\overline{\varepsilon}_k^{(n)}$ alors :

$$\overline{\varepsilon}_{2k}^{(n)} = a\, \varepsilon_{2k}^{(n)} + b \qquad \overline{\varepsilon}_{2k+1}^{(n)} = \varepsilon_{2k+1}^{(n)} / a$$

démonstration :

$\overline{\varepsilon}_0^{(n)} = a\, \varepsilon_0^{(n)} + b$ est vérifiée. On a :

$$\overline{\varepsilon}_1^{(n)} = \frac{\Delta\overline{\varepsilon}_0^{(n)}}{||\Delta\overline{\varepsilon}_0^{(n)}||^2} = \frac{a\,\Delta\varepsilon_0^{(n)}}{a^2||\Delta\varepsilon_0^{(n)}||^2} = \frac{\varepsilon_1^{(n)}}{a}$$

Supposons qu'elle le soit jusqu'aux colonnes 2k et 2k+1 et démontrons qu'elle est vraie pour les colonnes 2k+2 et 2k+3 :

$$\overline{\varepsilon}_{2k+2}^{(n)} = \overline{\varepsilon}_{2k}^{(n+1)} + \frac{\Delta\overline{\varepsilon}_{2k+1}^{(n)}}{||\Delta\overline{\varepsilon}_{2k+1}^{(n)}||^2}$$

$$= a\,\varepsilon_{2k}^{(n+1)} + b + \frac{\Delta\varepsilon_{2k+1}^{(n)}\ a^2}{a||\Delta\varepsilon_{2k+1}^{(n)}||^2}$$

$$= a(\varepsilon_{2k}^{(n+1)} + \frac{\Delta\varepsilon_{2k+1}^{(n)}}{||\Delta\varepsilon_{2k+1}^{(n)}||^2}) + b = a\,\varepsilon_{2k+1}^{(n)} + b$$

$$\overline{\varepsilon}_{2k+3}^{(n)} = \overline{\varepsilon}_{2k+1}^{(n+1)} + \frac{\Delta\overline{\varepsilon}_{2k+2}^{(n)}}{||\Delta\overline{\varepsilon}_{2k+2}^{(n)}||^2} = \frac{\varepsilon_{2k+1}^{(n+1)}}{a} + \frac{a\,\Delta\varepsilon_{2k+2}^{(n)}}{a^2||\Delta\varepsilon_{2k+2}^{(n)}||^2}$$

$$= \frac{\varepsilon_{2k+3}^{(n)}}{a}$$

Si a est remplacé par une matrice orthogonale A tous les vecteurs $\varepsilon_k^{(n)}$

sont multipliés par A.

Partant du théorème 91, Gekeler [83] a démontré le :

Théorème 92 :

Si on applique l'ε-algorithme vectoriel à la suite de vecteurs $\{x_n\}$ produits par $x_{n+1} = Ax_n + b$ avec x_0 donné où A est une matrice carrée réelle telle que I - A soit inversible alors :

$$\varepsilon_{2m}^{(n)} = x \qquad \text{pour } n = 0, 1, \ldots$$

où $x = (I - A)^{-1} b$ et où m est le degré du polynôme minimal de A pour le vecteur $x_0 - x$.

Démonstration : Soit $p(t) = \sum_{i=0}^{m} a_i t^i$ le polynôme minimal de A pour le vecteur $x_0 - x$.

On a, par définition du polynôme minimal d'une matrice pour un vecteur :

$$\left(\sum_{i=0}^{m} a_i A^i \right)(x_0 - x) = 0$$

Puisque la matrice I - A est inversible A ne possède pas la valeur propre $\lambda = 1$ et par conséquent p(t) ne possède pas un comme racine ; par conséquent on a :

$$p(1) = \sum_{i=0}^{m} a_i \neq 0$$

D'autre part on a $x = Ax + b$ d'où :

$$x_{n+1} - x = A(x_n - x)$$

et

$$x_k - x = A^k(x_0 - x) \quad \forall k \geq 0$$

D'où, en portant dans p(t) :

$$\sum_{i=0}^{m} a_i(x_i - x) = 0$$

ou encore $A^n \sum_{i=0}^{m} a_i(x_i - x) = \sum_{i=0}^{m} a_i(x_{n+i} - x) = 0 \ \forall n$

ce qui démontre, en utilisant le théorème 91, que

$$\varepsilon_{2m}^{(n)} = x \qquad \forall n \geq 0$$

On voit donc que l'ε-algorithme vectoriel fournit une méthode directe de résolution des systèmes d'équations linéaires. Le résultat du théorème précédent ainsi que celui du théorème 91 ont été généralisés par Brezinski [29]. Nous allons maintenant exposer ces résultats.

Théorème 93 :

Supposons que les conditions du théorème 92 soient vérifiées et que la matrice A soit singulière. Appelons r la multiplicité de la racine nulle pour le polynôme minimal de A pour le vecteur $x_0 - x$. On a alors :

$$\varepsilon_{2(m-r)}^{(n+r)} = x \qquad \text{pour } n = 0, 1, \ldots$$

démonstration : si p(t) admet la racine zéro avec la multiplicité r alors $a_0 = a_1 = \ldots = a_{r-1} = 0$. Par conséquent on a :

$$\sum_{i=r}^{m} a_i (x_{n+i} - x) = 0 \qquad \text{pour } n = 0, 1, \ldots$$

ce qui peut encore s'écrire :

$$\sum_{i=0}^{m-r} b_i (x_{n+r+i} - x) = 0 \qquad \text{pour } n = 0, 1, \ldots$$

avec $b_i = a_{r+i}$. D'où le résultat en appliquant le théorème 91.

Nous allons maintenant étudier ce qui se passe lorsque la matrice I-A est singulière

Théorème 94 :

Appliquons l'ε-algorithme vectoriel à la suite de vecteurs $\{x_n\}$ produits par $x_{n+1} = A x_n + b$ avec x_0 donné où A est une matrice carrée réelle telle que I - A soit singulière Soit x une solution du système x = Ax + b, soit m le degré du polynôme minimal de A pour le vecteur $x_0 - x$, q la multiplicité de la racine $\lambda = 1$ pour ce polynôme et r la multiplicité de la racine $\lambda = 0$ pour ce polynôme (avec éventuellement r = 0).

Si $b \in \mathrm{Im}(I - A)$ et si q = 1 alors $\varepsilon_{2(m-r)-2}^{(n+r)} = x$ pour n = 0, 1, ...

Si $b \in \mathrm{Im}(I - A)$ et si q = 2 alors $\varepsilon_{2(m-r)-3}^{(n+r)}$ = vecteur constant pour n = 0, 1, ...

Enfin si $b \notin \mathrm{Im}(I - A)$ soit m' le degré du polynôme minimal de A pour le vecteur Δx_0, q' et r' les multiplicités respectives des racines 1 et 0 pour ce polynôme minimal.

Si q' = 1 alors $\varepsilon_{2(m'-r')-1}^{(n+r')}$ = vecteur constant pour n = 0, 1, ...

Démonstration : Etudions d'abord le cas où $b \in Im(I-A)$ c'est-à-dire le cas où le système linéaire admet une infinité de solutions. Puisque $I-A$ est singulière $\lambda = 1$ est valeur propre de A et par conséquent racine de son polynôme minimal $\sum_{i=0}^{m} a_i t^i$. Supposons que ce soit une racine simple alors :

$$\sum_{i=0}^{m} a_i t^i = (1 - t) \sum_{i=0}^{m-1} b_i t^i \quad \text{avec} \quad \sum_{i=0}^{m-1} b_i \neq 0 \text{ puisque } \lambda = 1$$

n'est plus racine du polynôme $\sum_{i=0}^{m-1} b_i t^i$. On a :

$$(I - A) \sum_{i=0}^{m-1} b_i A^i (x_0 - x) = 0$$

d'où

$$(I - A) \sum_{i=0}^{m-1} b_i (x_i - x) = 0$$

ou encore
$$\sum_{i=0}^{m-1} b_i (x_i - x) = \sum_{i=0}^{m-1} b_i (x_{i+1} - x)$$

ce qui démontre que $\sum_{i=0}^{m-1} b_i x_{n+i}$ = vecteur constant pour $n = 0, 1, \ldots$ Puisque $\sum_{i=0}^{m-1} b_i \neq 0$ nous noterons $y \sum_{i=0}^{m-1} b_i$ ce vecteur constant. On aura donc d'après le théorème 91 :

$$\varepsilon_{2(m-1)}^{(n)} = y \quad \text{pour } n = 0, 1, \ldots$$

Démontrons maintenant que y vérifie $y = Ay + b$

on aura donc :

$$\sum_{i=0}^{m-1} b_i (x_{n+i} - y) = 0 \qquad n = 0, 1, \ldots$$

ou encore

$$(I - A) \sum_{i=0}^{m-1} b_i x_{n+i} = (I - A)\, y \sum_{i=0}^{m-1} b_i$$

or $(I - A) \sum_{i=0}^{m-1} b_i x_{n+i} = (I - A)\, x \sum_{i=0}^{m-1} b_i$ d'après ce qui précède, ce qui démontre

que $(I - A)\, y = (I - A)x = b$.

Si A est une matrice singulière on trouve immédiatement le premier résultat énoncé dans ce théorème en utilisant le théorème 93.

Considérons maintenant le cas q = 2, on a :

$$\sum_{i=0}^{m} a_i t^i = (1-t)^2 \sum_{i=0}^{m-2} c_i t^i \text{ avec } \sum_{i=0}^{m-2} c_i \neq 0 \text{ d'où}$$

$$(I - A)^2 \sum_{i=0}^{m-2} c_i (x_{n+i} - x) = 0 \quad \text{pour } n = 0, 1, \ldots$$

d'où

$$(I-A) \sum_{i=0}^{m-2} c_i (x_{n+i}-x) = (I-A) \sum_{i=0}^{m-2} c_i A(x_{n+i}-x) \quad n = 0, 1, \ldots$$

ce qui montre que :

$$(I - A) \sum_{i=0}^{m-2} c_i (x_{n+i} - x) = \text{vecteur constant } \forall n$$

donc $\quad (I - A) \sum_{i=0}^{m-2} c_i \Delta x_{n+i} = 0 \quad \text{pour } n = 0, 1, \ldots$

et $\quad \sum_{i=0}^{m-2} c_i \Delta x_{n+i} = \sum_{i=0}^{m-2} c_i A\Delta x_{n+i} \quad n = 0, 1, \ldots$

ce qui démontre que :

$$\sum_{i=0}^{m-2} c_i \Delta x_{n+i} = S \quad \forall n$$

où S est un vecteur constant.

Par conséquent, si on applique l'ε-algorithme vectoriel à la suite $\{\Delta x_n\}$ on aura :

$$\varepsilon_{2m-4}^{(n)} = S / \sum_{i=0}^{m-2} c_i \quad \forall n$$

d'après le théorème 91.

En utilisant une propriété de l'ε-algorithme vectoriel établie par Wynn [229], on sait que l'application de l'ε-algorithme vectoriel à la suite $\{x_n\}$ donnera $\varepsilon_{2m-3}^{(n)}$ = vecteur constant $\forall n$.

Etudions maintenant le cas où $b \notin \text{Im}(I - A)$. Dans ce cas le système $x = Ax + b$ n'admet pas de solution. Soit $\sum_{i=0}^{m'} a'_i t^i$ le polynôme minimal de A pour le vecteur Δx_0.

On a : $\Delta x_{n+1} = A\ \Delta x_n$

d'où $\Delta x_k = A^k\ \Delta x_0 \quad \forall k \geq 0$

et $\sum_{i=0}^{m'} a'_i\ A^i\ \Delta x_0 = 0$

si $q' = 1$ on a $\sum_{i=0}^{m'} a'_i\ t^i = (1-t) \sum_{i=0}^{m'-1} b'_i\ t^i$ avec $\sum_{i=0}^{m'-1} b'_i \neq 0$

Donc $(I - A) \sum_{i=0}^{m'-1} b'_i\ \Delta x_{n+i} = 0\ \forall n$

et par conséquent :

$$\sum_{i=0}^{m'-1} b'_i\ \Delta x_{n+i} = \text{vecteur constant } S\ \forall n$$

L'application de l'ε-algorithme vectoriel à $\{\Delta x_n\}$ donnera donc $\varepsilon_{2m'-2}^{(n)} = S / \sum_{i=0}^{m'-1} b'_i \quad \forall n$

d'après le théorème 91, et l'application de l'algorithme à $\{x_n\}$ donnera :

$$\varepsilon_{2m'-1}^{(n)} = \text{vecteur constant } \forall n$$

ce qui termine la démonstration du théorème 94.

REMARQUES :

1°) Dans le cas où $b \in \text{Im}\ (I-A)$ on ne peut rien dire si $q > 2$. Cependant, en utilisant d'autres résultats démontrés par Wynn [228] on peut émettre la conjecture que $\exists\ 1 \leq k \leq m$ tel que $\varepsilon_{2(m-k-r)+1}^{(n+r)}$ = vecteur constant pour $n = 0, 1, \ldots$ Des exemples numériques confirment cette conjecture. On peut émettre la même conjecture si $b \notin \text{Im}(I-A)$.

2°) Si k est le rang de $I-A$ on montre facilement que $m \leq k+1$

3°) Les résultats du théorème 94 peuvent être appliqués à la résolution de systèmes linéaires rectangulaires. Dans ce cas on complète la matrice avec des lignes ou des colonnes de zéro et on applique l'ε-algorithme vectoriel aux vecteurs produits à l'aide de la matrice ainsi complétée. Ainsi, dans le cas d'un système avec plus d'équations que d'inconnues si le rang est égal au nombre de colonnes de la matrice alors la solution du système est unique. L'utilisation de l'ε-algorithme vectoriel avec la matrice complétée évite d'avoir à chercher quelles sont les équations du système qui sont linéairement indépendantes.

Ceci peut être appliqué au calcul des vecteurs propres d'une matrice dont on connait les valeurs propres.

Nous allons maintenant considérer l'application de l'ε-algorithme vectoriel à des vecteurs produits par :

$$x_n = \sum_{i=1}^{k} A_i \, x_{n-i} + b$$

où b est un vecteur, où $x_0, \ldots, x_{k-1}$ sont donnés et où les A_i sont des matrices carrées réelles. Cette équation d'ordre k peut être transformée en un système de k équations du premier ordre de la façon suivante :

On pose $y_n = x_n - x$ puis

$$y_{n+1}^{(1)} = y_n$$

$$y_{n+1}^{(2)} = y_{n-1} = y_n^{(1)}$$

$$\text{-------------------}$$

$$y_{n+1}^{(k-1)} = y_{n-k+2} = y_n^{(k-2)}$$

$$y_{n+1} = A_1 \, y_n + A_2 \, y_n^{(1)} + \ldots + A_k \, y_n^{(k-1)}$$

D'où le système équivalent $Y_{n+1} = A \, Y_n$:

$$\begin{pmatrix} y_{n+1} \\ y_{n+1}^{(1)} \\ \vdots \\ y_{n+1}^{(k-1)} \end{pmatrix} = \begin{pmatrix} A_1 \; \ldots\ldots\ldots \; A_k \\ I \; \ldots\ldots\ldots \; 0 \\ \text{------------} \\ 0 \; \ldots\ldots\ldots \; I0 \end{pmatrix} \begin{pmatrix} y_n \\ y_n^{(1)} \\ \vdots \\ y_n^{(k-1)} \end{pmatrix}$$

Supposons que la matrice $B = I - \sum_{i=1}^{k} A_i$ soit inversible. Alors on déduit immédiatemen d'un résultat de Gantmacher [82] que $\lambda = 1$ n'est pas valeur propre de A et que, par conséquent, I - A est inversible. On peut donc appliquer les théorème 92 et 93. En utilisant le même résultat de Gantmacher on voit que si $\lambda = 1$ est valeur propre de A alors la matrice B est singulière. On peut donc, dans ce cas, appliquer le théorème

94 et par conséquent on a le résultat suivant [29] :

<u>Théorème 95</u> :

Appliquons l'ε-algorithme vectoriel à des vecteurs produits par :

$x_n = \sum_{i=1}^{k} A_i x_{n-i} + b$ où les A_i sont des matrices carrées réelles.

Posons $B = I - \sum_{i=1}^{k} A_i$ et

$$A = \begin{pmatrix} A_1 & \ldots & A_k \\ I & \ldots & 0 \\ \hline 0 & \ldots I & 0 \end{pmatrix}$$

soit m le degré du polynôme minimal de A pour le vecteur :

$$y_0 = \begin{pmatrix} x_k - x \\ \hline x_0 - x \end{pmatrix}$$

et r la multiplicité éventuelle de la racine $\lambda = 0$ pour ce polynôme minimal.

1°) Si B est inversible alors $\varepsilon_{2(m-r)}^{(n+r)} = x$ pour $n = 0, 1, \ldots$ où x est la solution unique du système $Bx = b$. De plus on a $k \leq m \leq kp$ où p est la dimension des vecteurs x_n et b.

2°) Si B est singulière appelons q la multiplicité de la racine $\lambda = 1$ pour le polynôme minimal précédent.

Si $b \in Im(B)$ et si $q = 1$ alors $\varepsilon_{2(m-r)-2}^{(n+r)} = x$ pour $n = 0, 1, \ldots$ où x est une solution de $Bx = b$.

Si $q = 2$ alors $\varepsilon_{2(m-r)-3}^{(n+r)}$ = vecteur constant pour $n = 0, 1, \ldots$

3°) Si B est singulière et si $b \notin Im(B)$ appelons m' le degré du polynôme minimal de A pour le vecteur Δy_0, r' et q' les multiplicités des racines $\lambda = 0$ et $\lambda = 1$ pour ce polynôme minimal. Si $q' = 1$ alors $\varepsilon_{2(m'-r')-1}^{(n+r')}$ = vecteur constant pour $n = 0, 1, \ldots$

Démonstration : la seule démonstration qu'il reste à faire est que $k \leqslant m \leqslant kp$. Il est évident que $m \leqslant kp$. Démontrons que $k \leqslant m$. Pour cela supposons que les polynômes minimaux de $A_1, \ldots, A_k$ soient du premier degré. Alors $A_i = \lambda_i I$ pour $i = 1, \ldots, k$ où λ_i est la valeur propre unique de A_i.

Considérons maintenant A comme une matrice k x k sur l'anneau des matrices p x p. Son polynôme minimal est :

$$\lambda^k - A_1 \lambda^{k-1} - \ldots - A_{k-1} \lambda - A_k$$

A annule son polynôme minimal, donc :

$$A^k - A_1 A^{k-1} - \ldots - A_{k-1} A - A_k = 0$$

où le produit $A_i A^{k-i}$ est le produit de chaque élément p x p de A^{k-i} par A_i au sens du produit matriciel. Remplaçons A_i par $\lambda_i I$ on obtient :

$$A^k - \lambda_1 A^{k-1} - \ldots - \lambda_k = 0$$

ce qui démontre que $k \leqslant m$

REMARQUE :

La première partie de ce théorème (B inversible) est une généralisation du théorème fondamental 91.

Une question qui se pose pour l'ε-algorithme vectoriel est de trouver toutes les suites de vecteurs qui vérifient le théorème 94. Au cas où ce théorème serait également une condition nécessaire on aurait ainsi obtenu la forme de toutes les suites pour lesquelle il existe k tel que $\varepsilon_{2k}^{(n)}$ = vecteur constant quelquesoit n. Une telle recherche nécessiterait la résolution complète du polynôme matriciel :

$$X + A_1 X^{k-1} + \ldots + A_k = 0$$

où l'inconnue X est une matrice. C'est un problème difficile car l'anneau des matrices p x p est non communicatif et non intègre. On peut cependant obtenir des résultats partiels [29] :

<u>Théorème 96</u> :

Si on applique l'ε-algorithme vectoriel à des vecteurs x_n produits par :

$$x_n = x + \sum_{i=1}^{k} \Lambda_i^n z_i$$

où les z_i et x sont des vecteurs et les Λ_i des matrices carrées réelles telles que $I - \Lambda_1, \ldots, I - \Lambda_k$ soient inversibles alors $\varepsilon_{2(m-r)}^{(n+r)} = x$ pour n = 0, 1, ... où m est le degré du polynôme minimal de

$$\Lambda = \begin{pmatrix} \Lambda_1 & \Lambda_2 - \Lambda_1 & \ldots\ldots & \Lambda_{k-1} - \Lambda_{k-2} & \Lambda_k - \Lambda_{k-1} \\ 0 & \Lambda_2 & \ldots\ldots & \Lambda_{k-1} - \Lambda_{k-2} & \Lambda_k - \Lambda_{k-1} \\ \hline 0 & 0 & \ldots\ldots & \Lambda_{k-1} & \Lambda_k - \Lambda_{k-1} \\ 0 & 0 & \ldots\ldots & 0 & \Lambda_k \end{pmatrix}$$

pour le vecteur

$$y = \begin{pmatrix} z_1 + \ldots + z_k \\ z_2 + \ldots + z_k \\ \hline z_k \end{pmatrix}$$

et r la multiplicité éventuelle de la racine $\lambda = 0$ pour ce polynôme minimal.

Démonstration : posons $y_n = x_n - x$ et $y_n^{(i)} = \sum_{j=i}^{k} \Lambda_j^n \; z_j$. On a $y_n = y_n^{(1)}$ et :

$$\begin{pmatrix} y_{n+1} \\ y_{n+1}^{(2)} \\ \vdots \\ y_{n+1}^{(k)} \end{pmatrix} = \Lambda \begin{pmatrix} y_n \\ y_n^{(2)} \\ \vdots \\ y_n^{(k)} \end{pmatrix}$$

que l'on démontre par induction mathématique :

$$y_{n+1}^{(k)} = \Lambda_k \; y_n^{(k)}$$

$$y_{n+1}^{(i)} = \Lambda_i \; y_n^{(i)} + \sum_{j=i+1}^{k} (\Lambda_j - \Lambda_{j-1}) \; y_n^{(j)} \quad \text{pour } i = 1, \ldots, k-1$$

en remplaçant $y_n^{(j)}$ par son expression et en réarrangeant l'ordre des termes dans les sommations on obtient

$$y_{n+1}^{(i)} = \sum_{j=i}^{k} \Lambda_j^{n+1} \; z_i$$

Les valeurs propres de Λ sont l'union des valeurs propres de $\Lambda_1, \ldots, \Lambda_k$. Donc si $\lambda = 1$ n'est pas valeur propre de $\Lambda_1, \ldots, \Lambda_k$ alors ce ne sera pas une valeur propre de Λ. Par conséquent $I - \Lambda$ sera inversible et l'on pourra appliquer le théorème 93 ce qui termine la démonstration.

REMARQUES :

1°) Dans le théorème précédent on a fait l'hypothèse que les matrices $I - \Lambda_1, \ldots, I - \Lambda_k$ étaient inversibles. Or on a vu au théorème 37 pour l'ε-algorithme scalaire que si l'équation aux différences admettait des racines nulles alors l'ε-algorithme donnait encore la réponse exacte.

La même propriété reste valable pour l'ε-algorithme vectoriel en prenant des matrices nilpotentes pour certaines matrices Λ_i. Des expériences numériques montrent également que si certains Λ_i^n sont multipliés par un polynôme en n ou par cos n b_i alors il existe m tel que $\varepsilon_{2m}^{(n)} = x \; \forall n$. Il reste encore à le démontrer.

2°) Si $x_n = x + \sum_{i=1}^{k} A_i \Lambda_i^n z_i$ où les A_i sont des matrices carrées réelles non singulières alors, avec les mêmes hypothèses, les conclusions du théorème 96 restent valables.

En effet posons $B_i = A_i \Lambda_i A_i^{-1}$ et $y_i = A_i z_i$ on a

$$B_i^n y_i = A_i \Lambda_i^n A_i^{-1} A_i z_i = A_i \Lambda_i^n z_i$$

3°) Si les vecteurs $\{x_n\}$ sont tels que

$$\sum_{i=0}^{k} a_i (x_{n+i} - x) = 0 \qquad \text{pour } n = 0, 1, \ldots$$

avec $a_k \neq 0$, les a_i réels et $\sum_{i=0}^{k} a_i \neq 0$

alors l'application de l'ε-algorithme scalaire aux suites de chacune des composantes des vecteurs x_n donne les composantes du vecteur x comme l'ε-algorithme vectoriel.

4°) Tous les résultats de ce paragraphe restent valables si la matrice correspondant à la matrice A du théorème 92 est une matrice complexe mais dont les coefficients du polynôme minimal par rapport au vecteur considéré sont des nombres réels.

L'utilisation pratique de l'ε-algorithme vectoriel est limitée par la propagation des erreurs dues à l'ordinateur lorsque deux vecteurs consécutifs d'une même colonne sont voisins.

Cordellier [66] a récemment obtenu des règles particulières pour éviter cette propagation ainsi que la règle singulière à utiliser lorsque deux vecteurs consécutifs d'une colonne sont strictement égaux. Ces règles particulières et singulières sont l'équivalent, pour l'ε-algorithme vectoriel, de celles obtenues par Wynn pour l'ε-algorithme scalaire (paragraphe IV-8).

Normalement, pour l'ε-algorithme vectoriel, la règle de la croix de la propriété 33, permet de calculer E dès que N,C,S et W sont connus :

$$E = C + [(N - C)^{-1} + (S - C)^{-1} - (W - C)^{-1}]^{-1}$$

Donnons la règle particulière de Cordellier qui est algébriquement équivalente à la règle de la croix. Nous avons auparavant besoin du :

Lemme 17ter : Si $a, b \in \mathcal{C}^p - \{0\}$ alors :

$$\|a^{-1} + b^{-1}\|^2 = \|a+b\|^2 \ / \ \|a\|^2 \ \|b\|^2$$

où $\|y\|^2 = (y,y)$.

Démonstration :

$$a^{-1} + b^{-1} = a \ / \ \|a\|^2 + b \ / \ \|b\|^2$$

$$\begin{aligned} \|a^{-1} + b^{-1}\|^2 &= (a \ / \ \|a\|^2 + b \ / \ \|b\|^2, \ a \ / \ \|a\|^2 + b \ / \ \|b\|^2) \\ &= 1/ \ \|a\|^2 + 2(a,b) \ / \ \|a\|^2 \ \|b\|^2 + 1 \ / \ \|b\|^2 \\ &= (a + b, \ a + b) \ / \ \|a\|^2 \ \|b\|^2 \end{aligned}$$

La règle particulière est maintenant donnée par le :

Théorème 96bis : Si les vecteurs N, S et W sont différents de C et si $(N-C)^{-1} + (S-C)^{-1} \neq (W-C)^{-1}$ alors E est donné par :

$E = (\gamma_N N + \gamma_S S - \gamma_W W + \lambda C) / (\gamma_N + \gamma_S - \gamma_W + \lambda)$ avec :

$\gamma_N = \|N-C\|^{-2}$, $\gamma_S = \|S-C\|^{-2}$, $\gamma_W = \|W-C\|^{-2}$

et

$\lambda = \gamma_N \gamma_W \|N-W\|^2 + \gamma_W \gamma_S \|W-S\|^2 - \gamma_S \gamma_N \|S-N\|^2.$

Démonstration : définissons μ par :

$$\mu = \|(N-C)^{-1} + (S-C)^{-1} - (W-C)^{-1}\|^2.$$

Puisque N, S et W sont différents de C alors μ est fini et non nul à cause de la condition du théorème. Un calcul simple donne :

$$\mu = \gamma_N + \gamma_S - \gamma_W + \|(N-C)^{-1} - (W-C)^{-1}\| + \|(S-C)^{-1} - (W-C)^{-1}\| - \|(N-C)^{-1} - (S-C)^{-1}\|$$

En utilisant trois fois la relation du lemme 17ter on trouve que :

$$\mu = \gamma_N + \gamma_S - \gamma_W + \lambda$$

Puisque $\mu \neq 0$ la relation de la croix peut s'écrire :

$$\begin{aligned} E &= C + [(N-C)^{-1} + (S-C)^{-1} - (W-C)^{-1}] / \mu \\ &= [(N-C)^{-1} + (S-C)^{-1} - (W-C)^{-1} + \mu C] / \mu \\ &= [\gamma_N(N-C) + \gamma_S(S-C) - \gamma_W(W-C) + \mu C] / \mu \\ &= [\gamma_N N + \gamma_S S - \gamma_W W + \lambda C] / \mu \end{aligned}$$

En faisant tendre C vers l'infini on obtient la règle singulière :

$$E = N + S - W$$

qui est exactement la même que celle obtenue par Wynn pour l'ε-algorithme scalaire.

Illustrons ceci par un exemple numérique. On considère les vecteurs $\{S_n\}$ de $\mathbb{R}^3$ générés par :

$$S_n = S_{n-4} / 2 \text{ pour } n = 4,5,\ldots$$

avec

$$S_0 = \begin{pmatrix} 1 \\ 0 \\ 0 \end{pmatrix} \qquad S_1 = \begin{pmatrix} 1 \\ 10^{-8} \\ 0 \end{pmatrix} \qquad S_2 = \begin{pmatrix} 0 \\ 0 \\ 0 \end{pmatrix} \qquad S_3 = \begin{pmatrix} 0 \\ 10^{-8} \\ 1 \end{pmatrix}$$

D'après le théorème 91 on doit théoriquement obtenir $\varepsilon_8^{(0)} = 0$. En pratique on trouve :

avec l'ε-algorithme de Wynn $\begin{pmatrix} 0,61 \\ -0,36 \\ 0,83 \end{pmatrix} 10^{-6}$

avec la règle particulière de Cordellier $\begin{pmatrix} 0,72 & 10^{-15} \\ 0,25 & 10^{-22} \\ 0,36 & 10^{-15} \end{pmatrix}$

IV - 4 Résolution de systèmes d'équations non linéaires par l'ε-algorithme vectoriel

Nous avons vu au paragraphe précédent que l'ε-algorithme vectoriel fournissait une méthode directe de résolution des systèmes d'équations linéaires.

Dans ce paragraphe nous allons montrer comment la même méthode appliquée à des systèmes d'équations non linéaires, fournit une méthode de résolution à convergence quadratique et cela sans avoir à effectuer ni calcul de dérivées ni inversion de matric[e] comme c'est malheureusement le cas pour la méthode de Newton. Cette méthode a été trouvée indépendemment et presque simultanément par Gekeler [83] et Brezinski [42 , 44

Soit à résoudre $x = F(x)$ où $F : \mathbb{R}^p \to \mathbb{R}^p$ et F différentiable au sens de Fréchet dans un voisinage d'une solution x que nous cherchons.

L'algorithme est le suivant :

x_0 donné

$(n+1)^{\text{ième}}$ itération $u_0 = x_n$

$u_k = F(u_{k-1})$ pour $k = 1, \ldots, 2m-r$

application de l'ε-algorithme aux vecteurs $u_0, \ldots, u_{2m-r}$ afin de calculer $\varepsilon_{2(m-r)}^{(r)}$

puis on prend :

$$x_{n+1} = \varepsilon_{2(m-r)}^{(r)}$$

m est le degré du polynôme minimal de $F'(x)$ pour le vecteur $x_n - x$ et r est la multiplicité éventuelle de la racine $\lambda = 0$ pour ce polynôme minimal ($r = 0$ si $F'(x)$ est inversible).

Le résultat fondamental est le suivant :

Théorème 97 :

Soit $F : \mathbb{R}^p \to \mathbb{R}^p$ telle qu'il existe $x \in \mathbb{R}^p$ qui vérifie $x = F(x)$, telle que F soit diffé rentiable au sens de Fréchet dans un voisinage de x et telle que $I - F'(x)$ soit inversible alors il existe un voisinage V de x tel que pour tout $x_0 \in V$ l'algorithme précédent converge vers x et ceci au moins quadratiquement c'est-à-dire que :

$$||x_{n+1} - x|| = O(||x_n - x||^2)$$

démonstration :

Si F est différentiable au sens de Fréchet au voisinage de x on a :

$$u_{k+1} - x = F'(x)(u_k - x) + O(||u_k - x||^2)$$

où la notation $O(||z_k||^2)$ désigne un vecteur y_k de $\mathbb{R}^p$ tel que $\forall k > K$

$||y_k|| \leqslant A\ ||z_k||^2$

Soit $p(t) = \sum_{i=0}^{m} a_i\, t^i$ le polynôme minimal de $F'(x)$ pour le vecteur $x_n - x$. Puisque $I - F'(x)$ est inversible on a $p(1) = \sum_{i=0}^{m} a_i \neq 0$

On a :

$$u_1 - x = F'(x)\ (u_0 - x) + O(||u_0 - x||^2)$$

et $u_k - x = [F'(x)]^k\ (u_0 - x) + O(||u_0 - x||^2)$

d'où en portant dans le polynôme minimal :

$$\sum_{i=0}^{m} a_i\ [F'(x)]^i\ (x_n - x) = \sum_{i=0}^{m} a_i(u_i - x) + O(||u_0 - x||^2) = 0$$

Par conséquent :

$$\sum_{i=0}^{m} a_i\ u_i = x \sum_{i=0}^{m} a_i + O\ (||x_n - x||^2)$$

puisque $u_0 = x_n$.

En utilisant les théorèmes 92, 93 ainsi que la continuité de l'ε-algorithme vectoriel on obtient :

$$\varepsilon_{2(m-r)}^{(r)} = x + O\ (||x_n - x||^2)$$

On pourra trouver la démonstration détaillée de ce théorème en [22] ou [83].

REMARQUES :

1°) si p = 1 l'algorithme proposé se réduit à la méthode de Steffesen :

$$x_{n+1} = \frac{F(F(x_n)).\ x_n - [F(x_n)]^2}{F(F(x_n)) - 2F(x_n) + x_n}$$

dont il possède les propriétés. Il peut donc être considéré comme sa généralisation à p dimensions.

2°) Si F est affine on retrouve le fait que $x_1 = x$ quelquesoit x_0. Ce n'est autre que le résultat du théorème 93.

3°) On ne suppose pas que les itérations de base $u_k = F(u_{k-1})$ convergent. Il ne faut cependant pas que ces itérations divergent trop rapidement car cela pourrait entraîner une instabilité numérique comme nous le verrons plus loin.

4°) En pratique m et r sont inconnus. On prendra donc $m = p$ et $r = 0$ pour effectuer les calculs. Si ce n'était pas le cas une utilisation en parallèle de l'ε-algorithme vectoriel (voir paragraphe IV - 8) permettrait de stopper les itérations de base et le calcul du tableau ε car deux vecteurs consécutifs deviendraient égaux dans ce tableau.

5°) A la place de l'ε-algorithme vectoriel on aurait pu utiliser l'ε-algorithme scalaire sur chaque composante des vecteurs u_k. L'organisation des calculs est plus simple avec l'ε-algorithme vectoriel.

6°) Il est impossible, sans calculer de dérivées ni inverser de matrices, de construire une méthode quadratique qui nécessite moins d'itérations de base. En ce sens la méthode proposée ici est optimale. Ulm [177] a proposé une extension de la méthode de Steffensen qui est à convergence quadratique mais qui nécessite plus d'évaluation de F ainsi que de inversions de matrices.

Henrici [110] a proposé une méthode quadratique qui nécessite seulement p+1 évaluations de F et une inversion de matrice par itération. Les essais numériques effectués avec la méthode d'Henrici montrent que celle-ci est numériquement instable [146].

Si l'on possède des informations supplémentaires sur la convergence de $u_{k+1} - x - F'(x)(u_k - x)$ vers zéro, on peut alors améliorer le résultat du théorème précédent :

<u>Théorème 98</u> : Si, pour tout y, appartenant à un voisinage de x on a

$F(y) - x - F'(x)(y - x) = O(||y - x||^a)$ avec $a > 1$ alors la suite générée par l'algorithme précédent vérifie :

$$||x_{n+1} - x|| = O(||x_n - x||^a)$$

Si $F(y) - x - F'(x)(y-x) = o(||y - x||^a)$ alors

$$||x_{n+1} - x|| = o(||x_n - x||^a)$$

La démonstration est immédiate à partir de celle du théorème 97. Ce résultat est une généralisation d'un théorème obtenu par Ostrowski [147] dans le cas $p = 1$.

Donnons un exemple numérique : soit à trouver la solution unique $x = -1$, $y = 1$ du système :

$$x = -\frac{y^4}{4} - \frac{3}{4}$$

$$y = -0.405\, e^{1-x^2} + 1.405$$

Les itérations de base convergent lentement car les valeurs propres du jacobien calculées à la solution valent ± 0.9.

En partant de $x_0 = y_0 = 0$ on obtient :

n	x_n	y_n
1	-0.85	0.87
2	-0.96	0.97
3	-0.9979	0.9978
4	-0.999989	0.999984
5	-0.9999999997	0.9999999991

Si on avait utilisé l'ε-algorithme scalaire on aurait obtenu :

n	x_n	y_n
1	-0.85	0.87
2	-0.97	0.97
3	-0.9980	0.9978
4	-0.9999905	0.999984
5	-0.9999999998	0.9999999991

Dans ce cas on avait $p = m = 2$.

Considérons maintenant le cas suivant :

$$x = \frac{y^2}{2} + x - \frac{1}{2}$$

$$y = \sin x + \sin (y-1) + 1$$

dont une solution est $x = 0$, $y = 1$. On a :

$$F'(x) = \begin{pmatrix} 1 & 1 \\ 1 & 1 \end{pmatrix}$$

et par conséquent $r = 1$. On prendra donc $x_{n+1} = \varepsilon_2^{(1)}$.

Partant de $x_0 = 1/2$ et $y_0 = -1$ on trouve :

n	x_n	y_n
1	0.28	1.22
2	0.15	0.907
3	$0.11\ 10^{-1}$	0.9928
4	$0.61\ 10^{-4}$	0.999957
5	$0.19\ 10^{-8}$	0.9999999983
6	$0.52\ 10^{-17}$	1.00000000000000000

Signalons deux utilisations particulièrement intéressantes de cet algorithme.

Quand on intègre une équation différentielle par une méthode de type Runge-Kutta explicite on sait que l'on ne peut pas augmenter le pas d'intégration à loisir pour des raisons de stabilité numérique. On est donc amené à s'orienter vers des méthodes de Runge-Kutta implicites. Si le système d'équations différentielles est non linéaire alors, à chaque étape, il faut résoudre un système d'équations non linéaires. L'utilisation de la méthode que nous venons de décrire permet d'effectuer cette résolution dans de bonnes conditions [7].

La seconde utilisation est la résolution des problèmes aux limites en plusieurs points pour des systèmes d'équations différentielles [48] : considérons en effet le système de p équations différentielles :

$$y'(t) = f(t, y(t))$$

avec les conditions aux limites en plusieurs points :

$$g(y(t_1), \ldots, y(t_k)) = 0$$

où $g : \mathbb{R}^k \to \mathbb{R}^p$.

Nous voulons transformer ce problème aux limites en un problème de conditions initiales Appelons $y(t,x)$ la solution de cette équation différentielle qui vérifie la condition $y(t_1) = x$. Par conséquent il nous faut trouver x tel que :

$$g(x, y(t_2, x), \ldots, y(t_k, x)) = 0$$

ce qui n'est autre que la résolution d'un système de p équations non linéaires à p inconnues. Ce système est résolu à l'aide de l'algorithme précédent. On trouvera dans

[162] des détails sur cette méthode et dans [222] une méthode pour estimer les paramètres dans les modèles mathématiques.
L'algorithme précédent permet de passer de x_n à x_{n+1} par une itération du type :

$$x_{n+1} = G(x_n) \qquad x_0 \text{ donné}$$

avec G différentiable au sens de Fréchet au voisinage de x. Nous allons étudier la propagation des erreurs d'arrondis dans un tel procédé. En effet à cause des erreurs d'arrondis on ne calcule pas exactement la suite $\{x_n\}$ mais la suite $\{\bar{x} = x_n + e_n\}$
Les itérations $x_{n+1} = G(x_n)$ peuvent s'écrire :

$$x_{n+1} = x_n + h\,(G(x_n) - x_n)$$

qui n'est autre que la méthode d'Euler avec h = 1 appliquée à l'équation différentielle :

$$\frac{dx(t)}{dt} = G(x(t)) - x(t)$$

$$x\,(0) = x_0$$

Nous allons donc, tout naturellement, utiliser la théorie de la propagation des erreurs d'arrondis mise au point pour les méthodes numériques d'intégration des équations différentielles [111,121,129] (A-stabilité). On a :

$$\bar{x}_{n+1} = \bar{x}_n + h(G(\bar{x}_n) - \bar{x}_n)$$

$$= x_n + e_n + h(G(x_n + e_n) - x_n - e_n)$$

Si G est Fréchet différentiable, alors :

$$G(x_n + e_n) = G(x_n) + G'(x_n)\,e_n + o\,(e_n)$$

où $o(e_n)$ dénote un vecteur dont la norme tend vers zéro plus vite que celle de e_n.
Ainsi on a : $\bar{x}_{n+1} = x_{n+1} + e_n + h(G'(x_n)\,e_n - e_n) + o(e_n)$

d'où

$$e_{n+1} = [I + h(G'(x_n) - I)]\,e_n + o\,(e_n)$$

Supposons que $o(e_n)$ puisse être négligé et que les valeurs propres des matrices $G'(x_n)$ soient indépendantes de n. Alors on sait qu'une condition nécessaire et suffisante pour que $\lim_{n\to\infty} e_n = 0$ est que le rayon spectral de la matrice $I + h(G'(x_n) - I)$ soit strictement inférieur à un. On dit, dans ce cas, que la méthode est numériquement stable : (A-stable).

Soient λ_i pour i = 1, ..., m les valeurs propres de $G'(x_n)$. Il faut donc que :

$$|1 + h(\lambda_i - 1)| < 1 \text{ pour } i = 1, \ldots, m$$

Soit C le disque ouvert de centre -1 et de rayon 1 :

$$C = \{z \in \mathbb{C} \mid |z+1| < 1\}$$

ainsi la condition de stabilité peut donc s'écrire :

$$h(\lambda_i - 1) \in C \qquad \text{pour } i = 1, \ldots, m$$

Il faudra donc remplacer les itérations $x_{n+1} = G(x_n)$ par $x_{n+1} = x_n + h\,(G(x_n) - x_n)$ où le pas h sera choisi afin que la condition de stabilité soit vérifiée. Il n'est possible de trouver un tel h que si les quantités Re (λ_i) i = 1, ..., m sont toutes de même signe. Si c'est impossible alors on choisira h afin que $h(\lambda - 1) \in C$ où λ est la valeur propre qui satisfait $|\lambda - 1| = \max_i |\lambda_i - 1|$.

On minimisera ainsi la propagation des erreurs d'arrondis. Appliquons cette théorie à notre méthode de résolution des systèmes d'équations non linéaires.

Puisque la méthode converge quadratiquement alors $G'(x) = 0$ et donc $\lambda_i = 0$ pour i = 1, ... m. Cette méthode est donc stable à condition que les itérations de base $u_k = F(u_{k-1})$ soient, elles aussi, stables. On remplacera donc ces itérations de base par les nouvelles itérations de base

$$u_k = u_{k-1} + h\,(F(u_{k-1}) - u_{k-1})$$

où h sera choisi afin de vérifier la condition de stabilité où les λ_i seront les valeurs propres de $F'(x)$.

Si la condition de stabilité n'est pas satisfaite alors l'instabilité numérique pourra prendre les formes :

- non convergence des itérations $x_{n+1} = G(x_n)$
- convergence de $x_{n+1} = G(x_n)$ mais perte du caractère quadratique de la convergence.

Du point de vue pratique les hypothèses que nous avons été amenés à faire ne sont qu'imparfaitement vérifiées. Il s'en suit que cette théorie ne sera qu'approximativement exacte. En particulier près de la frontière de C il pourra y avoir stabilité au dehors et instabilité dedans. Le caractère quadratique de la convergence sera plus affirmé si toutes les quantités $h\,(\lambda_i - 1)$ sont voisines du centre de C. Par contre si les hypothèses sont bien satisfaites les conclusions précédentes seront valables quelquesoit le point de départ x_0.

Reprenons par exemple le système

$$x = \begin{pmatrix} y \\ z \end{pmatrix} = \begin{pmatrix} -z^4/4 \quad -3/4 \\ -0.405\, e^{1-y^2} + 1.405 \end{pmatrix}$$

dont la solution unique est $y = -1$ et $z = 1$. La condition de stabilité est :

$$0 \leqslant h \leqslant 2/1.9$$

car $\lambda_1 = -0.9$ et $\lambda_2 = 0.9$

Partant de $y_0 = z_0 = 0$ on a les résultats suivants :

$h = -0.1$ convergence en 14 itérations ; non quadratique

$h = 0.1$ convergence en 9 itérations ; presque quadratique

$h = 0.5$ convergence en 7 itérations ; quadratique

$h = 1.1$ convergence en 6 itérations ; quadratique

$h = 2.3$ non convergence.

On trouvera les résultats détaillés en [45] ainsi qu'un exemple montrant que si les hypothèses faites ne sont absolument pas vérifiées, alors la théorie précédente ne s'applique pas.

V - 5 Calcul des valeurs propres d'une matrice par l'ε-algorithme vectoriel

La méthode de la puissance est une méthode très utilisée pour calculer la valeur propre de plus grand module d'une matrice. Nous allons voir comment l'ε-algorithme permet, dans certains cas, de calculer simultanément toutes les valeurs propres.

Soit A une matrice carrée réelle $p \times p$. Notons $\lambda_1, \ldots, \lambda_p$ ses valeurs propres et $v_1, \ldots, v_p$ les vecteurs propres correspondants. On supposera que :

$$\lambda_i \neq 1 \qquad i = 1, \ldots, p$$

$$|\lambda_1| > |\lambda_2| > \cdots > |\lambda_p| \geq 0$$

Si la première hypothèse n'est pas satisfaite, il suffira de multiplier la matrice par un scalaire ; la seconde condition est beaucoup plus contraignante mais des modifications appropriées des algorithmes devraient permettre de s'en affranchir. Si $\{u_n\}$ et $\{w_n\}$ sont deux suites de vecteurs de $\mathbb{R}^p$, la notation $u_n \sim w_n$ signifiera que :

$$\lim_{n\to\infty} (y,u_n) / (y_n,w_n) = 1$$

$$\forall\, y \neq o \in \mathbb{R}^p \text{ tel que } (y,v_i) \neq o \text{ pour } i = 1,\dots,p.$$

Soit x_o un vecteur de $\mathbb{R}^p$ tel que $(x_o,v_i) \neq o\ \forall\, i$ et construisons la suite des vecteurs $\{x_n\}$ par :

$$x_{n+1} = A\, x_n \qquad n = o, 1, \dots$$

Appliquons l'ε-algorithme vectoriel à cette suite $\{x_n\}$. On a le :

<u>Théorème 99</u> :

$$\varepsilon_{2k}^{(n)} \sim \sum_{i=k+1}^{p} \lambda_i^{n+k}\ z_i \qquad k = o,\dots,p-1$$

$$\varepsilon_{2k+1}^{(n)} \sim \frac{y_{k+1}^{-1}}{\lambda_{k+1}^{n+k}} \qquad k = o,\dots,p-1$$

avec $y_i = (\lambda_i - 1)\ z_i$.

Démonstration : puisque les vecteurs propres forment une base, on peut la supposer orthogonale et écrire que :

$$x_o = a_1 v_1 + \dots + a_p v_p.$$

La condition $(x_o,v_i) \neq o$ implique que $a_i \neq o$. Posons $z_i = a_i v_i$. On a :

$$\varepsilon_o^{(n)} = x_n = \sum_{i=1}^{p} \lambda_i^n\ z_i.$$

Il est facile de voir que :

$$\varepsilon_1^{(n)} \sim y_1^{-1} / \lambda_1^n.$$

En portant ces résultats dans les règles de l'ε-algorithme, une simple démonstration de récurrence permet d'établir le théorème. On en trouvera les détails dans [46].

On a donc l'algorithme suivant pour calculer toutes les valeurs propres de la matrice A :

1) Choisissons un vecteur arbitraire x_o tel que $(x_o, v_i) \neq o$ pour $i = 1, \ldots, p$

2) Effectuons les itérations $x_{n+1} = Ax_n$ pour $n = 0, 1, \ldots$

3) Appliquons l'ε-algorithme vectoriel à la suite $\{x_n\}$. Soient $\varepsilon_k^{(n)}$ les vecteurs ainsi obtenus.

4) Calculons les rapports :

$$a_k^{(n)} = (y, \varepsilon_{2k}^{(n+1)}) / (y, \varepsilon_{2k}^{(n)}) \quad \text{pour } k = o, \ldots, p-1 \text{ et } n = o, 1, \ldots$$

$$b_k^{(n)} = (y, \varepsilon_{2k+1}^{(n)}) / (y, \varepsilon_{2k+1}^{(n+1)}) \quad \text{pour } k = o, \ldots, p-1 \text{ et } n = o, 1, \ldots$$

où y est un vecteur arbitraire non nul, tel que $(y, v_i) \neq o$ pour tout i.

On tire immédiatement du théorème 99 que :

$$\lim_{n\to\infty} a_k^{(n)} = \lim_{n\to\infty} b_k^{(n)} = \lambda_{k+1} \qquad \text{pour } k = o, \ldots, p-1$$

$$\lim_{n\to\infty} \varepsilon_{2k}^{(n)} / (y, \varepsilon_{2k}^{(n)}) = \lim_{n\to\infty} (y, \varepsilon_{2k+1}^{(n)}) \varepsilon_{2k}^{(n)} = v_{k+1} \qquad \text{pour } k = o, \ldots, p-1$$

De plus, on montre que la vitesse de convergence est réglée par :

$$a_k^{(n)} = \lambda_{k+1} + O[(\lambda_{k+2} / \lambda_{k+1})^{n+k+1}] \quad \text{pour } n \to \infty.$$

Pour k fixé la suite $\{a_k^{(n)}\}$ est une suite de scalaires qui converge vers λ_{k+1} lorsque n tend vers l'infini. On peut donc essayer d'accélérer sa convergence en lui appliquant l'ε-algorithme scalaire ; on montre que cela est possible et l'on obtient alors :

$$\varepsilon_{2q}^{(n)} = \lambda_{k+1} + O[(\lambda_{k+q+2} / \lambda_{k+1})^{n+k}]$$

pour q et k fixés et n tendant vers l'infini.

Remarque : les résultats précédents restent valables si on utilise l'ε-algorithme normé décrit au paragraphe V-2 au lieu de l'ε-algorithme vectoriel.

Au lieu d'appliquer l'ε-algorithme vectoriel à la suite des vecteurs $\{x_n\}$ on peut appliquer l'ε-algorithme scalaire à la suite $\{(z,x_n)\}$ où z est un vecteur arbitraire tel que $(z,v_i) \neq o$ pour tout i. On considérera ensuite les rapports $a_k^{(n)} = \varepsilon_{2k}^{(n+1)}/\varepsilon_{2k}^{(n)}$ et $b_k^{(n)} = \varepsilon_{2k+1}^{(n)}/\varepsilon_{2k+1}^{(n+1)}$. Cette variante réduit considérablement l'encombrement mémoire de l'algorithme puisque l'on travaille sur une suite de scalaires au lieu de travailler sur une suite de vecteurs. Tous les résultats précédents restent valables sauf, évidemment, le fait qu'il est impossible d'obtenir les vecteurs propres de cette façon.

Au lieu d'utiliser l'ε-algorithme on peut utiliser l'application répétée du procédé Δ^2 d'Aitken.

Prenons la matrice

$$\begin{pmatrix} 3 & 12 & 30 \\ -6 & -27 & -66 \\ 4 & 16 & 37 \end{pmatrix}$$

dont les valeurs propres sont 9, 3 et 1. Avec $x_o = y = (1;o;o)^T$ on obtient :

$\{a_o^{(n)}\}$	$\{b_o^{(n)}\}$	$\{a_1^{(n)}\}$	$\{b_1^{(n)}\}$	$\{a_2^{(n)}\}$	$\{b_2^{(n)}\}$
3					
19	12.9				
11.4	10.1	-2.5			
9.7	9.3	5.2	2.93		
9.2	9.1	3.4	2.98	0.999999	
9.07	9.04	3.1	2.992	1.000000	0.5
9.02	9.01	3,04	2.997	0.999999	5.3
9.008	9.004	3.01	2.9991	1.000000	1.7
9.002	9.001	3.004	2,9997	0,999999	1.7

Si l'on accélère la convergence de $\{b_1^{(n)}\}$ à l'aide de l'ε-algorithme scalaire on trouve :

$\{\varepsilon_0^{(n)} = b_1^{(n)}\}$	$\{\varepsilon_2^{(n)}\}$	$\{\varepsilon_4^{(n)}\}$
2.93		
2.98	2.99992	
2.992	2.999991	2.999999989
2.997	2.9999990	2,9999999996
2.9991	2.99999989	
2,9997		

Remarque : cette méthode de calcul des valeurs propres d'une matrice est à relier à l'utilisation de l'algorithme q-d pour effectuer le même travail.

Une autre application du théorème 99 est l'accélération de la convergence des suites de vecteurs produits par relaxation. Considérons, en effet, les vecteurs $\{x_n\}$ générés par :

$$x_0 \text{ donné}$$

$$x_{n+1} = Ax_n + b$$

où A est une matrice carrée telle que I-A soit inversible et b un vecteur. Soient $\lambda_1,\ldots,\lambda_p$ les valeurs propres de A. Supposons que $|\lambda_1| > |\lambda_2| > \cdots > |\lambda_p|$ et posons $x = (I-A)^{-1}\,b$. Si nous appliquons l'ε-algorithme vectoriel à la suite des vecteurs $\{x_n\}$ alors le théorème 99 nous montre que :

$$\varepsilon_{2k}^{(n)} - x = 0(\lambda_{k+1}^n) \text{ pour } k = 0,\ldots,p-1.$$

Si la méthode de relaxation est convergente, c'est-à-dire si $\rho(A) = |\lambda_1| < 1$ alors chaque colonne de l'ε-algorithme converge plus vite que la précédente et l'on voit que l'accélération obtenue dépend de la proximité de deux valeurs propres consécutives.

Si la méthode de relaxation diverge et si $|\lambda_1| > \cdots |\lambda_i| > 1 > |\lambda_{i+1}| > \cdots > |\lambda_p|$ alors les suites $\{\varepsilon_{2k}^{(n)}\}$ divergeront pour $k = 0,\ldots,i-1$ et convergeront vers x pour

$k=i,\ldots,p-1$ et cela de plus en plus vite lorsque k augmentera.

Ainsi l'ε-algorithme vectoriel peut être utilisé pour accélérer la convergence des méthodes de relaxation qui convergent et pour induire la convergence, dans certains cas, de celles qui divergent.

On trouvera dans [76] les démonstrations de ces résultats ainsi qu'une étude d'autres procédés d'accélération de la convergence des méthodes de relaxation.

V - 6 L'ε-algorithme topologique

Dans ce qui précède on a vu que l'ε-algorithme scalaire est finalement un artifice commode pour mettre en oeuvre la transformation de Shanks qui est un rapport de déterminants. Par contre, la situation est différente pour l'ε-algorithme vectoriel ; il n'y a pas alors de rapport de déterminants et l'algorithme a été construit, on peut le dire, artificiellement à partir de la règle de l'ε-algorithme scalaire. Les propriétés connues de l'ε-algorithme vectoriel découlent toutes du théorème 91 dont la démonstration ne permet pas de comprendre la nature mathématique de l'algorithme ni la sorte de transformation qu'il sert à mettre en oeuvre. D'autre part, il est difficile d'obtenir de nouvelles propriétés sans finalement savoir ce que fait cet algorithme.

C'est pour pallier à ces inconvénients que l'ε-algorithme topologique a été défini [31]. La démarche qui a permis de l'obtenir est semblable à celle de Shanks pour la définition de la transformation puis à celle de Wynn pour trouver l'algorithme de mise en oeuvre.

Cet algorithme a été baptisé ε-algorithme topologique parce que la suite $\{S_n\}$ qu'il transforme est une suite d'éléments d'un espace vectoriel topologique E sur K ($\mathbb{R}$ ou $\mathbb{C}$).

Soit donc $\{S_n\}$ une suite d'éléments de E et $S \in E$. Supposons que la suite $\{S_n\}$ vérifie :

$$\sum_{i=0}^{k} a_i (S_{n+i} - S) = 0 \quad \forall n \text{ avec } \sum_{i=0}^{k} a_i \neq 0 \text{ et } a_i \in K \qquad (1)$$

on peut, sans restreindre la généralité, supposer que $\sum_{i=0}^{k} a_i = 1$.
La transformation de Shanks permet de transformer $\{S_n\}$ en une suite constante S.

Considérons le système :

$$\begin{array}{llll} a_o & + a_1 & + \ldots + a_k & = 1 \\ a_o S_n & + a_1 S_{n+1} & + \ldots + a_k S_{n+k} & = S \\ a_o S_{n+1} & + a_1 S_{n+2} & + \ldots + a_k S_{n+k+1} & = S \\ - - - & - - - & - - - & \\ a_o S_{n+k} & + a_1 S_{n+k+1} & + \ldots + a_k S_{n+2k} & = S \end{array} \qquad (2)$$

Ce système peut s'écrire :

$$\begin{array}{llll} a_o & + a_1 & + \ldots + a_k & = 1 \\ a_o \Delta S_n & + a_1 \Delta S_{n+1} & + \ldots + a_k \Delta S_{n+k} & = 0 \\ - - - & - - - & - - - & \\ a_o \Delta S_{n+k-1} & + a_1 \Delta S_{n+k} & + \ldots + a_k \Delta S_{n+2k-1} & = 0 \\ a_o S_n & + a_1 S_{n+1} & + \ldots + a_k S_{n+k} & = S \end{array} \qquad (3)$$

Dans tout ce qui suit quand Δ sera appliqué à des quantités avec deux indices il agira toujours soit sur l'indice n soit sur l'indice placé en position supérieure.

Considérons les k+1 premières équations de ce système et soit $y' \in E'$ dual topologique de E ; on a :

$$\begin{array}{l} a_o + a_1 + \ldots\ldots\ldots + a_k = 1 \\ a_o < y', \Delta S_n > + \ldots + a_k < y', \Delta S_{n+k} > = 0 \\ - - - - - - - - - - - - \\ a_o < y', \Delta S_{n+k-1} > + \ldots + a_k < y', \Delta S_{n+2k-1} > = 0 \end{array} \qquad (4)$$

où $<y',y>$ désigne la forme bilinéaire qui met E et E' en dualité.

On peut résoudre ce système linéaire à condition que son déterminant soit différent de zéro. On calculera ensuite S en utilisant la dernière relation du système (3). On a donc symboliquement :

$$S = \frac{\begin{vmatrix} S_n & \cdots\cdots\cdots & S_{n+k} \\ \langle y',\Delta S_n\rangle & \cdots\cdots\cdots & \langle y',\Delta S_{n+k}\rangle \\ - - - & - - - & - - - \\ \langle y',\Delta S_{n+k-1}\rangle & \cdots\cdots & \langle y',\Delta S_{n+2k-1}\rangle \end{vmatrix}}{\begin{vmatrix} 1 & \cdots\cdots\cdots & 1 \\ \langle y',\Delta S_n\rangle & \cdots\cdots\cdots & \langle y',\Delta S_{n+k}\rangle \\ - - - & - - - & - - - \\ \langle y',\Delta S_{n+k-1}\rangle & \cdots\cdots & \langle y',\Delta S_{n+2k-1}\rangle \end{vmatrix}} = e_k(S_n) \qquad (5)$$

Le déterminant généralisé qui se trouve au numérateur se développe de façon habituelle et désigne un élément de E. Si la suite $\{S_n\}$ ne vérifie pas une relation du type (1) alors (5) n'est pas égal à S mais à un élément de E que nous noterons $e_k(S_n)$. On transforme ainsi la suite $\{S_n\}$ en un ensemble de suites $\{e_k(S_n)\}$ pour différentes valeurs de k. Cette transformation généralise la transformation de Shanks.

Remarque 1 : pour passer du système (3) au système (4) on peut prendre un y' différent pour chaque équation de (3). (4) devient donc alors :

$$\begin{aligned} & a_o + \ldots\ldots + a_k = 1 \\ & a_o \langle y_1',\Delta S_n\rangle + \ldots\ldots + a_k \langle y_1',\Delta S_{n+k}\rangle = 0 \\ & - \\ & a_o \langle y_k', \Delta S_{n+k-1} + \ldots + a_k \langle y_k',\Delta S_{n+2k-1}\rangle = 0 \end{aligned}$$

Il est bien évident que les y_i' peuvent également dépendre de n. Cependant nous n'envisagerons pas ces deux cas par la suite car l'algorithme récursif de calcul de (5) étudié plus loin ne s'applique pas alors. y' sera donc toujours le même pour toutes les équations de (4) ; il ne dépendra ni de k ni de n.

Remarque 2 : Contrairement à ce qui se passe lorsque $E = \mathbb{C}$ le fait que (4) soit vérifié n'entraîne pas que (1) soit vérifié. C'est pour cette raison que la condition du théorème 100 est seulement suffisante.

Les propriétés de (5) sont les mêmes que celles de la transformation de Shanks habituelle :

<u>Théorème 100</u>: *Une condition suffisante pour que* $e_k(S_n) = S \quad \forall n > N$ *est que la suite* $\{S_n\}$ *vérifie la relation* $\sum_{i=0}^{k} a_i(S_{n+i}-S) = 0 \ \forall n > N$ *avec* $a_i \in K$ *et* $\sum_{i=0}^{k} a_i \neq 0$.

Cette propriété découle directement de la construction même du procédé. L'équation aux différences $\sum_{i=0}^{k} a_i(S_{n+i}-S) = 0$ peut être résolue dans E de même façon que lorsque $S_n \in \mathbb{R}$. On a donc un résultat analogue à celui démontré dans ce cas :

<u>Théorème 101</u>: *Une condition suffisante pour que* $e_k(S_n) = S \quad \forall n > N$ *est que :*

$$S_n = S + \sum_{i=1}^{p} A_i(n)\, r_i^n + \sum_{i=p+1}^{q} [B_i(n)\cos b_i n + C_i(n)\sin b_i n]\, e^{w_i n} + \sum_{i=0}^{m} c_i \delta_{in} \qquad \forall n > N.$$

r_i, w_i et b_i appartiennent à K et l'on a $r_i \neq 1$ pour $i=1,\ldots,p$.

A_i, B_i et C_i sont des polynômes en n dont les coefficients appartiennent à E. Les c_i appartiennent à E et δ_{in} est le symbole de Kronecker.

Si d_i désigne le degré de A_i plus un pour $i=1,\ldots,p$ et le plus grand des degrés de B_i et de C_i pour $i=p+1,\ldots,q$, on doit avoir :

$$m + \sum_{i=1}^{p} d_i + 2 \sum_{i=p+1}^{q} d_i = k-1$$

avec la convention que $m=-1$ s'il n'y a aucun terme en δ_{in}.

La transformation de Shanks généralisée est une transformation non linéaire de suite à suite ; cependant on a la

Propriété 36 :

$$e_k(aS_n+b) = a\, e_k(S_n) + b \qquad \forall\, n,k$$

$\forall\, a \neq 0 \in K$ *et* $\forall\, b \in E$.

La démonstration est évidente à partir de (5).

Propriété 37 : *Soit* $e_k(S_n)$ *l'élément de* E *obtenu en appliquant* (5) *à* $S_n, S_{n+1}, \ldots, S_{n+2k}$. *Si on applique* (5) *à* $u_n = S_{n+2k}$, $u_{n+1} = S_{n+2k-1}, \ldots$ $\ldots, u_{n+2k} = S_n$ *alors on obtient un élément de* E *généralement différent de* $e_k(S_n)$.

La démonstration est évidente en intervertissant les lignes et les colonnes dans (5). Cette propriété est l'inverse de celle démontrée par Gilewicz dans le cas scalaire où l'élément obtenu est identique.

Donnons maintenant une interprétation barycentrique de la généralisation de la transformation de Shanks que nous venons d'étudier, analogue à celle obtenue dans le cas scalaire (paragraphe III-4) :

d'après (2), (4) et (5) on a :

$$\sum_{i=0}^{k} a_i\, S_{p+i} = e_k(S_n) \sum_{i=0}^{k} a_i \qquad \text{pour } p=n,\ldots,n+k$$

avec $\sum_{i=0}^{k} a_i \neq 0$. $e_k(S_n)$ apparaît donc comme le barycentre des points $S_n, \ldots, S_{n+k}$ affectés des masses $a_o, \ldots, a_k$. Les masses $a_o, \ldots, a_k$ sont choisies de sorte que $e_k(S_n)$ soit également le barycentre de $(S_{n+1}, \ldots, S_{n+k+1})$, $\ldots$, $(S_{n+k}, \ldots, S_{n+2k})$ affectés des mêmes masses $a_o, \ldots, a_k$ (ces masses peuvent être ici négatives).

La propriété 36 provient tout simplement du fait que toute transformation affine transforme le barycentre en le barycentre des points transformés affectés des mêmes masses. Le fait que l'on puisse remplacer plusieurs points par leur barycentre affecté d'une masse égale à la somme de leurs masses nous fournira une méthode récursive de calcul de $e_k(S_n)$: ce sera l'ε-algorithme topologique.

Si $\sum_{i=0}^{k} a_i = 0$ alors tout point de E est barycentre, le déterminant intervenant au dénominateur de (5) est nul et le calcul de $e_k(S_n)$ ne peut pas alors être effectué.

Remarque : on voit que $\langle y', e_k(S_n)\rangle$ n'est autre que le résultat de la transformation habituelle de Shanks appliquée à la suite $\{\langle y', S_n\rangle\}$. Il est possible, à partir de la transformation (5), de donner une généralisation de la table de Padé.

Considérons la série de puissances formelle :

$$f(x) = \sum_{i=0}^{\infty} c_i x^i$$

avec $x \in K$ et $c_i \in E$.

Prenons comme suite $\{S_n\}$ la suite des sommes partielles de $f(x)$:

$$S_n = \sum_{i=0}^{n} c_i x^i .$$

D'après (5) on a :

$$e_k(S_n) = \frac{\begin{vmatrix} \sum_{i=0}^{n} c_i x^i & \cdots\cdots\cdots\cdots & \sum_{i=0}^{n+k} c_i x^i \\ x^{n+1}\langle y', c_{n+1}\rangle & \cdots\cdots & x^{n+k+1}\langle y', c_{n+k+1}\rangle \\ - - - - - - & - - - - - - & - - - - - - \\ x^{n+k}\langle y', c_{n+k}\rangle & \cdots\cdots & x^{n+2k}\langle y', c_{n+2k}\rangle \end{vmatrix}}{\begin{vmatrix} 1 & \cdots\cdots\cdots\cdots & 1 \\ x^{n+1}\langle y', c_{n+1}\rangle & \cdots\cdots & x^{n+k+1}\langle y', c_{n+k+1}\rangle \\ - - - - - - & - - - - - - & - - - - - - \\ x^{n+k}\langle y', c_{n+k}\rangle & \cdots\cdots & x^{n+2k}\langle y', c_{n+2k}\rangle \end{vmatrix}}$$

Multiplions la première colonne du numérateur et du dénominateur par x^k, la seconde par x^{k-1},..., la dernière par 1 ; on obtient :

$$e_k(S_n) = \frac{\begin{vmatrix} \sum\limits_{i=0}^{n} c_i x^{k+i} & & \sum\limits_{i=0}^{n+k} c_i x^i \\ x^{n+k+1}\langle y',c_{n+1}\rangle & \cdots\cdots\cdots & x^{n+k+1}\langle y',c_{n+k+1}\rangle \\ - - - - - - & - - - - - - & - - - - - - \\ x^{n+2k}\langle y',c_{n+k}\rangle & \cdots\cdots\cdots & x^{n+2k}\langle y',c_{n+2k}\rangle \end{vmatrix}}{\begin{vmatrix} x^k & \cdots\cdots\cdots & 1 \\ x^{n+k+1}\langle y',c_{n+1}\rangle & \cdots\cdots\cdots & x^{n+k+1}\langle y',c_{n+k+1}\rangle \\ - - - - - - & - - - - - - & - - - - - - \\ x^{n+2k}\langle y',c_{n+k}\rangle & \cdots\cdots\cdots & x^{n+2k}\langle y',c_{n+2k}\rangle \end{vmatrix}}$$

divisons maintenant les secondes lignes du numérateur et du dénominateur par x^{n+k+1}, les troisièmes par x^{n+k+2},..., les dernières par x^{n+2k}. On trouve :

$$e_k(S_n) = \frac{\begin{vmatrix} \sum\limits_{i=0}^{n} c_i x^{k+i} & \cdots\cdots & \sum\limits_{i=0}^{n+k} c_i x^i \\ \langle y',c_{n+1}\rangle & \cdots\cdots\cdots & \langle y',c_{n+k+1}\rangle \\ - - - - - & - - - - - & - - - - - \\ \langle y',c_{n+k}\rangle & \cdots\cdots\cdots & \langle y',c_{n+2k}\rangle \end{vmatrix}}{\begin{vmatrix} x^k & \cdots\cdots\cdots & 1 \\ \langle y',c_{n+1}\rangle & \cdots\cdots\cdots & \langle y',c_{n+k+1}\rangle \\ - - - - - & - - - - - & - - - - - \\ \langle y',c_{n+k}\rangle & & \langle y',c_{n+2k}\rangle \end{vmatrix}} \qquad (6)$$

On a :

$$\begin{vmatrix} \sum\limits_{i=0}^{n} c_i x^{k+i} & \cdots\cdots & \sum\limits_{i=0}^{n+k} c_i x^i \\ \langle y',c_{n+1}\rangle & \cdots\cdots\cdots & \langle y',c_{n+k+1}\rangle \\ - - - - - & - - - - - & - - - - - \\ \langle y',c_{n+k}\rangle & \cdots\cdots\cdots & \langle y',c_{n+2k}\rangle \end{vmatrix} - f(x) \begin{vmatrix} x^k & \cdots\cdots\cdots & 1 \\ \langle y',c_{n+1}\rangle & \cdots\cdots\cdots & \langle y',c_{n+k+1}\rangle \\ - - - - - & - - - - - & - - - - - \\ \langle y',c_{n+k}\rangle & \cdots\cdots\cdots & \langle y',c_{n+2k}\rangle \end{vmatrix} =$$

$$\begin{vmatrix} \sum_{i=n+1}^{\infty} c_i x^{k+i} & \cdots & \sum_{i=n+k+1}^{\infty} c_i x^i \\ \langle y', c_{n+1}\rangle & \cdots & \langle y', c_{n+k+1}\rangle \\ - - - & - - - & - - - \\ \langle y', c_{n+k}\rangle & \cdots & \langle y', c_{n+2k}\rangle \end{vmatrix} = Ax^{n+k+1} \quad \text{avec } A \in E \tag{7}$$

En examinant les relations (6) et (7) on voit que l'on peut considérer (5) comme une généralisation de la table de Padé de f(x) bien que la série située dans le membre de droite de (7) ne commence qu'avec un terme en x^{n+k+1}. Le numérateur de (6) est de degré n+k et son dénominateur de degré k. Nous noterons donc symboliquement :

$$e_k(S_n) = [n+k/k]$$

Pour rappeler que $e_k(S_n)$ est un approximant de la série de puissance f(x), nous le noterons quelques fois :

$$[n+k/k]_f(x)$$

Par analogie l'approximant de Padé généralisé sera défini par :

$$[p/q] = \frac{\begin{vmatrix} \sum_{i=0}^{p-q} c_i x^{q+i} & \cdots & \sum_{i=0}^{p} c_i x^i \\ \langle y', c_{p-q+1}\rangle & \cdots & \langle y', c_{p+1}\rangle \\ - - - & - - - & - - - \\ \langle y', c_p\rangle & \cdots & \langle y', c_{p+q}\rangle \end{vmatrix}}{\begin{vmatrix} x^q & \cdots & 1 \\ \langle y', c_{p-q+1}\rangle & \cdots & \langle y', c_{p+1}\rangle \\ - - - & - - - & - - - \\ \langle y', c_p\rangle & \cdots & \langle y', c_{p+q}\rangle \end{vmatrix}} = \frac{P(x)}{Q(x)} \tag{8}$$

où les c_i avec un indice négatif sont pris égaux à $0 \in E$. [p/q] pourra, dans la suite, être noté $[p/q]_f(x)$.

On voit, d'après (8), que l'on a :

$$[p/q] = \frac{\sum_{i=0}^{p} a_i x^i}{\sum_{i=0}^{q} b_i x^i} \quad \text{avec } a_i \in E \text{ et } b_i \in K$$

Le calcul des a_i et des b_i s'effectue comme pour la table de Padé ordinaire en écrivant que :

$$\sum_{i=0}^{\infty} c_i x^i - \frac{\sum_{i=0}^{p} a_i x^i}{\sum_{i=0}^{q} b_i x^i} = Ax^k$$

avec $A \in E$ et k le plus grand possible. En d'autres termes on veut déterminer les a_i et les b_i de sorte que :

$$(c_o+c_1x+\ldots)(b_o+b_1x+\ldots+b_qx^q) - (a_o+a_1x+\ldots+a_px^p) = Ax^{p+q+1}$$

ou encore :

$$(c_o+c_1x+\ldots)(b_o+\ldots+b_qx^q) = a_o+\ldots+a_px^p+0x^{p+1}+\ldots+0x^{p+q}+Ax^{p+q+1}$$

En identifiant les coefficients des termes de même degré en x on obtient :

$$\begin{array}{l} b_oc_{p+1} + b_1c_p + \ldots + b_qc_{p-q+1} = 0 \\ \text{- - - - - - - - - - - - - - - - - - - -} \\ b_oc_{p+q} + b_1c_{p+q-1}+\ldots+ b_qc_p = 0 \end{array}$$

avec la convention que $c_i = 0 \in E$ si $i < 0$. En prenant $b_0=1$ on trouve donc les b_i comme solution du système suivant de q équations à $q+1$ inconnues :

$$\begin{array}{l} b_o \langle y',c_{p+1}\rangle + b_1 \langle y',c_p\rangle + \ldots + b_q \langle y',c_{p-q+1}\rangle = 0 \\ \text{- -} \\ b_o \langle y',c_{p+q}\rangle + b_1 \langle y',c_{p+q-1}\rangle+\ldots+ b_q \langle y',c_p\rangle = 0 \end{array} \qquad (9)$$

Puis les $a_i \in E$ sont calculés à partir des p+1 relations :

$$
\begin{aligned}
& c_o b_o = a_o \\
& c_1 b_o + c_o b_1 = a_1 \\
& c_2 b_o + c_1 b_1 + c_o b_2 = a_2 \\
& - - - - - - - - - - - - - \\
& c_p b_o + c_{p-1} b_1 + \ldots + c_{p-q} b_q = a_p
\end{aligned}
\qquad (10)
$$

On voit que l'on a :

$$\langle y', [p/q] \rangle - \langle y', f(x) \rangle = 0(x^{p+q+1})$$

Par contre on a seulement :

$$[p/q] - f(x) = Ax^{p+1}$$

ceci tient au fait que $\langle y',y \rangle = 0$ n'entraîne pas que $y=0$. Nous considèrerons cependant (8) comme une généralisation de la table de Padé. (8) possède d'ailleurs les mêmes propriétés que les quotients de Padé ordinaires :

Théorème 102: Si P/Q est un approximant de Padé généralisé [p/q] de $f(x) = \sum_{i=0}^{\infty} c_i x^i$ *défini par (8), (9) et (10), alors on a :*

$$\langle y', f(x) \rangle \; Q(x) - \langle y', P(x) \rangle = \sum_{i=p+q+1}^{\infty} d_i x^i$$

$$\textit{avec} \quad d_i = \sum_{k=0}^{q} b_k \; \langle y', c_{i-k} \rangle$$

Démonstration : Elle est évidente ; c'est une simple identification de coefficients dans des séries de puissances. Ce résultat généralise un résultat classique de la table de Padé ordinaire [53].

Théorème 103: Une condition nécessaire et suffisante pour que [p/q] défini par (9) et (10) existe est que :

$$H_q^{(p-q+1)}(\langle y', c_{p-q+1} \rangle) \neq 0$$

où $H_k^{(n)}(u_n)$ *est le déterminant de Hankel défini de façon habituelle par :*

$$H_o^{(n)}(u_n) = 1$$

$$H_k^{(n)}(u_n) = \begin{vmatrix} u_n & \cdots\cdots & u_{n+k-1} \\ u_{n+1} & \cdots\cdots & u_{n+k} \\ - & - - - - - & - \\ u_{n+k-1} & \cdots\cdots & u_{n+2k-2} \end{vmatrix} \qquad \begin{matrix} n = 0,1,\ldots \\ k = 1,2,\ldots \end{matrix}$$

avec $u_n \in K \quad \forall\, n$.

Démonstration : C'est tout simplement la condition nécessaire et suffisante pour que le système (9) donnant les b_i admette une solution. Cette solution est d'ailleurs unique d'où le :

Théorème 104: S'il existe, [p/q], *défini par (9) et (10), est unique.*

Un certain nombre de propriétés de la table de Padé ordinaire restent encore valables ici :

Propriété 38 : Si $c_o = 0$ *alors :*

$$[p-1/q]_{f/x}(x) = [p/q]_f(x)/x$$

Propriété 39 : Soit $R_k(x) = \sum_{i=0}^{r} a_i x^i$ *avec* $a_i \in E$ *et* $x \in K$. *Alors si* $p \geq q+k$ *et* $r \leq k$:

$$[p/q]_{f+R_k}(x) = [p/q]_f(x) + R_k(x)$$

La démonstration de la première propriété découle immédiatement de (8). La seconde provient de la définition même des approximants de Padé (7) ; en effet posons :

$$[p/q]_{f+R_k}(x) = \frac{\sum_{i=0}^{p} a_i x^i}{\sum_{i=0}^{q} b_i x^i} \quad \text{avec } a_i \in E \quad b_i, x \in K$$

On a donc d'après (7) :

$$\frac{\sum_{i=0}^{p} a_i x^i}{\sum_{i=0}^{q} b_i x^i} - f(x) - R_k(x) = A\, x^{p+q+1} \qquad A \in E$$

ou encore

$$\frac{\sum_{i=0}^{p} a_i x^i - R_k(x) \sum_{i=0}^{q} b_i x^i}{\sum_{i=0}^{q} b_i x^i} - f(x) = A\, x^{p+q+1}$$

si $p \geq q+k$ alors le numérateur est de degré p. Par conséquent le rapport intervenant dans la relation précédente n'est autre que $[p/q]_f(x)$. D'où finalement :

$$[p/q]_{f+R_k}(x) - R_k(x) = [p/q]_f(x) \qquad \text{si} \quad p \geq q+k.$$

Remarque : prenons $p = n+k$ et $q = k$. On a $p-q=n$. Appliquons la propriété 39 :

$$[n+k/k]_{f-R_n}(x) = [n+k/k]_f(x) - R_n(x)$$

Prenons $f(x) = \sum_{i=0}^{\infty} c_i x^i$ et $R_n(x) = \sum_{i=0}^{n-1} c_i x^i$

on a $f(x) - R_n(x) = \sum_{i=n}^{\infty} c_i x^i = x^n \sum_{i=0}^{\infty} c_{n+i} x^i$

posons $f_n(x) = \sum_{i=0}^{\infty} c_{n+i} x^i$.

Par conséquent :

$$[n+k/k]_{x^n f_n}(x) = [n+k/k]_f(x) - R_n(x)$$

d'où, en divisant les deux membres par x^n :

$$[k/k]_{f_n}(x) = \frac{[n+k/k]_f(x) - R_n(x)}{x^n}$$

où $R_n(x)$ est la somme des n premiers termes de f(x). Cette relation permet donc de relier les approximants de Padé diagonaux [k/k] et les approximants non diagonaux [n+k/k].

Propriété 40 : Posons $y = \frac{x}{ax+b}$ et $g(x) = f(y)$ *alors*

$$[k/k]_f(y) = [k/k]_g(x)$$

La démonstration est identique à celle effectuée pour la table de Padé ordinaire.

Remarque : Lorsque E est un espace de Hilbert, les résultats de ce paragraphe sont à rapprocher de ceux obtenus par Wynn [217]. On comparera également au calcul d'approximants de Padé pour des matrices effectué par Rissanen [163].
Avant d'obtenir la règle de l'ε-algorithme topologique, il nous faut définir l'inverse d'un couple d'éléments et celui d'une série.

Soit $f(x) \in E$ une série de puissances formelle :

$$f(x) = \sum_{i=0}^{\infty} c_i x^i \qquad c_i \in E \qquad x \in K.$$

Nous voulons, comme pour la table de Padé ordinaire, définir son inverse. Il nous faut définir auparavant ce que l'on entend par inverse d'un élément $a \in E$ ou plutôt par inverse d'un couple $(a,b) \in E \times E'$. Cette notion est fondamentale pour pouvoir généraliser l'ε-algorithme.

Soit $a \in E$ et $b \in E'$ tels que $\langle b,a \rangle \neq 0$.

On appelle inverse du couple $(a,b) \in E \times E'$ le couple $(b^{-1},a^{-1}) \in E \times E'$ défini par :

$$a^{-1} = \frac{b}{\langle b,a \rangle} \qquad b^{-1} = \frac{a}{\langle b,a \rangle} \qquad a^{-1} \in E',\ b^{-1} \in E$$

On dira également par la suite que a^{-1} est l'inverse de a par rapport à b et réciproquement. Les propriétés de l'inverse de (a,b) sont les suivantes :

Propriété 41 : $\langle a^{-1},a\rangle = 1$

$$\langle b,b^{-1}\rangle = 1$$

$$\langle a^{-1},b^{-1}\rangle = 1/\langle b,a\rangle$$

$$(a^{-1})^{-1} = a \quad \text{et} \quad (b^{-1})^{-1} = b.$$

Exemple 1 : Soit (y'^{-1},d^{-1}) l'inverse de $(d,y') \in E \times E'$ où d est quelconque tel que $\langle y',d\rangle \neq 0$.

Soit $b \in E'$ tel que $\langle b,y'^{-1}\rangle \neq 0$; alors l'inverse de (y'^{-1},b) est $(b^{-1},(y'^{-1})^{-1})$ défini par :

$$(y'^{-1})^{-1} = \frac{b}{\langle b,y'^{-1}\rangle} \qquad b^{-1} = \frac{y'^{-1}}{\langle b,y'^{-1}\rangle}$$

Exemple 2 : Soit $(a,b) \in E \times E'$ et (b^{-1},a^{-1}) son inverse. Alors $a^{-1} \in E'$.

Considérons l'inverse de $(a,a^{-1}) \in E \times E'$:

$$a^{-1} = \frac{a^{-1}}{\langle a^{-1},a\rangle} \qquad (a^{-1})^{-1} = \frac{a}{\langle a^{-1},a\rangle}$$

On voit que la première relation entraîne que $\langle a^{-1},a\rangle = 1$ et que l'on a, par conséquent, $(a^{-1})^{-1} = a$.

Exemple 3 : Soit $(a,b) \in E \times E'$ et (b^{-1},a^{-1}) son inverse. On a $b^{-1} \in E$.

Considérons l'inverse de $(b^{-1},b) \in E \times E$:

$$(b^{-1})^{-1} = \frac{b}{\langle b,b^{-1}\rangle} \qquad b^{-1} = \frac{b^{-1}}{\langle b,b^{-1}\rangle}$$

On a $\langle b,b^{-1}\rangle = 1$ et par conséquent $(b^{-1})^{-1} = b$.

Remarque : Si E est un espace de Hilbert alors l'inverse du couple $(a,a) \in E \times E$ est (a^{-1},a^{-1}) avec :

$$a^{-1} = \frac{a}{\langle a,a\rangle}$$

où $<a,a>$ est le produit scalaire dans E. On voit que l'on retrouve dans ce cas la définition de l'inverse de $a \in \mathbb{R}^n$ utilisée par Wynn dans l'ε-algorithme vectoriel [242].

Nous pouvons maintenant définir l'inverse d'une série formelle. Pour cela utilisons les résultats précédents pour obtenir l'inverse du couple $(f(x),y') \in E \times E'$.

$$[f(x)]^{-1} = \frac{y'}{<y',f(x)>}$$

Nous allons chercher les coefficients de $[f(x)]^{-1}$ que nous mettrons sous la forme :

$$[f(x)]^{-1} = \sum_{i=0}^{\infty} d_i x^i \qquad \text{avec } d_i \in E' \quad \text{et} \quad x \in K$$

d'où :

$$[f(x)]^{-1} = \frac{y'}{<y', \sum_{i=0}^{\infty} c_i x^i>} = \sum_{i=0}^{\infty} d_i x^i$$

Posons $d_i = y'e_i$ avec $e_i \in K$. On a donc :

$$\frac{1}{<y', \sum_{i=0}^{\infty} c_i x^i>} = \sum_{i=0}^{\infty} e_i x^i$$

ce qui donne, comme dans le cas de l'inverse ordinaire d'une série de puissances formelle :

$$<y',c_o> e_o = 1$$

$$<y',c_o> e_1 + <y',c_1> e_o = 0$$

- - - - - - - - - - - - - - -

$$<y',c_o> e_k + <y',c_1> e_{k-1} + \ldots + <y',c_{k-1}> e_1 + <y',c_k> e_o = 0$$

- -

et permet de calculer les e_i à condition que $<y',c_o> \neq 0$.

D'après (8) on a :

$$[0/n] = \frac{a_o}{\sum_{i=0}^{n} b_i x^i} = \left[\sum_{i=0}^{n} d_i x^i \right]^{-1} = \frac{y'^{-1}}{< \sum_{i=0}^{n} d_i x^i , y'^{-1} >}$$

ou encore :

$$[0/n] = \frac{a_o}{\sum_{i=0}^{n} b_i x^i} = \frac{y'^{-1}}{\sum_{i=0}^{n} e_i x^i}$$ en utilisant le fait que $<y',y'^{-1}> = 1$

On obtient donc :

$$a_o = y'^{-1}$$

$$b_i = e_i \quad \text{pour} \quad i=0,\ldots,n$$

On pourra rapprocher l'inverse généralisé d'une série de puissances formelle tel que nous venons de le définir de celui donné par Wynn dans le cas où $E = \mathbb{R}^n$ [201].

L'inverse de f(x) étant défini de la façon précédente nous pouvons maintenant énoncer la :

<u>*Propriété 42*</u> : *Si* $<y',c_o> \neq 0$ *alors*

$$[q/p]_{f^{-1}}(x) = 1 / [p/q]_f(x)$$

Démonstration : On a :

$$1/[p/q]_f(x) = \left\{ \frac{\sum_{i=0}^{p} a_i x^i}{\sum_{i=0}^{q} b_i x^i} \right\}^{-1} = \frac{y' \sum_{i=0}^{q} b_i x^i}{<y', \sum_{i=0}^{p} a_i x^i>}$$

et $$f^{-1}(x) = \frac{y'}{<y', \sum_{i=0}^{\infty} c_i x^i>}$$

d'autre part :

$$\frac{\sum_{i=0}^{q} b_i x^i}{\langle y', \sum_{i=0}^{p} a_i x^i \rangle} - \frac{1}{\langle y', \sum_{i=0}^{\infty} c_i x^i \rangle} = \frac{\left(\sum_{i=0}^{\infty} \langle y', c_i \rangle x^i\right)\left(\sum_{i=0}^{q} b_i x^i\right) - \sum_{i=0}^{p} \langle y', a_i \rangle x^i}{\langle y', \sum_{i=0}^{p} a_i x^i \rangle \quad \langle y', \sum_{i=0}^{\infty} c_i x^i \rangle}$$

Le numérateur de cette expression est égal a :

$$\langle y', c_o \rangle\, b_o + \{\langle y', c_1 \rangle\, b_o + \langle y', c_o \rangle\, b_1\}\, x + \ldots + \{\langle y', c_p \rangle\, b_o + \ldots + \langle y', c_o \rangle b_p\}\, x^p$$

$$+ \{b_o\, \langle y', c_{p+1} \rangle + \ldots + b_q\, \langle y', c_{p-q+1} \rangle\} x^{p+1} + \ldots + \{b_o\, \langle y', c_{p+q} \rangle + \ldots + b_q\, \langle y', c_p \rangle\}\, x^{p+q}$$

$$+ Ax^{p+q+1} - \sum_{i=0}^{p} \langle y', a_i \rangle\, x^i$$

Or, d'après les relations (10), on a :

$$\langle y', c_o \rangle\, b_o = \langle y', a_o \rangle$$

$$\langle y', c_1 \rangle\, b_o + \langle y', c_o \rangle\, b_1 = \langle y', a_1 \rangle$$

- - - - - - - - - - - - - - - - - - -

$$\langle y', c_p \rangle\, b_o + \ldots + \langle y', c_o \rangle\, b_p = \langle y', a_p \rangle$$

D'autre part, d'après le système (9), on voit que les coefficients de $x^{p+1}, \ldots, x^{p+q}$ sont nuls. Par conséquent on a :

$$1/[p/q]_f(x) - f^{-1}(x) = A' x^{p+q+1} \qquad \text{avec} \qquad A' \in E'$$

ce qui démontre la propriété par définition des approximants de Padé puisque le numérateur de $1/[p/q]_f(x)$ est de degré q et que son dénominateur est de degré p.

On voit donc que la table de Padé généralisée que nous venons de définir possède les mêmes propriétés que la table de Padé ordinaire.

Venons-en maintenant à l'ε-algorithme topologique.

Le calcul des déterminants intervenant dans (5) est difficile dès que k devient élevé. Nous allons donc maintenant étudier un procédé récursif pour éviter le calcul de ces déterminants. Ce procédé sera une généralisation de l'ε-algorithme scalaire que l'on retrouve lorsque $E = \mathbb{R}$. Cet algorithme est basé sur le fait que les déterminants qui interviennent dans la relation (5) vérifient un certain nombre de propriétés analogues à celles vérifiées dans le cas scalaire.

Définissons d'abord les déterminants de Hankel généralisés.

Soit $\{u_n\}$ une suite d'éléments de E et y' un élément arbitraire de E'.

Nous poserons :

$$\tilde{H}_{k+1}^{(n)}(u_n) = \begin{vmatrix} u_n & \cdots\cdots\cdots\cdots & u_{n+k} \\ \langle y',\Delta u_n\rangle & \cdots\cdots & \langle y',\Delta u_{n+k}\rangle \\ - - - & - - - - - - & - - - \\ \langle y',\Delta u_{n+k}\rangle & \cdots & \langle y',\Delta u_{n+2k}\rangle \end{vmatrix} \quad \text{pour } k=0,1,\ldots$$

et nous appelerons $\tilde{H}_{k+1}^{(n)}(u_n)$ déterminant de Hankel généralisé ; on voit que c'est un élément de E défini par la combinaison linéaire de $u_n,\ldots,u_{n+k}$ obtenue en développant ce déterminant à l'aide des règles habituelles de calcul d'un déterminant.

Nous poserons :

$$H_{k+1}^{(n)}(u_n) = \langle y',\tilde{H}_{k+1}^{(n)}(u_n)\rangle$$

On voit que $H_{k+1}^{(n)}(u_n)$ n'est autre que le déterminant de Hankel classique de la suite scalaire $\{\langle y',u_n\rangle\}$.

Avec ces notations on voit que (5) s'écrit :

$$e_k(S_n) = \frac{\tilde{H}_{k+1}^{(n)}(S_n)}{H_k^{(n)}(\Delta^2 S_n)}$$

L'identité entre l'ε-algorithme scalaire et la transformation de Shanks repose sur le développement de Schweins du quotient de deux déterminants. La condensation d'un déterminant, les identités extentionnelles et le développement de Schweins s'obtiennent par combinaison linéaire des lignes ou des colonnes des déterminants mis en jeu [2]. Ces propriétés s'étendent donc immédiatement aux déterminants de Hankel généralisés que nous venons de définir. En particulier en effectuant un développement de Schweins on obtient la :

Propriété 43 :

$$e_{k+1}(S_n) - e_k(S_n) = - \frac{H_{k+1}^{(n)}(\Delta S_n)\ \tilde{H}_{k+1}^{(n)}(\Delta S_n)}{H_{k+1}^{(n)}(\Delta^2 S_n)\ H_k^{(n)}(\Delta^2 S_n)}$$

On voit que cette propriété est une généralisation d'une propriété bien connue dans le cas scalaire et que l'on retrouve immédiatement en écrivant que :

$$\langle y', e_{k+1}(S_n) - e_k(S_n)\rangle = - \frac{[H_{k+1}^{(n)}(\Delta S_n)]^2}{H_{k+1}^{(n)}(\Delta^2 S_n)\ H_k^{(n)}(\Delta^2 S_n)}$$

De la relation (5) et de la propriété 8 on tire la :

Propriété 44 :

$$H_k^{(n)}(\Delta^2 S_n)\ \tilde{H}_{k+2}^{(n)}(S_n) - H_{k+1}^{(n)}(\Delta^2 S_n)\ \tilde{H}_{k+1}^{(n)}(S_n) = - H_{k+1}^{(n)}(\Delta S_n)\ \tilde{H}_{k+1}^{(n)}(\Delta S_n)$$

Donnons maintenant les relations de l'ε-algorithme topologique.

Les quantités avec un indice inférieur pair sont des éléments de E ; celles avec un indice inférieur impair appartiennent à E' :

$$\varepsilon_{-1}^{(n)} = 0 \in E' \qquad \varepsilon_{o}^{(n)} = S_n \in E \qquad n=0,1,\ldots$$

$$\varepsilon_{2k+1}^{(n)} = \varepsilon_{2k-1}^{(n+1)} + [\Delta\varepsilon_{2k}^{(n)}]^{-1} \qquad n,k=0,1,\ldots \qquad (11')$$

avec $[\Delta\varepsilon_{2k}^{(n)}]^{-1} = \dfrac{y'}{<y',\Delta\varepsilon_{2k}^{(n)}>}$ et $y'^{-1} = \dfrac{\Delta\varepsilon_{2k}^{(n)}}{<y',\Delta\varepsilon_{2k}^{(n)}>}$ où $y' \in E'$

$$\varepsilon_{2k+2}^{(n)} = \varepsilon_{2k}^{(n+1)} + [\Delta\varepsilon_{2k+1}^{(n)}]^{-1} \qquad n,k=0,1,\ldots \qquad (11'')$$

avec $[\Delta\varepsilon_{2k+1}^{(n)}]^{-1} = \dfrac{y'^{-1}}{<\Delta\varepsilon_{2k+1}^{(n)},y'^{-1}>} = \dfrac{\Delta\varepsilon_{2k}^{(n)}}{<\Delta\varepsilon_{2k+1}^{(n)},\Delta\varepsilon_{2k}^{(n)}>}$

Appelons (11) l'ensemble des relations (11') et (11") qui définissent l'ε-algorithme topologique. D'après la définition de l'inverse d'un couple de E × E' donnée précédemment on voit que $\{[\Delta\varepsilon_{2k}^{(n)}]^{-1}\}^{-1} = \Delta\varepsilon_{2k}^{(n)}$. D'après le premier exemple de l'inverse d'un couple on a également $\{[\Delta\varepsilon_{2k+1}^{(n)}]^{-1}\}^{-1} = \Delta\varepsilon_{2k+1}^{(n)}$. Les exemples 2 et 3 montrent que :

$$<[\Delta\varepsilon_{2k}^{(n)}]^{-1}, \Delta\varepsilon_{2k}^{(n)}> = 1$$

$$<\Delta\varepsilon_{2k+1}^{(n)}, [\Delta\varepsilon_{2k+1}^{(n)}]^{-1}> = 1$$

Dans ce qui suit l'inverse de y'^{-1} sera toujours pris par rapport au couple $(y'^{-1},[\Delta\varepsilon_{2k}^{(n)}]^{-1}) \in E \times E'$ afin d'avoir $(y'^{-1})^{-1}=y'$.

Nous allons maintenant relier l'algorithme (11) avec la généralisation de la transformation de Shanks que nous avons exposée précédemment. Le résultat fondamental est le suivant :

Théorème 105 :

$$\varepsilon_{2k}^{(n)} = e_k(S_n) \text{ et } \varepsilon_{2k+1}^{(n)} = [e_k(\Delta S_n)]^{-1} = \frac{y'}{<y',e_k(\Delta S_n)>} \quad \text{pour } n,k=0,1,\ldots$$

Démonstration : On a $\varepsilon_0^{(n)} = e_0(S_n) = S_n$ et d'après (11') :

$$\varepsilon_1^{(n)} = \frac{y'}{\langle y',\Delta S_n\rangle} = \frac{y'}{\langle y',e_0(\Delta S_n)\rangle} = [e_0(\Delta S_n)]^{-1}$$

Supposons avoir démontré les relations du théorème jusqu'à $\varepsilon_{2k-1}^{(n)}$ et $\varepsilon_{2k}^{(n)}$; nous allons montrer qu'elles restent encore vraies pour $\varepsilon_{2k+1}^{(n)}$ et $\varepsilon_{2k+2}^{(n)}$ $\forall n$. D'après (11') on a :

$$\varepsilon_{2k+1}^{(n)} = \frac{y'}{\langle y',e_k(\Delta S_n)\rangle} = \frac{y'}{\langle y',e_{k-1}(\Delta S_{n+1})\rangle} + \frac{y'}{\langle y',\Delta\varepsilon_{2k}^{(n)}\rangle}$$

En utilisant (5) on voit que cette relation n'est autre que celle de l'ε-algorithme scalaire multipliée par $y' \in E'$. Elle est donc vérifiée si (11") est satisfaite. La démonstration de 11" est calquée sur celle de l'ε-algorithme scalaire et nous ne la donnerons pas ici.

Remarque : On voit que seules sont intéressantes les quantités avec un indice inférieur pair ; les quantités avec un indice inférieur impair ne représentent que des calculs intermédiaires.

On a les propriétés évidentes suivantes.

Propriété 45 :

$$e_{k+1}(S_n) = e_k(S_{n+1}) - \frac{\langle y',e_k(\Delta S_n)\rangle\langle y',e_k(\Delta S_{n+1})\rangle}{\langle y',\Delta e_k(S_n)\rangle\langle y',\Delta e_k(\Delta S_n)\rangle}\,\Delta e_k(S_n)$$

où l'opérateur Δ porte toujours sur l'indice n.

Propriété 46 :

$$[\varepsilon_{k+2}^{(n-1)} - \varepsilon_k^{(n)}]^{-1} - [\varepsilon_k^{(n)} - \varepsilon_{k-2}^{(n+1)}]^{-1} = [\Delta\varepsilon_k^{(n)}]^{-1} - [\Delta\varepsilon_k^{(n-1)}]^{-1}$$

La démonstration de cette propriété est analogue à celle effectuée par Wynn dans le cas de l'ε-algorithme scalaire . Elle résulte du fait que l'inverse de l'inverse d'un élément de E ou de E' est l'élément lui-même.

Lorsque k est pair cette relation devient :

$$<y',\varepsilon_{2k+2}^{(n-1)} - \varepsilon_{2k}^{(n)}>^{-1} - <y',\varepsilon_{2k}^{(n)} - \varepsilon_{2k-2}^{(n+1)}>^{-1} = <y',\Delta\varepsilon_{2k}^{(n)}>^{-1} - <y',\Delta\varepsilon_{2k}^{(n-1)}>^{-1}$$

Comme on le voit cette relation ne fait intervenir que des colonnes paires du tableau de l'ε-algorithme. Nous avons vu que l'ε-algorithme permettait de construire la moitié supérieure de la table de Padé. L'autre moitié peut être construite à l'aide de cette relation en partant des conditions aux limites :

$$[-1/q] = 0$$

$$<y',[p/-1]> = \infty$$

$$[p/0] = \sum_{i=0}^{p} c_i x^i$$

$$[0/q] = \left(\sum_{i=0}^{q} d_i x^i \right)^{-1}$$

où $\sum_{i=0}^{\infty} d_i x^i = [\sum_{i=0}^{\infty} c_i x^i]^{-1}$ où $c_i \in E$, $d_i \in E'$ et $x \in K$.

En termes de table de Padé, cette relation s'écrit :

$$([p/q+1] - [p/q])^{-1} - ([p/q] - [p/q-1])^{-1} = ([p+1/q] - [p/q])^{-1} - ([p/q] - [p-1/q])^{-1}$$

où l'inverse est pris par rapport à y'.

<u>*Propriété 47*</u> : *Si l'application de l'ε-algorithme (11) aux suites* $\{S_n\}$ *et* $\{aS_n + b\}$ avec $a \neq 0 \in K$ et $b \in E$ *fournit respectivement les éléments* $\varepsilon_k^{(n)}$ *et* $\bar{\varepsilon}_k^{(n)}$ *alors on a :*

$$\bar{\varepsilon}_{2k}^{(n)} = a\varepsilon_{2k}^{(n)} + b \quad et \quad \bar{\varepsilon}_{2k+1}^{(n)} = \varepsilon_{2k+1}^{(n)}/a$$

Démonstration : La relation sur les colonnes paires n'est autre que la propriété 1. La relation sur les colonnes impaires provient du théorème 6 et de la propriété 1.

Propriété 48 : Si on applique l'ε-algorithme topologique(11) à une suite $\{S_n\}$ d'éléments de E qui vérifie :

$$\sum_{i=0}^{k} a_i S_{n+i} = 0 \qquad \forall\, n > N$$

avec $\sum_{i=0}^{k} a_i \neq 0$, *alors :*

$$<\varepsilon_1^{(n)},\varepsilon_0^{(n)}> - <\varepsilon_1^{(n)},\varepsilon_2^{(n)}> + <\varepsilon_3^{(n)},\varepsilon_2^{(n)}> - \ldots + <\varepsilon_{2k-1}^{(n)},\varepsilon_{2k-2}^{(n)}> = - \sum_{i=1}^{k} i a_i / \sum_{i=0}^{k} a_i \quad \forall\, n>N.$$

Démonstration : on a, d'après (11) :

$$<\varepsilon_{2i+1}^{(n)} - \varepsilon_{2i-1}^{(n+1)}, \Delta\varepsilon_{2i}^{(n)}> = 1$$

$$<\Delta\varepsilon_{2i+1}^{(n)}, \varepsilon_{2i+2}^{(n)} - \varepsilon_{2i}^{(n+1)}>= 1$$

On a donc la même propriété que dans le cas de l'ε-algorithme scalaire mais où le produit ordinaire est remplacé par le produit de dualité. La démonstration reste donc la même que celle effectuée par Bauer [15] pour l'invariance de la somme quelque soit n et que celle de Wynn [195] pour la valeur numérique de la constante.

Remarque : Les relations (11) n'englobent pas les relations de l'ε-algorithme vectoriel défini par Wynn . En effet d'après cet algorithme on a :

$$\varepsilon_2^{(n)} = \{S_{n+2}(\Delta S_n,\Delta S_n) - 2S_{n+1}(\Delta S_{n+1},\Delta S_n) + S_n(\Delta S_{n+1},\Delta S_{n+1})\}/(\Delta^2 S_n,\Delta^2 S_n).$$

D'après (5) on voit ici que $e_1(S_n)$ est une combinaison de S_n et de S_{n+1} seulement.

Dans cet ε-algorithme vectoriel Wynn utilise comme définition de l'inverse y^{-1} de $y \in \mathbb{C}^p$: $y^{-1} = \bar{y}/(y,y)$. Dans [214], Wynn a émis la conjecture que l'on pouvait également utiliser la définition $y^{-1} = \bar{y}/(y,Dy)$ où D est une matrice symétrique et que la propriété du théorème 1 restait vraie. Greville [105] a montré que cette conjecture était fausse. On voit, à l'aide de l'étude précédente, que cela tient au fait que $(y,y^{-1}) \neq 1$ et au fait que $(y,y)(y^{-1},y^{-1}) \neq 1$. C'est pour la même raison que l'ε-algorithme normé (paragraphe V-2) ne vérifie pas la propriété du théorème 100 pour $k>1$.

Etudions maintenant la convergence de cet algorithme.

D'après (11") on a :

$$\varepsilon_{2k+2}^{(n)} = \varepsilon_{2k}^{(n+1)} + \frac{\Delta\varepsilon_{2k}^{(n)}}{<\Delta\varepsilon_{2k+1}^{(n)},\Delta\varepsilon_{2k}^{(n)}>}$$

$$= \varepsilon_{2k}^{(n+1)}\left\{1 + \frac{1}{<\Delta\varepsilon_{2k+1}^{(n)},\Delta\varepsilon_{2k}^{(n)}>}\right\} - \frac{\varepsilon_{2k}^{(n)}}{<\Delta\varepsilon_{2k+1}^{(n)},\Delta\varepsilon_{2k}^{(n)}>}$$

d'où immédiatement le :

Théorème 106 : *Supposons que* $\lim_{n\to\infty} \varepsilon_{2k}^{(n)} = S$. *Si* ∃ $\alpha' < 0 < \beta'$ *tels que*

$$<\Delta\varepsilon_{2k+1}^{(n)},\Delta\varepsilon_{2k}^{(n)}> \notin [\alpha',\beta'] \qquad \forall\, n > N$$

alors $\lim_{n\to\infty} \varepsilon_{2k+2}^{(n)} = S$.

Remarque 1 : Prenons $k=0$ dans le théorème . Alors la condition devient :

$$<\Delta\varepsilon_{1}^{(n)},\Delta S_n> = \frac{<y',\Delta S_n>}{<y',\Delta S_{n+1}>} - 1 \notin [\alpha',\beta']$$

ou, en d'autres termes, si ∃ $\alpha < 1 < \beta$ tels que :

$$\frac{<y',\Delta S_{n+1}>}{<y',\Delta S_n>} \notin [\alpha,\beta] \qquad \forall\, n > N$$

alors $\lim_{n\to\infty} \varepsilon_{2}^{(n)} = S$.

On retrouve ainsi un résultat analogue à ceux obtenus dans le cas de l'ε-algorithme ordinaire [38] et dans le cas de sa généralisation à un espace de Banach [40].

Théorème 107 : *Si* $\lim_{n\to\infty} \varepsilon_{2k}^{(n)} = S$, *si* $\lim_{n\to\infty} <\Delta\varepsilon_{2k+1}^{(n)},\Delta\varepsilon_{2k}^{(n)}> = a \neq 0$

et si $\lim_{n\to\infty} \frac{<z',\varepsilon_{2k}^{(n)}-S>}{<z',\varepsilon_{2k}^{(n+1)}-S>} = 1+a$ *avec* $z' \in E'$ *alors* $\lim_{n\to\infty} \varepsilon_{2k+2}^{(n)} = S$

et de plus :

$$\lim_{n\to\infty} \frac{<z',\varepsilon_{2k+2}^{(n)}-S>}{<z',\varepsilon_{2k}^{(n+1)}-S>} = 0$$

Démonstration : Il est évident, d'après le théorème 106, que $\lim_{n\to\infty} \varepsilon_{2k+2}^{(n)} = S$.

D'autre part on a :

$$<z',\varepsilon_{2k+2}^{(n)}-S> = <z',\varepsilon_{2k}^{(n+1)} - S> + \frac{<z',\varepsilon_{2k}^{(n+1)}-S> - <z',\varepsilon_{2k}^{(n)}-S>}{<\Delta\varepsilon_{2k+1}^{(n)},\Delta\varepsilon_{2k}^{(n)}>}$$

d'où :

$$\frac{<z',\varepsilon_{2k+2}^{(n)}-S>}{<z',\varepsilon_{2k}^{(n+1)}-S>} = 1 + \frac{1 - \dfrac{<z',\varepsilon_{2k}^{(n)}-S>}{<z',\varepsilon_{2k}^{(n+1)}-S>}}{<\Delta\varepsilon_{2k+1}^{(n)},\Delta\varepsilon_{2k}^{(n)}>}$$

Ce qui termine la démonstration du théorème.

Remarque 2 : Pour k=0 le théorème précédent nous donne :

si $$\lim_{n\to\infty} \frac{<z',S_{n+1}-S>}{<z',S_n - S>} = \lim_{n\to\infty} \frac{<y',\Delta S_{n+1}>}{<y',\Delta S_n>} = b \neq 1$$

alors $\lim_{n\to\infty} \varepsilon_2^{(n)} = S$ et, de plus :

$$\lim_{n\to\infty} \frac{<z',\varepsilon_2^{(n)}- S>}{<z',S_{n+1}-S>} = 0$$

On pourra de nouveau comparer avec les résultats de [40].

Si la suite $\{<y',S_n>\}$ est totalement monotone ou totalement oscillante alors les quantités $<y',\varepsilon_k^{(n)}>$ vérifient les inégalités qui ont été démontrées dans le cas scalaire et l'on a les résultats de convergence pour $\{<y',\varepsilon_{2k}^{(n)}>\}$ mais pas pour $\{\varepsilon_{2k}^{(n)}\}$.

Si la condition du théorème 107 n'est pas vérifiée, il est possible d'introduire dans l'algorithme un paramètre d'accélération, de caractériser sa valeur optimale et de construire un Θ-algorithme topologique en suivant la même démarche que dans le cas scalaire.

Dans la première généralisation de la transformation de Shanks que nous venons d'étudier $e_k(S_n)$ était calculé en résolvant le système (4) puis par $e_k(S_n) = a_0 S_n + \ldots + a_k S_{n+k}$. Il est évident, qu'au lieu d'utiliser la seconde des équations du système (2), on peut utiliser n'importe laquelle des autres équations et, en particulier, la dernière ; on obtiendra ainsi :

$$\tilde{e}_k(S_n) = a_0 S_{n+k} + \ldots + a_k S_{n+2k}$$

ce qui s'écrit encore :

$$\tilde{e}_k(S_n) = \frac{\begin{vmatrix} S_{n+k} & \cdots\cdots\cdots & S_{n+2k} \\ \langle y',\Delta S_n\rangle & \cdots\cdots & \langle y',\Delta S_{n+k}\rangle \\ - - - - & - - - - & - - - - \\ \langle y',\Delta S_{n+k-1}\rangle & \cdots & \langle y',\Delta S_{n+2k-1}\rangle \end{vmatrix}}{\begin{vmatrix} 1 & \cdots\cdots\cdots & 1 \\ \langle y',\Delta S_n\rangle & \cdots\cdots & \langle y',\Delta S_{n+k}\rangle \\ - - - - & - - - - & - - - - \\ \langle y',\Delta S_{n+k-1}\rangle & \cdots & \langle y',\Delta S_{n+2k-1}\rangle \end{vmatrix}} \qquad (12)$$

Pour cette seconde généralisation de la transformation de Shanks, le théorème 100 ainsi que les propriétés 36 et 37 sont vérifiées. Il est bien évident que (5) et (12) ne sont pas indépendants ; on a :

Propriété 49 : $\langle y',e_k(S_n)\rangle = \langle y',\tilde{e}_k(S_n)\rangle = e_k(\langle y',S_n\rangle)$

La démonstration est évidente par combinaison linéaire des lignes du numérateur de (12). Le dernier terme de cette double égalité représente la transformation habituelle de Shanks appliquée à la suite de scalaires $\langle y',S_n\rangle$. Cette propriété montre que l'on retrouve la transformation de Shanks ordinaire lorsque $E = \mathbb{R}$. On a également le résultat suivant qui est l'équivalent du résultat de Gilewicz déjà cité [90].

Propriété 50 : *Soient* $e_k(S_n)$ *et* $\tilde{e}_k(S_n)$ *les éléments de* E *obtenus en appliquant respectivement* (5) *et* (12) *aux* 2k+1 *éléments successifs* $S_n, S_{n+1}, \ldots, S_{n+2k}$. *Soient* $e_k(u_n)$ et $\tilde{e}_k(u_n)$ *les éléments de* E *obtenus en appliquant respectivement* (5) *et* (12) *aux* 2k+1 *éléments successifs* $u_n = S_{n+2k}$, $u_{n+1} = S_{n+2k-1}, \ldots, u_{n+2k} = S_n$. *Alors on a* :

$$e_k(S_n) = \tilde{e}_k(u_n)$$

et

$$e_k(u_n) = \tilde{e}_k(S_n)$$

La démonstration de cette propriété est évidente à partir de (5) et (12).

Si on applique (12) aux sommes partielles de la série formelle $f(x) = \sum_{i=0}^{\infty} c_i x^i$ avec $x \in K$ et $c_i \in E$ alors (12) nous fournit une seconde généralisation de la table de Padé. En effet en remplaçant S_n par sa valeur et en effectuant les mêmes transformations que précédemment on trouve que le numérateur de $\tilde{e}_k(S_n)$ est de degré n+2k par rapport à x et que son dénominateur est de degré k. On a :

$$\langle y', \tilde{e}_k(S_n)\rangle - \langle y', f(x)\rangle = O(x^{n+2k+1})$$

et $\tilde{e}_k(S_n) - f(x) = Ax^{n+2k+1}$ avec $A \in E$.

On notera donc symboliquement :

$$\tilde{e}_k(S_n) = [n+2k/k]$$

Par analogie l'approximant de Padé généralisé sera défini par :

$$[p/q] = \frac{\begin{vmatrix} \sum_{i=0}^{p-q} c_i x^{q+i} & \cdots\cdots & \sum_{i=0}^{p} c_i x^i \\ \langle y', c_{p-2q+1}\rangle & \cdots\cdots & \langle y', c_{p-q+1}\rangle \\ - - - & - - - & - - - \\ \langle y', c_{p-q}\rangle & & \langle y', c_p\rangle \end{vmatrix}}{\begin{vmatrix} x^q & \cdots\cdots & 1 \\ \langle y', c_{p-2q+1}\rangle & \cdots\cdots & \langle y', c_{p-q+1}\rangle \\ - - - & - - - & - - - \\ \langle y', c_{p-q}\rangle & \cdots\cdots & \langle y', c_p\rangle \end{vmatrix}} \qquad (13)$$

avec $c_i = 0 \in E_i$ si $i < 0$.

Remarque : D'après (13) on voit que $[p/q]$ n'est défini que pour $p-q+1 \geqslant 0$. On ne peut donc ainsi construire que la moitié de la table de Padé généralisée. Cette restriction sera toujours sous entendue dans la suite du paragraphe.

Les approximants sont de la forme :

$$[p/q] = \frac{\sum_{i=0}^{p} a_i x^i}{\sum_{i=0}^{q} b_i x^i} \qquad \text{avec } a_i \in E \text{ et } b_i \in \mathbb{C}$$

Le calcul des b_i s'effectue comme précédemment en résolvant le système :

$$\begin{aligned} b_0 \langle y',c_{p-q+1}\rangle + b_1 \langle y',c_{p-1}\rangle + \dots + b_q \langle y',c_{p-2q+1}\rangle &= 0 \\ \text{-------------------------------} & \\ b_0 \langle y',c_p\rangle \quad + b_1 \langle y',c_{p-1}\rangle + \dots + b_q \langle y',c_{p-q+1}\rangle &= 0 \end{aligned} \qquad (14)$$

avec $b_0=1$. Puis les $a_i \in E$ sont calculés à l'aide des relations (10).

Comme pour la généralisation de la table de Padé étudiée au second paragraphe, on voit que l'on a pour la même raison :

$$\langle y',[p/q]\rangle - \langle y',f(x)\rangle = 0(x^{p+q+1})$$

et $$[p/q] - f(x) = Ax^{p+1}$$

On a les résultats suivants :

Théorème 108 : Une condition nécessaire et suffisante pour que $[p/q]$ *défini par (14) et (10) existe est que :*

$$H_q^{(p-2q+1)}(\langle y',c_{p-2q+1}\rangle) \neq 0$$

Théorème 109 : S'il existe, $[p/q]$ *défini par (14) et (10), est unique.*

On peut calculer (12) au moyen d'un procédé récursif analogue à la généralisation (11) de l'ε-algorithme :

$$\varepsilon_{-1}^{(n)} = 0 \qquad \varepsilon_0^{(n)} = S_n \qquad n = 0,1,\ldots$$

$$\varepsilon_{2k+1}^{(n)} = \varepsilon_{2k-1}^{(n+1)} + \frac{y'}{\langle y', \Delta\varepsilon_{2k}^{(n)}\rangle} \qquad n,k=0,1,\ldots \qquad (15')$$

$$\varepsilon_{2k+2}^{(n)} = \varepsilon_{2k}^{(n+1)} + \frac{\Delta\varepsilon_{2k}^{(n+1)}}{\langle \Delta\varepsilon_{2k+1}^{(n)}, \Delta\varepsilon_{2k}^{(n+1)}\rangle} \qquad n,k=0,1,\ldots \qquad (15'')$$

où y'^{-1} qui intervient dans le calcul de $\varepsilon_{2k+2}^{(n)}$ est défini par :

$$y'^{-1} = \frac{\Delta\varepsilon_{2k}^{(n+1)}}{\langle y', \Delta\varepsilon_{2k}^{(n+1)}\rangle}$$

On a le résultat fondamental suivant :

Théorème 110: $\varepsilon_{2k}^{(n)} = \tilde{e}_k(S_n)$ et $\varepsilon_{2k+1}^{(n)} = [\tilde{e}_k(\Delta S_n)]^{-1} = \dfrac{y'}{\langle y', \tilde{e}_k(\Delta S_n)\rangle}$

Cette seconde généralisation de l'ε-algorithme possède des propriétés analogues à celles de la première généralisation (11) ; nous ne le retranscrirons pas ici.

Il faut remarquer que, dans le cas où $E = \mathbb{C}^p$, les relations (15) n'englobent pas les relations de l'ε-algorithme vectoriel. Cet algorithme ne semble pas devoir rentrer dans le cadre des généralisations que nous avons étudiées et ceci même en effectuant des combinaisons linéaires entre les diverses généralisations que l'on peut obtenir en utilisant les différentes équations du système (2) pour calculer $e_k(S_n)$.

On peut, pour cette seconde généralisation de l'ε-algorithme, démontrer des résultats de convergence analogues à ceux du paragraphe 5. Les théorèmes sur les suites totalement monotones ou totalement oscillantes sont encore vrais. Il est également possible d'introduire dans l'algorithme (15) un paramètre d'accélération de la convergence et de caractériser sa valeur optimale ; on peut aussi définir un Θ-algorithme en remplaçant la valeur optimale de ce paramètre d'accélération par son approximation. L'établissement de ces résultats est laissé au lecteur.

Supposons que la suite $\{S_n\}$ vérifie encore la relation (1). Comme précédemment on veut calculer S.

Soient $y'_1,..,y'_k \in E'$. Alors on a :

$$\begin{aligned}
&a_0 + a_1 + \ldots\ldots + a_k = 1 \\
&a_0\langle y'_1,\Delta S_n\rangle + \ldots + a_k\langle y'_1,\Delta S_{n+k}\rangle = 0 \\
&-\ -\ -\ -\ -\ -\ -\ -\ -\ -\ -\ - \\
&a_0\langle y'_k,\Delta S_n\rangle + \ldots + a_k\langle y'_k,\Delta S_{n+k}\rangle = 0
\end{aligned} \qquad (16)$$

Par conséquent, si le déterminant de ce système est différent de zéro, on obtient :

$$S = \frac{\begin{vmatrix} S_n & \cdots & S_{n+k} \\ \langle y'_1,\Delta S_n\rangle & \cdots & \langle y'_1,\Delta S_{n+k}\rangle \\ - - - & & - - - \\ \langle y'_k,\Delta S_n\rangle & \cdots & \langle y'_k,\Delta S_{n+k}\rangle \end{vmatrix}}{\begin{vmatrix} 1 & \cdots & 1 \\ \langle y'_1,\Delta S_n\rangle & \cdots & \langle y'_1,\Delta S_{n+k}\rangle \\ - - - & & - - - \\ \langle y'_k,\Delta S_n\rangle & \cdots & \langle y'_k,\Delta S_{n+k}\rangle \end{vmatrix}} = \bar{e}_k(S_n) \qquad (17)$$

On voit que l'on a immédiatement le :

Théorème 111 : Une condition nécessaire pour que le déterminant situé au dénominateur de (17) soit différent de zéro est que $y'_1,\ldots,y'_k$ soient linéairement indépendants.

Remarque 1 : Si E est de dimension p et si k>p alors la condition du théorème précédent n'est pas satisfaite. Dans ce cas on doit utiliser les généralisations (5) ou (12) étudiées dans les paragraphes précédents ; ceci est en particulier vrai lorsque $E = \mathbb{R}$: (17) est impossible à utiliser si $k > 1$; (5) et (12) se réduisent alors à la transformation de Shanks habituelle. On voit que cette remarque restreint singulièrement les possibilités d'utilisation de cette généralisation dans le cas vectoriel.

Remarque 2 : Dans le cas où (17) est applicable (c'est-à-dire lorsque son dénominateur est différent de zéro) on voit que seuls sont nécessaires les éléments $S_n,\ldots,S_{n+k+1}$ alors que l'utilisation de (5) ou (12) demande la connaissance de $S_n,\ldots,S_{n+2k}$.

Pour cette troisième généralisation $\bar{e}_k(S_n)$ de la transformation de Shanks le théorème 100 ainsi que les propriétés 36 et 37 restent vérifiées.

Considérons de nouveau la série formelle :

$$f(x) = \sum_{i=0}^{\infty} c_i x^i$$

avec $x \in K$ et $c_i \in E$. Si l'on prend comme suite $\{S_n\}$ les sommes partielles de $f(x)$:

$$S_n = \sum_{i=0}^{n} c_i x^i$$

alors (17) nous fournit une troisième généralisation de la table de Padé définie par :

$$[p/q] = \frac{\begin{vmatrix} \sum_{i=0}^{p-q} c_i x^{q+i} & \cdots\cdots & \sum_{i=0}^{p} c_i x^i \\ \langle y_1', c_{p-q+1}\rangle & \cdots\cdots & \langle y_1', c_{p+1}\rangle \\ - - - & - - - & - - - \\ \langle y_k', c_{p-q+1}\rangle & \cdots\cdots & \langle y_k', c_{p+1}\rangle \end{vmatrix}}{\begin{vmatrix} x^q & \cdots\cdots & 1 \\ \langle y_1', c_{p-q+1}\rangle & \cdots\cdots & \langle y_1', c_{p+1}\rangle \\ - - - & - - - & - - - \\ \langle y_k', c_{p-q+1}\rangle & \cdots\cdots & \langle y_k', c_{p+1}\rangle \end{vmatrix}}$$

avec la convention que $c_i = 0 \in E$ si $i < 0$. Pour que ces approximants soient définis il faut donc que : $p \geqslant q-1$; cette restriction sera par conséquent toujours sous entendue par la suite.

Les approximants sont de la forme :

$$[p/q] = \frac{\sum_{i=0}^{p} a_i x^i}{\sum_{i=0}^{q} b_i x^i} \quad \text{avec } a_i \in E \quad \text{et} \quad b_i \in \mathbb{C}$$

Le calcul des a_i et des b_i s'effectue comme pour la première généralisation de la table de Padé étudiée précédemment. En prenant $b_0=1$ on trouve les b_i comme solution du système :

$$b_0 \langle y_1', c_{p+1} \rangle + b_1 \langle y_1', c_p \rangle + \ldots + b_q \langle y_1', c_{p-q+1} \rangle = 0$$

$$- -$$

$$b_0 \langle y_q', c_{p+1} \rangle + b_1 \langle y_q', c_p \rangle + \ldots + b_q \langle y_q', c_{p-q+1} \rangle = 0$$

Puis les a_i sont calculés à partir des relations (10).

On a les mêmes résultats théoriques que pour les première et seconde généralisations de la table de Padé : théorèmes 101,102, et 103, propriétés 38, 39 et 40.

Remarque : Puisque l'on doit avoir $p \geqslant q-1$ seule la moitié de cette table de Padé est définie. Il est par conséquent inutile d'étudier dans ce cas la série inverse de $f(x)$.

Le calcul effectif de $\bar{e}_k(S_n)$ à partir de (17) est difficile dès que k vaut 4 ou 5. Il est donc nécessaire d'éviter ce calcul à l'aide d'un algorithme analogue aux généralisations (11) et (15) de l'ε-algorithme. Un tel algorithme n'a pas encore été obtenu. Il est cependant toujours possible de calculer numériquement $\bar{e}_k(S_n)$ en résolvant le système (16) dont les inconnues sont $a_0,\ldots,a_k$ puis en écrivant que :

$$\bar{e}_k(S_n) = a_0 S_n + a_1 S_{n+1} + \ldots + a_k S_{n+k}$$

Cette façon de procéder est à rapprocher d'une méthode utilisée par Henrici [110, paragraphe 5-9 page 115] pour résoudre les systèmes d'équations non linéaires par un procédé qui étend à plusieurs dimensions la formule de Steffensen (dans ce cas il faut prendre comme y'_i dans (17) le $i^{ème}$ vecteur de base de $\mathbb{R}^k$). On pourra également comparer la méthode d'Henrici avec celle proposée au paragraphe V-4 qui est basée sur l'utilisation de l'ε-algorithme vectoriel.

L'utilisation d'une relation du type (17) est liée à la méthode des moments [184]. La connexion qui existe entre cette méthode et les procédés d'accélération de la convergence a été mise en évidence par Germain-Bonne [88] ; ce nouvel aspect de ces méthodes semble d'ailleurs devoir se développer. On voit également la liaison qui existe entre (17) et le procédé d'orthogonalisation de Schmidt lorsque E est un espace de Hilbert et que $y'_1 = \Delta S_n, \ldots, y'_k = \Delta S_{n+k-1}$.

Dans les applications, c'est évidemment le cas vectoriel qui est le plus important. Tous les résultats des paragraphes V-3, V-4 et V-5 sont encore valables pour les deux ε-algorithmes topologiques puisque leurs démonstrations reposent uniquement sur le théorème de Wynn-McLeod. L'avantage des ε-algorithmes topologiques est que l'on connaît les $e_k(S_n)$ sous forme d'un rapport de déterminants ; cela nous a permis d'obtenir un certain nombre de résultats qui restent encore à démontrer dans le cas de l'ε-algorithme vectoriel. Il a d'autre part été possible de rattacher le premier ε-algorithme topologique à la méthode des moments et par là à la méthode du gradient conjugué et à la théorie des polynômes orthogonaux. Cette étude n'est pas encore achevée actuellement mais elle devrait fournir un cadre théorique à l'ε-algorithme et à la table de Padé et permettre d'englober tous les résultats connus.

Remarque 1: Dans l'ε-algorithme vectoriel utilisé par Wynn l'inverse d'un vecteur $y \in \mathbb{C}^p$ est définit par $y^{-1} = \bar{y}/(y,y)$. Dans cet algorithme, si l'on remplace, dans y^{-1}, $\bar{y}$ par y alors les vecteurs d'indice inférieur pair restent inchangés alors que les vecteurs d'indice inférieur impair sont remplacés par leurs conjugués. On remarquera, que dans les algorithmes que nous venons d'étudier, le conjugué d'un élément de E n'apparait jamais.

Remarque 2:Les algorithmes (11) et (15) peuvent être utilisés de façon similaire à celle décrite au paragraphe V-5 pour calculer les valeurs propres d'un endomorphisme de E lorsque celui-ci est normal et compact. Les algorithmes (11) et (15) peuvent également être utilisés pour résoudre des équations de point fixe de la forme x = Ax + b où A est un endomorphisme de E normal, compact et de rang fini. Une telle méthode est à rapprocher de l'utilisation par Chisholm [55] de la table de Padé ordinaire pour résoudre certaines équations intégrales provenant de problèmes de mécanique quantique.

Les résultats présentés ici montrent que l'on peut généraliser de cette façon tous les algorithmes d'accélération de la convergence. Il n'y a en particulier aucune difficulté pour donner les règles des procédés p et q ainsi que des première et seconde généralisation de l'ε-algorithme et le ρ-algorithme . La suite auxiliaire $\{x_n\}$ qui intervient dans ces algorithmes est toujours une suite d'éléments de K.

Les procédés linéaires de sommation et, en particulier, l'extrapolation polynomiale de Richardson se généralisent immédiatement puisque la notion d'inverse d'un élément n'y intervient pas. On voit d'ailleurs que la base de la définition de l'ε-algorithme généralisé que nous avons donnée ici est la notion d'inverse d'un couple d'éléments l'un appartenant à E et l'autre à E'. On remarquera aussi que l'ensemble des généralisations que nous venons d'exposer dépend d'un élément arbitraire y' ∈ E (ou d'une suite de tels éléments) ; il se pose dont le problème du choix optimal de cet élément y' (ou de cette suite d'éléments).

CHAPITRE VI

ALGORITHMES DE PREDICTION CONTINUE

Jusqu'à présent le problème auquel nous nous sommes intéressés était celui de l'estimation de la limite S de la suite convergente $\{S_n\}$ à partir des quantités S_n, ΔS_n, $\Delta^2 S_n$, ..., $\Delta^k S_n$ pour une certaine valeur de n. Dans ce chapitre nous étudierons le problème suivant : étant donnée $f : \mathbb{R} \to \mathbb{R}$ continue, suffisamment dérivable et telle que $\lim_{t\to\infty} f(t) = S$ existe et soit finie on veut estimer S à partir des quantités f(t), $f'(t)$, $f''(t)$, ..., $f^{(k)}(t)$ pour une certaine valeur de t.

On voit la ressemblance que présentent ces deux problèmes ; pour traiter le second problème on utilisera donc des formes spécialement adaptées des algorithmes que nous connaissons pour les suites : ce seront les formes confluentes des ε et ρ-algorithmes, etc ... Nous n'étudierons ici que la première forme confluente de l'ε-algorithme, la forme confluente du procédé d'Overholt et le développement en série de Taylor.

VI - 1 La première forme confluente de l'ε-algorithme

Cet algorithme a été obtenu par Wynn [239] de la façon suivante : dans la règle de l'ε-algorithme scalaire on remplace la variable discrète n par la variable continue $t = a + n.\Delta t$, puis $\varepsilon_{2k+1}^{(n)}$ par $\varepsilon_{2k+1}(t) / \Delta t$ et $\varepsilon_{2k}^{(n)}$ par $\varepsilon_{2k}(t)$ et ensuite on fait tendre Δt vers zéro. D'où la première forme confluente de l'ε-algorithme donc les règles sont les suivantes :

$$\varepsilon_{-1}(t) = 0 \qquad \varepsilon_0(t) = f(t)$$

$$\varepsilon_{k+1}(t) = \varepsilon_{k-1}(t) + \frac{1}{\varepsilon_k'(t)}$$

On peut démontrer, pour cet algorithme, des propriétés analogues à celle de l'ε-algorithme

Propriété 51 : Si l'application de la première forme confluente de l'ε-algorithme aux fonctions f et af + b, où a et b sont des constantes avec $a \neq 0$, fournit respectivement les fonctions ε_k et $\bar{\varepsilon}_k$ alors :

$$\bar{\varepsilon}_{2k}(t) = a\,\varepsilon_{2k}(t) + b \qquad \bar{\varepsilon}_{2k+1}(t) = \varepsilon_{2k+1}(t)\,/\,a$$

démonstration : on a $\bar{\varepsilon}_0(t) = a\,\varepsilon_0(t) + b$ et $\bar{\varepsilon}'_0(t) = a\,\varepsilon'_0(t)$.

Donc

$$\bar{\varepsilon}_1(t) = \frac{1}{\bar{\varepsilon}'_0(t)} = \varepsilon_1(t)\,/\,a.$$

Supposons que la propriété soit vraie jusqu'aux fonctions d'indices 2k et 2k+1 et démontrons qu'elle est vraie pour les fonctions d'indices 2k+2 et 2k+3. On a :

$$\bar{\varepsilon}_{2k+2}(t) = \bar{\varepsilon}_{2k}(t) + \frac{1}{\bar{\varepsilon}'_{2k+1}(t)} = a\,\varepsilon_{2k}(t) + b + \frac{a}{\varepsilon'_{2k+1}(t)}$$

$$= a\,\varepsilon_{2k+2}(t) + b$$

De même :

$$\bar{\varepsilon}_{2k+3}(t) = \bar{\varepsilon}_{2k+1} + \frac{1}{\bar{\varepsilon}'_{2k+2}(t)}$$

$$= \frac{\varepsilon_{2k+1}(t)}{a} + \frac{1}{a\,\varepsilon'_{2k+2}(t)} = \frac{\varepsilon_{2k+3}(t)}{a}$$

définition 23 : On appelle déterminants fonctionnels de Hankel les déterminants

$$H_0^{(n)}(t) = 1$$

$$H_k^{(n)}(t) = \begin{vmatrix} f^{(n)}(t) & f^{(n+1)}(t) & \dots & f^{(n+k-1)}(t) \\ \text{-----} & \text{-----} & \text{-----} & \text{-----} \\ f^{(n+k-1)}(t) & f^{(n+k)}(t) & \dots & f^{(n+2k-2)}(t) \end{vmatrix}$$

Wynn [240] a démontré que l'on avait la :

propriété 52 :

$$\varepsilon_{2k}(t) = \frac{H_{k+1}^{(0)}(t)}{H_k^{(2)}(t)} \qquad \varepsilon_{2k+1}(t) = \frac{H_k^{(3)}(t)}{H_{k+1}^{(1)}(t)}$$

la démonstration est analogue à celle effectuée pour l'ε-algorithme scalaire en utilisant un développement de Schweins. On notera également $\varepsilon_{2k}(t) = e_k(f,t)$.

Propriété 53 :

$$\varepsilon_{2k+2}(t) - \varepsilon_{2k}(t) = - \frac{[H_{k+1}^{(1)}(t)]^2}{H_k^{(2)}(t)\ H_{k+1}^{(2)}(t)}$$

Nous allons maintenant chercher les conditions que doit vérifier f pour que $\varepsilon_{2k}(t) = S$ $\forall t > T$. On a le :

Théorème 112 :

Une condition nécessaire et suffisante pour que $\varepsilon_{2k}(t) = S\ \forall t > T$ est que f(t) vérifie :

$$\sum_{i=0}^{k} a_i\, f^{(i)}(t) = a_0 S \qquad \forall t > T \text{ avec } a_0 \neq 0$$

démonstration : démontrons que la condition est nécessaire ; d'après la propriété 52 on doit avoir :

$$S = \frac{H_{k+1}^{(0)}(t)}{H_k^{(2)}(t)} \qquad \forall t > T$$

or

$$H_k^{(2)}(t) = \begin{vmatrix} f''(t) & \cdots & f^{(k+1)}(t) \\ \hline f^{(k+1)}(t) & \cdots & f^{(2k)}(t) \end{vmatrix} = \begin{vmatrix} 1 & f'(t) & \cdots & f^{(k)}(t) \\ 0 & f''(t) & \cdots & f^{(k+1)}(t) \\ \hline 0 & f^{(k+1)}(t) & \cdots & f^{(2k)}(t) \end{vmatrix}$$

d'où

$$\begin{vmatrix} f(t) - S & f'(t) & \cdots & f^{(k)}(t) \\ f'(t) & f''(t) & \cdots & f^{(k+1)}(t) \\ \hline f^{(k)}(t) & f^{(k+1)}(t) & \cdots & f^{(2k)}(t) \end{vmatrix} = 0$$

Ce déterminant est nul si et seulement s'il existe $a_0, \ldots, a_k$ non tous nuls tels que :

$$\begin{aligned} &a_0(f(t) - S) + a_1 f'(t) + \ldots + a_k f^{(k)}(t) = 0 \\ &a_0 f'(t) + a_1 f''(t) + \ldots + a_k f^{(k+1)}(t) = 0 \\ &\text{-------------------------------} \\ &a_0 f^{(k)}(t) + a_1 f^{(k+1)}(t) + \ldots + a_k f^{(2k)}(t) = 0 \end{aligned}$$

ce qui démontre le théorème

D'où, immédiatement en résolvant cette équation différentielle :

Théorème 113 : une condition nécessaire et suffisante pour que $\varepsilon_{2k}(t) = S \; \forall t > T$ est que :

$$f(t) = S + \sum_{i=1}^{p} A_i(t)\, e^{r_i t} + \sum_{i=p+1}^{q} [B_i(t) \cos b_i t + C_i(t) \sin b_i t]\, e^{r_i t} \quad \text{pour } t > T \text{ avec}$$

$r_i \neq 0$ pour $i=1,\ldots,p$.

A_i, B_i et C_i sont des polynômes en t tels que si d_i est égal au degré de A_i plus un pour $i = 1, \ldots, p$ et au plus grand des degrés de B_i et de C_i plus un pour $i = p+1, \ldots, q$ on ait :

$$\sum_{i=1}^{p} d_i + 2 \sum_{i=p+1}^{q} d_i = k$$

La démonstration de ce théorème est évidente. On écrit tout simplement que f est la solution de l'équation différentielle du théorème 112.

Propriété 54 : Supposons que l'on applique la première forme confluente de l'ε-algorithme a $f(t) = S + \sum_{i=0}^{k} a_i\, e^{\lambda_i t}$

Si $S \neq 0$ et $Re(\lambda_1) > Re(\lambda_2) > \ldots > Re(\lambda_k) > 0$ alors

$$\lim_{t\to\infty} \frac{\varepsilon'_{2i}(t)}{\varepsilon_{2i}(t)} = \lambda_{i+1} \qquad i = 0, \ldots, k-1$$

$$\lim_{t\to\infty} \frac{\varepsilon'_{2i+1}(t)}{\varepsilon_{2i+1}(t)} = -\lambda_{i+1} \qquad i = 0, \ldots, k-1$$

Si $S = 0$ et $Re(\lambda_1) > Re(\lambda_2) > \ldots > Re(\lambda_k)$ alors les mêmes conclusions restent vraies.

Wynn [230] a démontré également la :

Propriété 55 : Si on applique la première forme confluente de l'ε-algorithme à une fonction f qui vérifie :

$$\sum_{i=0}^{k} a_i\, f^{(i)}(t) = 0$$

alors

$$\sum_{i=0}^{2k-2} (-1)^i\, \varepsilon_i(t)\, \varepsilon_{i+1}(t) = -\, a_1/a_0 \qquad \forall t$$

Nous ne donnerons pas la démonstration de ce résultat.

Propriété 56 : on a les relations suivantes :

$$H_{k+2}^{(n-1)}(t).\ H_k^{(n+1)}(t) + [H_{k+1}^{(n)}(t)]^2 = H_{k+1}^{(n-1)}(t)\ H_{k+1}^{(n+1)}(t)$$

$$\varepsilon'_{2k+1}(t) = -\frac{H_k^{(2)}(t)\ H_{k+1}^{(2)}(t)}{[H_{k+1}^{(1)}(t)]^2}$$

$$\varepsilon'_{2k}(t) = \frac{H_k^{(1)}(t)\ H_{k+1}^{(1)}(t)}{[H_k^{(2)}(t)]^2}$$

démonstration : d'après la propriété 53 on a :

$$\frac{H_{k+2}^{(0)}}{H_{k+1}^{(2)}} - \frac{H_{k+1}^{(0)}}{H_k^{(2)}} = -\frac{[H_{k+1}^{(1)}]^2}{H_k^{(2)}\ H_{k+1}^{(2)}}$$

ce qui donne la première propriété en remplaçant f par $f^{(n-1)}$.

On a :

$$\varepsilon_{2k+2}(t) - \varepsilon_{2k}(t) = \frac{1}{\varepsilon'_{2k+1}(t)} = -\frac{[H_{k+1}^{(1)}]^2}{H_k^{(2)}\ H_{k+1}^{(2)}}$$

Ce qui démontre la seconde propriété.

D'après ce qui précède on a :

$$H_{k+2}^{(1)}\ H_k^{(3)} - H_{k+1}^{(1)}\ H_{k+1}^{(3)} = -[H_{k+1}^{(2)}]^2$$

divisons par $H_{k+1}^{(1)}.\ H_{k+2}^{(1)}$; on obtient :

$$\frac{H_k^{(3)}}{H_{k+1}^{(1)}} - \frac{H_{k+1}^{(3)}}{H_{k+2}^{(1)}} = -\frac{[H_{k+1}^{(2)}]^2}{H_{k+1}^{(1)}\ H_{k+2}^{(1)}}$$

ou encore, d'après la propriété 52 :

$$\varepsilon_{2k+1}(t) - \varepsilon_{2k+3}(t) = -\frac{[H_{k+1}^{(2)}]^2}{H_{k+1}^{(1)}\ H_{k+2}^{(1)}} = -\frac{1}{\varepsilon'_{2k+2}(t)}$$

on peut donner pour la première forme confluente de l'ε-algorithme une interprétation analogue à celle donnée au paragraphe III.4 pour l'ε-algorithme scalaire.

L'application de la première forme confluente de l'ε-algorithme à $f(t)$, $f'(t)$, ..., $f^{(2k)}(t)$ revient à résoudre le système :

$$a_0 f(t) + a_1 f'(t) + \ldots + a_k f^{(k)}(t) = c \neq 0$$

$$a_0 f'(t) + a_1 f''(t) + \ldots + a_k f^{(k+1)}(t) = 0$$

$$a_0 f^{(k)}(t) + a_1 f^{(k+1)}(t) + \ldots + a_k f^{(2k)}(t) = 0$$

puis à calculer $\frac{c}{a_0} = \varepsilon_{2k}(t)$. En effet on voit immédiatement que :

$$a_0 = \frac{c\, H_k^{(2)}(t)}{H_{k+1}^{(0)}(t)}$$

d'où

$$\varepsilon_{2k}(t) = \frac{H_{k+1}^{(0)}(t)}{H_k^{(2)}(t)} = \frac{c}{a_0}$$

on voit que si c est remplacé par ac où a est une constante non nulle alors a_0 est remplacé par a a_0. Par conséquent $\varepsilon_{2t}(t)$ est indépendant de c. On voit également que la condition $H_k^{(2)}(t) \neq 0$ est équivalente à la condition $a_0 \neq 0$

VI - 2 Etude de la convergence

Les théorèmes de convergence pour la première forme confluente de l'ε-algorithme ne sont pas encore nombreux. On a cependant les résultats suivants [20] :

Théorème 114 :

Si $\lim_{t\to\infty} \varepsilon_{2k}(t) = S$, si $\varepsilon'_{2k-1}(t) \leq 0$, $\forall\, t > T$ et si $\varepsilon''_{2k}(t) \geq 0$, $\forall\, t > T$ alors il existe une suite strictement croissante $\{t_n\}$ tendant vers l'infini telle que $\lim_{n\to\infty} \varepsilon_{2k+2}(t_n) = S$.

démonstration : elle est calquée sur celle du théorème 54. Remarquons d'abord que $\varepsilon'_{2k}(t) \leq 0$ puisque $\varepsilon''_{2k}(t) \geq 0$ et que $\lim_{t\to\infty} \varepsilon'_{2k}(t) = 0$. Posons $u_0 = \varepsilon_{2k}(t)$ et $u_{p+1} = \varepsilon_{2k}(t+(p+1)k) - \varepsilon_{2k}(t+ph)$. On démontre qu'il n'existe pas $A < 0$ fini tel que $A < u_{n+1}^{-1} - u_n^{-1} \leqslant 0$, $\forall$ n. Or $u_{n+1} = h\varepsilon'_{2k}(t+\Theta h)$ avec $\Theta \in [n,n+1]$. Le fait que

$\varepsilon''_{2k}(t) \geq 0$, $\forall\, t > T$ entraîne que ε'_{2k} est une fonction croissante de t, $\forall\, t > T$;

donc $0 \geq h\varepsilon'_{2k}(t+(n+1)h) \geq u_{n+1} \geq h\varepsilon'_{2k}(t+nh)$. Posons $x = t+nh$, on a :

$$\frac{1}{h\varepsilon'_{2k}(x+h)} \leq \frac{1}{u_{n+1}} \leq \frac{1}{h\varepsilon'_{2k}(x)} \leq 0 \quad \text{d'où}$$

$$\frac{1}{h}\left[\frac{1}{\varepsilon'_{2k}(x+h)} - \frac{1}{\varepsilon'_{2k}(x-h)}\right] \leq \frac{1}{h\,\varepsilon'_{2k}(x+h)} - \frac{1}{u_n} \leq \frac{1}{u_{n+1}} - \frac{1}{u_n} \leq 0$$

car $1/u_n \leq 1/h\varepsilon'_{2k}(x-h)$.

Donc $\exists\, t_n \in [x-h,x+h]$ tel que :

$$\left(\frac{1}{\varepsilon'_{2k}(t_n)}\right)' = \frac{1}{2h}\left[\frac{1}{\varepsilon'_{2k}(x+h)} - \frac{1}{\varepsilon'_{2k}(x-h)}\right]$$

Lorsqu'on fait tendre n vers l'infini $\{(1/\varepsilon'_{2k}(t_n))'\}$ n'est pas borné inférieurement. Comme cette propriété est vraie $\forall\, t > T$, la suite des abscisses $\{t_n\}$ tend vers l'infini. Or on a :

$$\varepsilon_{2k+2}(t) = \varepsilon_{2k}(t) + \frac{1}{\varepsilon'_{2k+1}(t)} = \varepsilon_{2k}(t) + \frac{1}{\varepsilon'_{2k-1}(t)+\left(\frac{1}{\varepsilon'_{2k}(t)}\right)'}$$

Puisque $\varepsilon'_{2k-1}(t) \leqslant 0$ la suite $\{\varepsilon'_{2k+1}(t_n)\}$ est négative et non bornée inférieurement lorsque n tend vers l'infini, ce qui termine la démonstration.

Remarque : si $\lim_{t\to\infty} \varepsilon_{2k-2}(t) = S$ alors $\lim_{t\to\infty} \varepsilon'_{2k-1}(t) = -\infty$ et donc, sous les mêmes hypothèses, $\lim_{t\to\infty} \varepsilon'_{2k+1}(t) = -\infty$ et $\lim_{t\to\infty} \varepsilon_{2k+2}(t) = S$.

Nous allons maintenant étudier la convergence de la première forme confluente de l'ε-algorithme pour les fonctions totalement monotones. On a d'abord la :

<u>Définition 24</u> : on dit que f est une fonction totalement monotone de la variable t si :

$$(-1)^k\, f^{(k)}(t) \geqslant 0 \qquad \forall t > T \text{ et } \forall k \geqslant 0$$

<u>Propriété 57</u> : si f est totalement monotone alors $\lim_{t\to\infty} f(t)$ existe et est finie.

démonstration : $f'(t) \leq 0$ entraîne $0 \leq f(t+h) \leq f(t)$ $\forall t > T$ et $\forall h$.

On a la propriété suivante [188]:

<u>Propriété 58</u> : si f est totalement monotone alors $H_k^{(0)}(t) \geq 0$ $\forall t > T$

<u>Lemme 18</u> : si on applique la première forme confluente de l'ε-algorithme à une fonction f totalement monotone alors :

$$\varepsilon_{2k}(t) \geq 0 \quad \text{et} \quad \varepsilon_{2k+1}(t) \leq 0 \qquad \forall t > T$$

démonstration : si f est totalement monotone alors $(-1)^k f^{(k)}$ est aussi totalement monotone et par conséquent, d'après la propriété 58 on a :

$$(-1)^{hk} H_k^{(n)}(t) \geq 0$$

ce qui démontre le lemme en utilisant la propriété 52.

Remarque : si f et g sont deux fonctions totalement monotones alors, en utilisant une propriété donnée par Dieudonné [72, problème 17, p.51], on a :

$$e_k(f+g,t) \geq e_k(f,t) + e_k(g,t) \geq 0 \quad \forall\, k,t$$

<u>Théorème 115</u> :

Appliquons la première forme confluente de l'ε-algorithme à une fonction f telle que $\lim_{t\to\infty} f(t) = S$ existe et soit finie. S'il existe deux constantes $a \neq 0$ et b telles que la fonction af + b soit totalement monotone alors :

$$\lim_{t\to\infty} \varepsilon_{2k}(t) = \lim_{t\to\infty} f(t) \text{ pour } k = 0, 1, \ldots$$

démonstration : d'après les propriétés 53 et 54 on a

$$\varepsilon_{2k+2}(t) - \varepsilon_{2k}(t) \leq 0 \qquad \forall t > T$$

d'où, en utilisant le lemme 18 :

$$0 \leq \varepsilon_{2k+2}(t) \leq \varepsilon_{2k}(t) \quad \forall t > T$$

La fonction f - S est totalement monotone, par conséquent en utilisant la propriété 51 :

$$0 \leq \varepsilon_{2k+2}(t) - S \leq \varepsilon_{2k}(t) - S$$

la convergence de $\varepsilon_{2k}(t)$ vers S lorsque t tend vers l'infini entraîne donc celle de $\varepsilon_{2k+2}(t)$ ce qui démontre le théorème en utilisant de nouveau la propriété 51.

On a également le :

Théorème 116 :

Si on applique la première forme confluente de l'ε-algorithme à une fonction f totalement monotone alors :

$$\varepsilon_{2k+1}(t) < \varepsilon_{2k-1}(t) \leq 0 \qquad \forall t > T$$

$$\text{et} \lim_{t\to\infty} \varepsilon_{2k+1}(t) = -\infty \qquad \text{pour } k = 0, 1, \ldots$$

démonstration : d'après les propriétés 56 et 58 on a :

$$\varepsilon'_{2k}(t) = \frac{H_k^{(1)}(t).\, H_{k+1}^{(1)}(t)}{[H_k^{(2)}(t)]^2} \leq 0$$

d'où

$$\varepsilon_{2k+1}(t) - \varepsilon_{2k-1}(t) = \frac{1}{\varepsilon'_{2k}(t)} \quad \text{et } \varepsilon_{2k+1}(t) \leq \varepsilon_{2k-1}(t) \leq 0$$

en utilisant le lemme 18.

D'autre part $\varepsilon_1(t) = 1 / f'(t)$ d'où

$$\lim_{t\to\infty} \varepsilon_1(t) = \lim_{t\to\infty} \varepsilon_{2k+1}(t) = -\infty \text{ pour } k = 0, 1, \ldots$$

puisque $f'(t) \leq 0 \quad \forall t > T$

Exemple : $f(t) = \int_1^t \frac{e^{-x}}{x}\, dx$. On a $\lim_{t\to\infty} f(t) = S = 0.21983934...$La fonction $S - f(t)$ est totalement monotone $\forall t \geq 1$. On obtient :

t	f(t)	$\varepsilon_2(t)$	$\varepsilon_4(t)$	$\varepsilon_6(t)$
1	0	0.18393972	0.21021682	0.21639967
3	0.20633555	0.21878232	0.21932348	0.21937502
5	0.21823564	0.21935863	0.21938252	0.21938381
6	0.21902385	0.21937796	0.21938367	0.21938392

On remarque que l'on a bien :

$$\lim_{t\to\infty} \varepsilon_{2k}(t) = S$$

$$0 \leq \varepsilon_{2k+2}(t) - S \leq \varepsilon_{2k}(t) - S$$

$$0 \leq \varepsilon_{2k}(t_2) - S \leq \varepsilon_{2k}(t_1) - S \qquad \forall t_2 > t_1$$

Le théorème 115 n'assure pas une convergence de $\varepsilon_{2k+2}(t)$ plus rapide que celle de $\varepsilon_{2k}(t)$. Prenons, en effet, $f(t) = 1 / t$. On trouve que $\varepsilon_{2k}(t) = 1 / (k+1)t$ et par conséquent on a :

$$\frac{\varepsilon'_{2k+2}(t)}{\varepsilon'_{2k}(t)} = \frac{k+1}{k+2}$$

Le fait que $\lim_{k\to\infty} \varepsilon_{2k}(t) = S \ \forall\, t$ reste à démontrer.

VI - 3 Le problème de l'accélération de la convergence

En terminant le paragraphe précédent nous avons vu un exemple où il n'y avait pas accélération de la convergence. Donnons d'abord la :

définition 25 : Soient f et g deux fonctions dérivables telles que $\lim_{t\to\infty} g(t)$ et $\lim_{t\to\infty} f(t)$ existent et soient finies. On dit que f(t) converge plus vite que g(t) si :

$$\lim_{t\to\infty} \frac{f'(t)}{g'(t)} = 0$$

En d'autres termes $f' = o(g')$.

Dans ce paragraphe nous allons étudier l'introduction d'un facteur d'accélération dans les algorithmes de façon analogue à ce qui a été fait au paragraphe IV-5 pour les formes discrètes des algorithmes. Afin d'homogénéïser les notations nous étudierons les algorithmes confluents de la forme :

$$\theta_{-1}(t) = 0 \qquad \theta_0(t) = f(t)$$

$$\theta_{k+1}(t) = \theta_{k-1}(t) + D_k(t)$$

ainsi pour $D_k(t) = 1 / \theta'_k(t)$ on retrouve la première forme confluente de l'ε-algorithme. Pour $D_k(t) = (k+1) / \theta'_k(t)$ on retrouve la première forme confluente du ρ-algorithme [239].

On a d'abord le :

Théorème 117 :

Supposons que $\lim_{t\to\infty} \theta_{2k}(t) = \lim_{t\to\infty} \theta_{2k+2}(t)$; alors une condition nécessaire et suffisante pour que $\theta_{2k+2}(t)$ converge plus vite que $\theta_{2k}(t)$ est que :

$$\lim_{t\to\infty} \frac{D'_{2k+1}(t)}{\theta'_{2k}(t)} = -1$$

La démonstration est évidente ; elle est laissée en exercice. Si cette condition n'est pas vérifiée alors on introduit dans l'algorithme un facteur d'accélération de la convergence w_k :

$$\theta_{2k+1}(t) = \theta_{2k-1}(t) + D_{2k}(t)$$

$$\theta_{2k+2}(t) = \theta_{2k}(t) + w_k\, D_{2k+1}(t)$$

Le choix optimal de w_k est caractérisé par le :

Théorème 118 :

Supposons que $\lim_{t\to\infty} \theta_{2k}(t) = \lim_{k\to\infty} \theta_{2k+2}(t)$. Une condition nécessaire et suffisante pour que $\theta_{2k+2}(t)$ converge plus vite que $\theta_{2k}(t)$ est de prendre :

$$w_k = -\lim_{t\to\infty} \frac{\theta'_{2k}(t)}{D'_{2k+1}(t)}$$

démonstration :

$$\theta'_{2k+2}(t) = \theta'_{2k}(t) + w_k\, D'_{2k+1}(t)$$

d'où

$$\lim_{t\to\infty} \frac{\theta'_{2k+2}(t)}{\theta'_{2k}(t)} = 0 = 1 + w_k \lim_{t\to\infty} \frac{D'_{2k+1}(t)}{\theta'_{2k}(t)}$$

le reste de la démonstration est évident.

Considérons par exemple $D_k(t) = 1/\theta'_k(t)$ c'est-à-dire la première forme confluente de l'ε-algorithme et $f(t) = 1 + 1/t$.

On trouve $w_0 = +2$ d'où :

$$\theta_2(t) = f(t) + 2\, D_1(t)$$

qui n'est autre que la première forme confluente du ρ-algorithme. Ce facteur w_k apparait donc comme le lien entre les premières formes confluentes de l'ε et du ρ-algorithme.

Comme pour l'ε-algorithme discret on peut donner une interprétation de ces résultats en considérant les $w_k \, D_{2k+1}(t)$ comme les termes successifs d'un développement asymptotique.

Soit f une fonction de t telle que $\lim_{t\to\infty} f(t) = S$ existe et soit finie. Soit G l'ensemble des $D_{2k+1}(t)$. Supposons que $D_{2k+1}(t) = o\,(D_{2k-1}(t))$. Alors, s'il vérifie cette propriété, l'ensemble G est une échelle de comparaison. Cherchons le développement asymptotique de $S - f(t)$ par rapport à G au voisinage de $+\infty$. Si un tel développement existe jusqu'à l'ordre k on aura :

$$S - f(t) = \sum_{i=0}^{k} w_{i-1} \, D_{2i-1}(t) + o(D_{2k-1}(t))$$

Le problème est de trouver les coefficients w_i de ce développement asymptotique. On a :

$$\theta_{2k}(t) = f(t) + \sum_{i=1}^{k} w_{i-1} \, D_{2i-1}(t)$$

d'où :

$$S = \theta_{2k}(t) + o\,(D_{2k-1}(t))$$

$$= \theta_{2k-2}(t) + w_{k-1} \, D_{2k-1}(t) + o(D_{2k-1}(t))$$

On va choisir w_{k-1} de façon à avoir :

$$0 = \theta'_{2k-2}(t) + w_{k-1} \, D'_{2k-1}(t) + o\,(D'_{2k-1}(t))$$

ce qui entraînera $S = \theta_{2k}(t) + o\,(D_{2k-1}(t))$ en supposant, ce qui est effectivement vérifié que les intégrales :

$$\int_t^\infty \theta'_{2k-2}(t)\,dt \qquad \text{et} \qquad \int_t^\infty D'_{2k-1}(t)\,dt$$

sont convergentes. On a donc :

$$w_{k-1} \, D'_{2k-1}(t) = -\,\theta'_{2k-2}(t) + o\,(D'_{2k-1}(t))$$

d'où

$$w_{k-1} = -\lim_{t\to\infty} \frac{\theta'_{2k-2}(t)}{D'_{2k-1}(t)}$$

ce qui n'est autre que le choix optimal du théorème 118. Le théorème 117 apparait donc comme une condition nécessaire et suffisante pour que $w_k = 1$. Le fait que le choix optimal du théorème 118 fournisse le meilleur algorithme signifie simplement que w_k est le coefficient de $D_{2k+1}(t)$ dans le développement asymptotique de $S - f(t)$ par rapport à G au voisinage de $+\infty$. Ce choix optimal de w_k est le seul pour lequel on ait

$$S - \theta_{2k}(t) = 0\,(D_{2k-1}(t))$$

d'où encore :

$$S - \theta_{2k-2}(t) = w_{k-1}\, D_{2k-1}(t) + o\,(D_{2k-1}(t))$$

et $S - \theta_{2k-2}(t) \sim w_{k-1}\, D_{2k-1}(t)$

$$S - \theta_{2k-2}(t) = 0\,(D_{2k-1}(t))$$

ce qui fournit une estimation de l'erreur.

REMARQUE :

Si $f(t) = \int_a^t g(x)\,dx$ on aura

$$\varepsilon_2(t) = \int_a^t g(x)\,dx - \frac{g^2(x)}{g'(t)} \quad \text{d'où}$$

$$\int_t^\infty g(x)\,dx \sim \frac{g^2(t)}{g'(t)}$$

d'après ce qui précède. On retrouve ainsi un résultat connu sur la partie principale d'une primitive [72].

Nous avions vu pour l'ε-algorithme discret qu'il n'avait pas été possible de trouver un théorème analogue au théorème 35 pour caractériser l'ensemble des suites pour lesquelles on obtenait le résultat exact quand on introduisait les w_k. Il n'en est pas de même ici. On a le :

Théorème 119 :

Une condition nécessaire et suffisante pour que $\theta_{2k}(t) = S \;\forall t > T$ est que la fonction f vérifie l'équation différentielle :

$$S = \sum_{i=0}^{k} a_i \frac{H_{i+1}^{(0)}(t)}{H_i^{(2)}(t)} \quad \text{avec} \quad \sum_{i=0}^{k} a_i = 1 \quad \forall t > T$$

démonstration : on a

$$\theta_{2k}(t) = f(t) + \sum_{i=1}^{k} w_{i-1}\, D_{2i-1}(t)$$

or $D_{2i-1}(t) = \theta_{2i}(t) - \theta_{2i-2}(t)$

d'où :

$$\theta_{2k}(t) = f(t) + \sum_{i=1}^{k} w_{i-1}\, [\theta_{2i}(t) - \theta_{2i-2}(t)]$$

Posons $a_0 = 1 - w_0$

$a_i = w_{i-1} - w_i$ pour $i = 1, \ldots, k-1$

$a_k = w_{k-1}$

on a :

$$S = \theta_{2k}(t) = \sum_{i=0}^{k} a_i\, \theta_{2i}(t) \qquad \text{avec} \sum_{i=0}^{k} a_i = 1$$

d'où la condition nécessaire en utilisant la propriété 52. La condition suffisante est évidente.

Cette équation différentielle est difficile à résoudre pour k quelconque. Cependant pour $k = 1$ cette résolution est possible. On a le :

Théorème 120 :

Une condition nécessaire et suffisante pour que $\theta_2(t) = S \;\; \forall t > T$ est que $f(t) = S + c_1\, e^{-c_2 t}$

ou que $f(t) = [(1 - w_0)\, c_1 t + c_2]^{1/(1-w_0)} + S \;\; \forall t > T.$

démonstration : on doit avoir pour $t > T$

$$S = f(t) - w_0\, f'^2(t) \,/\, f''(t)$$

ou encore

$$g(t) = w_0\, g'^2(t) \,/\, g''(t) \text{ en posant } g(t) = f(t) - S.$$

Ceci peut s'écrire :

$$\frac{g}{g'} = w_0 \frac{g'}{g''} \quad \text{ou encore} \quad \frac{g'}{g} = \frac{1}{w_0} \frac{g''}{g'}$$

d'où en intégrant :

$$\text{Log } g = \frac{1}{w_0} \text{ Log } g' + c$$

ce qui donne :

$$g(t) = c\,[g'(t)]^{1/w_0}$$

ou encore $g^{-w_0}(t)\, dg(t) = c$

si $w_0 = 1$ on a $g(t) = c_1 e^{-c_2 t}$

si $w_0 \neq 1$ on a $g(t) = [(1-w_0)\, c_1 t + c_2]^{1/(1-w_0)}$

La condition suffisante est immédiate. En effet si $f(t) = S + c_1 e^{-c_2 t}$ on trouve $w_0=1$ et si $f(t) = S + [a\, c_1 t + c_2]^{1/a}$ on trouve que $w_0 = 1-a$.

Nous allons essayer d'élargir encore la classe des fonctions pour lesquelles $\varepsilon_{2k}(t) = S$ $\forall t > T$. Pour cela, au lieu d'introduire un paramètre d'accélération w_k nous allons utiliser une fonction d'accélération $w_k(t)$. D'où l'algorithme :

$$\theta_{2k+1}(t) = \theta_{2k-1}(t) + D_{2k}(t)$$

$$\theta_{2k+2}(t) = \theta_{2k}(t) + w_k(t)\, D_{2k+1}(t)$$

on démontrerait comme précédemment le :

<u>Théorème 121</u> :

une condition nécessaire et suffisante pour que $\theta_{2k}(t) = S$ $\forall t > T$ est que la fonction f vérifie l'équation différentielle :

$$S = \sum_{i=0}^{k} a_i(t) \frac{H_{i+1}^{(0)}(t)}{H_i^{(2)}(t)} \quad \text{avec} \quad \sum_{i=0}^{k} a_i(t) = 1 \ \forall t > T$$

La démonstration est analogue à celle du théorème 119. Elle est laissée en exercice.

Pour k = 1 on peut résoudre cette équation différentielle :

<u>Théorème 122</u> :

Une condition nécessaire et suffisante pour que $\theta_2(t) = S$ $\forall t > T$ est que :

$$f(t) = S + c_1 .\ \exp \int \frac{dt}{c_2 + t - \Omega_0(t)} \quad \forall t > T$$

où $\Omega_0(t)$ est une primitive de $w_0(t)$.

démonstration : la condition est nécessaire. En effet on doit avoir :

$$w_0 \frac{g'}{g} = \frac{g''}{g} \text{ avec } g(t) = f(t) - S$$

cherchons les solutions de la forme $g(t) = e^{z(t)}$. On obtient pour $z(t)$ l'équation différentielle :

$$w_0(t) - 1 = z''(t) \;/\; z'^2(t) \text{ d'où } - \frac{1}{z'(t)} = - c_2 - t + \Omega_0(t) \text{ où } \Omega_0(t) \text{ est une primitive}$$

de $w_0(t)$. En intégrant une nouvelle fois on obtient :

$$z(t) = \int \frac{dt}{c_2 + t - \Omega_0(t)}$$

La condition suffisante se démontre en portant $f(t)$ dans l'équation différentielle.

<u>Théorème 123</u> :

une condition nécessaire et suffisante pour que $\theta_2(t)$ = constante $\forall t > T$ est que $w_0(t)$ vérifie l'équation différentielle :

$$f''^2 - w'_0 \, f' \, f'' - w_0(2f''^2 - f' \, f''') = 0 \quad \forall t > T$$

démonstration : si $\theta_2(t)$ = constante alors $\theta'_2(t) = 0$ d'où la condition du théorème. Réciproquement si cette condition est vérifiée alors $\theta'_2(t) = 0$ donc $\theta_2(t)$ = constante. Si l'on compare cet algorithme avec l'algorithme où $w_k(t) = w_k$ = constante on voit que, pour $k = 1$ la première méthode converge plus vite que la seconde si :

$$\lim_{t \to \infty} \frac{f''^2 - w'_0(t) \, f' \, f'' - w_0(t)(2f''^2 - f' \, f''')}{f''^2 - w_0 \, (2f''^2 - f' \, f''')} = 0$$

Dans l'algorithme avec w_k = constante on voit que la valeur optimale donnée par le théorème 118 est difficile à obtenir car elle fait intervenir le calcul d'une limite. D'autre part dans l'algorithme avec une fonction d'accélération $w_k(t)$ on voit qu'il est difficile d'effectuer un choix intéressant pour cette fonction. Ces deux raisons conduisent donc naturellement à l'idée de prendre :

$$w_k(t) = - \frac{\theta'_{2k}(t)}{D'_{2k+1}(t)}$$

or $D_{2k+1}(t) = 1 \;/\; \theta'_{2k+1}(t)$ d'où :

$$w_k(t) = \frac{\theta'_{2k}(t) \, \theta'^2_{2k+1}(t)}{\theta''_{2k+1}(t)}$$

On obtient ainsi le nouvel algorithme :

$$\theta_{-1}(t) = 0 \qquad \theta_0(t) = f(t)$$

$$\theta_{2k+1}(t) = \theta_{2k-1}(t) + 1 \,/\, \theta'_{2k}(t)$$

$$\theta_{2k+2}(t) = \theta_{2k}(t) + \theta'_{2k}(t)\ \theta'_{2k+1}(t) \,/\, \theta''_{2k+1}(t)$$

La démonstration du théorème suivant est évidente :

Théorème 124 :

une condition suffisante pour que $\lim_{t\to\infty} \theta_{2k}(t) = \lim_{t\to\infty} \theta_{2k+2}(t)$ est que :

$$\lim_{t\to\infty} \frac{\theta''_{2k+1}(t)}{\theta'_{2k+1}(t)} \neq 0$$

Théorème 125 :

Supposons que $\lim_{t\to\infty} \theta_{2k+2}(t) = \lim_{t\to\infty} \theta_{2k}(t) = S$. Alors si $\lim_{t\to\infty} w_k(t)$ existe et est finie

alors $\theta_{2k+2}(t)$ converge plus vite que $\theta_{2k}(t)$ en ce sens que :

$$\lim_{t\to\infty} \frac{\theta_{2k+2}(t) - S}{\theta_{2k}(t) - S} = 0$$

démonstration : on a :

$$\theta_{2k+2}(t) - S = \theta_{2k}(t) - S + w_k(t)\ D_{2k+1}(t)$$

$$\frac{\theta_{2k+2}(t) - S}{\theta_{2k}(t) - S} = 1 + w_k(t)\ \frac{D_{2k+1}(t)}{\theta_{2k}(t) - S}$$

La quantité $\lim_{t\to\infty} \dfrac{\theta_{2k}(t) - S}{D_{2k+1}(t)}$ se présente sous la forme indéterminée $\frac{0}{0}$.

On applique la règle de l'Hospital :

$$\lim_{t\to\infty} \frac{\theta_{2k}(t) - S}{D_{2k+1}(t)} = \lim_{t\to\infty} \frac{\theta'_{2k}(t)}{D'_{2k+1}(t)} = - \lim_{t\to\infty} w_k(t)$$

d'où :

$$\lim_{t\to\infty} \frac{\theta_{2k+2}(t) - S}{\theta_{2k}(t) - S} = 0$$

ce qui termine la démonstration du théorème.

VI - 4 FORME CONFLUENTE DE L'ε-ALGORITHME TOPOLOGIQUE

Il est possible de définir une forme confluente pour l'ε-algorithme topologique qui a été étudié au paragraphe V-6.

Soit E un espace vectoriel topologique séparé sur K ($\mathbb{R}$ ou $\mathbb{C}$) et soit E' son dual topologique.

Soit, d'autre part, f une application de $\mathbb{R}$ dans E. Nous supposerons que f est différentiable autant de fois qu'il sera nécessaire et nous désignerons par $D^k f$ les dérivées successives de f.

Considérons maintenant le premier ε-algorithme topologique. Dans les règles de cet algorithme , remplaçons n par $t = a+nh$, $\varepsilon_{2k}^{(n)}$ par $\varepsilon_{2k}(t)$ et $\varepsilon_{2k+1}^{(n)}$ par $\varepsilon_{2k+1}(t)/h$ puis faisons tendre h vers zéro. On obtient immédiatement les règles de la forme confluente de l'ε-algorithme topologique [26] :

$$\varepsilon_{-1}(t) = o \in E \qquad \varepsilon_o(t) = f(t)$$

$$\varepsilon_{2k+1}(t) = \varepsilon_{2k-1}(t) + \frac{y'}{\langle y', D\varepsilon_{2k}(t)\rangle}$$

$$k = o,1,\ldots$$

$$\varepsilon_{2k+2}(t) = \varepsilon_{2k}(t) + \frac{D\varepsilon_{2k}(t)}{\langle D\varepsilon_{2k+1}(t), D\varepsilon_{2k}(t)\rangle}$$

où y' est un élément arbitraire non nul de E' et où $\langle\ ,\ \rangle$ désigne la forme bilinéaire qui met E et E' en dualité.

Remarque : si l'on effectue le même changement de variable dans le second ε-algorithme topologique on obtient la même forme confluente.

On voit que le calcul de $\varepsilon_{2k}(t)$ à l'aide de la forme confluente de l'ε-algorithme topologique nécessite la connaissance de $f(t), Df(t),\ldots, D^{2k}f(t)$. D'autre part l'application qui fait passer de ces valeurs à $\varepsilon_{2k}(t)$ est non linéaire ; on a cependant la :

propriété 59: Si l'application de la forme confluente de l'ε-algorithme topologique à f et à af + b ou a est un scalaire non nul et où b $\in$ E fournit respectivement les fonctions ε_k et $\bar{\varepsilon}_k$ alors :

$$\bar{\varepsilon}_{2k}(t) = a\varepsilon_{2k}(t) + b \text{ et } \bar{\varepsilon}_{2k+1}(t) = \varepsilon_{2k+1}(t)/a$$

La démonstration est évidente à partir des règles de l'algorithme.

Nous allons maintenant donner les éléments $\varepsilon_k(t)$ sous forme de rapports de déterminants.

Définition 26: On appelle déterminants fonctionnels de Hankel généralisés les élément $\tilde{H}_k^{(n)}(t)$ de E :

$$\tilde{H}_k^{(n)}(t) = \begin{vmatrix} D^n f(t) \ldots\ldots\ldots\ldots D^{n+k-1} f(t) \\ \langle y', D^{n+1} f(t)\rangle \ldots\ldots \langle y', D^{n+k} f(t)\rangle \\ \ldots\ldots\ldots\ldots\ldots\ldots\ldots\ldots \\ \langle y', D^{n+k-1} f(t)\rangle \ldots\ldots \langle y', D^{n+2k-2} f(t)\rangle \end{vmatrix}$$

Cet élément $\tilde{H}_k^{(n)}(t)$ désigne l'élément de E obtenu en développant ce déterminant généralisé de façon habituelle.

Nous poserons :

$$H_k^{(n)}(t) = \langle y', \tilde{H}_k^{(n)}(t)\rangle$$

avec la convention que $H_o^{(n)}(t) = 1$ pour tout n et tout t. $H_k^{(n)}(t)$ est déterminant fonctionnel de Hankel habituel.

propriété 60:

$$\varepsilon_{2k}(t) = \frac{\tilde{H}_{k+1}^{(o)}(t)}{H_k^{(2)}(t)} \qquad \varepsilon_{2k+1}(t) = \frac{H_k^{(3)}(t)}{H_{k+1}^{(1)}(t)}\, y'$$

La démonstration peut être effectuée directement en utilisant un développement de Schweins comme on l'avait fait pour établir les règles des ε-algorithmes topologiques. On peut aussi, plus simplement, partir de la relation de la première transformation de Shanks topologique :

$$\varepsilon_{2k}^{(n)} = \frac{\begin{vmatrix} S_n & \dots & S_{n+k} \\ \langle y', \Delta S_n \rangle & \dots & \langle y', \Delta S_{n+k} \rangle \\ \dots & \dots & \dots \\ \langle y', \Delta S_{n+k-1} \rangle & \dots & \langle y', \Delta S_{n+2k-1} \rangle \end{vmatrix}}{\begin{vmatrix} 1 & \dots & 1 \\ \langle y', \Delta S_n \rangle & \dots & \langle y', \Delta S_{n+k} \rangle \\ \dots & \dots & \dots \\ \langle y', \Delta S_{n+k-1} \rangle & \dots & \langle y', \Delta S_{n+2k-1} \rangle \end{vmatrix}}$$

En effectuant des différences de lignes et de colonnes on trouve que :

$$\varepsilon_{2k}^{(n)} = \frac{\begin{vmatrix} S_n & \Delta S_n & \dots & \Delta^k S_n \\ \langle y', \Delta S_n \rangle & \langle y', \Delta^2 S_n \rangle & \dots & \langle y', \Delta^{k+1} S_n \rangle \\ \dots & \dots & \dots & \dots \\ \langle y', \Delta^k S_n \rangle & \langle y', \Delta^{k+1} S_n \rangle & \dots & \langle y', \Delta^{2k} S_n \rangle \end{vmatrix}}{\begin{vmatrix} \langle y', \Delta^2 S_n \rangle & \dots & \langle y', \Delta^{k+1} S_n \rangle \\ \dots & \dots & \dots \\ \langle y', \Delta^{k+1} S_n \rangle & \dots & \langle y', \Delta^{2k} S_n \rangle \end{vmatrix}}$$

Pour le numérateur on divise la première ligne par h, la seconde par $h^2, \dots$, la dernière par h^{k+1} ; puis on multiplie la première colonne par h, la seconde par 1, la troisième par 1/h,...., la dernière par $1/h^{k-1}$. On effectue une transformation semblable pour le dénominateur.

Comme pour l'établissement des règles de la forme confluente de l'ε-algorithme topologique on remplace maintenant S_n par f(t) puis on fait tendre h vers zéro. En utilisant $\lim_{h \to 0} \Delta^p f(t)/h^p = D^p f(t)$ on obtient immédiatement la première relation de la propriété 60. La seconde relation découle de :

$$\varepsilon_{2k+1}^{(n)} = \frac{y'}{\langle y', e_k(\Delta S_n) \rangle}$$

En effectuant un travail analogue on démontrerait de même la :

propriété 61:

$$\varepsilon_{2k+2}(t) - \varepsilon_{2k}(t) = -\frac{H_{k+1}^{(1)}(t)\ \tilde{H}_{k+1}^{(1)}(t)}{H_k^{(2)}(t)\ H_{k+1}^{(2)}(t)}$$

Nous allons maintenant chercher les conditions que doit vérifier f pour que $\varepsilon_{2k}(t) = S \quad \forall t>T$. On a le :

Théorème 126: Une condition suffisante pour que $\varepsilon_{2k}(t) = S\ \forall t>T$ est que, pour tout $t>T$, f vérifie :

$$\sum_{i=o}^{k} a_i\ D^i f(t) = a_o S$$

où $S \in E$, $a_i \in K$ et $a_o \neq o$.

démonstration : si la condition du théorème est vérifiée alors on a :

$$a_o(f(t)-S) + a_1\ Df(t) + \dots\dots\dots\dots + a_k D^k f(t) = o$$

$$a_o Df(t) + a_1 D^2 f(t) + \dots\dots\dots\dots + a_k D^{k+1} f(t) = o$$

$$\dots\dots\dots\dots\dots\dots\dots\dots\dots\dots\dots\dots\dots\dots\dots$$

$$a_o D^k f(t) + a_1 D^{k+1} f(t) + \dots\dots\dots + a_k D^{2k} f(t) = o$$

Puisque $a_o \neq o$ on peut supposer sans restreindre la généralité que $a_o = 1$. Soit y' un élément arbitraire non nul de E'. Alors, d'après le système précédent on a :

$$a_o = 1$$

$$a_o \langle y', Df(t)\rangle + \dots\dots\dots\dots\dots\dots + a_k \langle y', D^{k+1} f(t)\rangle = o$$

$$\dots\dots\dots\dots\dots\dots\dots\dots\dots\dots\dots\dots\dots\dots\dots$$

$$a_o \langle y', D^k f(t)\rangle + \dots\dots\dots\dots\dots\dots + a_k \langle y', D^{2k} f(t)\rangle = o$$

Le déterminant de ce système est égal à $H_k^{(2)}(t)$. Nous le supposerons différent de zéro. Résolvons ce système puis calculons S en utilisant la relation :

$$S = f(t) + a_1 \, Df(t) + \dots\dots\dots + a_k \, D^k f(t)$$

on obtient :

$$\tilde{S} = \frac{H_{k+1}^{(o)}(t)}{H_k^{(2)}(t)} \qquad \forall t > T$$

ce qui termine la démonstration du théorème d'après la propriété 60.

remarque : contrairement à ce qui se passe pour la première forme confluente de l'ε-algorithme scalaire, la condition du théorème 126 n'est que suffisante. Cela tient au fait que $\langle y', y \rangle = o$ n'entraine pas obligatoirement $y = o$.

Une conséquence du théorème 126 est le :

Théorème 127: une condition suffisante pour que $\varepsilon_{2k}(t) = S$ $\forall t > T$ est que :

$$f(t) = S + \sum_{i=1}^{p} A_i(t)\, e^{r_i t} + \sum_{i=p+1}^{q} \left[B_i(t) \cos b_i t + C_i(t) \sin b_i t \right] e^{r_i t}$$

pour $t > T$ avec $r_i \neq o$ pour $i = 1, \dots, p$. A_i, B_i et C_i sont des polynômes de la variable réelle t dont les coefficients sont des éléments de E et tels que si d_i est égal au degré de A_i plus un pour $i = 1, \dots, p$ et au plus grand des degrés de B_i et de C_i plus un pour $i = p+1, \dots, q$ on ait :

$$\sum_{i=1}^{p} d_i + 2 \sum_{i=p+1}^{q} d_i = k$$

La démonstration de ce théorème est immédiate. Elle exprime tout simplement le fait qu'une telle fonction f vérifie l'équation différentielle du théorème 126 pour tout $t > T$.

on a les relations suivantes :

<u>propriété 62</u>:

$$H_k^{(n+1)}(t)\,\tilde{H}_{k+2}^{(n-1)}(t) - H_{k+1}^{(n+1)}(t)\,\tilde{H}_{k+1}^{(n-1)}(t) = - H_{k+1}^{(n)}(t)\,\tilde{H}_{k+1}^{(n)}(t)$$

$$\frac{D\varepsilon_{2k}(t)}{<D\varepsilon_{2k+1}(t),\ D\varepsilon_{2k}(t)>} = - \frac{H_{k+1}^{(1)}(t)\,\tilde{H}_{k+1}^{(1)}(t)}{H_k^{(2)}(t)\,H_{k+1}^{(2)}(t)}$$

$$\frac{1}{<y',\ D\varepsilon_{2k}(t)>} = \frac{\left[H_k^{(2)}(t)\right]^2}{H_k^{(1)}(t)\,H_{k+1}^{(1)}(t)}$$

démonstration : d'après les propriétés 60 et 61 on a :

$$\frac{\tilde{H}_{k+2}^{(0)}(t)}{H_{k+1}^{(2)}(t)} - \frac{\tilde{H}_{k+1}^{(0)}(t)}{H_k^{(2)}(t)} = - \frac{H_{k+1}^{(1)}(t)\,\tilde{H}_{k+1}^{(1)}(t)}{H_{k+1}^{(2)}(t)\,H_k^{(2)}(t)}$$

ce qui démontre la première des relations en remplaçant f(t) par $D^{n-1}f(t)$. La seconde des relations découle immédiatement de la propriété 61 et de la définition des règles de l'algorithme.

Dans la première des relations de la propriété 62 faisons n = 2 et divisons les deux membres de l'égalité ainsi obtenue par $H_{k+1}^{(1)}\ H_{k+2}^{(1)}$. On obtient la troisième relation en multipliant scalairement par y' et en utilisant les règles de l'algorithme.

On peut donner, pour la forme confluente de l'ε-algorithme topologique, une interprétation barycentrique analogue à celles donnés dans les cas discrets et confluent :

l'application de la forme confluente de l'ε-algorithme topologique à f(t), Df(t),..., $D^{2k}f(t)$ revient à résoudre le système :

$$a_0 = 1$$

$$a_0 <y',\ Df(t)> + \ldots\ldots + a_k <y',\ D^{k+1}f(t)> = 0$$

$$\ldots\ldots\ldots\ldots\ldots\ldots\ldots\ldots\ldots\ldots\ldots\ldots$$

$$a_0 <y',\ D^k f(t)> + \ldots\ldots + a_k <y',\ D^{2k}f(t)> = 0$$

puis à calculer :

$$\varepsilon_{2k}(t) = a_o f(t) + a_1 Df(t) + \ldots\ldots + a_k D^k f(t)$$

Les théorèmes de convergence pour la première forme confluente de l'ε-algorithme scalaire ne sont pas encore nombreux.

Cependant certains résultats démontrés dans le cas scalaire restent vérifiés si l'on remarque que les quantités $\langle y', \varepsilon_k(t)\rangle$ sont égales à celles obtenues en appliquant la forme confluente de l'ε-algorithme scalaire à $\langle y', f(t)\rangle$. Ou a donc les :

<u>Théorème 128</u> : si $\lim_{t\to\infty} \langle y', \varepsilon_{2k}(t)\rangle = S$, si $\langle y', D\varepsilon_{2k-1}(t)\rangle \leq 0$ $\forall\ t > T$ et si $\langle y', D^2\varepsilon_{2k}(t)\rangle \geq 0\ \forall\ t > T$ alors il existe une suite strictement croissante $\{t_n\}$ tendant vers l'infini telle que $\lim_{n\to\infty} \langle y', \varepsilon_{2k+2}(t_n)\rangle = S$.

<u>Théorème 129</u>:

S'il existe $a \neq o \in K$ et $b \in E$ tels que :

$(-1)^k \langle y', D^k\ (af(t) + b)\rangle \geqslant o\ \forall t > T$ et $\forall k \geqslant o$

alors

$$\lim_{t\to\infty} \langle y', \varepsilon_{2k}(t)\rangle = \lim_{t\to\infty} \langle y', f(t)\rangle\ \forall k \geqslant o$$

Ecrivons maintenant les règles de l'algorithme sous la forme condensée :

$$\varepsilon_{-1}(t) = o \qquad \varepsilon_o(t) = f(t)$$

$$\varepsilon_{k+1}(t) = \varepsilon_{k-1}(t) + D_k(t)$$

avec $D_{2k}(t) = y'/\langle y', D\varepsilon_{2k}(t)\rangle$

et $D_{2k+1}(t) = D\varepsilon_{2k}(t)/\langle D\varepsilon_{2k+1}(t), D\varepsilon_{2k}(t)\rangle$

<u>Définiton 27</u>: Soit $z' \neq o \in E'$. On dira que ε_{2k+2} converge plus vite que ε_{2k} par rapport à z' si :

$$\lim_{t\to\infty} \frac{\langle z', D\varepsilon_{2k+2}(t)\rangle}{\langle z', D\varepsilon_{2k}(t)\rangle} = o$$

Théorème 130: Supposons que $\lim_{t\to\infty} <z', \varepsilon_{2k+2}(t)> = \lim_{t\to\infty} <z', \varepsilon_{2k}(t)>$.
Une condition nécessaire et suffisante pour que ε_{2k+2} converge plus vite que ε_{2k} par rapport à z' est que :

$$\lim_{t\to\infty} \frac{<z', DD_{2k+1}(t)>}{<z', D\varepsilon_{2k}(t)>} = -1$$

Si cette condition n'est pas vérifiée alors on peut introduire un paramètre d'accélération dans les règles de l'algorithme comme cela a été fait pour l'ε-algorithme scalaire et sa forme confluente.
Les règles de l'algorithme deviennent alors :

$$\varepsilon_{-1}(t) = o \qquad \varepsilon_o(t) = f(t)$$

$$\varepsilon_{2k+1}(t) = \varepsilon_{2k-1} + D_{2k}(t)$$

$$\varepsilon_{2k+2}(t) = \varepsilon_{2k}(t) + w_k D_{2k+1}(t)$$

Le choix optimal du paramètre w_k est caractérisé par le :

Théorème 131: Supposons que $\lim_{t\to\infty} <z', \varepsilon_{2k}(t)> = \lim_{t\to\infty} <z', \varepsilon_{2k+2}(t)>$.
Une condition nécessaire et suffisante pour que ε_{2k+2} converge plus vite que ε_{2k} par rapport à z' est de prendre :

$$w_k = - \lim_{t\to\infty} \frac{<z', D\varepsilon_{2k}(t)>}{<z', DD_{2k+1}(t)>}$$

On voit qu'en pratique le calcul de la valeur optimale de w_k est difficile parce qu'il fait intervenir la limite d'une expression. On remplacera donc w_k par l'expression elle même sans en prendre la limite. On obtient ainsi la forme confluente du θ-algorithme généralisé :

$$\theta_{-1}(t) = o \qquad \theta_o(t) = f(t)$$

$$\theta_{2k+1}(t) = \theta_{2k-1}(t) + D_{2k}(t)$$

$$\theta_{2k+2}(t) = \theta_{2k}(t) + w_k(t)\, D_{2k+1}(t)$$

avec

$$w_k(t) = - \frac{<z', D\theta_{2k}(t)>}{<z', DD_{2k+1}(t)>}$$

$$D_{2k}(t) = y'/\langle y', D\theta_{2k}(t)\rangle$$

$$D_{2k+1}(t) = D\theta_{2k}(t)/\langle D\theta_{2k+1}(t), D\theta_{2k}(t)\rangle$$

Pour cet algorithme on a le :

Théorème 132 : supposons que $\lim_{t\to\infty} \langle z', \theta_{2k+2}(t)\rangle = \lim_{t\to\infty} \langle z', \theta_{2k}(t)\rangle = S$.

Si $-\lim_{t\to\infty} w_k(t)$ et $\lim_{t\to\infty} \frac{\langle z', \theta_{2k}(t)\rangle - S}{\langle z', D_{2k+1}(t)\rangle}$ existent et sont égales alors :

$$\lim_{t\to\infty} \frac{\langle z', \theta_{2k+2}(t)\rangle - S}{\langle z', \theta_{2k}(t)\rangle - S} = o$$

Les applications pratiques des formes confluentes de l'ε-algorithme et du Θ-algorithme topologiques restent encore à trouver.

L'ε-algorithme topologique est relié à la méthode du gradient conjugué et à la méthode des moments [184] de la façon suivante.

Soit à résoudre le système de p équations linéaires à p inconnues $Bx = b$ où la matrice B est symétrique définie positive et soit $\{x_k\}$ la suite des vecteurs obtenus en appliquant la méthode du gradient conjugué à ce système en partant de $x_o = o$.

D'un autre côté, posons $A = I-B$ et appliquons l'ε-algorithme topologique à la suite $\{S_n\}$ produite par :

$$S_o = o$$
$$S_{n+1} = AS_n + b$$

avec $y' = b$.

Enfin appelons $\{v_k\}$ la suite des vecteurs obtenus par la méthode des moments. Alors on peut montrer que :

$$\varepsilon_{2k}^{(o)} = x_k = v_k \qquad \text{pour } k = 0,\ldots,p$$

et l'on a :

$$\varepsilon_{2p}^{(o)} = x_p = v_p = B^{-1}b.$$

La théorie des polynômes orthogonaux peut également être utilisée pour comprendre l'ε-algorithme topologique. Soit $P_k(x) = a_o + \ldots + a_k x^k$ le polynôme orthogonal de degré k par rapport à la suite $\{c_n = \langle y, \Delta S_n \rangle\}$; alors on montre que l'ε-algorithme topologique est tel que :

$$\varepsilon_{2k}^{(o)} = (a_o S_o + \ldots + a_k S_k) \,/\, P_k(1)$$

L'ε-algorithme topologique peut également être relié à la méthode des moments, à celle de Lanczos et au gradient biconjugué dans le cas où la matrice B est quelconque.

VI - 5 Le développement en série de Taylor

Il est évident que, par un changement de variable, on peut transformer le problème du calcul de $\lim_{t\to\infty} f(t)$ en celui du calcul de $\lim_{t\to o} f(t)$.

On va supposer que f se comporte comme un polynôme en t au voisinage de zéro et on va essayer de définir une forme confluente de la méthode de Richardson que nous avons utilisé dans le cas discret (paragraphe III-3) :

$$T_0^{(n)} = S_n$$

$$T_{k+1}^{(n)} = \frac{x_{n+k+1}\, T_k^{(n)} - x_n\, T_k^{(n+1)}}{x_{n+k+1} - x_n} \qquad k, n = 0, 1, \ldots$$

Remplaçons la variable discrète n par la variable continue $t = a + n.\Delta t$ et $x_{n+p} = t + p.\Delta t$.

On obtient :

$$T_{k+1}^{(n)} = T_k^{(n)} + \frac{x_n\,(T_k^{(n)} - T_k^{(n+1)})}{x_{n+k+1} - x_n}$$

d'où :

$$T_{k+1}(t) = T_k(t) + \frac{t}{k+1}\;\frac{T_k(t) - T_k(t+\Delta t)}{\Delta t}$$

Faisons tendre Δt vers zéro, on obtient la forme confluente du procédé de Richardson :

$$T_0(t) = f(t)$$

$$T_{k+1}(t) = T_k(t) - \frac{t}{k+1} T'_k(t) \qquad k = 0, 1, \ldots$$

On a

$$T_1(t) = f(t) - t\, f'(t)$$

$$T_2(t) = f(t) - t\, f'(t) - \frac{t}{2} (f'(t) - t\, f''(t) - f'(t))$$

d'oü

$$T_2(t) = f(t) - t\, f'(t) + \frac{t^2}{2!} f''(t)$$

Supposons que $T_k(t) = f(t) - t\, f'(t) + \ldots + \frac{(-1)}{k!} t^k f^{(k)}(t)$

on démontre facilement que :

$$T_{k+1}(t) = f(t) - t\, f'(t) + \ldots + \frac{(-1)^{k+1}}{(k+1)!} t^{k+1} f^{(k+1)}(t)$$

ceci n'est autre que le développement en série de Taylor de f au voisinage de zéro.

Dans la pratique, l'utilisation du développement de Taylor peut rendre des services.

Prenons, par exemple, $f(t) = e^{-t}$. On obtient pour $t = 0.25$:

$$T_0(t) = 0.77$$

$$T_1(t) = 0.97$$

$$T_2(t) = 0.997$$

$$T_3(t) = 0.9998$$

$$T_4(t) = 0.999993$$

$$T_5(t) = 0.9999997$$

VI - 6 Forme confluente du procédé d'Overholt

Il est possible de construire, pour les fonctions, un procédé analogue au procédé d'Overholt pour les suites (paragraphe IV-1). Soit f une fonction telle que :

$$f(t) - S = \sum_{i>0} a_i \, [f'(t)]^i$$

Le problème est de trouver des estimations de S d'ordre de plus en plus élevé à partir de f(t), f'(t) ... On a :

$$f' = f'' \sum_i i\, a_i\, f'^{\,i-1}$$

Posons $V_0(t) = f(t)$ et $V_1(t) = V_0(t) - \frac{f'(t)}{f''(t)} V'_1(t)$.

On a :

$$V_1 = f - \sum_i i\, a_i\, f'^i = S + \sum_{i=1} (1-i)\, a_i\, f'_i = S + \sum_{i=2} (1-i)\, a_i\, f'_i$$

$$V'_1 = f'' \sum_{i=2} i\,(1-i)\, a_i\, f'^{i-1}$$

Posons $V_2(t) = V_1(t) - \frac{f'(t)}{f''(t)} \frac{V_1(t)}{2}$

On trouve que :

$$V_2(t) = S + \sum_{i=3} \left(1 - \frac{i}{2}\right)(1-i)\, a_i\, f'^i$$

cela suggère donc que la forme confluente du procédé d'Overholt est donnée par la règle :

$$V_0(t) = f(t)$$

$$V_{k+1}(t) = V_k(t) - \frac{f'(t)}{f''(t)} \frac{V'_k(t)}{k+1} \qquad k = 0, 1, \ldots$$

Si l'on suppose que :

$$V_k(t) = S + \sum_{i=k+1} \left(1 - \frac{i}{k}\right)\left(1 - \frac{i}{k-1}\right) \ldots (1-i)\, a_i\, f'^i$$

on trouve facilement que $V_{k+1}(t)$ vérifie la même relation où k est remplacé par k+1.

On a le :

<u>Théorème 133</u> :

Une condition suffisante pour que $\lim_{t\to\infty} V_k(t) = \lim_{t\to\infty} f(t)\ \forall k$ est que :

$$\lim_{t\to\infty} \frac{f''(t)}{f'(t)} \neq 0$$

démonstration : puisque $\lim_{t\to\infty} f(t) = S$ alors $\lim_{t\to\infty} V_0(t) = S$.

Par conséquent $\lim_{t\to\infty} V_1(t) = S$ si la condition du théorème est remplie. On peut faire le même raisonnement de proche en proche pour tout k.

Pour ce procédé on a :

$$V_1(t) = f(t) - \frac{f'^2(t)}{f''(t)} = \varepsilon_2(t)$$

$$V_2(t) = f - \frac{f'^2}{2f''^3} (f''^2 + f' f''')$$

Malgré sa linéarité ce procédé est plus difficile à mettre en oeuvre que la première forme confluente de l'ε-algorithme car pour ce dernier on utilise, pour effectuer les calculs, la relation de récurrence qui existe entre les déterminants fonctionnels de Hankel.

VI - 7 Transformation rationnelle d'une fonction

On peut définir pour les fonctions [39] des transformations rationnelles analogues à celles définies par Pennacchi pour les suites (paragraphe IV - 7).

Définition 28 : On appelle transformation rationnelle d'ordre p et de degré m l'application $T_{p,m}$ qui à la fonction f fait correspondre la fonction $V_{p,m}$ définie par :

$$V_{p,m}(t) = f(t) + \frac{P_m\,(f'(t), \ldots, f^{(p)}(t))}{Q_{m-1}\,(f'(t), \ldots, f^{(p)}(t))}$$

où P_m, Q_{m-1} sont des polynômes homogènes de degrés respectifs m et m-1 par rapport aux variables $f'(t), \ldots, f^{(p)}(t)$. On posera $R_m = P_m \,/\, Q_{m-1}$ et $R_m \equiv 0$ si $f'(t) = \ldots = f^{(p)}(t) = 0$ pour $m > 1$.

L'application $T_{p,m}$ possède les propriétés suivantes :

Propriété 63 :

- $T_{p,m}\,[a\,f(t) + b] = a\,T_{p,m}\,[f(t)] + b$
- $T_{p,m}\,[a] = a$
- les puissances successives d'une transformation rationnelle ne sont pas en général des transformations rationnelles.

Définition 29 : On dit que f est régulière si :

- $\lim_{t\to\infty} f(t) = S$ existe et est finie
- $\exists T : \forall t > T \quad f'(t) \neq 0$
- $\lim_{t\to\infty} \frac{f'(t)}{f(t) - S} = \rho \neq 0$

Théorème 134 :

Si f est régulière et si

$$Q_{m-1}(1, \rho, \ldots, \rho^{p-1}) \neq 0$$

alors $\lim_{t\to\infty} V_{p,m}(t) = S$

démonstration : on pose pour $t > T$ $\rho_k(t) = f^{(k+1)}(t) / f^{(k)}(t)$ pour $k = 1, \ldots, p-1$.

On a donc :

$$f^{(k)}(t) / f'(t) = \rho_1(t) \ldots \rho_{k-1}(t)$$

d'où $V_{p,m}(t) = f(t) + f'(t) R_m(1, \rho_1, \ldots, \rho_1 \ldots \rho_{p-1})$

on pose $\sigma_0(t) = 1$ et $\sigma_i(t) = \rho_1(t) \ldots \rho_i(t)$ pour $i = 1, \ldots, p-1$ ce qui donne :

$$V_{p,m}(t) = f(t) + f'(t) R_m(1, \sigma_1, \ldots, \sigma_{p-1})$$

d'après la règle de l'Hospital on a :

$$\lim_{t\to\infty} \rho_k(t) = \rho \qquad k = 1, \ldots, p-1$$

et par conséquent :

$$\lim_{t\to\infty} V_{p,m}(t) = \lim_{t\to\infty} f(t) + R_m(1, \rho, \ldots, \rho^{p-1}) \lim_{t\to\infty} f'(t)$$

d'où $\lim_{t\to\infty} V_{p,m}(t) = S$ si $Q_{m-1}(1, \rho, \ldots, \rho^{p-1}) \neq 0$ puisque $\lim_{t\to\infty} f'(t) = 0$; ce qui démontre le théorème.

Pour l'accélération de la convergence nous utiliserons la :

Définition 30 : On dit que $T_{p,m}$ accélère la convergence si :

$$\lim_{t\to\infty} \frac{V_{p,m}(t) - S}{f(t) - S} = 0$$

Théorème 135 :

Une condition nécessaire et suffisante pour que $T_{p,m}$ accélère la convergence est que :

$$R_m(1, \rho, \ldots, \rho^{p-1}) = -1 / \rho$$

démonstration :

$$V_{p,m}(t) - S = f(t) - S + f'(t) R_m(\sigma_0, \ldots, \sigma_{p-1})$$

$$\frac{V_{p,m}(t) - S}{f(t) - S} = 1 + \frac{f'(t)}{f(t) - S} R_m(\sigma_0, \ldots, \sigma_{p-1})$$

d'où :

$$0 = 1 + \rho\, R_m\, (1, \rho, \ldots, \rho^{p-1})$$

ce qui démontre le théorème.

On obtient également des résultats analogues à ceux donnés par Pennacchi pour les transformations rationnelles de suites. Les démonstrations sont laissées en exercices :

Théorème 136 :

$T_{1,m}$ et $T_{p,1}$ ne peuvent pas accélérer la convergence de toute fonction régulière.

Théorème 137 :

Il existe une et une seule transformation $T_{2,2}$ qui accélère la convergence de toute fonction régulière. Cette transformation est donnée par :

$$V_{2,2}(t) = f(t) - f'^2(t) \,/\, f''(t)$$

on voit que $V_{2,2}(t) = \varepsilon_2(t)$.

Définition 31 : on dit que $T_{p,m}$ et $T_{q,k}$ sont équivalentes si :

$$T_{p,m}\,[f(t)] = T_{q,k}\,[f(t)]$$

on peut démontrer les :

Théorème 138 :

Pour m > 2, toute transformation $T_{2,m}$ qui accélère la convergence est toujours équivalente à $T_{2,2}$.

Théorème 139 :

il existe une transformation unique d'ordre 2 qui accélère la convergence : c'est $\varepsilon_2(t)$.

VI - 8 Applications

Dans ce paragraphe nous allons donner quelques applications de la première forme confluente de l'ε-algorithme et de celle du ρ-algorithme.

Il est évident que la forme confluente de l'ε-algorithme est difficile à mettre en oeuvre directement sur ordinateur ; il faudrait en effet disposer d'un compilateur capable de dériver formellement les fonctions ε_k. On réalise la mise en oeuvre effective en utilisant l'une des relations :

$$\varepsilon_{2k}(t) = H_{k+1}^{(0)}(t)/H_k^{(2)}(t)$$

ou

$$\varepsilon_{2k}(t) = \varepsilon_{2k-2}(t) - \frac{[H_k^{(1)}(t)]^2}{H_{k-1}^{(2)}(t).H_k^{(2)}(t)}$$

avec $\varepsilon_o(t) = f(t)$. On calcule les déterminants fonctionnels de Hankel à l'aide de leur relation de récurrence :

$$H_{k+2}^{(n-1)}(t).H_k^{(n+1)}(t) + [H_{k+1}^{(n)}(t)]^2 = H_{k+1}^{(n-1)}(t).H_{k+1}^{(n+1)}(t)$$

en partant des conditions initiales :

$$H_o^{(n)}(t) = 1 \text{ et } H_1^{(n)}(t) = f^{(n)}(t) \text{ pour } n = 0,1,\ldots$$

On trouvera dans [19,22,35] des programmes FORTRAN. Signalons qu'il est utile d'avoir à sa disposition un programme de dérivation formelle pour calculer les $f^{(n)}(t)$.

Au lieu d'utiliser la relation de récurrence des déterminants de Hankel fonctionnels on peut également se servir du ω-algorithme qui a été spécialement mis au point par Wynn [238] dans ce but et qui est plus économique :

Les règles du ω-algorithme sont les suivantes :

$$\omega_{-1}^{(n)} = 0 \qquad \omega_0^{(n)} = f^{(n)}(t) \qquad \text{pour } n = 0,1,\ldots$$

$$\omega_{2k+1}^{(n)} = \omega_{2k-1}^{(n+1)} + \omega_{2k}^{(n)} / \omega_{2k}^{(n+1)}$$

$$\omega_{2k+2}^{(n)} = \omega_{2k}^{(n+1)}(\omega_{2k+1}^{(n)} - \omega_{2k+1}^{(n+1)}) \qquad n,k = 0,1,\ldots$$

En utilisant de nouveau le développement de Schweins, Wynn a démontré que l'on avait :

$$\omega_{2k}^{(n)} = H_{k+1}^{(n)}(t) / H_k^{(n+2)}(t)$$

et par conséquent :

$$\omega_{2k}^{(0)} = \varepsilon_{2k}(t)$$

Dans la mise en oeuvre de cet algorithme, il peut évidemment se produire une division par zéro. Wynn a donné une forme particulière du ω-algorithme qu'il faut utiliser lorsque les $\bar{n}$ premières dérivées de f en t sont nulles. On trouvera dans [238] des programmes ALGOL et dans [19] des programmes FORTRAN.

Remarque : l'étude d'un algorithme similaire au ω-algorithme pour mettre en oeuvre la forme confluente de l'ε-algorithme topologique reste à faire.

La mise en oeuvre de la forme confluente du ρ-algorithme qui est :

$$\rho_{-1}(t) = 0 \qquad \rho_0(t) = f(t) \text{ et } \rho_{k+1}(t) = \rho_{k-1}(t) + (k+1)/\rho_k'(t)$$

s'effectue à l'aide des mêmes relations que celles du ω-algorithme, seules les initialisations changent. Cet algorithme est le ω'-algorithme que l'on initialise avec :

$$\omega_0^{(n)'} = f^{(n)}(t) \;/\; n\,! \qquad\qquad \text{pour } n = 0,1,\ldots$$

Wynn a démontré que l'on avait alors :

$$\omega_{2k}^{(0)'} = \rho_{2k}(t)$$

et que, comme pour la forme confluente de l'ε-algorithme, les quantités $\rho_{2k}(t)$ s'expriment sous forme d'un rapport de deux déterminants :

$$\rho_{2k}(t) = H_{k+1}^{(0)'}(t) \;/\; H_k^{(2)'}(t)$$

où $H_k^{(n)'}(t)$ est le déterminant obtenu en remplaçant $f^{(i)}(t)$ par $f^{(i)}(t)/i!$ pour tout i dans $H_k^{(n)}(t)$.

On trouvera dans [19] de nombreuses applications au calcul des intégrales impropres et dans [224] la théorie de la convergence de telles méthodes d'intégration.

Soit donc à calculer $I = \int_a^\infty g(x)\,dx$.

Posons $f(t) = \int_a^t g(x)\,dx$. On aura donc $f^{(n)}(t) = g^{(n-1)}(t)$ pour $n = 1, 2, \ldots$ Les quantités $\varepsilon_{2k}(t)$ seront donc des approximations de I. Inversement si l'on connait $\int_a^\infty g(x)\,dx$ on pourra déduire des $\varepsilon_{2k}(t)$ des approximations de $\int_a^t g(x)\,dx$.

La première forme confluente de l'ε-algorithme apparait donc ainsi comme un moyen pour passer d'un intervalle d'intégration fini à un intervalle semi-infini et inversement. C'est une propriété que ne possèdent pas les transformations G de Gray, Atchison, Clark et Schucany [98 à 103] qui fournissent des approximations de $\lim_{t\to\infty} f(t)$ connaissant $f(t)$ et $f(t+k)$ ou $f(t)$ et $f(kt)$.

La valeur de $f(t) = \int_a^t g(x)\,dx$ sera calculée si possible par intégration directe ; dans le cas contraire on l'estimera à l'aide d'une formule de quadrature numérique. Donnons trois exemples :

1°) Passage d'un intervalle d'intégration fini à un intervalle semi-infini

$$\Gamma(\tfrac{1}{2}, t^2) = \sqrt{\pi}\,(1 - \text{erf } t) = \int_{t^2}^\infty \frac{e^{-x}}{\sqrt{x}}\,dx.$$

Prenons $f(t) = \int_0^{t^2} \frac{e^{-x}}{\sqrt{x}}\,dx$. Nous aurons :

$$\int_0^\infty \frac{e^{-x}}{\sqrt{x}}\,dx = \Gamma(\tfrac{1}{2}, 0) = \sqrt{\pi} \simeq \varepsilon_{2k}(t)$$

On aura donc :

$$\text{erf } t \simeq 1 - \frac{1}{\sqrt{\pi}} \sum_{i=1}^{k} D_{2i-1}(t) = E_k(t)$$

On obtient les résultats suivants :

$t = 1$ $D_1(t) = -0.13836917$ $D_3(t) = -0.\ 415651176.10^{-1}$ $D_5(t) = -0.30309000\ 10^{-2}$

$E_1(t) = 0.86163084$ $E_2(t) = 0.84706566$ $E_3(t) = 0.84403476$

$\text{erf}(1) = 0.84270079$

$t = 1.5$ $D_1(t) = -0.32435536.10^{-1}$ $D_3(t) = -0.12909670.10^{-2}$ $D_5(t) = -0.13913150.10^{-3}$

$E_1(t) = 0.96756448$ $E_2(t) = 0.96627350$ $E_3(t) = 0.96613437$

$\text{erf}(1.5) = 0.96610515$

$t = 2$ $D_1(t) = -0.45926636.10^{-2}$ $D_3(t) = -0.79872380.10^{-4}$ $D_5(t) = -0.46709767\ .10^{-5}$

$E_1(t) = 0.99540734$ $E_2(t) = 0.99532747$ $E_3(t) = 0.99532279$

$\text{erf}(2) = 0.99532227.$

On observe une très nette amélioration de la précision quand t augmente. On remarque également que $D_3(t)$ et $D_5(t)$ sont une bonne approximation de l'erreur sur $E_1(t)$ et $E_2(t)$ et que l'erreur sur $E_3(t)$ est petite devant $D_5(t)$ comme cela avait été mis en évidence au paragraphe VI-3.

Sur cet exemple si l'on effectue les calculs analytiquement on s'aperçoit que l'on retrouve exactement les convergents successifs de la fraction continue obtenue par Levy-Soussan [132].

2°) Passage d'un intervalle d'intégration semi-infini à un intervalle fini. On veut calculer $\Gamma(x) = \int_0^\infty t^{x-1} e^{-t} dt$. Prenons x = 2 et $f(t) = \int_0^t x e^{-x} dx$. On a $\Gamma(2) = 1$ et f(2) est calculé par une formule de quadrature dont la précision est de 10^{-5}. On obtient :

t	$\varepsilon_2(t)$	$\varepsilon_4(t)$
6	1.0049	0.99980
8	1.00056	0.999986
10	1.000068	0.9999989
12	1.0000086	0.99999991
14	1.0000011	1.00000002
16	1.0000004	1.00000005

3°) $J_0(x) = \frac{2}{\pi} \int_0^\infty \sin(x \operatorname{ch} u)\, du$. Prenons $f(t) = \frac{2}{\pi} \int_0^t \sin(x \operatorname{ch} u)\, du$ et t = 7.

f(t) est calculé avec une précision de 10^{-8}.

x	$J_0(x)$	$\varepsilon_2(7)$	$\varepsilon_4(7)$	$\varepsilon_6(7)$
0.1	0.9975	1.0774	0.9983	0.9983
0.3	0.97762	0.96849	0.97754	0.97752
0.5	0.93846	0.94195	0.93849	0.93851

Sur la liaison entre la première forme confluente de l'ε-algorithme et les intégrales définies on pourra consulter [218].

La première forme confluente de l'ε-algorithme peut être appliquée à la résolution d'une équation g(x) = 0.

Posons y = g(x) ; on a $x = g^{-1}(y)$, résoudre g(x) = 0 revient donc à chercher $\lim_{y \to 0} g^{-1}(y)$.

En posant $y = 1/t$ et $f(t) = g^{-1}(1/t)$ on voit encore que résoudre $g(x) = 0$ revient à calculer $\lim_{t\to\infty} f(t)$.

Appliquons à f la première forme confluente de l'ε-algorithme. On obtient :

$$\varepsilon_2(t) = x - w_0 \frac{g(t)\, g'(t)}{2g'^2(t) - g(t)\, g''(t)}$$

si l'on suppose que la racine est une racine simple on trouve que $w_0 = 2$, ce qui nous donne la méthode itérative suivante :

$$x_{n+1} = x_n - 2 \frac{g(x_n)\, g'(x_n)}{2g'^2(x_n) - g(x_n)\, g''(x_n)}$$

On retrouve ainsi une méthode itérative connue : la méthode de Schröder. C'est une méthode d'ordre trois. Soit par exemple à résoudre $x = e^{-x}$ dont la racine unique est $x = 0.56714329$ ... Avec $x_0 = 0$ on trouve

	Méthode de Newton	Méthode de Schröder
x_1	0.506	0.571
x_2	0.5603	0.56714329

Si la racine est multiple alors on ne sait plus calculer w_0. On remplacera donc comme précédemment w_0 par :

$$w_0(t) = - \frac{g'(t)}{D_1'(t)}$$

ce qui donne la méthode itérative :

$$x_{n+1} = x_n - \frac{gg'(2g'^2 - gg'')}{2g'^4 - 2gg'^2 g'' + g^2 g' g''' - g^2 g''^2}$$

où toutes les fonctions sont calculées en donnant la valeur x_n à la variable. Soit à résoudre $(x - 1)^6 = 0$ en partant de $x_0 = - 2$ à l'aide de cette méthode ; on obtient $x_1 = 1.0000001$ alors que la méthode de Schröder n'est plus que du premier ordre.

Donnons maintenant une application de la forme confluente du ρ-algorithme à l'intégration des équations différentielles [43].

Soit à intégrer l'équation différentielle :

$$y' = f(x,y)$$
$$y(x_0) = y_0$$

avec les hypothèses habituelles sur f. On a :

$$y(x+h) - y(x) = \lim_{t\to\infty} \int_x^{x+h-1/t} f(u,y(u))\, du$$

où h est un paramètre positif arbitraire. D'où l'idée de poser :

$$g(t) = y(x) + \int_x^{x+h-1/t} f(u,y(u))\, du = y(x+h-1/t)$$

Appliquons la forme confluente du ρ-algorithme à cette fonction g. On obtient :

$$\rho_2(t) = y(x+h-1/t) - 2g'^2(t)/g''(t)$$

avec $g'(t) = y'(x+h-1/t)/t^2$

$g''(t) = y''(x+h-1/t)/t^4 - 2y'(x+h-1/t)/t^3$

Puisque $\lim_{t\to\infty} g(t) = y(x+h)$ ceci nous donne l'idée de prendre $\rho_2(t)$ comme approximation de $y(x+h)$. En donnant à t la valeur 1/h on obtient le schéma d'intégration suivant :

y_0 donné

$$y_{n+1} = y_n + 2h \frac{y_n'^2}{2y_n' - hy_n''}$$

où y_n' et y_n'' sont les valeurs respectives de $y'(x)$ et $y''(x)$ obtenues en donnant à x la valeur x_n et à y la valeur approchée y_n dans les relations :

$$y'(x) = f(x,y) \qquad \text{et} \qquad y''(x) = \frac{\partial f(x,y)}{\partial x} + \frac{\partial f(x,y)}{\partial y} f(x,y)$$

Cette méthode est une méthode à pas séparés de la forme

$$y_{n+1} = y_n + h\phi(x_n,y_n,h) \qquad \text{avec} \qquad \phi(x,y,h) = \frac{2y'^2}{2y'-hy''}$$

Il est bien évident que ϕ n'est définie que si $2y' - hy'' \neq 0$. Si $y' = y'' = 0$ on prendra $\phi(x,y,h) = 0$. Si $y'' \neq 0$ et $y' = 0$ on posera également $\phi(x,y,h) = 0$.

On démontre que cette méthode est consistante avec l'équation différentielle. De plus si f et f' vérifient une condition de Lipschitz par rapport à leur seconde variable et si $f'(x,y) = O(f(x,y))$ pour tout x appartenant à l'intervalle d'intégration et pour tout y tel que $f(x,y) \neq 0$ alors la méthode est stable. Elle est donc convergence et l'on démontre quelle est du second ordre c'est-à-dire que :

$$y_n - y(x_n) = O(h^2)$$

Mais l'intérêt principal de cette méthode est d'être A-stable au sens de Dahlquist [68] c'est-à-dire que si l'on intègre l'équation différentielle $y' = -\lambda y$ avec $\mathrm{Re}\lambda > 0$ on a $\lim_{n\to\infty} y_n = 0$.

Soit par exemple à intégrer $y' = -10y$ avec $y(0) = 1$. On obtient respectivement avec cette méthode et avec la méthode de Runge-Kutta classique d'ordre 2, les erreurs relatives suivantes :

	x	méthode A-stable	Runge-Kutta
h = 0,01	0,3	$0,25\ 10^{-2}$	$-\ 0,54\ 10^{-2}$
	0,6	$0,50\ 10^{-2}$	$-\ 0,11\ 10^{-1}$
	1,0	$0,83\ 10^{-2}$	$-\ 0,18\ 10^{-1}$
h = 0,04	0,6	$0,79\ 10^{-1}$	$-\ 0,24$
	1,0	0,13	$-\ 0,43$
h = 0,16	0,96	0,97	$-\ 0,15\ 10^{4}$

Remarque 1 : il est théoriquement possible d'obtenir des méthodes d'ordre plus élevé en utilisant ρ_{2k} au lieu de ρ_2 mais il est évident que la méthode devient rapidement d'une utilisation trop difficile puisqu'il faut commencer par dériver l'équation différentielle à intégrer.

Remarque 2 : les méthodes explicites et A-stables sont d'un grand intérêt pratique pour l'intégration des équations différentielles. De nombreuses études ont été faites sur ce sujet depuis un certain temps [165,166,167,178]. Signalons que la méthode précédente peut également être obtenue à partir des approximants de Padé de e^{-x} [73,74,125].

Remarque 3 : la généralisation aux systèmes d'équations différentielles n'est pas encore résolue actuellement.

Remarque 4 : dans la méthode précédente on peut remplacer hy''_n par son approximation $y'_n - y'_{n-1}$. On obtient alors une méthode à pas liés dont l'étude reste à terminer.

Cette méthode peut être appliquée au calcul des intégrales définies [193]. Soit, en effet, à calculer :

$$I = \int_a^b f(x)\,dx$$

ce calcul est équivalent à intégrer l'équation différentielle :

$$y' = f(x)$$
$$y(a) = 0$$

On a bien évidemment $y(b) = I$ et la méthode précédente se simplifie puisque f ne dépend pas de y. Dans ce cas on on :

$$y_0 = 0$$

$$y_{n+1} = y_n + 2h \frac{f^2(x_n)}{2f(x_n) - hf'(x_n)} \qquad (1)$$

Si l'on remplace $hf'(x_n)$ par son approximation $f(x_{n+1}) - f(x_n)$ on obtient :

$$y_0 = 0$$

$$y_{n+1} = y_n + 2h \frac{f^2(x_n)}{3f(x_n) - f(x_{n+1})} \qquad (2)$$

On peut comparer les méthodes (1) et (2) à la méthode des trapézes (T) et à la méthode de Simpson (S) à nombre égal d'évaluations de fonctions (une évaluation supplémentaire est nécessaire pour la méthode (1)). Les exemples suivants sont empruntés à Wuytack [194] qui a étudié trés complètement cette méthode de calcul des intégrales définies.

Soit à calculer $I = \int_0^1 e^x\,dx = 1,7182818...$

On obtient :

évaluations de fonctions	T	S	(1)	(2)
5	1,727	1,7183	1,73	1,76
15	1,719	1,7182821	1,72	1,72
25	1,7185	1,7182819	1,719	1,719
35	1,7184	1,7182818	1,7187	1,7188
45	1,7183	1,7182818	1,7185	1,7186

Soit maintenant à calculer $I = \int_0^1 \frac{e^x}{(3-e^x)^2} dx = 3,0496468...$

évaluations de fonctions	T	S	(1)	(2)
5	5,3	4,07	3,17	- 63,7
15	3,3	3,10	3,057	76,9
25	3,15	3,059	3,052	3,65
35	3,09	3,052	3,0507	3,27
45	3,08	3,0507	3,0502	3,17

Si f est la dérivée d'une fraction rationnelle dont numérateur et dénominateur sont des polynômes du premier degré alors (1) fournit le résultat exact. La méthode (2) semble souffrir d'une certaine instabilité. Si f possède un pôle à l'extérieur de l'intervalle d'intégration mais au voisinage de l'une de ces bornes les méthodes (1) et (2) donnent des résultats mailleurs que les méthodes classiques ; cela tient au fait que f est mieux représentée alors par une fraction rationnelle que par un polynôme. Pour les fonctions bien "lisses" les méthodes classiques donnent de meilleurs résultats que les méthodes (1) et (2). De telles méthodes d'intégration semblent cependant très intéressantes mais beaucoup de travail reste encore à faire sur ce sujet.

CHAPITRE VII

LES FRACTIONS CONTINUES

VII-1 - Définitions et propriétés

Considérons l'expression suivante :

$$b_0 + \cfrac{a_1}{b_1 + \cfrac{a_2}{b_2 + \cfrac{a_3}{b_3 + \cfrac{a_4}{\ddots}}}} \qquad (1)$$

Pour des raisons typographiques évidentes (1) sera écrite sous la forme :

$$b_0 + \frac{a_1}{b_1 +} \; \frac{a_2}{b_2 +} \; \frac{a_3}{b_3 +} \cdots \qquad (2)$$

ou sous la forme :

$$b_0 + \left.\frac{a_1}{|b_1}\right. + \left.\frac{a_2}{|b_2}\right. + \left.\frac{a_3}{|b_3}\right. + \cdots \qquad (3)$$

Nous utiliserons cette dernière forme car elle nous parait la plus claire.

L'expression (1) (ou les formes équivalentes (2) et (3)) est appelée une fraction continue.

Voyons quelle signification on peut donner à une telle fraction continue.

Donnons d'abord quelques définitions :

a_k et b_k s'appellent respectivement $k^{ième}$ numérateur partiel et $k^{ième}$ dénominateur partiel. Le rapport a_k/b_k est le $k^{ième}$ quotient partiel et la quantité :

$$C_n = b_0 + \left.\frac{a_1}{|b_1}\right. + \left.\frac{a_2}{|b_2}\right. + \cdots + \left.\frac{a_n}{|b_n}\right. \qquad (4)$$

s'appelle le $n^{ième}$ convergent (ou approximant) de la fraction continue (1).
Le nombre C_n ne peut évidemment être défini que si aucun des dénominateurs rencontrés dans les divisions successives ne s'annule.

Si tous les convergents C_n sont définis, sauf peut-être un nombre fini d'entre eux, et si la quantité :

$$C = \lim_{n\to\infty} C_n \tag{5}$$

existe alors nous écrirons :

$$C = b_0 + \frac{a_1|}{|b_1} + \frac{a_2|}{|b_2} + \dots \tag{6}$$

On dit dans ce cas que la fraction continue (1) est convergente et qu'elle a C comme valeur. Elle sera dite divergente dans le cas contraire.
Le concept de fraction continue est important en théorie de l'approximation.
Considérons par exemple la fraction continue suivante qui a été étudiée par Gauss :

$$\frac{z|}{|1} - \frac{z^2|}{|3} - \frac{z^2|}{|5} - \dots - \frac{z^2|}{|2n-1} - \dots \tag{7}$$

Cette formule est valable pour toute valeur de la variable complexe z et sa valeur, qui dépend de z ainsi que ses convergents, est $C(z) = \text{tg } z$.

Calculons par exemple $\text{tg } \frac{\Pi}{4} = 1$ à l'aide des convergents successifs de cette fraction continue ; on obtient :

$C_1(\frac{\Pi}{4}) = 0.78$ $\qquad$ $C_2(\frac{\Pi}{4}) = 0.988$

$C_3(\frac{\Pi}{4}) = 0.99978$ $\qquad$ $C_4(\frac{\Pi}{4}) = 0.9999978$

$C_5(\frac{\Pi}{4}) = 0.999999986$ $\qquad$ $C_6(\frac{\Pi}{4}) = 0.999999999941$

On voit ainsi que l'utilisation des fractions continues fournit des approximations précises ; elles sont d'ailleurs utilisées pour le calcul de nombreuses fonctions mathématiques standard sur ordinateur [108,132].

Le calcul effectif des convergents successifs d'une fraction continue peut s'effectuer de deux façons différentes. Nous ne donnerons la première que pour mémoire car elle n'est pas utilisée en pratique :

$$D_0 = b_n$$
$$D_{k+1} = b_{n-k-1} + \frac{a_{n-k}}{D_k} \qquad k = 0,\ldots,n-1$$

On aura :

$$D_n = C_n$$

La démonstration est évidente. On voit que l'on calcule la suite des dénominateurs de (4) en partant du $n^{\text{ième}}$ quotient partiel pour arriver au premier.

Thèorème 140 :

Posons $C_n = A_n/B_n$. On peut calculer A_n et B_n récursivement à l'aide des relations :

$$\left.\begin{aligned} A_k &= b_k A_{k-1} + a_k A_{k-2} \\ B_k &= b_k B_{k-1} + a_k B_{k-2} \end{aligned}\right\} \quad \text{pour } k = 1,2,\ldots \qquad (8)$$

en partant des conditions initiales :

$$A_0 = b_0 \qquad A_{-1} = 1$$
$$B_0 = 1 \qquad B_{-1} = 0$$

démonstration [220] : elle se fait par récurrence. Pour C_1 on a

$$C_1 = b_0 + \frac{a_1|}{|b_1} = \frac{b_0 b_1 + a_1}{b_1}$$

Les relations de récurrence donnent :

$$A_1 = b_1 A_0 + a_1 A_{-1} = b_0 b_1 + a_1$$
$$B_1 = b_1 B_0 + a_1 B_{-1} = b_1$$

Supposons que les relations sont vérifiées jusqu'à $k=n$ et démontrons qu'elles sont encore vraies pour $k=n+1$.

En effet C_{n+1} est obtenu à partir de C_n en remplaçant simplement b_n par $b_n + \frac{a_{n+1}|}{|b_{n+1}}$; d'où d'après (8) :

$$A_{n+1} = (b_n + \frac{a_{n+1}}{b_{n+1}}) A_{n-1} + a_n A_{n-2}$$

$$= \frac{b_{n+1}(b_n A_{n-1} + a_n A_{n-2}) + a_{n+1} A_{n-1}}{b_{n+1}}$$

et de même :

$$B_{n+1} = \frac{b_{n+1}(b_n B_{n-1} + a_n B_{n-2}) + a_{n+1} B_{n-1}}{b_{n+1}}$$

on a donc :

$$A_{n+1} = \frac{b_{n+1} A_n + a_{n+1} A_{n-1}}{b_{n+1}} \quad \text{et} \quad B_{n+1} = \frac{b_{n+1} B_n + a_{n+1} B_{n-1}}{b_{n+1}} \tag{9}$$

Par conséquent :

$$C_{n+1} = \frac{A_{n+1}}{B_{n+1}} = \frac{b_{n+1} A_n + a_{n+1} A_{n-1}}{b_{n+1} B_n + a_{n+1} B_{n-1}}$$

ce qui termine la démonstration.

Voyons maintenant la relation qui existe entre C_{n-1} et C_n. On a le :

Théorème 141: deux convergents successifs de la fraction continue (1) sont reliés par :

$$\frac{A_n}{B_n} - \frac{A_{n-1}}{B_{n-1}} = (-1)^{n-1} \frac{a_1 a_2 \ldots . a_n}{B_n B_{n-1}} \qquad n = 1,2,\ldots .$$

démonstration : il suffit de montrer que :

$$A_n B_{n-1} - A_{n-1} B_n = (-1)^{n-1} a_1 a_2 \ldots a_n$$

ce qui peut être fait par récurrence. Pour n=1 on a :

$$A_1B_0 - A_0B_1 = a_1$$

Supposons que la formule est vraie jusqu'à l'indice n et démontrons qu'elle reste valable pour n+1. On a pour n+1 :

$$B_nA_{n+1} - A_nB_{n+1} = B_n(b_{n+1}A_n + a_{n+1}A_{n-1})$$

$$- A_n(b_{n+1}B_n + a_{n+1}B_{n-1})$$

d'après le théorème (1). D'où :

$$B_nA_{n+1} - A_nB_{n+1} = - a_{n+1}(B_{n-1}A_n - A_{n-1}B_n)$$

$$= (-1)^n a_1a_2\dots a_{n+1}$$

ce qui termine la démonstration.

Théorème 142: On peut exprimer le $n^{\text{ième}}$ convergent de la fraction continue (1) sous forme de la somme finie :

$$\frac{A_n}{B_n} = b_0 + \frac{a_1}{B_0B_1} - \frac{a_1a_2}{B_1B_2} + \frac{a_1a_2a_3}{B_2B_3} - \dots +(-1)^{n+1}\frac{a_1a_2\dots a_n}{B_{n-1}B_n}$$

démonstration : elle découle de l'utilisation du théorème 141 dans la relation :

$$\frac{A_n}{B_n} = (\frac{A_n}{B_n} - \frac{A_{n-1}}{B_{n-1}}) +(\frac{A_{n-1}}{B_{n-1}} - \frac{A_{n-2}}{B_{n-2}}) +\dots+ (\frac{A_1}{B_1} - \frac{A_0}{B_0}) + \frac{A_0}{B_0}$$

remarque : supposons que $a_i \neq 0$ pour $i=1,\dots,n$ et que $a_{n+1} = 0$.

D'après la relation du théorème 141 on a

$$\frac{A_{n+1}}{B_{n+1}} - \frac{A_n}{B_n} = 0$$

et par conséquent $C_p = C_n$ pour $p = n+1, n+2, \ldots$

On dit dans ce cas que la fraction continue est d'ordre fini n. Elle est égale à C_n.

Le théorème 140 nous a montré comment l'on pouvait exprimer les convergents d'une fraction continue en fonction de ses éléments a_n et b_n. Exprimons maintenant les éléments en fonction des convergents. Pour cela supposons que $a_i \neq 0$ pour tout i ; alors les relations du théorème 140 nous donnent immédiatement :

$$a_n = -\frac{A_n B_{n-1} - A_{n-1} B_n}{A_{n-1} B_{n-2} - A_{n-2} B_{n-1}}$$

$$b_n = \frac{A_n B_{n-2} - B_n A_{n-2}}{A_{n-1} B_{n-2} - A_{n-2} B_{n-1}} \tag{10}$$

avec $b_0 = C_0$, $b_1 = 1$ et $a_1 = C_1 - C_0$

ou encore :

$$a_n = \frac{B_n}{B_{n-2}} \frac{C_{n-1} - C_n}{C_{n-1} - C_{n-2}}$$

$$b_n = \frac{B_n}{B_{n-1}} \frac{C_n - C_{n-2}}{C_{n-1} - C_{n-2}} \tag{11}$$

avec $b_0 = C_0$, $b_1 = 1$ et $a_1 = C_1 - C_0$.

Puisque le rapport $C_n = A_n/B_n$ n'est déterminé qu'à un facteur multiplicatif près, on voit que les B_n peuvent être pris arbitrairement. On obtient ainsi des fractions continues ayant même suite de convergents mais ayant des éléments a_n et b_n différents : on dit, dans ce cas, que les fractions continues ainsi obtenues sont équivalentes. En particulier on peut choisir les B_n de sorte que $B_n/B_{n-1} = 1$ pour tout n.

Soit C une fraction continue :

$$C = b_0 + \frac{a_1|}{|b_1} + \frac{a_2|}{|b_2} + \dots$$

et soit $d_1, d_2, \dots$ des nombres non nuls. Alors, on montre facilement que la fraction continue :

$$C' = b_0 + \frac{d_1a_1|}{|d_1b_1} + \frac{d_1d_2a_2|}{|d_2b_2} + \dots + \frac{d_{n-1}d_na_n|}{|d_nb_n} + \dots$$

est équivalente à la fraction continue C. De plus on a :

$$A'_k = d_1d_2\dots d_kA_k \quad \text{et} \quad B'_k = d_1d_2\dots d_kB_k$$

Toutes les fractions continues équivalentes à C peuvent être obtenues de cette manière.

VII-2 - Transformation d'une série en fraction continue.

Considérons la série :

$$S = u_0 + u_1 + \dots$$

et appelons S_n ses sommes partielles :

$$S_n = \sum_{i=0}^{n} u_i \qquad n=0,1,\dots$$

On veut lui associer une fraction continue :

$$C = b_0 + \frac{a_1|}{|b_1} + \frac{a_2|}{|b_2} + \dots$$

telle que :

$$C_n = S_n \quad \text{pour } n = 0,1,2\dots$$

Puisque l'on connait les convergents successifs S_n de cette fraction continue on peut obtenir immédiatement ses éléments a_n et b_n à l'aide des relations (11) étudiées au paragraphe précédent.

Pour $n \geq 2$ on a donc :

$$a_n = \frac{S_{n-1} - S_n}{S_{n-1} - S_{n-2}} = - \frac{u_n}{u_{n-1}}$$

$$b_n = \frac{S_n - S_{n-2}}{S_{n-1} - S_{n-2}} = 1 + \frac{u_n}{u_{n-1}} \qquad (12)$$

avec

$$b_0 = S_0 = u_0$$

$$\frac{a_1}{b_1} = S_1 - S_0 = u_1$$

d'où la fraction continue :

$$C = u_0 + \frac{u_1|}{|1} - \frac{\frac{u_2}{u_1}|}{|1 + \frac{u_2}{u_1}} - \dots - \frac{\frac{u_n}{u_{n-1}}|}{|1 + \frac{u_n}{u_{n-1}}} - \dots \qquad (13)$$

Réciproquement, une fraction continue correspond à la série de terme général :

$$u_n = C_n - C_{n-1}.$$

Si nous prenons, comme série particulière, la série de puissances :

$$S = c_0 + c_1 x + c_2 x^2 + \dots.$$

on obtient, en faisant $u_n = c_n x^n$ dans (13) :

$$S = c_0 + \frac{c_1 x|}{|1} - \frac{\frac{c_2}{c_1}x|}{|1 + \frac{c_2}{c_1}x} - \dots - \frac{\frac{c_n}{c_{n-1}}x|}{|1 + \frac{c_n}{c_{n-1}}x} - \dots \qquad (14)$$

exemple : on a ainsi :

$$\mathrm{Log}(1+x) = \frac{x}{1} - \frac{x^2}{2} + \frac{x^3}{3} - \ldots + (-1)^{n-1}\frac{x^n}{n} + \ldots$$

$$= \frac{x\,|}{|\,1} + \frac{\frac{x}{2}\,|}{|\,1-\frac{x}{2}} + \ldots + \frac{\frac{n-1}{n}x\,|}{|\,1-\frac{n-1}{n}x} + \ldots$$

$$= \frac{x\,|}{|\,1} + \frac{1^2x\,|}{|\,2-x} + \ldots + \frac{(n-1)^2\,x\,|}{|\,n-(n-1)x} + \ldots$$

Etant donnée une suite $\{S_n\}$ qui converge vers S on peut lui associer la série :

$$u_0 + u_1 + \ldots$$

avec $u_0 = S_0$, $u_1 = \Delta S_0, \ldots,$ $u_n = \Delta S_{n-1}, \ldots$

Les sommes partielles de cette série sont égales aux termes S_n de la suite et cette série converge vers S.
Exprimée à l'aide de la suite initiale $\{S_n\}$la fraction continue (13) s'écrit :

$$S_0 + \frac{\Delta S_0\,|}{|\,1} - \frac{\frac{\Delta S_1}{\Delta S_0}\,|}{|\,1+\frac{\Delta S_1}{\Delta S_0}} - \ldots - \frac{\frac{\Delta S_{n+1}}{\Delta S_n}\,|}{|\,1+\frac{\Delta S_{n+1}}{\Delta S_n}} - \ldots \qquad (15)$$

Si tous les convergents de cette fraction continue existent sauf un nombre fini d'entre eux alors la fraction continue est convergente et sa valeur est S puisque $C_n = S_n$ et que $\lim\limits_{n\to\infty} S_n = S$.

VII-3 - Contraction d'une fraction continue

Soit $\{C_n\}$ la suite des convergents successifs d'une fraction continue C et soit $\{C_{p_n}\}$ une suite extraite de $\{C_n\}$. Considérons la fraction continue C', d'éléments a'_n et b'_n, dont les convergents successifs C'_n sont égaux à C_{p_n} : on a effectué une contraction de la fraction continue C en la fraction continue C'.

D'après (11) on voit que l'on a :

$$a'_n = \frac{C_{p_{n-1}} - C_{p_n}}{C_{p_{n-1}} - C_{p_{n-2}}}$$

$$\tag{16}$$

$$b'_n = \frac{C_{p_n} - C_{p_{n-2}}}{C_{p_{n-1}} - C_{p_{n-2}}}$$

ainsi que $b'_1 = 1$, $a'_1 = C_{p_1} - C_{p_0}$ et $b'_0 = C_{p_0}$

Considérons en détail le cas où $p_n = 2n$; on a :

$$a'_n = \frac{C_{2n-2} - C_{2n}}{C_{2n-2} - C_{2n-4}}$$

$$b'_n = \frac{C_{2n} - C_{2n-4}}{C_{2n-2} - C_{2n-4}}$$

or $C'_n = A'_n/B'_n = C_{2n} = A_{2n}/B_{2n}$; d'où :

$$A_{2n} = b_{2n} A_{2n-1} + a_{2n} A_{2n-2}$$

$$A_{2n-1} = b_{2n-1} A_{2n-2} + a_{2n-1} A_{2n-3} \qquad (17)$$

$$A_{2n-2} = b_{2n-2} A_{2n-3} + a_{2n-2} A_{2n-4}$$

Multiplions la première de ces égalités par b_{2n-2}, la seconde par $b_{2n}\, b_{2n-2}$, la dernière par $-a_{2n-1}\, b_{2n}$ et faisons la somme ; il vient :

$$b_{2n-2}\, A_{2n} = (a_{2n} b_{2n-2} + b_{2n}\, b_{2n-1}\, b_{2n-2} + a_{2n-1}\, b_{2n})\, A_{2n-2}$$

$$- a_{2n-1}\, a_{2n-2}\, b_{2n}\, A_{2n-4}$$

d'où :

$$A'_n = \frac{a_{2n}\, b_{2n-2} + b_{2n}\, b_{2n-1}\, b_{2n-2} + a_{2n-1}\, b_{2n}}{b_{2n-2}}\, A'_{n-1} - \frac{a_{2n-1}\, a_{2n-2}\, b_{2n}}{b_{2n-2}}\, A'_{n-2}$$

et une relation analogue pour les B'_n. On a donc :

$$a'_n = - \frac{a_{2n-1}\, a_{2n-2}\, b_{2n}}{b_{2n-2}}$$

$$b'_n = \frac{a_{2n}\, b_{2n-2} + b_{2n}\, b_{2n-1}\, b_{2n-2} + a_{2n-1}\, b_{2n}}{b_{2n-2}} \tag{18}$$

avec $b'_0 = b_0$, $b'_1 = 1$ et $a'_1 = C_2 - C_0 = a_1 b_2/(b_1 b_2 + a_2)$.

Nous venons donc d'effectuer la contraction de la fraction continue

$C = b_0 + \frac{a_1|}{|b_1} + \ldots$ en la fraction continue $C' = b'_0 + \frac{a'_1|}{|b'_1} + \ldots$

où les éléments a'_n et b'_n sont donnés par les relations (18).

VII-4 - Fractions continues associée et correspondante

Considérons la fraction continue :

$$C^{(0)} = b_0 + \frac{a_1 x|}{|1} + \frac{a_2 x|}{|1} + \ldots \tag{19}$$

On voit, en utilisant les relations de récurrence du théorème 1 que $A^{(0)}_{2k-1}$, $A^{(0)}_{2k}$ et $B^{(0)}_{2k}$ sont des polynômes de degré k en x et que $B^{(0)}_{2k-1}$ est un polynôme de degré k-1 en x.

D'autre part, d'après le théorème 2, on a :

$$C_k^{(0)} - C_{k-1}^{(0)} = (-1)^{k-1} \frac{a_1 a_2 \ldots a_k}{B_k^{(0)} B_{k-1}^{(0)}} x^k$$

$$= (-1)^{k-1} \frac{a_1 a_2 \ldots a_k}{b_0 + \ldots} x^k \qquad (20)$$

Ceci montre que les développements de $C_k^{(0)}$ et de $C_{k-1}^{(0)}$ en puissances croissantes de x ont leurs k premiers termes identiques.

Considérons maintenant une série formelle :

$$f(x) = \sum_{i=0}^{\infty} c_i x^i \qquad (21)$$

Il est possible de choisir b_0, a_1, a_2,... de telle sorte que $C_k^{(0)}$ possède un développement en puissances croissantes de x identique à celui de f(x) jusqu'au terme de degré k compris. On dit que la fraction continue (19) est la fraction continue correspondante à la série (21).

En effectuant une contraction de la fraction continue correspondante, par la méthode exposée au paragraphe précédent, on obtient une fraction continue dont le développement du $k^{\text{ième}}$ convergent en puissances croissantes de x est analogue à celui de f(x) jusqu'au terme de degré 2k compris ; cette fraction continue s'appelle la fraction continue associée à la série (21).
Nous verrons plus loin comment l'on obtient les nombres a_k à partir des coefficients de la série.
Examinons maintenant la connexion entre les fractions continues associée et correspondante et la table de Padé.

Connexion avec l' ε-algorithme

D'après ce que l'on vient de voir, $C_{2k}^{(0)}$ est le rapport de deux polynômes de degré k en x et qui, de plus, possède la propriété :

$$C_{2k}^{(0)} - f(x) = 0(x^{2k+1})$$

Cette propriété n'est autre que la propriété fondamentale de l'approximant de Padé $[k/k]$. Par conséquent, d'après la propriété d'unicité de ceux-ci et la connexion entre l' ε-algorithme et la table de Padé, on a :

$$C_{2k}^{(0)} = [k/k] = \varepsilon_{2k}^{(0)}$$

L'approximant $C_{2k}^{(0)}$ de la fraction continue correspondante à la série est égal à l'approximant $D_{k}^{(0)}$ de la fraction continue associée.

De même $C_{2k+1}^{(0)}$ est le rapport d'un polynôme de degré k+1 en x sur un polynôme de degré k et qui vérifie :

$$C_{2k+1}^{(0)} - f(x) = 0(x^{2k+2})$$

On a donc également :

$$C_{2k+1}^{(0)} = [k+1/k] = \varepsilon_{2k}^{(1)}$$

Considérons maintenant la série (21) dans laquelle on a groupé les n+1 premiers termes. On peut écrire :

$$f(x) = (c_0 + c_1 x + \ldots + c_n x^n) + x^n(c_{n+1} x + c_{n+2} x^2 + \ldots) \quad (22)$$

Considérons également la fraction continue $C^{(n)}$ correspondante à cette série. Soient $C_k^{(n)}$ ses approximants successifs. $A_{2k-1}^{(n)}$ et $A_{2k}^{(n)}$ sont des polynômes de degré n+k en x, $B_{2k}^{(n)}$ est de degré k et $B_{2k-1}^{(n)}$ est de degré k-1.

Si l'on effectue la division de $A_{2k}^{(n)}$ par $B_{2k}^{(n)}$ suivant les puissances croissantes de x on retrouve les k+1 premiers termes de (22) c'est-à-dire que l'on a , puisque x^n est en facteur :

$$C_{2k}^{(n)} - f(x) = O(x^{n+2k+1})$$

On a donc par conséquent :

$$C_{2k}^{(n)} = [n+k/k] = \varepsilon_{2k}^{(n)} = D_k^{(n)}$$

et de même on aurait :

$$C_{2k+1}^{(n)} = [n+k+1/k] = \varepsilon_{2k}^{(n+1)}$$

On voit que ces différentes fractions continues correspondantes sont reliées par la relations :

$$C_{2k}^{(n+1)} = C_{2k+1}^{(n)} \qquad n,k=0,1,\ldots$$

L' ε-algorithme apparait donc ainsi comme une méthode pour transformer les sommes partielles d'une série en les approximants successifs de diverses fractions continues correspondantes et associées à la série.

Ces diverses fractions continues correspondantes se distinguent par le choix du premier terme (celui correspondant à b_0 dans (19)). Appelons $C^{(n)}$ la fraction continue correspondante à la série (22) de premier terme $c_0 + c_1 x + \ldots + c_n x^n$ et écrivons $C^{(n)}$ sous la forme suivante qui sera plus adaptée à la suite de l'exposé :

$$C^{(n)} = c_0+c_1 x + \ldots + c_n x^n + \frac{c_{n+1}x^{n+1}|}{|1} - \frac{q_1^{(n+1)}x|}{|1}$$

$$- \frac{e_1^{(n+1)}x|}{|1} - \ldots - \frac{e_{k-1}^{(n+1)}x|}{|1} - \frac{q_k^{(n+1)}x|}{|1} - \frac{e_k^{(n+1)}x|}{|1} - \ldots \qquad (23)$$

Calcul des éléments de la fraction continue correspondante

Nous allons maintenant donner les expressions des éléments $e_k^{(n)}$ et $q_k^{(n)}$ de ces diverses fractions continues correspondantes. Nous verrons au paragraphe 4.3 un algorithme récursif pour les calculer.

Puisque les coefficients des polynômes $A_{2k}^{(n)}$ et $B_{2k}^{(n)}$ ne sont que des intermédiaires de calcul nous écrirons, pour simplifier les notations :

$$B_{2k}^{(n)} = b_0 + b_1 x + \ldots + b_k x^k$$

On a vu, dans l'étude de la table de Padé, que les coefficients b_i sont solutions du système :

$$c_{n+1}b_k + c_{n+2}b_{k-1} + \ldots + c_{n+k}\, b_1 + c_{n+k+1}\, b_0 = 0$$

- -

$$c_{n+k}b_k + c_{n+k+1}\, b_{k-1} + \ldots + c_{n+2k-1}b_1 + c_{n+2k}\, b_0 = 0$$

avec $b_0 = 1$.

On a donc :

$$b_k = (-1)^k \frac{H_k^{(n+2)}(c_{n+2})}{H_k^{(n+1)}(c_{n+1})}$$

D'autre part, en utilisant les relations de récurrence du théorème 140 et les notations de (23) on voit facilement que :

$$b_k = (-1)^k q_1^{(n+1)} q_2^{(n+1)} \ldots q_k^{(n+1)}$$

d'où finalement :

$$q_1^{(n+1)} \ldots q_k^{(n+1)} = \frac{H_k^{(n+2)}(c_{n+2})}{H_k^{(n+1)}(c_{n+1})}$$

et par conséquent :

$$q_k^{(n)} = \frac{H_k^{(n+1)}(c_{n+1})\, H_{k-1}^{(n)}(c_n)}{H_k^{(n)}(c_n)\, H_{k-1}^{(n+1)}(c_{n+1})} \tag{24}$$

Les quantités $e_k^{(n)}$ sont déterminées de façon tout à fait analogue. On trouve que

$$e_k^{(n)} = \frac{H_{k+1}^{(n)}(c_n)\, H_{k-1}^{(n+1)}(c_{n+1})}{H_k^{(n)}(c_n)\, H_k^{(n+1)}(c_{n+1})} \tag{25}$$

L'algorithme Q D

Il existe, comme c'est le cas pour l'ε-algorithme, un algorithme récursif qui évite le calcul effectif des déterminants de Hankel qui interviennent dans les expressions de $q_k^{(n)}$ et de $e_k^{(n)}$: c'est l'algorithme QD de Rutishauser [109,110,164].

$$q_1^{(n)} = c_{n+1}/c_n \qquad e_0^{(n)} = 0 \qquad n = 0,1,\ldots$$

$$q_{k+1}^{(n)}\, e_k^{(n)} = q_k^{(n+1)}\, e_k^{(n+1)} \qquad k = 1,2,\ldots \quad n = 0,1,\ldots \tag{26}$$

$$q_k^{(n)} + e_k^{(n)} = q_k^{(n+1)} + e_{k-1}^{(n+1)}$$

En utilisant (24) et (25) on voit immédiatement que la première des relations (26) est satisfaite. La seconde des relations (26) se démontre de façon analogue à partir de (24), de (25) et de la relation de récurrence entre déterminants de Hankel.

Les nombres $e_k^{(n)}$ et $q_k^{(n)}$ sont placés dans un tableau à double entrée :

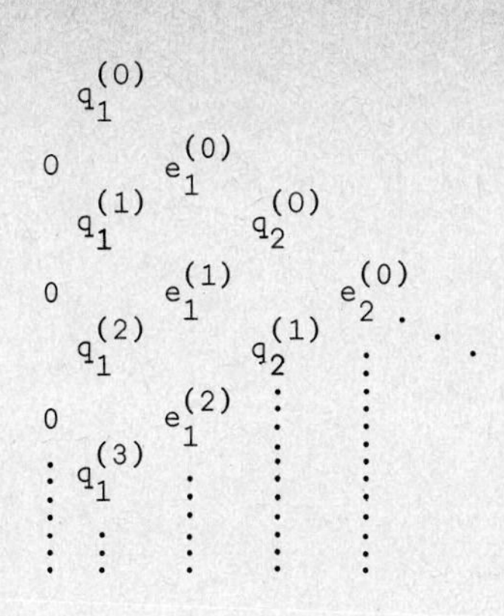

Ils sont calculés à l'aide des relations (26) en allant de gauche et de haut en bas dans ce tableau, à partir des valeurs initiales $e_0^{(n)}$ et $q_1^{(n)}$ $\forall n$.

Il est facile de voir les règles de l'algorithme QD peuvent aussi se déduire de la relation $C_{2k}^{(n+1)} = C_{2k+1}^{(n)}$

Liaison entre l'algorithme QD et l'ε-algorithme

Nous allons, dans ce paragraphe, relier les quantités $q_k^{(n)}$ et $e_k^{(n)}$ calculées à l'aide des relations (26) en partant des conditions initiales $q_1^{(n)} = c_{n+1}/c_n$ aux quantités $\varepsilon_{2k}^{(n)}$ obtenues avec l' ε-algorithme et les conditions initiales $\varepsilon_0^{(n)} = \sum_{i=0}^{n} c_i x^i$.

Utilisons les relations (11) et le fait que les coefficients de la fraction continue sont donnés par :

$$a_1 = c_{n+1} x^{n+1}$$

$$\left.\begin{aligned} a_{2k} &= - q_k^{(n+1)} x \\ a_{2k+1} &= - e_k^{(n+1)} x \end{aligned}\right\} \quad k = 1,2,\ldots$$

On obtient alors :

$$- q_{k+1}^{(n+1)} x = \frac{B_{2k+2}^{(n)}}{B_{2k}^{(n)}} \quad \frac{C_{2k+1}^{(n)} - C_{2k+2}^{(n)}}{C_{2k+1}^{(n)} - C_{2k}^{(n)}}$$

$$- e_{k+1}^{(n+1)} x = \frac{B_{2k+3}^{(n)}}{B_{2k+1}^{(n)}} \quad \frac{C_{2k+2}^{(n)} - C_{2k+3}^{(n)}}{C_{2k+2}^{(n)} - C_{2k+1}^{(n)}}$$

or $\quad C_{2k}^{(n)} = \varepsilon_{2k}^{(n)}$ et $C_{2k+1}^{(n)} = \varepsilon_{2k}^{(n+1)}$; d'où :

$$q_{k+1}^{(n+1)} x = \frac{B_{2k+2}^{(n)}}{B_{2k}^{(n)}} \quad \frac{\varepsilon_{2k+2}^{(n)} - \varepsilon_{2k}^{(n+1)}}{\varepsilon_{2k}^{(n+1)} - \varepsilon_{2k}^{(n)}} .$$

$$e_{k+1}^{(n+1)} x = \frac{B_{2k+3}^{(n)}}{B_{2k+1}^{(n)}} \quad \frac{\varepsilon_{2k+2}^{(n+1)} - \varepsilon_{2k+2}^{(n)}}{\varepsilon_{2k+2}^{(n)} - \varepsilon_{2k}^{(n+1)}}$$

Utilisons de nouveau les relations (11) et le fait que tous les dénominateurs partiels sont égaux à un. On trouve que :

$$B_{2k+1}^{(n)} \left[C_{2k+1}^{(n)} - C_{2k}^{(n)} \right] = B_{2k+2}^{(n)} \left[C_{2k+2}^{(n)} - C_{2k}^{(n)} \right]$$

$$B_{2k+2}^{(n)} \left[C_{2k+2}^{(n)} - C_{2k+1}^{(n)} \right] = B_{2k+3}^{(n)} \left[C_{2k+3}^{(n)} - C_{2k+1}^{(n)} \right]$$

ou encore :

$$B_{2k+1}^{(n)} \left[\varepsilon_{2k}^{(n+1)} - \varepsilon_{2k}^{(n)} \right] = B_{2k+2}^{(n)} \left[\varepsilon_{2k+2}^{(n)} - \varepsilon_{2k}^{(n)} \right]$$

$$B_{2k+2}^{(n)} \left[\varepsilon_{2k+2}^{(n)} - \varepsilon_{2k}^{(n+1)} \right] = B_{2k+3}^{(n)} \left[\varepsilon_{2k+2}^{(n+1)} - \varepsilon_{2k}^{(n+1)} \right]$$

Par conséquent on obtient :

$$\frac{B_{2k+3}^{(n)}}{B_{2k+1}^{(n)}} = \frac{\varepsilon_{2k+2}^{(n)} - \varepsilon_{2k}^{(n+1)}}{\varepsilon_{2k+2}^{(n+1)} - \varepsilon_{2k}^{(n+1)}} \; \frac{\varepsilon_{2k}^{(n+1)} - \varepsilon_{2k}^{(n)}}{\varepsilon_{2k+2}^{(n)} - \varepsilon_{2k}^{(n)}}$$

et de même :

$$\frac{B_{2k+2}^{(n)}}{B_{2k}^{(n)}} = \frac{\varepsilon_{2k}^{(n+1)} - \varepsilon_{2k}^{(n)}}{\varepsilon_{2k+2}^{(n)} - \varepsilon_{2k}^{(n)}} \; \frac{\varepsilon_{2k}^{(n)} - \varepsilon_{2k-2}^{(n+1)}}{\varepsilon_{2k}^{(n+1)} - \varepsilon_{2k-2}^{(n+1)}}$$

d'où finalement :

$$e_{k+1}^{(n+1)} x = \frac{\varepsilon_{2k}^{(n+1)} - \varepsilon_{2k}^{(n)}}{\varepsilon_{2k+2}^{(n)} - \varepsilon_{2k}^{(n)}} \; \frac{\varepsilon_{2k+2}^{(n+1)} - \varepsilon_{2k+2}^{(n)}}{\varepsilon_{2k+2}^{(n+1)} - \varepsilon_{2k}^{(n+1)}} \qquad k = 0,1,\ldots$$

$$q_{k+1}^{(n+1)} x = \frac{\varepsilon_{2k}^{(n)} - \varepsilon_{2k-2}^{(n+1)}}{\varepsilon_{2k}^{(n+1)} - \varepsilon_{2k-2}^{(n+1)}} \; \frac{\varepsilon_{2k+2}^{(n)} - \varepsilon_{2k}^{(n+1)}}{\varepsilon_{2k+2}^{(n)} - \varepsilon_{2k}^{(n)}} \qquad k = 1,2,\ldots$$

On vérifiera que les relations (26) sont satisfaites. On voit qu'en faisant des rapports de telles quantités on obtient des constantes indépendantes de x ; par exemple $e_{k+1}^{(n+1)} / q_{k+1}^{(n+1)}$ est indépendant de x.

Un tel invariant a également été obtenu par Baker [8] qui démontre que le rapport suivant est indépendant de x :

$$\frac{\left[\varepsilon_{2k}^{(n+1)} - \varepsilon_{2k}^{(n)}\right] \left[\varepsilon_{2k+2}^{(n+1)} - \varepsilon_{2k+2}^{(n)}\right]}{\left[\varepsilon_{2k}^{(n)} - \varepsilon_{2k+2}^{(n+1)}\right] \left[\varepsilon_{2k}^{(n+1)} - \varepsilon_{2k+2}^{(n)}\right]}$$

Nous venons donc de voir comment calculer les quantités $e_k^{(n)}$ et $q_k^{(n)}$ à partir des $\varepsilon_{2k}^{(n)}$.

On sait également calculer les $\varepsilon_{2k}^{(n)}$ à partir des $e_k^{(n)}$ et des $q_k^{(n)}$ en écrivant que $\varepsilon_{2k}^{(n)} = C_{2k}^{(n)}$.

D'autre part comme pour l' ε-algorithme les relations (26) relient des quantités situées aux quatre sommets d'un losange.

Mais la liaison qui existe entre l' ε-algorithme et l'algorithme QD ne s'arrête pas là.Bauer [15,16,17] a mis en évidence un algorithme qui semble être l'algorithme de base ainsi que l'algorithme qui leur sert de lien.

VII-5 - Les algorithmes de losange

Il est bien évident, d'après ce qui précède, qu'il existe une connexion entre l' ε-algorithme de Wynn et l'algorithme QD de Rutishauser. Cette connexion a été trouvée par Bauer ; elle est basée sur la décomposition g d'une fraction continue [17]. Le lien qui existe entre ces deux algorithmes est le η-algorithme [16] . Cet algorithme peut être utilisé à la place de l'ε-algorithme dans toutes les applications ; son avantage est qu'il semble être plus stable numériquement dans certains cas [93]. Tous ces algorithmes relient des quantités situées aux quatre sommets d'un losange ; c'est ce qui leur donne leur nom.

La décomposition g

posons :

$$q_k^{(n)} = g_{2k-2}^{(n)} (C - g_{2k-1}^{(n)})$$

$$e_k^{(n)} = g_{2k-1}^{(n)} (1 - g_{2k}^{(n)})$$

où C est un nombre arbitraire.

Portons ces expressions dans les règles de l'algorithme QD. On voit facilement que ces relations (26) sont satisfaites si les quantités $g_k^{(n)}$ vérifient :

$$g_{2k-1}^{(n)} \, g_{2k}^{(n)} = g_{2k-2}^{(n+1)} \, g_{2k-1}^{(n+1)}$$

$$(1 - g_{2k}^{(n)})(C-g_{2k+1}^{(n)}) = (C - g_{2k-1}^{(n+1)})(1-g_{2k}^{(n+1)}) \qquad (27)$$

pour $k = 1,2,\ldots$ et $n = 0,1,\ldots$ .Ces relations (27) sont initialisées avec les conditions :

$$g_0^{(n)} = 1 \qquad g_1^{(n)} = C - c_{n+1}/c_n \qquad \text{pour } n = 0,1,\ldots$$

Il est évident que la fraction continue correspondante (23) peut être exprimée à l'aide de ces quantités $g_k^{(n)}$ au lieu des quantités $e_k^{(n)}$ et $q_k^{(n)}$.
On obtient alors ce qu'on appelle la décomposition g de la fraction continue.

le η-algorithme

Posons maintenant :

$$\eta_k^{(n)} = c_n \prod_{i=1}^{k} r_i^{(n)} \qquad \text{avec} \quad r_{2i}^{(n)} = \frac{1-g_{2i}^{(n)}}{g_{2i}^{(n)}}$$

$$\text{et} \qquad r_{2i+1}^{(n)} = \frac{C-g_{2i+1}^{(n)}}{g_{2i+1}^{(n)}}$$

On vérifiera facilement que :

$$1 + r_{2k}^{(n)} = 1/g_{2k}^{(n)} \qquad 1+r_{2k+1}^{(n)} = C/g_{2k+1}^{(n)}$$

$$1+1/r_{2k}^{(n)} = 1/(1-g_{2k}^{(n)}) \qquad 1+1/r_{2k+1}^{(n)} = C/(C-g_{2k+1}^{(n)})$$

En remplaçant dans les relations (27) du g-algorithme on trouve que les quantités $r_k^{(n)}$ satisfont :

$$(1+r_{2k-1}^{(n)})(1+r_{2k}^{(n)}) = (1+r_{2k-2}^{(n+1)})(1+r_{2k-1}^{(n+1)})$$

$$(1+1/r_{2k}^{(n)})(1+1/r_{2k+1}^{(n)}) = (1+1/r_{2k-1}^{(n+1)})(1+1/r_{2k}^{(n+1)}) \qquad (28)$$

avec les conditions initiales :

$$r_0^{(n)} = 0 \qquad \text{et} \quad r_1^{(n)} = \frac{c_{n+1}}{c\, c_n - c_{n+1}}$$

L'établissement des règles de l' η-algorithme est un peu plus compliqué. On a :

$$\eta_{2k-1}^{(n)} + \eta_{2k}^{(n)} = c_n \, (1+r_{2k}^{(n)}) \prod_{i=1}^{2k-1} r_i^{(n)}$$

$$\eta_{2k-2}^{(n+1)} + \eta_{2k-1}^{(n+1)} = c_{n+1} \, (1+r_{2k-1}^{(n)}) \prod_{i=1}^{2k-2} r_i^{(n+1)}$$

On a également :

$$c\, c_n = c_{n+1}(1+1/r_1^{(n)})$$

et l'on peut écrire :

$$r_k^{(n)} = (1+r_k^{(n)}) \frac{1}{1+1/r_k^{(n)}}$$

En portant dans les relations (28) on obtient les règles de l'η-algorithme :

$$C(\eta_{2k-1}^{(n)} + \eta_{2k}^{(n)}) = \eta_{2k-2}^{(n+1)} + \eta_{2k-1}^{(n+1)}$$

$$\frac{1}{C}\left(\frac{1}{\eta_{2k}^{(n)}} + \frac{1}{\eta_{2k+1}^{(n)}}\right) = \frac{1}{\eta_{2k-1}^{(n+1)}} + \frac{1}{\eta_{2k}^{(n+1)}}$$

avec les conditions initiales :

$$\frac{1}{\eta_{-1}^{(n)}} = 0 \qquad \text{et} \qquad \eta_0^{(n)} = c_n$$

On montre également que le η-algorithme et l'algorithme QD sont reliés par :

$$\frac{C}{q_k^{(n)}} = (1 + \eta_{2k-2}^{(n)} / \eta_{2k-3}^{(n)})(1 + \eta_{2k-2}^{(n)} / \eta_{2k-1}^{(n)})$$

$$\frac{C}{e_k^{(n)}} = (1 + \eta_{2k-1}^{(n)} / \eta_{2k-2}^{(n)})(1 + \eta_{2k-1}^{(n)} / \eta_{2k}^{(n)})$$

L' ε-algorithme

Posons maintenant :

$$\varepsilon_{2k}^{(n)} = \sum_{i=0}^{n} \frac{\eta_0^{(i)}}{C^i} + \frac{1}{C^{n+1}} \sum_{i=0}^{2k-1} \eta_i^{(n+1)}$$

$$\varepsilon_{2k+1}^{(n)} = C^{n+1} \sum_{i=0}^{2k} \frac{1}{\eta_i^{(n+1)}}$$

On montre, en utilisant la premières des règles (29) de l' η-algorithme, que l'on a :

$$\varepsilon_{2k}^{(n+1)} - \varepsilon_{2k}^{(n)} = \eta_{2k}^{(n+1)} / C^{n+1}$$

$$\varepsilon_{2k+2}^{(n)} - \varepsilon_{2k}^{(n+1)} = \eta_{2k+1}^{(n+1)} / C^{n+1}$$

Avec la seconde des relations (29) on trouve de même que :

$$\varepsilon_{2k+1}^{(n+1)} - \varepsilon_{2k+1}^{(n)} = C^{n+1} / \eta_{2k+1}^{(n+1)}$$

$$\varepsilon_{2k+1}^{(n)} - \varepsilon_{2k-1}^{(n+1)} = C^{n+1} / \eta_{2k}^{(n+1)}$$

On obtient par conséquent :

$$(\varepsilon_k^{(n+1)} - \varepsilon_k^{(n)})(\varepsilon_{k+1}^{(n)} - \varepsilon_{k-1}^{(n+1)}) = 1$$

qui n'est autre que la règle habituelle de l' ε-algorithme. On voit que cette règle est indépendante de C. Les conditions initiales sont obtenues directement à partir de celles de l' η-algorithme :

$$\varepsilon_{-1}^{(n)} = 0 \qquad \varepsilon_0^{(n)} = \sum_{i=0}^{n} \frac{c_i}{C^i} = \sum_{i=0}^{n} c_i\, x^i$$

en posant $x = 1/C$.

On retrouve donc ainsi les résultats classiques de l' ε-algorithme.

VII-6 - Quelques résultats de convergence

L'importance des fractions continues dans la théorie de l'ε-algorithme et de la table de Padé se manifeste surtout dans les résultats de convergence ; en effet la convergence des diagonales du tableau ε et de la moitié supérieure de la table de Padé revient à la convergence de la fraction continue associée à la série.

Nous ne donnerons ici que quelques résultats ; pour des compléments on pourra se reporter à [122,123,143,152,185,197,220].

Convergence des fractions continues correspondantes

Considérons une série :

$$c_0 + c_1 x \text{ et } c_2 x^2 + \ldots$$

et sa fraction continue correspondante :

$$\frac{1|}{|1} - \frac{a_2 x|}{|1} - \frac{a_3 x|}{|1} - \ldots$$

Le premier problème qui se pose est le suivant : si la série et la fraction continue convergent ont-elles mêmes limites ? La réponse est donnée par un théorème dû à Van Vleck [183] :

Théorème 143: Si la fraction continue correspondante converge uniformément pour $|x| \leq M$ alors la série a un rayon de convergence au moins égal à M et sa somme est égale à la valeur de la fraction continue correspondante.

Démonstration : Soit $C_k(x)$ le $k^{\text{ième}}$ approximant de la fraction continue correspondante à la série.

Si la suite $\{C_k(x)\}$ converge uniformément pour $x \leq M$ alors il existe K tel que $\forall k > K$ le développement de $C_k(x)$ suivant les puissances croissantes de x converge pour $|x| \leq M$. Posons :

$$u_1(x) = C_k(x)$$

$$u_i(x) = C_{k+i-1}(x) - C_{k+i-2}(x) \qquad i = 2,3,\ldots$$

On a :

$$\sum_{i=1}^{\infty} u_i(x) = \lim_{k\to\infty} C_k(x) = u(x)$$

uniformément pour $x \leq M$, ou u est une fonction analytique pour $|x| < M$.

D'après le théorème de Weierstrass la série des dérivées $n^{\text{ièmes}}$ converge vers $u^{(n)}(x)$ pour $x < M$:

$$\sum_{i=1}^{\infty} u_i^{(n)}(x) = u^{(n)}(x)$$

Puisque le développement de $C_k(x)$ en puissances croissantes de x est identique à la série depuis le premier terme jusqu'au $k^{\text{ième}}$ terme inclus, on a : en faisant tendre k vers l'infini.

$$\lim_{k\to\infty} C_k(0) = u^{(n)}(0) = \sum_{i=0}^{\infty} u_i^{(n)}(0) = n!\, c_n \qquad n=0,1,\ldots$$

d'où pour $|x| < M$:

$$u(x) = \sum_{i=0}^{\infty} \frac{u^{(i)}(0)}{i!} x^i = \sum_{i=0}^{\infty} c_i x^i$$

ce qui démontre le théorème.

Donnons maintenant un résultat de convergence uniforme pour la fraction continue correspondante [185]:

<u>Théorème 144</u> : Si $|a_k| \leq M$ pour $k = 2,3,\ldots$ alors la fraction continue correspondante converge uniformément pour $|x| \leq 1/4M$.

Démonstration : d'après le théorème 142 on a :

$$\sum_{k=1}^{\infty} \left(\frac{A_k}{B_k} - \frac{A_{k-1}}{B_{k-1}}\right) = 1 + \sum_{k=1}^{\infty} \rho_1 \rho_2 \ldots \rho_k \tag{30}$$

avec $$\rho_k = \frac{a_{k+1} \times B_{k-1}}{B_{k+1}}$$

Démontrons maintenant le résultat intermédiaire suivant :

S'il existe des nombres $r_n \geq 0$ tels que :

$$r_n|1-a_n x-a_{n+1}x| \geq r_n r_{n-2}|a_n x|+|a_{n+1}x| \quad n = 1,2,... \tag{31}$$

avec $a_1 = 0$, $r_0 = r_{-1} = 0$ alors tous les dénominateurs partiels B_k de la fraction continue correspondante sont différents de zéro et, de plus, $|\rho_k| \leq r_k$ pour $k = 1,2,...$

Pour $k = 1$ et 2 on a :

$$r_1|1-a_2x| \geq |a_2x|$$

$$r_2|1-a_2x-a_3x| \geq |a_3x|$$

Par conséquent :

$$B_2 = 1-a_2x \neq 0$$

$$B_3 = 1-a_2x-a_3x \neq 0$$

et

$$|\rho_1| = \left|\frac{a_2x}{1-a_2x}\right| \leq r_1$$

$$|\rho_2| = \left|\frac{a_3x}{1-a_2x-a_3x}\right| \leq r_2$$

Démontrons, par récurrence, que cela est vrai pour tout k.

Il faut distinguer deux cas : si $a_{k+2} = 0$ alors $B_{k+2} = B_{k+1}$ et :

$$|\rho_{k+1}| = \left|\frac{a_{k+2} \times B_k}{B_{k+2}}\right| = 0 \leq r_{k+1}$$

Si $a_{k+2} \neq 0$ en prenant $n = k+1$ dans (31) on voit que $r_{k+1} > 0$. De plus :

$$B_{k+2} = (1-a_{k+1}\, x - a_{k+2}\, x)\, B_k - a_k\, a_{k+1}\, x^2\, B_{k-2}$$

ce qui entraine que :

$$\left|\frac{B_{k+2}}{a_{k+2} x\, B_k}\right| = \left|\frac{1-a_{k+1}\, x - a_{k+2}\, x}{a_{k+2}\, x} - \frac{a_{k+1}}{a_{k+2}}\ \frac{a_k\, x\, B_{k-2}}{B_k}\right|$$

$$\geqslant \left|\frac{1 - a_{k+1}\, x - a_{k+2}\, x}{a_{k+2}\, x}\right| - \left|\frac{a_{k+1}}{a_{k+2}}\right|\ r_{k-1} \geqslant \frac{1}{r_{k+1}} > 0$$

Par conséquent $B_{k+2} \neq 0$ et $|\rho_{k+1}| \leqslant r_{k+1}$; ce qui termine la démonstration de ce résultat intermédiaire.

On vérifiera facilement que si $|a_k x| \leqslant \frac{1}{4}$ alors l'inégalité (31) est satisfaite avec $r_k = k/(k+2)$ pour $k = 1,2,\ldots$

De plus, puisque $|\rho_k| \leqslant r_k$, $1 + \sum_{k=1}^{\infty} r_1\, r_2 \ldots r_k$ est un majorant de la série (30). On a :

$$1 + \sum_{k=1}^{\infty} r_1\, r_2 \ldots r_k = 1 + \sum_{k=1}^{\infty} \frac{2}{(k+1)(k+2)} = 2$$

Par conséquent la série (30) converge uniformément pour $|a_k x| \leqslant 1/4$ $k = 2,3,\ldots$ et son module est inférieur ou égal à deux ; ceci démontre donc la convergence uniforme de la fraction continue correspondante sous les conditions du théorème puisque la série (30) converge et est égale à la valeur de la fraction continue.

Les fractions continues correspondantes aux séries de Stieltjes

Soit la série formelle :

$$f(x) = \sum_{i=0}^{\infty} (-1)^i c_i x^i$$

On dit que cette série est une série de Stieltjes [171] si :

$$c_i = \int_0^{\infty} t^i \, dg(t) \qquad i=0,1,... \tag{32}$$

où g est une fonction bornée non décroissante dans $[0,+\infty)$.

Si la fonction g est donnée et si toutes les intégrales (32) existent alors les nombres c_i sont déterminés de façon unique : ce sont les moments de la fonction g. Réciproquement si la suite $\{c_n\}$ est donnée le problème de la construction de la fonction g s'appelle le problème des moments de Stieltjes. On trouvera dans Widder [188] des conditions pour que la solution de ce problème existe. Une de ces conditions, qui a été démontrée par Carleman [52], est que :

$$H_k^{(0)}(c_0) > 0 \quad \text{et} \quad H_k^{(1)}(c_1) > 0 \qquad \text{pour } k = 0,1,...$$

Considérons la fonction F définie par :

$$F(x) = \int_0^{\infty} \frac{dg(t)}{1 + xt}$$

et supposons que la série $f(x)$ converge pour $|x| < R$.

Alors pour $|x| < R$ $f(x) = F(x)$; en effet on a :

$$f(x) = \sum_{i=0}^{\infty} (-1)^i x^i \int_0^{\infty} t^i \, dg(t)$$

$f(x)$ étant uniformément convergente dans $|x| < R$ on peut intervertir l'intégration et la sommation, d'où :

$$f(x) = \int_0^{\infty} \left\{ \sum_{i=0}^{\infty} (-1)^i (xt)^i \right\} dg(t)$$

et par conséquent :

$$f(x) = \int_0^\infty \frac{dg(t)}{1 + xt} = F(x)$$

Pour $|x| > R$ alors $F(x)$ est le prolongement analytique de la série $f(x)$. On dira $f(x)$ est le développement formel de $F(x)$.

La fonction F ainsi définie est analytique dans tout domaine ouvert borné du plan complexe ne contenant aucun point du demi axe réel $(-\infty,0]$.

Sur cette question on pourra consulter [152]. Une discussion complète est également donnée dans [211] et [218].

Considérons maintenant le cas où la fonction g est constante pour $x > b$. On a alors :

$$c_i = \int_0^b t^i \, dg(t) \qquad i = 0,1,\ldots$$

et
$$F(x) = \int_0^b \frac{dg(t)}{1 + xt}$$

L'étude de la convergence de la fraction continue correspondante à $f(x)$ a été faite par Markov [137] dans le cas ou $0< b< \infty$ et où g est une fonction bornée non décroissante sur $[0,b]$. Le résultat de Markov est le suivant :

Théorème 145 La fraction continue correspondante à la série $f(x) = \sum_{i=0}^{\infty} (-1)^i c_i x^i$ avec :

$$c_i = \int_0^b t^i \, dg(t) \qquad i = 0,1,\ldots \qquad 0<b<\infty$$

où g est une fonction bornée non décroissante sur $[0,b]$, converge uniformément vers $F(x) = \int_0^b dg(t)/(1+xt)$ pour tout x appartenant à un ouvert borné du plan complexe ne contenant aucun point de $(-\infty, -b^{-1}]$

remarque : $F(x)$ est analytique dans $|x| < b^{-1}$ et par conséquent $f(x)$ converge pour $|x| < b^{-1}$.

Si g est telle que $g(b-0) \neq g(b)$ alors Wynn [211] a démontré la convergence de $f(x)$ pour $x = b^{-1}$ ainsi que la convergence uniforme de toutes les diagonales du tableau de l' ε-algorithme vers $F(x)$ pour tout x appartenant à un ouvert borné du plan complexe ne contenant aucun point de $(-\infty, -b^{-1}]$.

Dans le même article Wynn a également étudié la convergence des colonnes du tableau de l' ε-algorithme.

Si la suite des coefficients $\{c_n\}$ est une suite totalement monotone alors on est dans un cas particulier d'application du théorème de Markov et du théorème de Wynn qui en découle. On sait en effet que l'on a :

$$c_i = \int_0^1 t^i \, dg(t) \qquad i = 0, 1, \ldots$$

où g est une fonction bornée non décroissante sur $[0,1]$.

Si $0 \leqslant x \leqslant 1$ alors la suite $\{x^n\}$ est totalement monotone. Le produit terme à terme de deux suites totalement monotones étant une suite totalement monotone (voir par exemple [185]) il en résulte que $\{c_n x^n\}$ est totalement monotone et que la suite $\{u_n = S_n - S\}$ avec $S_n = \sum_{i=0}^{n} (-1)^i c_i x^i$ est totalement oscillante. On peut donc utiliser également les résultats de III-6 pour démontrer la convergence des colonnes du tableau de l'ε-algorithme vers $F(x)$.

Si maintenant $-1 < x \leqslant 0$ alors $\{x^n\}$ et $\{c_n x^n\}$ sont des suites totalement oscillantes. Dans ce cas $\{u_n = S - S_n\}$ est totalement monotone et les résultats de III-6 démontrent la convergence des diagonales et des colonnes du tableau de l' ε-algorithme vers $F(x)$.

Il faut remarquer que les résultats de III-6 ne sont que des résultats de convergence ponctuelle et non pas de convergence uniforme.

Pour les séries de Stieltjes dont les coefficients forment une suite totalement monotone, on peut, grâce à certaines inégalités sur les déterminants de Hankel démontrées par Wynn [234] , obtenir de nombreuses inégalités entre approximants de Padé pour $x \in]-R, R[$. La majorité de ces inégalités peut être démontrées uniquement à l'aide de propriétés algébriques des approximants de Padé [28].

Wynn [237] a montré sur des exemples numériques que si $\{c_n\}$ est une suite totalement monotone alors l' ε-algorithme est un procédé très puissant d'accélération de la convergence. Considérons, par exemple, la série formelle :

$$f(x) = \sum_{i=0}^{\infty} (-1)^i \frac{x^i}{i+1}$$

on a donc $$c_i = \frac{1}{i+1} = \int_0^1 t^i \, dt$$

$$\text{et} \quad F(x) = \int_0^1 \frac{dt}{1+xt} = \frac{1}{x} \, \text{Log}(1+x)$$

On est par conséquent dans les conditions d'applications des théorèmes de Markov et de Wynn. On sait que la série converge pour $x \in]-1,1]$. Appliquons l' ε-algorithme aux sommes partielles de la série $x\,f(x)$ pour $x = 1$.

On obtient :

$$\varepsilon_0^{(0)} = 1$$

$$\varepsilon_2^{(0)} = 0,7$$

$$\varepsilon_4^{(0)} = 0,6933... \qquad \text{Log } 2 = 0,6931471805...$$

$$\varepsilon_6^{(0)} = 0,69315...$$

$$\varepsilon_8^{(0)} = 0,6931473...$$

$$\varepsilon_{10}^{(0)} = 0,69314718...$$

alors que $S_{10} = 0,7365...$

Lorsque $x=2$ la série $f(x)$ diverge et la fraction continue correspondante doit cependant converger vers $\text{Log}(1+x)$ qui est le prolongement analytique de la série. Dans ce cas on obtient :

$$\varepsilon_0^{(0)} = 2$$

$$\varepsilon_0^{(0)} = 1,14...$$

$$\varepsilon_4^{(0)} = 1,101...$$

$$\varepsilon_6^{(0)} = 1,0988.... \qquad\qquad \text{Log } 3 = 1,0986122886681...$$

$$\varepsilon_8^{(0)} = 1,098625...$$

$$\varepsilon_{10}^{(0)} = 1,098613...$$

$$\varepsilon_{20}^{(0)} = 1,0986122886698...$$

alors que $S_{10} = 121,35...$ et $S_{20} = 65504,6...$

remarque 1 : le cas où $\{c_n\}$ est une suite totalement monotone et où l'on considère la série $f(x) = \sum_{i=0}^{\infty} (-1)^i \, c_i \, x^i$ pour $0 \leqslant x \leqslant 1$ est analogue à celui où l'on considère la série $v(x) = \sum_{i=0}^{\infty} c_i \, x^i$ pour $-1 \leqslant x \leqslant 0$. Inversement le cas de $f(x)$ pour $-1 \leqslant x \leqslant 0$ est identique au cas de $v(x)$ pour $0 \leqslant x \leqslant 1$. On peut également relier aux deux cas précédents tous les cas où la suite $\{c_n\}$ est totalement oscillante.

remarque 2 : sur l'exemple précédent on pourra consulter [92].

VII-7 - Les fractions continues d'interpolation

Soit f une fonction dont on connait la valeur aux abscisses distinctes $x_0, x_1, ...$ On sait que l'on peut définir les différences réciproques de f par [142] :

$$\rho_0(x_k) = f(x_k) \qquad k = 0, 1, \ldots$$

$$\rho_1(x_0,x_1) = \frac{x_0 - x_1}{\rho_0(x_0) - \rho_0(x_1)}$$

$$\rho_2(x_0,x_1,x_2) = \frac{x_0 - x_2}{\rho_1(x_0,x_1) - \rho_1(x_1,x_2)} + \rho_0(x_1) \qquad \text{etc...}$$

Ces différences réciproques ne sont autres que les quantités $\rho_k^{(0)}$ rencontrées dans le ρ-algorithme. En remplaçant x_0 par x_n, x_1 par $x_{n+1}, \ldots$ on obtiendrait de même les quantités notées $\rho_k^{(n)}$.

La formule d'interpolation de Thiele

A partir de ces différences réciproques, il est possible de développer $f(x)$ en fraction continue. C'est la formule d'interpolation de Thiele [174]; d'après ce qui précède on a :

$$\rho_1(x,x_0) = \frac{x - x_0}{f(x) - f(x_0)}$$

d'où :

$$f(x) = f(x_0) + \frac{x - x_0}{\rho_1(x,x_0)}$$

or on a, en utilisant la notation $\rho_k^{(n)} = \rho_k(x_n, x_{n+1}, \ldots, x_{n+k})$:

$$\rho_1(x,x_0) = \rho_1^{(0)} + \frac{x - x_1}{\rho_2(x,x_0,x_1) - \rho_0^{(0)}}$$

$$\rho_{k-1}(x,x_0,\ldots,x_{k-2}) = \rho_{k-1}^{(0)} + \frac{x - x_{k-1}}{\rho_k(x,x_0,\ldots,x_{k-1}) - \rho_{k-2}^{(0)}}$$

d'où finalement :

$$f(x) = \rho_0^{(0)} + \frac{x-x_0|}{|\rho_1^{(0)}} + \frac{x-x_1|}{|\rho_2^{(0)}-\rho_0^{(0)}} + \frac{x-x_2|}{|\rho_3^{(0)} - \rho_1^{(0)}} + \ldots$$

ou encore :

$$f(x) = \alpha_0 + \frac{x-x_0|}{|\alpha_1} + \frac{x-x_1|}{|\alpha_2} + \ldots \qquad (33)$$

avec $\alpha_0 = \rho_0^{(0)}$, $\alpha_k = \rho_k^{(0)} - \rho_{k-2}^{(0)}$ pour $k=1,\ldots$

et la convention $\rho_{-1}^{(n)} = 0$ pour tout n.

Appelons $C_k(x)$ le $k^{\text{ième}}$ approximant de cette fraction continue pour $k = 0,1,\ldots$

On a la propriété fondamentale :

$$C_k(x_i) = f(x_i) \quad \text{pour} \quad i=0,\ldots,k$$

Cette propriété découle du fait que, si $x=x_i$, le $i^{\text{ème}}$ numérateur partiel de la fraction continue est nul ; par conséquent la fraction continue ne comporte qu'un nombre fini de termes et l'on a :

$$C_k(x_i) = \alpha_0 + \frac{x_i-x_0|}{|\alpha_1} + \ldots + \frac{x_i-x_{i-1}|}{|\alpha_i} \qquad i=0,\ldots,k$$

d'après la construction même de cette fraction continue on a l'identité :

$$f(x_i) = \alpha_0 + \frac{x_i - x_0|}{|\alpha_1} + \ldots + \frac{x_i - x_{i-1}|}{|\alpha_i} + \frac{x_i - x_i|}{|\rho_{i+1}(x_i, x_0, \ldots, x_i) - \rho_{i-1}^{(0)}}$$

ce qui démontre la propriété fondamentale.

De façon plus générale on peut considérer la fraction continue :

$$f(x) = \alpha_0^{(n)} + \frac{x - x_n|}{|\alpha_1^{(n)}} + \frac{x - x_{n+1}|}{|\alpha_2^{(n)}} + \ldots \tag{34}$$

avec $\alpha_0^{(n)} = \rho_0^{(n)}$, $\alpha_k^{(n)} = \rho_k^{(n)} - \rho_{k-2}^{(n)}$ pour $k=1,\ldots$ et $\rho_{-1}^{(n)} = 0$

Appelons $C_k^{(n)}(x)$ le $k^{\text{ième}}$ convergent de cette fraction continue pour $k = 0,1,\ldots$

On a de même :

$$C_k^{(n)}(x_i) = f(x_i) \qquad i = n, \ldots, n+k$$

L'utilisation des différences réciproques permet donc de développer $f(x)$ en fraction continue. Cette fraction continue prend, aux abscisses x_i utilisées dans la construction des différences réciproques, la même valeur que la fonction f.

En utilisant les relations de récurrence du théorème 140 on montre facilement que :

$$A_{2k-1}^{(n)}(x) = \sum_{i=0}^{k} a_i^{(2k-1)} x^i$$

$$A_{2k}^{(n)}(x) = \sum_{i=0}^{k} a_i^{(2k)} x^i$$

$$B_{2k-1}^{(n)}(x) = \sum_{i=0}^{k-1} b_i^{(2k-1)} x^i$$

$$B_{2k}^{(n)}(x) = \sum_{i=0}^{k} b_i^{(2k)} x^i$$

avec

$$A_0^{(n)}(x) = a_0^{(0)} = \alpha_0^{(n)}$$

$$B_0^{(n)}(x) = b_0^{(0)} = 1$$

$$A_{-1}^{(n)}(x) = a_0^{(-1)} = 1$$

$$B_{-1}^{(n)}(x) = b_0^{(-1)} = 0$$

En identifiant les coefficients de termes de même degré on montre également que les coefficients $a_i^{(k)}$ et $b_i^{(k)}$ sont obtenues à l'aide des relations suivantes :

$$a_0^{(2k-1)} = \alpha_{2k-1}^{(n)} a_0^{(2k-2)} - x_{2k-2+n} a_0^{(2k-3)}$$

$$a_i^{(2k-1)} = \alpha_{2k-1}^{(n)} a_i^{(2k-2)} - x_{2k-2+n} a_i^{(2k-3)} + a_i^{(2k-3)} \qquad i=1,\dots,k-1$$

$$a_k^{(2k-1)} = a_{k-1}^{(2k-3)}$$

puis

$$a_0^{(2k)} = \alpha_{2k}^{(n)} a_0^{(2k-1)} - x_{2k-1+n} a_0^{(2k-2)}$$

$$a_i^{(2k)} = \alpha_{2k}^{(n)} a_i^{(2k-1)} - x_{2k-1+n} a_i^{(2k-2)} + a_{i-1}^{(2k-2)} \qquad i=1,\dots,k-1$$

$$a_k^{(2k)} = \alpha_{2k}^{(n)} a_k^{(2k-1)} + a_{k-1}^{(2k-2)}$$

et de même :

$$b_0^{(2k-1)} = \alpha_{2k-1}^{(n)} b_0^{(2k-2)} - x_{2k-2+n} b_0^{(2k-3)}$$

$$b_i^{(2k-1)} = \alpha_{2k-1}^{(n)} b_i^{(2k-2)} - x_{2k-2+n} b_i^{(2k-3)} + b_{i-1}^{(2k-3)} \qquad i=1,\dots,k-2$$

$$b_{k-1}^{(2k-1)} = \alpha_{2k-1}^{(n)} b_{k-1}^{(2k-2)} + b_{k-2}^{(2k-3)}$$

puis :

$$b_0^{(2k)} = \alpha_{2k}^{(n)} b_0^{(2k-1)} - x_{2k-1+n} b_0^{(2k-2)}$$

$$b_i^{(2k)} = \alpha_{2k}^{(n)} b_i^{(2k-1)} - x_{2k-1+n} b_i^{(2k-2)} + b_{i-1}^{(2k-2)} \qquad i=1,\dots,k-1$$

$$b_k^{(2k)} = b_{k-1}^{(2k-2)}$$

On voit également en utilisant ces relations que :

$$a_k^{(2k-1)} = 1 \qquad b_{k-1}^{(2k-1)} = \rho_{2k-1}^{(n)}$$

$$a_k^{(2k)} = \rho_{2k}^{(n)} \qquad b_k^{(2k)} = 1$$

D'après ce qui précède on peut donc considérer l'utilisation des différences réciproques comme une méthode pour construire des fractions rationnelles d'interpolation. On voit que les fractions rationnelles ainsi obtenues sont soit le rapport de deux polynômes de degré k soit le rapport d'un polynôme de degré k sur un polynôme de degré k-1. Différents auteurs ont étudié le problème général de l'interpolation (ou de l'extrapolation) par une fraction rationnelle où numérateur et dénominateur sont de degré quelconque. Nous y reviendrons plus loin (VII-8 On trouvera dans [34] le cas où la fraction rationnelle est le rapport de deux polynômes généralisés de la forme :

$$a_0 + a_1 \varphi(x) + \dots + a_k \left[\varphi(x)\right]^k$$

où φ est une fonction arbitraire donnée.

Notons enfin que les fractions continues d'interpolation sont également appelées approximants de Padé de type II. On trouvera dans [13,14,245] des résultats et des applications les concernant.

Il faut signaler une application de l'interpolation par une fraction rationnelle qui peut être intéressante : c'est l'inversion de la transformée de Laplace.

Soit f une fonction réelle ou complexe de la variable réelle t. On appelle transformée de Laplace de f, la fonction F définie par :

$$F(p) = \int_o^\infty e^{-pt} f(t) dt$$

Nous supposerons naturehlement que cette intégrale existe. Le problème de l'inversion de la transformée de Laplace consiste à trouver f lorsque F est connue. L'idée de base de la méthode est simple : on commence par remplacer F par une fraction rationnelle de la variable p (il faut que le degré du dénominateur soit supérieur à celui du numérateur puisque $\lim_{p\to\infty} F(p) = o$). On décompose ensuite cette fraction rationnelle en éléments simples puis on inverse chacun de ces éléments.

Si F est connue pour un certain nombre de valeurs de p alors on peut construire une fraction rationnelle d'interpolation en utilisant le procédé qui vient d'être décrit. L'idée de cette méthode est due à Fouquart [78,79] qui a obtenu de très bons résultats numériques.

Si F est connue par son développement en série alors on peut utiliser les approximants de Padé pour trouver cette fraction rationnelle. Cette idée est due à Longman [134] à qui l'on doit également une méthode d'inversion d'une fraction rationnelle qui ne nécessite pas la décomposition en éléments simples [135] (recherche des racines d'un polynôme). Sur cette question voir également [131].

Ces deux méthodes sont très bonnes numériquement mais beaucoup de travail théorique reste encore à faire pour les étudier complètement.

Etude de l'erreur

On peut écrire :

$$f(x) = C_k^{(n)}(x) + R_k^{(n)}(x)$$

$R_k^{(n)}(x)$ représente l'erreur faite en remplaçant $f(x)$ par le convergent $C_k^{(n)}(x)$.

Le but de ce paragraphe est de donner l'expression de $R_k^{(n)}(x)$ sous l'hypothèse que f est k+1 fois différentiable dans le plus petit intervalle $[a,b]$ contenant $x, x_n, \dots, x_{n+k}$.

Supposons que, dans $[a,b]$, f ait les pôles $\alpha_1, \alpha_2, \dots, \alpha_j$ de multiplicités $r_1, r_2, \dots, r_j$ avec $r_1 + \dots + r_j = m$.

Supposons de plus qu'aucun de ces pôles ne coïncide avec l'un des points d'interpolation $x_n, \dots, x_{n+k}$, qu'en dehors des pôles f admette une dérivée $(k+1)^{\text{ième}}$ bornée et que le degré de $B_k^{(n)}$ est plus grand ou égal à m. Posons :

$$\phi(x) = (x-\alpha_1)^{r_1}(x-\alpha_2)^{r_2} \dots (x-\alpha_j)^{r_j}$$

Alors $f(x)\ \phi(x)$ est borné en tout point de $[a,b]$.

Soit ψ un polynôme tel que $Q(x) = \phi(x)\ \psi(x)$ soit de même degré que $B_k^{(n)}$.

Posons :

$$R_k^{(n)}(x) = g(x) \frac{(x-x_n)\dots\dots(x-x_{n+k})}{B_k^{(n)}(x)\ Q(x)}$$

et considérons :

$$f(t) - \frac{A_k^{(n)}(t)}{B_k^{(n)}(t)} - g(x) \frac{(t-x_n)\dots(t-x_{n+k})}{B_k^{(n)}(t)\ Q(t)}$$

Cette quantité s'annule en $t=x_n, \dots, x_{n+k}$ et en $t=x$.

On pose donc :

$$\omega(t) = f(t)B_k^{(n)}(t)Q(t) - A_k^{(n)}(t)Q(t) - g(x)(t-x_n)\dots(t-x_{n+k})$$

ω s'annule en $x, x_n, \dots, x_{n+k} \in [a,b]$. D'après le théorème de Rolle ω' s'annule $k+1$ fois dans $[a,b]$, ω'' k fois dans $[a,b], \dots$ et $\omega^{(k+1)}$ s'annule une fois dans $[a,b]$; soit ξ cette abscisse.

Si $k = 2p$ alors Q est de degré p ainsi que $A_k^{(n)}$. Si $k = 2p-1$ alors Q est de degré $p-1$ et $A_k^{(n)}$ est de degré p. Par conséquent $Q(t)\, A_k^{(n)}(t)$ est de degré k et sa dérivée $(k+1)^{\text{ième}}$ est identiquement nulle. On a donc :

$$g(x) = \frac{1}{(k+1)!} \frac{d^{k+1}}{d\xi^{k+1}} \left[f(\xi)\, B_k^{(n)}(\xi)\, Q(\xi)\right]$$

d'où finalement :

$$R_k^{(n)}(x) = \frac{(x-x_n)\dots(x-x_{n+k})}{(k+1)!\, B_{k(x)}^{(n)} Q(x)} \frac{d^{k+1}}{d\xi^{k+1}} \left[f(\xi)\, B_k^{(n)}(\xi) Q(\xi)\right] \qquad (35)$$

Si f n'a pas de pôles dans $[a,b]$ on peut prendre $Q(x) = B_k^{(n)}(x)$; l'erreur est alors donnée par :

$$R_k^{(n)}(x) = \frac{(x-x_n)\dots(x-x_{n+k})}{(k+1)!\left[B_k^{(n)}(x)\right]^2} \frac{d^{k+1}}{d\xi^{k+1}} \{f(\xi) \left[B_k^{(n)}(\xi)\right]^2\} \qquad (36)$$

Le cas confluent

Dans les paragraphes précédents nous avons supposé que les abscisses $x_0, x_1, \dots$ qui interviennent dans les différences réciproques étaient deux à deux distinctes. Nous allons maintenant étudier ce qui se passe lorsque toutes ces abscisses coïncident.

Posons $x_n = t+nh$ pour $n = 0,1 \dots$

Alors les différences réciproques s'écrivent :

$$\rho_0(x_k) = f(x_k)$$

$$\rho_1(x_0,x_1) = \frac{h}{\rho_0(x_1)-\rho_0(x_0)}$$

$$\rho_2(x_0,x_1,x_2) = \rho_0(x_1) + \frac{2h}{\rho_1(x_1,x_2)-\rho_1(x_0,x_1)} \qquad \text{etc...}$$

Faisons maintenant tendre h vers zéro. On obtient la première forme confluente du ρ-algorithme qui a été donnée par Wynn [239] :

$$\rho_{-1}(t) = 0 \qquad \rho_0(t) = f(t)$$

$$\rho_{k+1}(t) = \rho_{k-1}(t) + \frac{k+1}{\rho'_k(t)} \qquad k = 0,1,...$$

On a donc en posant $x = t + h$:

$$f(t+h) = f(t) + \frac{h|}{|\alpha_1} + \frac{h|}{|\alpha_2} + \dots \tag{37}$$

avec $\alpha_k = \rho_k(t) - \rho_{k-2}(t)$ pour $k = 1,2,...$

C'est ce qu'on appelle le développement de Thiele d'une fonction. Le développement de Taylor donne le développement d'une fonction en série tandis que celui de Thiele le donne sous forme de fraction continue. De même que le développement de Taylor se termine lorsque la fonction est un polynôme, celui de Thiele se termine lorsque la fonction est une fraction rationnelle dont numérateur et dénominateur sont de mêmes degrés ou dont le degré du dénominateur est inférieur de un à celui du numérateur. L'erreur est donnée par :

$$f(t+h) = C_k(t+h) + R_k(t+h)$$

et

$$R_k(t+h) = \frac{h^{k+1}}{(k+1)!} \frac{1}{B_k(t+h)Q(t+h)} \frac{d^{k+1}}{d\xi^{k+1}} \left[f(\xi)B_k(\xi)Q(\xi)\right]$$

avec $\xi \in [x,x+h]$.

Remplaçons maintenant t par 0 et h par x ; il vient :

$$f(x) = f(0) + \frac{x|}{|\alpha_1} + \frac{x|}{|\alpha_2} + \ldots \tag{38}$$

avec $\alpha_k = \rho_k(0) - \rho_{k-2}(0)$ pour $k = 1,2,\ldots$

Soient $C_k(x)$ les approximants successifs de cette fraction continue : $C_k(x) = A_k(x)/B_k(x)$. A_{2k-1}, A_{2k} et B_{2k} sont des polynômes de degré k en x tandis que B_{2k-1} est de degré k-1.

Voyons ce que devient la propriété d'interpolation pour la fraction continue lorsque toutes les abscisses coïncident en zéro. Cette propriété d'interpolation peut encore s'écrire :

$$\Delta^p C_k(x_0) = \Delta^p f(x_0) \quad p = 0,\ldots,k$$

En faisant tendre h vers 0 on voit que l'on a :

$$\frac{d^p}{dt^p} C_k(t) = f^{(p)}(t) \quad p = 0,\ldots,k$$

Supposons maintenant que f soit un développement en série formelle :

$$f(t) = \sum_{i=0}^{\infty} c_i t^i$$

On a :

$$f^{(p)}(0) = p!\, c_p$$

D'autre part on peut développer C_k suivant les puissances croissantes de t ; d'où :

$$C_k(t) = \sum_{i=0}^{\infty} c_i' \, t^i$$

et par conséquent :

$$f^{(p)}(0) = p! \; c_p = C_k^{(p)}(0) = p! \; c_p' \qquad \text{pour } p = 0,\dots,k$$

Ce qui montre que :

$$c_p = c_p' \quad \text{pour } p=0,\dots,k$$

Le résultat fondamental auquel on aboutit est par conséquent le suivant :

$$C_k(x) - f(x) = O(x^{k+1})$$

Les convergents successifs de cette fraction continue sont donc les approximants de Padé de f puisque ceux-ci sont uniques lorsqu'ils existent :

$$C_{2k}(x) = [k/k] = \varepsilon_{2k}^{(0)}$$

$$C_{2k-1}(x) = [k/k-1] = \varepsilon_{2k-2}^{(1)}$$

De plus on a l'expression de l'erreur :

$$f(x) - C_k(x) = \frac{x^{k+1}}{(k+1)! \; B_k(x)Q(x)} \; \frac{d^{k+1}}{dt^{k+1}} \left[f(t) \; B_k(t)Q(t)\right] \tag{39}$$

avec $t \in [0,x]$. Cette relation donne par conséquent la constante qui intervient dans la notation $O(x^{k+1})$. Si f n'a pas de pôle dans $[0,x]$ alors :

$$f(x)-C_k(x) = \frac{x^{k+1}}{(k+1)!\left[B_k(x)\right]^2} \; \frac{d^{k+1}}{dt^{k+1}} \{f(t) \left[B_k(t)\right]^2\} \tag{40}$$

Une autre conséquence de ceci est que la fraction continue précédente (38) est équivalente à la fraction continue (19) ou (23) pour n=0 obtenue à l'aide de l'algorithme QD de Rutishauser ; en d'autres termes il existe une connexion entre l'algorithme QD et la forme confluente du ρ-algorithme. C'est ce que nous allons maintenant exposer :

On a vu que $c_p = f^{(p)}(0)/p!$. Il est facile de voir [238] que si on applique la première forme confluente du ρ-algorithme à la fonction f donnée par son développement en série alors :

$$\rho_{2k}(0) = \frac{H_{k+1}^{(0)}(c_0)}{H_k^{(2)}(c_2)}$$

$$k=0,1,\ldots$$

$$\rho_{2k+1}(0) = \frac{H_k^{(3)}(c_3)}{H_{k+1}^{(1)}(c_1)}$$

Ecrivons maintenant que la fraction continue (38) est équivalente à la fraction continue (23) dans laquelle on a pris $n=0$. On doit donc (voir paragraphe VII-1) trouver les nombres $d_1, d_2, \ldots$ tels que :

$$\alpha_1 d_1 = 1$$

$$d_1 = c_1$$

$$\alpha_2 d_2 = 1$$

$$d_1 d_2 = -q_1^{(1)}$$

....

$$\alpha_{2k} d_{2k} = 1$$

$$d_{2k-1} d_{2k} = -q_k^{(1)}$$

$$\alpha_{2k+1} d_{2k+1} = 1$$

$$d_{2k} d_{2k+1} = -e_k^{(1)}$$

....

d'où finalement les relations :

$$- \alpha_{2k}\, \alpha_{2k+1}\, e_k^{(1)} = 1$$

$$- \alpha_{2k}\, \alpha_{2k-1} q_k^{(1)} = 1 \qquad (41)$$

En utilisant la définition des quantités α_k, $\rho_k(0)$, $q_k^{(1)}$ et $e_k^{(1)}$ ainsi que la relation de récurrence entre les déterminants de Hankel, on vérifiera facilement que les relations précédentes sont satisfaites.

En utilisant le fait que $\rho_k(0) - \rho_{k-2}(0) = k/\rho'_{k-1}(0)$ les relations précédentes peuvent également s'écrire :

$$- 2k(2k+1)\, e_k^{(1)} = \rho'_{2k-1}(0) \quad \rho'_{2k}(0)$$

$$- 2k(2k-1)\, q_k^{(1)} = \rho'_{2k-1}(0) \quad \rho'_{2k-2}(0) \qquad (42)$$

Ces relations sont a rapprocher de celles du ω'-algorithme défini par Wynn et utilisées pour le calcul des intégrales impropres
lement comparer ces relations à celles de la généralisation de l'algorithme QD donnée par Wuytack [190] qui est également utilisée pour construire des fractions continues d'interpolation.

Les relations (41) sont reliées à celles établies au paragraphe VII-4. de la façon suivante :
d'après les relations (11) on peut exprimer les éléments d'une fraction continue en fonction de ses approximants successifs ; on a :

$$x = \frac{B_{2k+1}(x)}{B_{2k-1}(x)} \quad \frac{C_{2k}(x)-C_{2k+1}(x)}{C_{2k}(x)-C_{2k-1}(x)}$$

$$\alpha_{2k} = \frac{B_{2k}(x)}{B_{2k-1}(x)} \quad \frac{C_{2k}(x)-C_{2k-2}(x)}{C_{2k-1}(x)-C_{2k-2}(x)}$$

$$\alpha_{2k+1} = \frac{B_{2k+1}(x)}{B_{2k}(x)} \quad \frac{C_{2k+1}(x)-C_{2k-1}(x)}{C_{2k}(x) \;\; -C_{2k-1}(x)}$$

Dans (41) remplaçons α_{2k} et α_{2k+1} par leurs expressions et multiplions par x : on retrouve immédiatement les relations du paragraphe VII-4.
On peut également lier directement l'ε-algorithme et la première forme confluente du ρ-algorithme. A partir des relations du paragraphe VII-4 et de (41) on obtient :

$$\frac{\varepsilon_{2k}^{(1)}-\varepsilon_{2k}^{(0)}}{\varepsilon_{2k+2}^{(0)}-\varepsilon_{2k}^{(0)}} \quad \frac{\varepsilon_{2k+2}^{(1)}-\varepsilon_{2k+2}^{(0)}}{\varepsilon_{2k+2}^{(1)}-\varepsilon_{2k}^{(1)}} = - \frac{x}{[\rho_{2k+2}(0)-\rho_{2k}(0)]\,[\rho_{2k+3}(0)-\rho_{2k+1}(0)]}$$

$$\frac{\varepsilon_{2k}^{(0)}-\varepsilon_{2k-2}^{(1)}}{\varepsilon_{2k}^{(1)}-\varepsilon_{2k-2}^{(1)}}\;\frac{\varepsilon_{2k+2}^{(0)}-\varepsilon_{2k}^{(1)}}{\varepsilon_{2k+2}^{(0)}-\varepsilon_{2k}^{(0)}} = -\frac{x}{\left[\rho_{2k+4}(0)-\rho_{2k+2}(0)\right]\left[\rho_{2k+3}(0)-\rho_{2k+1}(0)\right]}$$

lorsque : $\varepsilon_0^{(n)} = \sum_{i=0}^{n} c_i\, x^i$

$$f(t) = \sum_{i=0}^{\infty} c_i\, x^i$$

$$\rho_0(0) = f(0) = c_0$$

VII-8 - L'interpolation d'Hermite rationnelle

Le problème de l'interpolation d'Hermite par des fractions rationnelles ne conduit pas directement à des algorithmes d'accélération de la convergence. Cependant, comme nous allons le voir, il existe un lien très étroit avec ceux-ci et c'est la raison pour laquelle nous allons en parler brièvement.

Le problème de l'interpolation d'Hermite consiste à chercher une fraction rationnelle r telle que :

$$r^{(i)}(x_j) = y_j^{(i)}$$

pour $i=0,\ldots,n_j$ et $j=0,\ldots,m$ où les abscisses d'interpolation x_j sont données ainsi que les nombres $y_j^{(i)}$.

On voit que, formulé de cette façon, ce problème contient certains de ceux que nous avons étudiés précédemment. En effet si $m=0$ ce problème n'est rien d'autre que celui des approximants de Padé. Par contre si $n_j = 0$ pour tout j c'est le problème de l'interpolation par une fraction rationnelle dont numérateur et dénominateur sont de degré quelconque ; il généralise donc le problème d'interpolation étudié au paragraphe précédent à l'aide du ρ-algorithme.

Si maintenant le degré du dénominateur de r est nul et si $m=0$ la solution du problème est donnée par le développement en série de

Taylor (paragraphe VI-5) tandis que si n_j = o pour tout j c'est le procédé de Neville-Aitken qui résoud la question (paragraphe II-3).

Les premières études sur le problème d'interpolation d'Hermite sont dues à Cauchy qui a traité le cas où tous les n_j sont nuls. Jacobi [116] trouva ensuite des expressions avec des déterminants pour ces fractions rationnelles d'interpolation. Il fallut attendre Kronecker [124] et Thiele [174] pour voir apparaître les premiers algorithmes de calcul. Plus tard, l'étude de ce problème fut reprise, parallèlement à celle des approximants de Padé ; on trouve alors chronologiquement les contributions de Wynn [203], Thacher et Tukey [173], Stoer [172], Larkin [126,127] et Meinguet [140].

Ces toutes dernières années enfin, les bases mêmes de ce problème ont été réexaminées en détail. Wuytack [191,192] a étudié l'existence et la construction de la table des fractions rationnelles d'interpolation ; il a démontré certaines propriétés de cette table ainsi que des relations entre ses éléments. Il a prouvé l'existence de fractions continues dont les convergents successifs forment les éléments de cette table. Dans un autre article, Wuytack [190] a donné un algorithme qui généralise l'algorithme q-d et qui permet de calculer les numérateurs et dénominateurs partiels de ces fractions continues. On trouvera des applications à l'accélération de la convergence dans [189].

Dans sa thèse en 1974, Warner [187] s'est livré à une étude très complète de cette table d'interpolation rationnelle et a obtenu un nombre considérable de relations algébriques la concernant. Il a notamment généralisé la règle de la croix de Wynn (propriété 17). Il s'est livré à une étude systématique des algorithmes de calcul existants et en a proposé un nouveau basé sur la règle de la croix généralisée. Il a enfin également étudié la convergence de ces fractions rationnelles d'interpolation. On pourra consulter également [186].

En 1976, Claessens [62] dans sa thèse a étudié la structure en blocs de la table d'interpolation rationnelle et unifié les définitions de normalité (abscence de blocs) données par Wuytack et par Warner qui différaient quelque peu. En généralisant les déterminants de Hankel ainsi que la notion de bigradient [113] il a donné des formules des interpolants rationnels faisant intervenir les déterminants. Claessens a également étudié les algorithmes qui permettent de construire la table d'interpolation rationnelle à l'aide des fractions continues. Il a notamment simplifié la généralisation de l'algorithme q-d donnée par Wuytack et dont nous avons parlé plus haut. Enfin il a établi la connexion avec les travaux de Barnsley [11,12] ainsi qu'avec l'algorithme d'Euclide pour trouver le p.g.c.d. de deux polynômes. D'autres résultats sont donnés dans [81].

Tous ces travaux récents, trop longs et trop complexes pour être présentés ici, permettent une bonne connaissance théorique du problème d'interpolation d'Hermite par des fractions rationnelles. Il existe de nombreux algorithmes pour construire ces fractions rationnelles. Les applications à l'analyse numérique restent encore à trouver.

CONCLUSION

On voit que les méthodes d'accélération de la convergence sont intéressantes à deux points de vue : leur étude en elle-même n'est pas dénuée d'intérêt (c'est du moins une opinion personnelle que personne n'est obligé de partager) ; en second lieu, il est indiscutable que leur emploi dans les méthodes itératives ainsi que les diverses applications que nous avons examinées sont d'un grand intérêt pratique.

De nombreuses questions concernant ces méthodes restent encore à étudier. Nous les avons signalées au fur et à mesure de l'exposé. L'équipe d'analyse numérique de Lille étudie en ce moment un certain nombre de ces problèmes : inversion de la transformée de Laplace, lissage par des sommes d'exponentielles, calcul des fonctions spéciales, étude du Θ-algorithme, étude de la stabilité numérique et utilisation des moments modifiés, propriétés des suites de moments, théorie des polynômes orthogonaux, méthodes de quadratures, formalisation des procédés d'accélération de la convergence et étude de nouvelles méthodes.

La question la plus importante qui reste à résoudre est celle du choix du procédé d'accélération de la convergence le mieux adapté à la suite à accélérer. Comme nous l'avons vu on peut répondre à cette question dans des cas particuliers. A l'heure actuelle l'algorithme qui semble donner de bons résultats pour le plus grand nombre de suites est le Θ-algorithme. Une fois choisi l'algorithme, il faudra, bien évidemment, s'assurer qu'il converge et qu'il accélère la convergence. Il faudra aussi savoir le mettre en oeuvre correctement. Enfin... le champ des applications reste encore à élargir mais le plus difficile est finalement de convaincre ceux qui utilisent des méthodes itératives d'employer systématiquement des méthodes d'accélération de la convergence. Ne désespérons pas : la méthode de Romberg a mis quelque trente ans à s'imposer. Pour être complet sur ce sujet, il faut signaler qu'un certain nombre de résultats connus n'ont pas été donnés ici ni que certains points n'ont même pas été abordés. Ce sont les suivants :

- Formalisme des procédés d'extrapolation [22]
- Connexion entre l'ε-algorithme et les polynômes orthogonaux [4,5,179,180,181,182,205]
- Equations aux dérivées partielles associées aux algorithmes de losange [202,204,221]
- Interprétation géométrique de l'ε-algorithme [65]
- Certaines propriétés de la table de Padé [9,10,96,97]
- Généralisations du ρ-algorithme [50,51,127,189]
- Stabilité numérique et propagation des erreurs d'arrondis [232,233]
- Etude de la seconde forme confluente de l'ε-algorithme [22,198,199]
- La transformation G [159, 98 à 104]
- Approximants de Padé en plusieurs points [11,12]
- Approximation quadratique [169]
- Approximants de Padé à plusieurs variables [54,64]
- Liaison avec la théorie de l'approximation [70]
- Approximants de Padé pour les séries de Laurent [95]

D'autres questions relatives à ce sujet ont été passées sous silence et cela même dans la liste précédente puisque, par exemple, moins de la moitié des articles publiés par Wynn sont cités dans la bibliographie (une bibliographie plus étendue est donnée dans [37]). On peut d'ailleurs compléter la lecture de ce qui précède par celle d'un livre que Wynn doit faire paraître prochainement (The ε-algorithm, Birkhauser-Verlag). Que ceux que le sujet intéresse ne se découragent pas pour autant : je pense que ce livre contient l'essentiel de ce qu'il faut savoir sur la question. Du moins je l'espère.

RÉFÉRENCES

[1] *A.C. AITKEN* - On Bernoulli's numerical solution of algebraic equations - Proc. Roy. Soc. Edinburgh, 46 (1926) 289-305.

[2] *A.C. AITKEN* - Determinants and matrices - Oliver and Boyd, 1951.

[3] *N.I. AKHIEZER* - The classical moment problem - Oliver and Boyd, London, 1965.

[4] *G.D. ALLEN, C.K. CHUI, W.R. MADYCH, F.J. NARCOWICH, P.W. SMITH* - Padé approximation and gaussian quadrature - Bull. Austral. Math. Soc., 11 (1974) 63-69.

[5] *G.D. ALLEN, C.K. CHUI, W.R. MADYCH, F.J. NARCOWICH, P.W. SMITH* - Padé approximation and orthogonal polynomials - Bull. Austral. Math. Soc., 10 (1974) 263-270.

[6] *G.D. ALLEN, C.K. CHUI, W.R. MADYCH, F.J. NARCOWICH, P.W. SMITH* - Padé approximation of Stieltjes series - J. Approx. Theory, 14 (1975) 302-316.

[7] *R. ALT* - Méthodes A-stables pour l'intégration des systèmes différentiels mal conditionnés - Thèse 3ème cycle, Paris, 1971.

[8] *G.A. BAKER Jr* - The Padé approximant method and some related generalization - in "The Padé approximant in theoretical physics", G.A. Baker Jr. and J.L. Gammel eds., Academic Press, New York, 1970.

[9] *G.A. BAKER Jr* - Essential of Padé approximants - Academic Press, New York, 1975.

[10] *G.A. BAKER Jr, J.L. GAMMEL eds.* - The Padé approximant in theoretical physics - Academic Press, New York, 1972.

[11] *M. BARNSLEY* - The bounding properties of the multipoint Padé approximant to a series of Stieltjes - Rocky Mountains J. Math., 4 (1974) 331-334.

[12] *M. BARNSLEY* - The bounding properties of the multipoint Padé approximant to a series of Stieltjes on the real line - J. Math. Phys., 14 (1973) 299-313.

[13] *J.L. BASDEVANT* - Padé approximants - in "Methods in subnuclear physics", vol. IV, Gordon and Breach, London, 1970.

[14] *J.L. BASDEVANT* - The Padé approximation and its physical applications - Fort. der Physik, 20 (1972) 283-331.

[15] *F.L. BAUER* - Connections between the q-d algorithm of Rutishauser and the ε-algorithm of Wynn - Deutsche Forschungsgemeinschaft Tech. Rep. Ba/106, 1957.

[16] *F.L. BAUER* - Nonlinear sequence transformations - in "Approximation of functions", Garabedian ed., Elsevier, New York, 1965.

[17] *F.L. BAUER* - The g-algorithm - SIAM J., 8 (1960) 1-17.

[18] *N. BOURBAKI* - Fonctions d'une variable réelle (chapitre 5) - Hermann, Paris, 1951.

[19] *L.C. BREAUX* - A numerical study of the application of acceleration techniques and prediction algorithms to numerical integration - M. Sc. Thesis, Louisiana State Univ., New Orleans, 1971.

[20] *C. BREZINSKI* - Convergence d'une forme confluente de l'ε-algorithme - C.R. Acad. Sc. Paris, 273 A (1971) 582-585.

[21] *C. BREZINSKI* - L'ε-algorithme et les suites totalement monotones et oscillantes - C.R. Acad. Sc. Paris, 276 A (1973) 305-308.

[22] *C. BREZINSKI* - Méthodes d'accélération de la convergence en analyse numérique - Thèse, Univ. de Grenoble, 1971.

[23] *C. BREZINSKI* - Application du ρ-algorithme à la quadrature numérique - C.R. Acad. Sc. Paris, 270 A (1970) 1252-1253.

[24] *C. BREZINSKI* - Etudes sur les ε et ρ-algorithmes - Numer. Math., 17 (1971) 153-162.

[25] *C. BREZINSKI* - Résultats sur les procédés de sommations et l'ε-algorithme - RIRO, R3 (1970) 147-153.

[26] *C. BREZINSKI* - Forme confluente de l'ε-algorithme topologique - Numer. Math., 23 (1975) 363-370.

[27] *C. BREZINSKI* - Computation of Padé approximants and continued fractions - J. Comp. Appl. Math., 2 (1976) 113-123.

[28] *C. BREZINSKI* - Séries de Stieltjes et approximants de Padé - Colloque Euromech 58, Toulon, 12-14 mai 1975.

[29] *C. BREZINSKI* - Some results in the theory of the vector ε-algorithm - Linear Algebra, 8 (1974) 77-86.

[30] *C. BREZINSKI* - Some results and applications about the vector ε-algorithm - Rocky Mountains J. Math., 4 (1974) 335-338.

[31] *C. BREZINSKI* - Généralisations de la transformation de Shanks, de la table de Padé et de l'ε-algorithme - Calcolo 12 (1975) 317-360.

[32] *C. BREZINSKI* - Comparaison de suites convergentes - RIRO, R2 (1971) 95-99.

[33] *C. BREZINSKI* - Limiting relationships and comparison theorems for sequences - Rend. Circ. Mat. Palermo, à paraître.

[34] *C. BREZINSKI* - Généralisation des extrapolations polynomiales et rationnelles - RAIRO, R1 (1972) 61-66.

[35] *C. BREZINSKI* - Méthodes numériques générales pour l'accélération de la convergence - à paraître.

[36] *C. BREZINSKI* - Génération de suites totalement monotones et oscillantes - C.R. Acad. Sc. Paris, 280 A (1975) 729-731.

[37] *C. BREZINSKI* - A bibliography on Padé approximation and some related matters. dans "Padé approximants method and its applications to mechanics", Lecture Notes in Physics 47, H. Cabannes ed., Springer-Verlag.

[38] *C. BREZINSKI* - Accélération de suites à convergence logarithmique - C.R. Acad. Sc. Paris, 273 A (1971) 727-730.

[39] *C. BREZINSKI* - Transformation rationnelle d'une fonction - C.R. Acad. Sc. Paris, 273 A (1971) 772-774.

[40] *C. BREZINSKI* - Accélération de la convergence de suites dans un espace de Banach - C.R. Acad. Sc. Paris, 278 A (1974) 351-354.

[41] *C. BREZINSKI* - Conditions d'application et de convergence de procédés d'extrapolation - Numer. Math., 20 (1972) 64-79.

[42] *C. BREZINSKI* - Application de l'ε-algorithme à la résolution des systèmes non linéaires - C.R. Acad. Sc. Paris, 271 A (1970) 1174-1177.

[43] *C. BREZINSKI* - Intégration des systèmes différentiels à l'aide du ρ-algorithme - C.R. Acad. Sc. Paris, 278 A (1974) 875-878.

[44] *C. BREZINSKI* - Sur un algorithme de résolution des systèmes non linéaires - C.R. Acad. Sc. Paris, 272 A (1971) 145-148.

[45] *C. BREZINSKI* - Numerical stability of a quadratic method for solving systems of non linear equations - Computing, 14 (1975) 205-211.

[46] *C. BREZINSKI* - Computation of the eigenelements of a matrix by the ε-algorithm - Linear Algebra, 11 (1975) 7-20.

[47] *C. BREZINSKI, M. CROUZEIX* - Remarques sur le procédé Δ^2 d'Aitken - C.R. Acad. Sc. Paris, 270 A (1970) 896-898.

[48] *C. BREZINSKI, A.C. RIEU* - The solution of systems of equations using the ε-algorithm, and an application to boundary value problems - Math. Comp., 28 (1974) 731-741.

[49] *T.J. BROMWICH* - An introduction to the theory of infinite series - Macmillan, London, 1949, 2d ed.

[50] *R. BULIRSCH, J. STOER* - Fehlerabschatzungen und extrapolation mit rationalen funktionen bei verfahren von Richardson-typus - Numer. Math., 6 (1964) 413-427.

[51] *R. BULIRSCH, J. STOER* - Numerical quadrature by extrapolation - Numer. Math., 9 (1967) 271-278.

[52] *T. CARLEMAN* - Les fonctions quasi-analytiques - Gauthier-Villars, Paris, 1923.

[53] *E.W. CHENEY* - Introduction to approximation theory - McGraw-Hill, (1966).

[54] *J.S.R. CHISHOLM* - Rational approximants defined from double power series - Math. Comp., 27 (1973) 841-848.

[55] *J.S.R. CHISHOLM* - Padé approximants and linear integral equations - in "The Padé approximant in theoretical physics", G.A. Baker Jr. and J.L. Gammel eds., Academic Press, New York, 1970.

[56] *J.S.R. CHISHOLM* - Application of Padé approximation to numerical integration - Rocky Mountains J. Math., 4 (1974) 159-168.

[57] *J.S.R. CHISHOLM* - Padé approximation of single variable integrals - Colloquium on computational methods in theoretical physics, Marseille, 1970.

[58] *J.S.R. CHISHOLM* - Accelerated convergence of sequences of quadrature approximants - second colloquium on computational methods in theoretical physics, Marseille, 1971.

[59] *J.S.R. CHISHOLM, A.C. GENZ, G.E. ROWLANDS* - Accelerated convergence of sequences of quadrature approximation - J. Comp. Phys., 10 (1972) 284-307.

[60] *C.K. CHUI* - Recent results on Padé approximants and related problems - Approximation theory conference, Austin, 1976.

[61] *G. CLAESSENS* - A new look at the Padé table and the different methods for computing its elements - J. Comp. Appl. Math., 1 (1975) 141-151.

[62] *G. CLAESSENS* - Some aspects of the rational Hermite interpolation table and its applications - Thèse, Univ. d'Anvers, 1976.

[63] *W.D. CLARK* - Infinite series transformations and their applications - Thesis, University of Texas, 1967.

[64] *A.K. COMMON, P.R. GRAVES-MORRIS* - Some properties of Chisholm approximants - J. Inst. Maths. Applics., 13 (1974) 229-232.

[65] *F. CORDELLIER* - Interprétation géométrique d'une étape de l'ε-algorithme - Publ. 40, Labo. de Calcul, Univ. de Lille, 1973.

[66] *F. CORDELLIER* - Particular rules for the vector ε-algorithm. Numer. Math., 27 (1977) 203-207.

[67] *F. CORDELLIER* - Détermination des suites que le Θ-algorithme transforme en une suite constante - à paraître.

[68] *G. DAHLQUIST* - Convergence and stability in the numerical integration of ordinary differential equations - Math. Scand., 4 (1956) 33-53.

[69] *P.J. DAVIS, P. RABINOWITZ* - Numerical integration - Blaisdell, Waltham, 1967.

[70] *J. DELLA DORA* - Approximation non archimédienne - Colloque d'analyse numérique, Port-Bail, 1976.

[71] *J. DIEUDONNE* - Fondements de l'analyse moderne - Gauthier-Villars, Paris, 1967.

[72] *J. DIEUDONNE* - Calcul infinitésimal - Hermann, Paris, 1968.

[73] *B.L. EHLE* - A-stable methods and Padé approximantions to the exponential - SIAM J. Math. Anal., 4 (1973) 671-680.

[74] *B.L. EHLE* - On Padé approximantions to the exponential function and A-stable methods for the numerical solution of initial value problems - Research rep. CSRR 2010, dept. of AACS, Univ. of Waterloo, Ontario, 1969.

[75] *C. ESPINOZA* - Applications de l'ε-algorithme à des suites non scalaires et comparaison de quelques résultats numériques obtenus avec les ε, ρ et Θ-algorithmes - Mémoire de DEA, Lille, 1975.

[76] *C. ESPINOZA* - Accélération de la convergence des méthodes de relaxation - Thèse 3ème cycle, à paraître.

[77] *V.N. FADDEEVA* - Computational methods of linear algebra - Dover, New York, 1959.

[78] *Y. FOUQUART* - Utilisation des approximants de Padé pour l'étude des largeurs équivalentes des raies formées en atmosphère diffusante - J. Quant. Spectrosc. Radiat. Transfert, 14 (1974) 497-508.

[79] *Y. FOUQUART* - Contribution à l'étude des spectres réfléchis par les atmosphères planétaires diffusantes. Application à Vénus - Thèse, Lille, 1975.

[80] *L. FOX* - Romberg integration for a class of singular integrands - Computer J., 10 (1967) 87-93.

[81] *M.A. GALLUCCI, W.B. JONES* - Rational approximations corresponding to Newton series - J. Approx. Theory, 17 (1976) 366-392.

[82] *F.R. GANTMACHER* - The theory of matrices, vol.1 - Chelsea, New York, 1960.

[83] *E. GEKELER* - On the solution of systems of equations by the epsilon algorithm of Wynn - Math. Comp., 26 (1972) 427-436.

[84] *A. GENZ* - Applications of the ε-algorithm to quadrature problems - in "Padé approximants and their applications", P.R. Graves-Morris ed., Academic Press, New York, 1973.

[85] *B. GERMAIN-BONNE* - Transformations de suites - RAIRO, R1 (1973) 84-90.

[86] *B. GERMAIN-BONNE* - Transformations non linéaires de suites - Séminaire d'analyse numérique, Lille, 28 mars 1973.

[87] *B. GERMAIN-BONNE* - Accélération de la convergence d'une suite par extrapolation - Colloque d'analyse numérique d'Anglet, juin 1971.

[88] *B. GERMAIN-BONNE* - Etude de quelques problèmes d'accélération de convergence - Publication 65, Laboratoire de Calcul, Université de Lille, 1976.

[89] *J. GILEWICZ* - Totally monotonic and totally positive sequences for the Padé approximation method - Rapport 74/P. 619, CPT-CNRS, Marseille.

[90] *J. GILEWICZ* - Thèse (à paraître).

[91] *W.B. GRAGG* - On extrapolation algorithms for ordinary initial value problems - SIAM J. Numer. Anal., 2 (1965) 384-403.

[92] *W.B. GRAGG* - Truncation error bounds for g-fractions - Numer. Math., 11 (1968) 370-379.

[93] *W.B. GRAGG* - The Padé table and its relation to certain algorithms of numerical analysis - SIAM Rev., 14 (1972) 1-62.

[94] *W.B. GRAGG* - Matrix interpretations and applications of the continued fraction algorithm - Rocky Mountains J. Math., 4 (1974) 213-226.

[95] *W.B. GRAGG, G.D. JOHNSON* - The Laurent - Padé table - Proceedings IFIP Congress, North-Holland, (1974) 632-637.

[96] *P.R. GRAVES-MORRIS ed.* - Padé approximants and their applications - Academic Press, New York, 1973.

[97] *P.R. GRAVES-MORRIS ed.* - Padé approximants - The institute of physics, London, 1973.

[98] *H.L. GRAY, T.A. ATCHISON* - A note on the G-transformation - J. Res. NBS, 72 B (1968) 29-31.

[99] *H.L. GRAY, T.A. ATCHISON* - Applications of the G and B transforms to Laplace transform - Proceedings ACM National conference, 1968.

[100] *H.L. GRAY, T.A. ATCHISON* - Nonlinear transformations related to the evaluation of improper integrals - SIAM J. Numer. Anal., 4 (1967) 363-371 et 5 (1968) 451-459.

[101] *H.L. GRAY, T.A. ATCHISON* - The generalized G-transform - Math. of Comp., 22 (1968) 595-606.

[102] *H.L. GRAY, W.D. CLARK* - On a class of nonlinear transformations and their applications to the evaluation of infinite series - J. Res. NBS, 73 B (1969) 251-274.

[103] *H.L. GRAY, W.R. SCHUCANY* - Some limiting cases of the G-transformation - Math. of Comp., 23 (1969) 849-859.

[104] *H.L. GRAY, T.A. ATCHISON, G.V. Mc WILLIAMS* - Higher order G transformations - SIAM J. Numer. Anal., 8 (1971) 365-381.

[105] *T.N.E. GREVILLE* - On some conjectures of P. Wynn concerning the ε-algorithm - MRC Technical summary report 877, Madison, 1968.

[106] *A.O. GUELFOND* - Calcul des différences finies - Dunod, Paris, 1963.

[107] *G.H. HARDY* - Divergent series - Clarendon Press, Oxford, 1949.

[108] *C. HASTING Jr.* - Approximations for digital computers - Princeton University Press, 1955.

[109] *P. HENRICI* - The quotient-difference algorithm - NBS appl. Math. series, 49 (1958) 23-46.

[110] *P. HENRICI* - Elements of numerical analysis - Wiley, 1964.

[111] *P. HENRICI* - Error propagation for difference methods - John Wiley and sons, Wiley, 1963.

[112] *A.S. HOUSEHOLDER* - The numerical treatment of a single nonlinear equation - Mc Graw-Hill, New York, 1970.

[113] *A.S. HOUSEHOLDER, G.W. STEWART* - Bigradients, Hankel determinants and the Padé table - in "Constructive aspects of the fundamental theorem of algebra", B. Dejon and P. Henrici eds., Academic Press, New York, 1969.

[114] *D.B. HUNTER* - The numerical evaluation of Cauchy principal values of integrals by Romberg integration - Numer. Math., 21 (1973) 185-192.

[115] *C.G.J. JACOBI* - De fractione continue, in quam integrale $\int_x^\infty e^{-x^2} dx$ evoldere licet - J. für die reine u. angew. math., 12 (1834), 346-347.

[116] *C.G.I. JACOBI* - Uber die Darstellung einer Reice gegebener Werte durch eine gebrochene Rationale Funktion - J. Reine u. angew. Math., 30 (1846) 127-156.

[117] *W.B. JONES, W.J. THRON* - On convergence of Padé approximants - SIAM J. Math. Anal., 6 (1975) 9-16.

[118] *D.C. JOYCE, W.J. THRON* - Survey of extrapolation processes in numerical analysis - SIAM Rev., 13 (1971) 435-490.

[119] *D.K. KAHANER* - Numerical quadrature by the ε-algorithm - Math. Comp., 26 (1972) 689-694.

[120] *L.V. KANTOROVITCH, G.P. AKILOV* - Functional analysis in normed spaces - Pergamon Press, 1964.

[121] *S.M. KEATHLEY, T.J. AIRD* - Stability theory of multistep methods - NASA - TN - D - 3976.

[122] *A. Ya. KHINTCHINE* - Continued fractions - P. Noordhoff, Groningen, 1963.

[123] *A.N. KHOVANSKII* - The application of continued fractions and their generalizations to problems in approximation theory - P. Noordhoff, Groningen, 1963.

[124] *L. KRONECKER* - Zur Theorie der Elimination einer Variabeln aus swei algeraischen Gelichungen - Montasber. Konigl. Preuss. Akad. Wiss. Berlin (1881) 535-600.

[125] *J.D. LAMBERT* - Nonlinear methods for stiff systems of ordinary differential equations - dans "Conference on the numerical solution of differential equations", Lecture Notes in Mathematics 363, Springer-Verlag, 1974.

[126] *F.M. LARKIN* - A class of methods for tabular interpolation - Proc. Cambridge Phil. Soc., 63 (1967) 1101-1114.

[127] *F.M. LARKIN* - Some techniques for rational interpolation - Computer J., 10 (1967) 178-187.

[128] *P.J. LAURENT* - Etude de procédés d'extrapolation en analyse numérique - thèse, Grenoble, 1964.

[129] *R.N. LEA* - On the stability on numerical solutions of ordinary differential equations - NASA - TN - D - 3760.

[130] *D. LEVIN* - Development of non-linear transformations for improving convergence of sequences - Intern. J. Comp. Math., B3 (1973) 371-388.

[131] *D. LEVIN* - Numerical inversion of the Laplace transform by accelerating the convergence of Bromwich's integral - J. Comp. Appl. Math., 1 (1975) 247-250.

[132] *G. LEVY-SOUSSAN* - Application des fractions continues à la programmation de quelques fonctions remarquables - Thèse 3ème cycle, Grenoble, 1962.

[133] *I.M. LONGMAN* - Computation of the Padé table - Intern. J. Comp. Math., 3B (1971) 53-64.

[134] *I.M. LONGMAN* - Numerical Laplace transform inversion of a function arising in viscoelasticity - J. Comp. Phys., 10 (1972) 224-231.

[135] *I.M. LONGMAN, M. SHARIR* - Laplace transform inversion of rational functions - Geophys. J. R. astr. Soc., 25 (1971) 299-305.

[136] *J.N. LYNESS, B.W. NINHAM* - Numerical quadrature and asymptotic expansions - Math. Comp., 21 (1967) 162-178.

[137] *A. MARKOV* - Deux démonstrations de la convergence de certaines fractions continues - Acta Math., 19 (1895) 93-104.

[138] *I. MARX* - Remark concerning a nonlinear sequence to sequence transformation - J. Math. Phys., 42 (1963) 334-335.

[139] *J.B. McLEOD* - A note on the ε-algorithm - Computing, 7 (1971) 17-24.

[140] *J. MEINGUET* - On the solubility of the Cauchy interpolation problem - dans "Approximation theory", A. Talbot ed., Academic Press, 1970.

[141] *S.E. MIKELADZE* - Numerical methods of mathematical analysis - AEC - TR - 4285.

[142] *L.M. MILNE-THOMSON* - The calculus of finite differences - Macmillan, London, 1965.

[143] *P. MONTEL* - Leçons sur les récurrences et leurs applications - Gauthier-Villars, 1957.

[144] *R. DE MONTESSUS DE BALLORE* - Sur les fractions continues algébriques - Bull. Soc. Math. de France, 30 (1902) 28-36.

[145] *E.H. MOORE* - On the reciprocal of the general algebraic matrix - Bull. Amer. Math. Soc., 26 (1920) 394-395.

[146] *J.M. ORTEGA, W.C. RHEINBOLDT* - Iterative solution of nonlinear equations in several variables - Academic Press, New York, 1970.

[147] *A.M. OSTROWSKI* - Solution of equation and systems of equations - Academic Press, 1966.

[148] *K.J. OVERHOLT* - Extended Aitken acceleration - BIT, 5 (1965) 122-132.

[149] *H. PADE* - Sur la représentation approchée d'une fonction par des fractions rationnelles - Ann. Ec. Norm. Sup., 9 (1892) 1-93.

[150] *R. PENNACCHI* - La transformazioni razionali di una successione - Calcolo, 5 (1968) 37-50.

[151] *R. PENROSE* - A generalized inverse for matrices - Proc. Cambridge Phil. Soc., 51 (1955) 406-413.

[152] *O. PERRON* - Die Lehre von dem Kettenbrüchen - Chelsea Pub. Co., New York, 1950.

[153] *D. PETIT* - Etude de la transformation G - Mémoire de DEA, Lille, 1975.

[154] *D. PETIT* - Etude de certains procédés d'accélération de la convergence - Thèse 3ème cycle, Lille, à paraître.

[155] *A. PEYERIMHOFF* - Lectures on summability - Springer-Verlag, 1969.

[156] *R. PIESSENS* - Numerical evaluation of Cauchy principal values of integrals - BIT, 10 (1970) 476-480.

[157] *C. PISOT, M. ZAMANSKY* - Mathématiques Générales - Dunod, Paris, 1966.

[158] Procédures Algol en analyse numérique, tome 2 - CNRS, Paris, 1972.

[159] *W.C. PYE, T.A. ATCHISON* - An algorithm for the computation of higher order G-transformation - SIAM J. Numer. Anal., 10 (1973) 1-7.

[160] *L.D. PYLE* - A generalized inverse ε-algorithm for constructing intersection projection matrices, with applications - Numer. Math., 10 (1967) 86-102.

[161] *L.F. RICHARDSON* - The deferred approach to the limit - Trans. Phil. Roy. Soc., 226 (1927) 261-299.

[162] *A.C. RIEU* - Contribution à la résolution des problèmes différentiels à condition en deux ou plusieurs points - Thèse 3ème cycle, Paris, 1973.

[163] *J. RISSANEN* - Recursive evaluation of Padé approximants for matrix sequences - IBM J. Res. Develop., (juillet 1972) 401-406.

[164] *H. RUTISHAUSER* - Der quotienten, differenzen algorithms - Birkhauser Verlag, 1957.

[165] *E.B. SAFF, R.S. VARGA* - Convergence of Padé approximants to e^{-z} on unbounded sets - J. Approx. Theory, 13 (1975) 470- 488.

[166] *E.B. SAFF, R.S. VARGA* - On the zeros and poles of Padé approximants to e^{z} - Numer. Math., 25 (1975) 1-14.

[167] *E.B. SAFF, R.S. VARGA, W.C. NI* - Geometric convergence of rational approximations to e^{-z} in infinite sectors - Numer. Math., 26 (1976) 211-225.

[168] *J.R. SCHMIDT* - On the numerical solution of linear simultaneous equations by an iterative method - Phil. Mag., 7 (1951) 369-383.

[169] *R.E. SHAFER* - On quadratic approximantion - SIAM J. Numer. Anal., 11 (1974) 447-460.

[170] *D. SHANKS* - Non linear transformations of divergent and slowly convergent series - J. Math. Phys., 34 (1955) 1-42.

[171] *T.J. STIELTJES* - Recherches sur les fractions continues - Ann. Fac. Sci. Univ. Toulouse, 8 (1894) 1-122.

[172] *J. STOER* - Uber zwei algorithmen zur interpolation mit rationalen funktionen - Numer. Math., 3 (1961) 285-304.

[173] *H.C. THACHER Jr., J.W. TUKEY* - Rational interpolation made easy by recursive algorithm - manuscript non publié, 1960.

[174] *T.N. THIELE* - Interpolationsrechnung - Teubner, 1909.

[175] *W.F. TRENCH* - An algorithm for the inversion of finite Hankel matrices - SIAM J. Appl. Math., 13 (1965) 1102-1107.

[176] *R.P. TUCKER* - Remark concerning a paper by Imanuel Mark - J. Math. Phys., 45 (1966) 233-234.

[177] *S.Y. ULM* - Extension of Steffensen's method for solving operator equations - USSR Comp. Math. Phys., 4 (1964) 159-165.

[178] *C. UNDERHILL, A. WRAGG* - Convergence properties of Padé approximants to exp (z) and their derivatives - J. Inst. Maths. Applics., 11 (1973) 361-367.

[179] *A. VAN DER SLUIS* - General orthogonal polynomials - Thèse, Univ. d'Utrecht, 1956.

[180] *H. VAN ROSSUM* - A theory of orthogonal polynomials based on the Padé table - Van Gorcum, Assen, 1953.

[181] *H. VAN ROSSUM* - Contiguous orthogonal systems - Koninkl. Nederl. Akad. Wet., 63 A (1960) 323-332.

[182] *H. VAN ROSSUM* - Systems of orthogonal and quasi orthogonal polynomials connected with the Padé table - Koninkl. Nederl. Akad. Wet. 58 A (1955) 517-534 et 675-682.

[183] *E.B. VAN VLECK* - On the convergence of the continued fraction of Gauss and other continued fractions - Ann. Math., 3 (1901) 1-18.

[184] *Yu. VOROBYEV* - Method of moments in appleid mathematics - Gordon and Breach, New York, 1965.

[185] *H.S. WALL* - The analytic theory of continued fractions - Van Nostrand, New York, 1948.

[186] *D.D. WARNER* - An extension of Saff's theorem on the convergence of interpolating rational functions - J. Approx. Theory, à paraître.

[187] *D.D. WARNER* - Hermite interpolation with rational functions - Thèse, Univ. de Californie, 1974.

[188] *D.V. WIDDER* - The Laplace transform - Princeton University Press, 1946.

[189] *L. WUYTACK* - A new technique for rational extrapolation to the limit - Numer. Math., 17 (1971) 215-221.

[190] *L. WUYTACK* - An algorithm for rational interpolation similar to the qd-algorithm - Numer. Math., 20 (1973) 418-424.

[191] *L. WUYTACK* - On some aspects of the rational interpolation problem - SIAM J. Numer. Anal., 11 (1974) 52-60.

[192] *L. WUYTACK* - On the osculatory rational interpolation problem - Math. Comp., 29 (1975) 837-843.

[193] *L. WUYTACK* - Numerical integration by using nonlinear techniques - J. Comp. Appl. Math., à paraître.

[194] *L. WUYTACK* - The use of Padé approximation in numerical integration - dans "Padé approximants method and its applications to mechanics", Lecture Notes in Physics 47, H. Cabannes ed., Springer-Verlag.

[195] *P. WYNN* - Upon an invariant associated with the epsilon algorithm - MRC Technical summary report 675 (1966).

[196] *P. WYNN* - Sur les suites totalement monotones - C.R. Acad. Sc. Paris, 275 A (1972) 1065-1068.

[197] *P. WYNN* - The numerical efficiency of certain continued fraction expansions - Konink1. Nederl. Akad. Wet., 65 A (1962) 127-148.

[198] *P. WYNN* - On a connection between the first and the second confluent form of the ε-algorithm - Nieuw. Arch. Wisk., 11 (1963) 19-21.

[199] *P. WYNN* - Upon a second confluent form of the ε-algorithm - Proc. Glasgow Math. Soc., 5 (1962) 160-165.

[200] *P. WYNN* - Upon the inverse of formal power series over certain algebras - Centre de recherches mathématiques, Université de Montréal, 1970.

[201] *P. WYNN* - Upon the generalized inverse of a formal power series with vector valued coefficients - Compositio Math., 23 (1971) 453-460.

[202] *P. WYNN* - Sur l'équation aux dérivées partielles de la surface de Padé - C.R. Acad. Sc. Paris, 278 A (1974) 847-850.

[203] *P. WYNN* - Uber finen interpolations - algorithmus und gewise andere formeln, die in der theorie der interpolation durch rationale funktionen bestehen - Numer. Math., 2 (1961) 151-182.

[204] *P. WYNN* - Difference - differential recursions for Padé quotients - Proc. London Math. Soc., 23 (1971) 283-300.

[205] *P. WYNN* - A general system of orthogonal polynomials - Quart. J. Math., 18 ser. 2 (1967) 69-81.

[206] *P. WYNN* - Some recent developments in the theories of continued fractions and the Padé table - Rocky Mountains J. Math., 4 (1974) 297-324.

[207] *P. WYNN* - Upon the diagonal sequences of the Padé table - MRC Technical summary report 660, Madison, 1966.

[208] *P. WYNN* - Extremal properties of Padé quotients - Acta Math. Acad. Sci. Hungaricae, 25 (1974) 291-298.

[209] *P. WYNN* - Upon a convergence result in the theory of the Padé table - Trans. Amer. Math. Soc., 165 (1972) 239 - 249.

[210] *P. WYNN* - Zur theorie der mit gewissen speziellen funktionen verknüpften Padèschen tafeln - Math. Z., 109 (1969) 66-70.

[211] *P. WYNN* - Upon the Padé table derived from a Stieltjes series - SIAM J. Numer. Anal., 5 (1968) 805-834.

[212] *P. WYNN* - Upon systems of recursions which obtain among the quotients of the Padé table - Numer. Math., 8 (1966) 264-269.

[213] *P. WYNN* - L'ε-algoritmo e la tavola di Padé - Rend. di Mat. Roma, 20 (1961) 403-408.

[214] *P. WYNN* - Upon a conjecture concerning a method for solving linear equations, and certain other matters - MRC technical summary report 626, Madison, 1966.

[215] *P. WYNN* - Continued fractions whose coefficients obey a noncommutative law of multiplication - Arch. Rat. Mech. Anal., 12 (1963) 273-312.

[216] *P. WYNN* - A note on the convergence of certain noncommutative continued fractions - MRC technical summary report 750, Madison, 1967.

[217] *P. WYNN* - Vector continued fractions - Linear Algebra, 1 (1968) 357-395.

[218] *P. WYNN* - Upon the definition of an integral as the limit of a continued fraction - Arch. Rat. Mech. Anal., 28 (1968) 83-148.

[219] *P. WYNN* - An arsenal of Algol procedures for the evaluation of continued fractions and for effecting the epsilon algorithm - Chiffres, 9 (1966) 327-362.

[220] *P. WYNN* - Four lectures on the numerical application of continued fractions - CIME summer school lectures, 1965.

[221] *P. WYNN* - Partial differential equations associated with certain nonlinear algorithms - ZAMP, 15 (1964) 273-289.

[222] *P. WYNN* - A numerical method for estimating parameters in mathematical models - Centre de recherches mathématiques, Univ. de Montréal, rep. CRM, 443, 1974.

[223] *P. WYNN* - On a device for computing the $e_m(S_n)$ transformation - MTAC, 10 (1956) 91-96.

[224] *P. WYNN* - A convergence theory of some methods of integration - J. Reine Angew. Math., 285 (1976) 181-208.

[225] *P. WYNN* - Acceleration techniques in numerical analysis with particular reference to problems in one independant variable - Proc. IFIP Congress, North Holland, (1962) 149-156.

[226] *P. WYNN* - A note on programming repeated application of the ε-algorithm - Chiffres, 8 (1965) 23-62.

[227] *P. WYNN* - The rational approximation of functions which are formally defined by a power series expansion - Math. Comp., 14 (1960) 147-186.

[228] *P. WYNN* - The abstract theory of the epsilon algorithm - Centre de recherches mathématiques n°74, Univ. de Montréal, 1971.

[229] *P. WYNN* - Upon a hierarchy of epsilon arrays - Louisiana State Univ., New Orleans, techn. rep. 46, 1970.

[230] *P. WYNN* - Invariants associated with the epsilon algorithm and its first confluent form - Rend. Circ. Mat. Palermo, (2) 21 (1972) 31-41.

[231] *P. WYNN* - Singular rules for certain nonlinear algorithms - BIT, 3 (1963) 175-195.

[232] *P. WYNN* - A sufficient condition for the instability of the ε-algorithm - Nieuw. Arch. Wisk., 3 (1961) 117-119.

[233] *P. WYNN* - On the propagation of error in certain nonlinear algorithms - Numer. Math., 1 (1959) 142-149.

[234] *P. WYNN* - On the convergence and stability of the epsilon algorithm - SIAM J. Numer. Anal., 3 (1966) 91-122.

[235] *P. WYNN* - Hierarchies of arrays and function sequences associated with the epsilon algorithm and its first confluent form - Rend. Mat. Roma, 5 (1972) 819-852.

[236] *P. WYNN* - Accélération de la convergence de séries d'opérateurs en analyse numérique - C.R. Acad. Sc. Paris, 276 A (1973) 803-806.

[237] *P. WYNN* - Transformations de séries à l'aide de l'ε-algorithme - C.R. Acad. Sc. Paris, 275 A (1972) 1351-1353.

[238] *P. WYNN* - Upon some continuous prediction algorithms - Calcolo, 9 (1972) 197-234 and 235-278.

[239] *P. WYNN* - Confluent forms of certain nonlinear algorithms - Arch. Math., 11 (1960) 223-234.

[240] *P. WYNN* - A note on a confluent form of the ε-algorithm - Arch. Math., 11 (1960) 237-240.

[241] *P. WYNN* - On a procrustean technique for the numerical transformation of slowly convergent sequences and series - Proc. Camb. Phil. Soc., 52 (1956) 663-671.

[242] *P. WYNN* - Acceleration techniques for iteraded vector and matrix problems - Math. Comp., 16 (1962) 301-322.

[243] *K. YOSIDA* - Functional analysis - Springer-Verlag, 1968.

[244] *M. ZAMANSKY* - Introduction à l'algèbre et à l'analyse modernes - Dunod, Paris, 1967.

[245] *J. ZINN-JUSTIN* - Strong interactions dynamics with Padé approximants - Phys. Lett., 1 C (1971) 55-102.

INDEX

Vol. 460: O. Loos, Jordan Pairs. XVI, 218 pages. 1975.

Vol. 461: Computational Mechanics. Proceedings 1974. Edited by J. T. Oden. VII, 328 pages. 1975.

Vol. 462: P. Gérardin, Construction de Séries Discrètes p-adiques. »Sur les séries discrètes non ramifiées des groupes réductifs déployés p-adiques«. III, 180 pages. 1975.

Vol. 463: H.-H. Kuo, Gaussian Measures in Banach Spaces. VI, 224 pages. 1975.

Vol. 464: C. Rockland, Hypoellipticity and Eigenvalue Asymptotics. III, 171 pages. 1975.

Vol. 465: Séminaire de Probabilités IX. Proceedings 1973/74. Edité par P. A. Meyer. IV, 589 pages. 1975.

Vol. 466: Non-Commutative Harmonic Analysis. Proceedings 1974. Edited by J. Carmona, J. Dixmier and M. Vergne. VI, 231 pages. 1975.

Vol. 467: M. R. Essén, The Cos $\pi\lambda$ Theorem. With a paper by Christer Borell. VII, 112 pages. 1975.

Vol. 468: Dynamical Systems – Warwick 1974. Proceedings 1973/74. Edited by A. Manning. X, 405 pages. 1975.

Vol. 469: E. Binz, Continuous Convergence on C(X). IX, 140 pages. 1975.

Vol. 470: R. Bowen, Equilibrium States and the Ergodic Theory of Anosov Diffeomorphisms. III, 108 pages. 1975.

Vol. 471: R. S. Hamilton, Harmonic Maps of Manifolds with Boundary. III, 168 pages. 1975.

Vol. 472: Probability-Winter School. Proceedings 1975. Edited by Z. Ciesielski, K. Urbanik, and W. A. Woyczyński. VI, 283 pages. 1975.

Vol. 473: D. Burghelea, R. Lashof, and M. Rothenberg, Groups of Automorphisms of Manifolds. (with an appendix by E. Pedersen) VII, 156 pages. 1975.

Vol. 474: Séminaire Pierre Lelong (Analyse) Année 1973/74. Edité par P. Lelong. VI, 182 pages. 1975.

Vol. 475: Répartition Modulo 1. Actes du Colloque de Marseille-Luminy, 4 au 7 Juin 1974. Edité par G. Rauzy. V, 258 pages. 1975. 1975.

Vol. 476: Modular Functions of One Variable IV. Proceedings 1972. Edited by B. J. Birch and W. Kuyk. V, 151 pages. 1975.

Vol. 477: Optimization and Optimal Control. Proceedings 1974. Edited by R. Bulirsch, W. Oettli, and J. Stoer. VII, 294 pages. 1975.

Vol. 478: G. Schober, Univalent Functions – Selected Topics. V, 200 pages. 1975.

Vol. 479: S. D. Fisher and J. W. Jerome, Minimum Norm Extremals in Function Spaces. With Applications to Classical and Modern Analysis. VIII, 209 pages. 1975.

Vol. 480: X. M. Fernique, J. P. Conze et J. Gani, Ecole d'Eté de Probabilités de Saint-Flour IV-1974. Edité par P.-L. Hennequin. XI, 293 pages. 1975.

Vol. 481: M. de Guzmán, Differentiation of Integrals in R^n. XII, 226 pages. 1975.

Vol. 482: Fonctions de Plusieurs Variables Complexes II. Séminaire François Norguet 1974–1975. IX, 367 pages. 1975.

Vol. 483: R. D. M. Accola, Riemann Surfaces, Theta Functions, and Abelian Automorphisms Groups. III, 105 pages. 1975.

Vol. 484: Differential Topology and Geometry. Proceedings 1974. Edited by G. P. Joubert, R. P. Moussu, and R. H. Roussarie. IX, 287 pages. 1975.

Vol. 485: J. Diestel, Geometry of Banach Spaces – Selected Topics. XI, 282 pages. 1975.

Vol. 486: S. Stratila and D. Voiculescu, Representations of AF-Algebras and of the Group $U(\infty)$. IX, 169 pages. 1975.

Vol. 487: H. M. Reimann und T. Rychener, Funktionen beschränkter mittlerer Oszillation. VI, 141 Seiten. 1975.

Vol. 488: Representations of Algebras, Ottawa 1974. Proceedings 1974. Edited by V. Dlab and P. Gabriel. XII, 378 pages. 1975.

Vol. 489: J. Bair and R. Fourneau, Etude Géométrique des Espaces Vectoriels. Une Introduction. VII, 185 pages. 1975.

Vol. 490: The Geometry of Metric and Linear Spaces. Proceedings 1974. Edited by L. M. Kelly. X, 244 pages. 1975.

Vol. 491: K. A. Broughan, Invariants for Real-Generated Uniform Topological and Algebraic Categories. X, 197 pages. 1975.

Vol. 492: Infinitary Logic: In Memoriam Carol Karp. Edited by D. W. Kueker. VI, 206 pages. 1975.

Vol. 493: F. W. Kamber and P. Tondeur, Foliated Bundles and Characteristic Classes. XIII, 208 pages. 1975.

Vol. 494: A Cornea and G. Licea. Order and Potential Resolvent Families of Kernels. IV, 154 pages. 1975.

Vol. 495: A. Kerber, Representations of Permutation Groups II. V, 175 pages. 1975.

Vol. 496: L. H. Hodgkin and V. P. Snaith, Topics in K-Theory. Two Independent Contributions. III, 294 pages. 1975.

Vol. 497: Analyse Harmonique sur les Groupes de Lie. Proceedings 1973–75. Edité par P. Eymard et al. VI, 710 pages. 1975.

Vol. 498: Model Theory and Algebra. A Memorial Tribute to Abraham Robinson. Edited by D. H. Saracino and V. B. Weispfenning. X, 463 pages. 1975.

Vol. 499: Logic Conference, Kiel 1974. Proceedings. Edited by G. H. Müller, A. Oberschelp, and K. Potthoff. V, 651 pages 1975.

Vol. 500: Proof Theory Symposion, Kiel 1974. Proceedings. Edited by J. Diller and G. H. Müller. VIII, 383 pages. 1975.

Vol. 501: Spline Functions, Karlsruhe 1975. Proceedings. Edited by K. Böhmer, G. Meinardus, and W. Schempp. VI, 421 pages. 1976.

Vol. 502: János Galambos, Representations of Real Numbers by Infinite Series. VI, 146 pages. 1976.

Vol. 503: Applications of Methods of Functional Analysis to Problems in Mechanics. Proceedings 1975. Edited by P. Germain and B. Nayroles. XIX, 531 pages. 1976.

Vol. 504: S. Lang and H. F. Trotter, Frobenius Distributions in GL_2-Extensions. III, 274 pages. 1976.

Vol. 505: Advances in Complex Function Theory. Proceedings 1973/74. Edited by W. E. Kirwan and L. Zalcman. VIII, 203 pages. 1976.

Vol. 506: Numerical Analysis, Dundee 1975. Proceedings. Edited by G. A. Watson. X, 201 pages. 1976.

Vol. 507: M. C. Reed, Abstract Non-Linear Wave Equations. VI, 128 pages. 1976.

Vol. 508: E. Seneta, Regularly Varying Functions. V, 112 pages. 1976.

Vol. 509: D. E. Blair, Contact Manifolds in Riemannian Geometry. VI, 146 pages. 1976.

Vol. 510: V. Poènaru, Singularités C^∞ en Présence de Symétrie. V, 174 pages. 1976.

Vol. 511: Séminaire de Probabilités X. Proceedings 1974/75. Edité par P. A. Meyer. VI, 593 pages. 1976.

Vol. 512: Spaces of Analytic Functions, Kristiansand, Norway 1975. Proceedings. Edited by O. B. Bekken, B. K. Øksendal, and A. Stray. VIII, 204 pages. 1976.

Vol. 513: R. B. Warfield, Jr. Nilpotent Groups. VIII, 115 pages. 1976.

Vol. 514: Séminaire Bourbaki vol. 1974/75. Exposés 453 – 470. IV, 276 pages. 1976.

Vol. 515: Bäcklund Transformations. Nashville, Tennessee 1974. Proceedings. Edited by R. M. Miura. VIII, 295 pages. 1976.

Vol. 516: M. L. Silverstein, Boundary Theory for Symmetric Markov Processes. XVI, 314 pages. 1976.

Vol. 517: S. Glasner, Proximal Flows. VIII, 153 pages. 1976.

Vol. 518: Séminaire de Théorie du Potentiel, Proceedings Paris 1972–1974. Edité par F. Hirsch et G. Mokobodzki. VI, 275 pages. 1976.

Vol. 519: J. Schmets, Espaces de Fonctions Continues. XII, 150 pages. 1976.

Vol. 520: R. H. Farrell, Techniques of Multivariate Calculation. X, 337 pages. 1976.

Vol. 521: G. Cherlin, Model Theoretic Algebra – Selected Topics. IV, 234 pages. 1976.

Vol. 522: C. O. Bloom and N. D. Kazarinoff, Short Wave Radiation Problems in Inhomogeneous Media: Asymptotic Solutions. V. 104 pages. 1976.

Vol. 523: S. A. Albeverio and R. J. Høegh-Krohn, Mathematical Theory of Feynman Path Integrals. IV, 139 pages. 1976.

Vol. 524: Séminaire Pierre Lelong (Analyse) Année 1974/75. Edité par P. Lelong. V, 222 pages. 1976.

Vol. 525: Structural Stability, the Theory of Catastrophes, and Applications in the Sciences. Proceedings 1975. Edited by P. Hilton. VI, 408 pages. 1976.

Vol. 526: Probability in Banach Spaces. Proceedings 1975. Edited by A. Beck. VI, 290 pages. 1976.

Vol. 527: M. Denker, Ch. Grillenberger, and K. Sigmund, Ergodic Theory on Compact Spaces. IV, 360 pages. 1976.

Vol. 528: J. E. Humphreys, Ordinary and Modular Representations of Chevalley Groups. III, 127 pages. 1976.

Vol. 529: J. Grandell, Doubly Stochastic Poisson Processes. X, 234 pages. 1976.

Vol. 530: S. S. Gelbart, Weil's Representation and the Spectrum of the Metaplectic Group. VII, 140 pages. 1976.

Vol. 531: Y.-C. Wong, The Topology of Uniform Convergence on Order-Bounded Sets. VI, 163 pages. 1976.

Vol. 532: Théorie Ergodique. Proceedings 1973/1974. Edité par J.-P. Conze and M. S. Keane. VIII, 227 pages. 1976.

Vol. 533: F. R. Cohen, T. J. Lada, and J. P. May, The Homology of Iterated Loop Spaces. IX, 490 pages. 1976.

Vol. 534: C. Preston, Random Fields. V, 200 pages. 1976.

Vol. 535: Singularités d'Applications Differentiables. Plans-sur-Bex. 1975. Edité par O. Burlet et F. Ronga. V, 253 pages. 1976.

Vol. 536: W. M. Schmidt, Equations over Finite Fields. An Elementary Approach. IX, 267 pages. 1976.

Vol. 537: Set Theory and Hierarchy Theory. Bierutowice, Poland 1975. A Memorial Tribute to Andrzej Mostowski. Edited by W. Marek, M. Srebrny and A. Zarach. XIII, 345 pages. 1976.

Vol. 538: G. Fischer, Complex Analytic Geometry. VII, 201 pages. 1976.

Vol. 539: A. Badrikian, J. F. C. Kingman et J. Kuelbs, Ecole d'Eté de Probabilités de Saint Flour V-1975. Edité par P.-L. Hennequin. IX, 314 pages. 1976.

Vol. 540: Categorical Topology, Proceedings 1975. Edited by E. Binz and H. Herrlich. XV, 719 pages. 1976

Vol. 541: Measure Theory, Oberwolfach 1975. Proceedings. Edited by A. Bellow and D. Kölzow. XIV, 430 pages. 1976.

Vol. 542: D. A. Edwards and H. M. Hastings, Čech and Steenrod Homotopy Theories with Applications to Geometric Topology. VII, 296 pages. 1976.

Vol. 543: Nonlinear Operators and the Calculus of Variations, Bruxelles 1975. Edited by J. P. Gossez, E. J. Lami Dozo, J. Mawhin, and L. Waelbroeck, VII, 237 pages. 1976.

Vol. 544: Robert P. Langlands, On the Functional Equations Satisfied by Eisenstein Series. VII, 337 pages. 1976.

Vol. 545: Noncommutative Ring Theory. Kent State 1975. Edited by J. H. Cozzens and F. L. Sandomierski. V, 212 pages. 1976.

Vol. 546: K. Mahler, Lectures on Transcendental Numbers. Edited and Completed by B. Diviš and W. J. Le Veque. XXI, 254 pages. 1976.

Vol. 547: A. Mukherjea and N. A. Tserpes, Measures on Topological Semigroups: Convolution Products and Random Walks. V, 197 pages. 1976.

Vol. 548: D. A. Hejhal, The Selberg Trace Formula for PSL (2, ℝ). Volume I. VI, 516 pages. 1976.

Vol. 549: Brauer Groups, Evanston 1975. Proceedings. Edited by D. Zelinsky. V, 187 pages. 1976.

Vol. 550: Proceedings of the Third Japan – USSR Symposium on Probability Theory. Edited by G. Maruyama and J. V. Prokhorov. VI, 722 pages. 1976.

Vol. 551: Algebraic K-Theory, Evanston 1976. Proceedings. Edited by M. R. Stein. XI, 409 pages. 1976.

Vol. 552: C. G. Gibson, K. Wirthmüller, A. A. du Plessis and E. J. N. Looijenga. Topological Stability of Smooth Mappings. V, 155 pages. 1976.

Vol. 553: M. Petrich, Categories of Algebraic Systems. Vector and Projective Spaces, Semigroups, Rings and Lattices. VIII, 217 pages. 1976.

Vol. 554: J. D. H. Smith, Mal'cev Varieties. VIII, 158 pages. 1976.

Vol. 555: M. Ishida, The Genus Fields of Algebraic Number Fields. VII, 116 pages. 1976.

Vol. 556: Approximation Theory. Bonn 1976. Proceedings. Edited by R. Schaback and K. Scherer. VII, 466 pages. 1976.

Vol. 557: W. Iberkleid and T. Petrie, Smooth S^1 Manifolds. III, 163 pages. 1976.

Vol. 558: B. Weisfeiler, On Construction and Identification of Graphs. XIV, 237 pages. 1976.

Vol. 559: J.-P. Caubet, Le Mouvement Brownien Relativiste. IX, 212 pages. 1976.

Vol. 560: Combinatorial Mathematics, IV, Proceedings 1975. Edited by L. R. A. Casse and W. D. Wallis. VII, 249 pages. 1976.

Vol. 561: Function Theoretic Methods for Partial Differential Equations. Darmstadt 1976. Proceedings. Edited by V. E. Meister, N. Weck and W. L. Wendland. XVIII, 520 pages. 1976.

Vol. 562: R. W. Goodman, Nilpotent Lie Groups: Structure and Applications to Analysis. X, 210 pages. 1976.

Vol. 563: Séminaire de Théorie du Potentiel. Paris, No. 2. Proceedings 1975–1976. Edited by F. Hirsch and G. Mokobodzki. VI, 292 pages. 1976.

Vol. 564: Ordinary and Partial Differential Equations, Dundee 1976. Proceedings. Edited by W. N. Everitt and B. D. Sleeman. XVIII, 551 pages. 1976.

Vol. 565: Turbulence and Navier Stokes Equations. Proceedings 1975. Edited by R. Temam. IX, 194 pages. 1976.

Vol. 566: Empirical Distributions and Processes. Oberwolfach 1976. Proceedings. Edited by P. Gaenssler and P. Révész. VII, 146 pages. 1976.

Vol. 567: Séminaire Bourbaki vol. 1975/76. Exposés 471–488. IV, 303 pages. 1977.

Vol. 568: R. E. Gaines and J. L. Mawhin, Coincidence Degree, and Nonlinear Differential Equations. V, 262 pages. 1977.

Vol. 569: Cohomologie Etale SGA $4\frac{1}{2}$. Séminaire de Géométrie Algébrique du Bois-Marie. Edité par P. Deligne. V, 312 pages. 1977.

Vol. 570: Differential Geometrical Methods in Mathematical Physics, Bonn 1975. Proceedings. Edited by K. Bleuler and A. Reetz. VIII, 576 pages. 1977.

Vol. 571: Constructive Theory of Functions of Several Variables, Oberwolfach 1976. Proceedings. Edited by W. Schempp and K. Zeller. VI, 290 pages. 1977

Vol. 572: Sparse Matrix Techniques, Copenhagen 1976. Edited by V. A. Barker. V, 184 pages. 1977.

Vol. 573: Group Theory, Canberra 1975. Proceedings. Edited by R. A. Bryce, J. Cossey and M. F. Newman. VII, 146 pages. 1977.

Vol. 574: J. Moldestad, Computations in Higher Types. IV, 203 pages. 1977.

Vol. 575: K-Theory and Operator Algebras, Athens, Georgia 1975. Edited by B. B. Morrel and I. M. Singer. VI, 191 pages. 1977.

Vol. 576: V. S. Varadarajan, Harmonic Analysis on Real Reductive Groups. VI, 521 pages. 1977.

Vol. 577: J. P. May, E_∞ Ring Spaces and E_∞ Ring Spectra. IV, 268 pages. 1977.

Vol. 579: Combinatoire et Représentation du Groupe Symétrique, Strasbourg 1976. Proceedings 1976. Edité par D. Foata. IV, 339 pages. 1977.

Vol. 580: C. Castaing and M. Valadier, Convex Analysis and Measurable Multifunctions. VIII, 278 pages. 1977.

8480-77-15
5-07